D1259305

MANAGING HAZARDOUS AIR POLLUTANTS
State of the Art

Edited by

Winston Chow
Electric Power Research Institute
Palo Alto, California

Katherine K. Connor
Decision Focus Incorporated
Mountain View, California

Contributing Editors

Peter Mueller
Ronald Wyzga
Donald Porcella
Leonard Levin
Ramsay Chang

EPRI TR-101890

Library of Congress Cataloging-in-Publication Data

Managing hazardous air pollutants : state of the art / edited by
Winston Chow, Katherine K. Connor.
p. cm.
"Conference on Managing Hazardous Air Pollutants keynote address"—CIP galley.
Includes bibliographical references and index.
ISBN 0-87371-866-6
1. Air quality management—Congresses. 2. Hazardous wastes—
Management—Congresses. I. Chow, Winston. II. Connor, Katherine K. III. Conference on Managing Hazardous Air Pollutants.
TD881.M28 1993
363.73′92—dc20 92-43825
CIP

Direct all inquiries to CRC Press, Inc., 2000 Corporate Blvd., N.W., Boca Raton, Florida 33431.

Lewis Publishers is an imprint of CRC Press

No claim to original U.S. Government works
International Standard Book Number 0-87371-866-6
Library of Congress Card Number 92-43825
Printed in the United States of America 1 2 3 4 5 6 7 8 9 0
Printed on acid-free paper

Acknowledgments

CONFERENCE STEERING COMMITTEE

Lori J. Adams, Conference Coordinator, Electric Power Research Institute
Ramsay Chang, Control Technologies, Electric Power Research Institute
Winston Chow, Conference Chair, Electric Power Research Institute
Katherine K. Connor, Conference Program Director, Decision Focus Inc.
Leonard Levin, Risk Issues, Electric Power Research Institute
Peter K. Mueller, Atmospheric Concerns, Electric Power Research Institute
Donald B. Porcella, Ecological Effects, Electric Power Research Institute
Ronald E. Wyzga, Health Effects, Electric Power Research Institute

COOPERATING ORGANIZATIONS AND THEIR REPRESENTATIVES

British Coal
Peter W. Sage

Canadian Electrical Association
William P. Peel

Electricité de France
Gerard Maffiolo

IEA Coal Research
Irene M. Smith

PowerGen
Hugh E. Evans

U.S. Department of Energy
Charles E. Schmidt

Preface

This volume contains the proceedings from the first international conference, Managing Hazardous Air Pollutants: State of the Art held in Washington, D.C., November 4–6, 1991. The conference was sponsored by the Electric Power Research Institute (EPRI) in cooperation with a number of other organizations involved in energy research and development in the U.S. and abroad. The goal of the conference was to stimulate the exchange of information regarding hazardous air pollutants, or "air toxics", and provide a forum for assessing the current state of knowledge and potential future research needs. Over 200 delegates representing a broad range of U.S. and foreign utilities, governments, consulting firms, equipment suppliers, and academic and research organizations attended the conference.

The conference was organized in response to the growing needs of electric utilities, in the U.S. and abroad, to understand and manage their emissions of air toxics. New regulations, including the Clean Air Act Amendments of 1990, and heightened public concerns have made it imperative to understand the magnitude and nature of potential health and environmental risks due to air toxic emissions and to develop successful strategies for communicating with a range of constituencies about risks and risk management options. However, this task is particularly difficult given the considerable uncertainty that characterizes most, if not all, of the components of an air toxics risk analysis — including emissions, transport and fate, exposure and dose, health and environmental effects, risk management options, and total risks and costs. This conference was designed around the risk analysis framework as an appropriate context for integrating and evaluating current air toxics research addressing the issues in each of the individual components.

The conference consisted of six technical sessions and a poster session aligned with the major inputs to an air toxics risk analysis. Each session comprises a single chapter in this volume. Chapter 1 contains material presented at the opening General Session. This chapter provides an introduction and overview of the subject, with papers describing the range of national and international approaches to regulating air toxics and synopses of utility and government-sponsored air toxics research activities. Chapter 2 includes information on a range of research efforts designed to develop sampling and analytical methods and characterize plant emissions. Chapter 3 focuses on what is known about the fate of air toxic emissions in the atmosphere and includes papers addressing a broad range of stack and atmospheric monitoring programs and their results. The likely impact of human and ecosystem exposures to ambient levels of air toxics is addressed in Chapter 4. The methodology and results of several air toxics risk assessments are contained in Chapter 5 and the design and effectiveness of a range of potential control strategies are included in Chapter 6. Chapter 7 includes the papers presented at the conference Poster Session. Finally, Chapter 8 contains a synthesis of the conference which highlights some of the major themes and summarizes key insights and areas where further research is necessary.

The information presented at this conference and in these proceedings represents the state of the art in 1991. As the results from ongoing and new research activities become available, the information contained in this document will need to be updated, as appropriate. It is our plan and hope that succeeding conferences will provide a forum for the kind of international cooperation and communication that will continue to enhance our understanding of the difficult technical issues surrounding air toxics risks and facilitate the development of economic and effective risk management strategies.

We would like to express our gratitude to the many individuals that submitted and presented papers at the conference and to the chairs and co-chairs of the technical sessions. We appreciate the time and energy you invested in preparing and presenting high-quality materials and ensuring that the sessions ran smoothly. Thanks also go to Ian Torrens and Stephen Peck at EPRI for their ongoing support and participation in this effort. Special appreciation goes to Lori Adams at EPRI for her diligence and attention to detail throughout the planning, organizing, and conduct of the conference and to Lori Lehmann and Sheila Mattaliano for their contributions to organizing the conference and producing this document. We are also indebted to the individuals and organizations listed in the following pages for their substantial contributions to the success of this conference.

Winston Chow
Katherine K. Connor

Winston Chow is a Program Manager in the Environmental Control Systems Department, Generation and Storage Division of the Electric Power Research Institute (EPRI), Palo Alto, California. In his position, he manages research and development projects covering cooling system heat rate and condenser performance improvements, water quality control, environmental monitoring, and hazardous/toxic substances management. He concurrently serves as deputy manager of the Waste and Water Management Program at EPRI. In 1988, he was part of a 4-person planning team which developed a 20-year strategic plan for corporate research on power systems environmental control.

Before joining EPRI in 1979, Winston spent 7 years with Bechtel Corporation as a Mechanical Engineering Supervisor, where he supervised power plant design. He holds Bachelor and Master degrees in Chemical Engineering, and a Master of Business Administration degree. As a Registered Professional Engineer, he is licensed to practice chemical and mechanical engineering in the State of California.

Winston is a member of Beta Gamma Sigma, the American Institute of Chemical Engineers, the Air and Waste Management Association, the Water Pollution Control Federation, and the National Society of Professional Engineers. During his career, he has contributed more than 35 papers, articles, and publications to scientific, technical, and business forums.

Katherine Kahle Connor is a Senior Associate and manager of Decision Focus Incorporated's air toxics risk assessment and risk management consulting. Ms. Connor has over 10 years of experience performing environmental risk analyses for industry and government. She has developed decision frameworks to help clients evaluate and communicate the environmental and business risks associated with a wide range of issues including air toxics, petroleum-contaminated soils, spent solvents, boiler cleaning waste, and others. She has led major projects to develop and apply quantitative risk analysis methods to emission sources ranging from electric power plants to university and research laboratories.

Prior to joining Decision Focus, Ms. Connor worked as an environmental engineer at Bechtel Corporation, developing waste and water management plans for a variety of industrial facilities, and as a business analyst at Sohio Petroleum Company, where she conducted engineering-economic evaluations for enhanced recovery projects at Prudhoe Bay. She received her B.S. in Applied Earth Science and M.S. in Environmental Engineering from Stanford University and her M.B.A. from Harvard Business School.

List of Contributors

David J. Akers
CQ Inc.
One Quality Center
Homer City, Pennsylvania 15748

Michael T. Alberts
Radian Corporation
10389 Old Placerville Road
Sacramento, California 95827

Donald J. Ames
Stationary Source Division
California Air Resources Board
2020 L Street
Sacramento, California 95814

Michael St. J. Arnold
British Coal Corporation
Coal Research Establishment
Cheltenham, Glos
GL52 4RZ United Kingdom

Daniel C. Baker
Westhollow Research Center
Shell Development Company
3333 Highway 6 South
Houston, Texas 77082-3101

Albert M. Beim
Institute of Ecological Toxicology
P.O. Box 48
Baikalsk, Irkutsk Region
665914 USSR

Michael B. Berkenpas
Center for Energy and
Environmental Studies
Carnegie Mellon University
Pittsburgh, Pennsylvania 15213

John D. Bignell
Warren Spring Laboratory
Stevenage, Herts,
SG1 2NG United Kingdom

Nicolas S. Bloom
Frontier Geosciences
414 Pontius North
Seattle, Washington 98109

Mary Anna Bogle
Environmental Sciences Division
Oak Ridge National Laboratory
Oak Ridge, Tennessee 37831

Roger P. Brower
Versar, Inc.
9200 Rumsey Road
Columbia, Maryland 21045

Thomas D. Brown
U.S. DOE/Pittsburgh Energy
Technology Center
P.O. Box 10940
Pittsburgh, Pennsylvania 15236

Andrew Burnette
Radian Corporation
8501 North MoPac Boulevard
Austin, Texas 78759

W. V. Bush
Westhollow Research Center
Shell Development Company
3333 Highway 6 South
Houston, Texas 77082-3101

Sally A. Campbell
S.A. Campbell, Associates
10714 Midsummer Lane
Columbia, Maryland 21044

Ramsay Chang
Electric Power Research Institute
3412 Hillview Avenue
Palo Alto, California 94303

Winston Chow
Electric Power Research Institute
3412 Hillview Avenue
Palo Alto, California 94303

Jane C. Chuang
Battelle Memorial Institute
505 King Avenue
Columbus, Ohio 43201-2693

Lee B. Clarke
IEA Coal Research
Gemini House
10-18 Putney Hill
London, SW15 6AA, England

Yoram Cohen
Department of Chemical Engineering; and
National Center for Intermedia Transport Research
University of California, Los Angeles
Los Angeles, California 90024

Katherine Connor
Decision Focus Inc.
650 Castro Street, Suite 300
Mountain View, California 94041

Vincent B. Conrad
CONSOL Inc.
Research and Development
4000 Brownsville Road
Library, Pennsylvania 15129

Elpida Constantinou
ENSR Consulting and Engineering
1320 Harbor Bay Parkway, Suite 210
Alameda, California 94501

John A. Cooper
Chester Environmental
12242 S.W. Garden Place
Tigard, Oregon 97223

Keith E. Curtis
Environmental Sciences Department
Research Division
Ontario Hydro
Toronto, Ontario, Canada M8Z-584

Stephen G. Dawes
British Coal Corporation
Coal Research Establishment
Cheltenham, Glos
GL52 4RZ United Kingdom

Michael DePhillips
Biomedical and Environmental Assessment Group
Department of Applied Science
Brookhaven National Laboratory
Upton, New York 11973

Matthew S. DeVito
CONSOL Inc.
Research & Development
4000 Brownsville Road
Library, Pennsylvania 15129

Ronald J. Dickson
Radian Corporation
10389 Old Placerville Road
Sacramento, California 95827

A. Gwen Eklund
Radian Corporation
8501 North MoPac Boulevard
Austin, Texas 78759

Hugh E. Evans
Environmental Branch
PowerGen plc U.K.
Solihull, West Midlands,
B91 2JN United Kingdom

Vasilis M. Fthenakis
Biomedical and Environmental Assessment Group
Department of Applied Science
Brookhaven National Laboratory
Upton, New York 11973

Robert W. Garber
Tennessee Valley Authority
219 Chemical Engineering Bldg.
Muscle Shoals, Alabama 35660

Robert J. Gemmill
Assessment Branch
British Coal Corporation
Coal Research Establishment
Cheltenham, Glos
GL52 4RZ, United Kingdom

Jeroen Gerritsen
Versar, Inc.
9200 Rumsey Road
Columbia, Maryland 21045

William Gleason
Occidental Chemical Corporation
Niagara Falls, New York

Lawrence S. Goldstein
Electric Power Research Institute
3412 Hillview Avenue
Palo Alto, California 94304

Lawrence Gratt
IWG Corp.
San Diego, California 92101

Elena I. Grosheva
Institute of Ecological Toxicology
P.O. Box 48
Baikalsk, Irkutsk Region
665914 USSR

Allyn Hemenway
Office of Fossil Energy
U.S. Department of Energy
Washington, D.C. 20585

Jeffrey B. Hicks
Radian Corporation
10389 Old Placerville Road
Sacramento, California 95827

Lynn M. Hildemann
Department of Civil Engineering
Stanford University
Stanford, California 94305

Krishna N. Krishnamurthy
Environment Support Department
Thermal Integration and Services Division
Ontario Hydro
Toronto, Ontario, Canada M8Z-554

William S. Kyte
PowerGen plc U.K.
Solihull, West Midlands,
B91 2JN United Kingdom

John S. Lagarias
California Air Resources Board
2020 L Street
Sacramento, California 95814

Dennis L. Laudal
Energy and Environmental Research Center
University of North Dakota
Box 8213, University Station
Grand Forks, North Dakota 58202

John Lengyel, Jr.
CONSOL Inc.
Research and Development
4000 Brownsville Road
Library, Pennsylvania 15129

Leonard Levin
Electric Power Research Institute
3412 Hillview Avenue
Palo Alto, California 94303

Ken Lionarons
Paradise Robotics
59 Skyline Drive
Glen Mills, Pennsylvania 19342

K. R. Loos
Westhollow Research Center
Shell Development Company
3333 Highway 6 South
Houston, Texas 77082-3101

William V. Loscutoff
Monitoring and Laboratory Division
California Air Resources Board
1309 T Street
Sacramento, California 95814

Gordon Maller
Radian Corporation
8501 North MoPac Boulevard
Austin, Texas 78759

Robert M. Mann
Radian Corporation
8501 North MoPac Boulevard
Austin, Texas 78759

Senichi Masuda
Masuda Research, Inc.
Kaneyasu Building 6F1
2-40-11, Hongo, Bunkyo-ku
Tokyo 114, Japan

Andrew R. McFarland
Department of Mechanical Engineering
Texas A&M University
EPB Bldg., Spence Street
College Station, Texas 77843-3123

Douglas G. McKinney
Air and Energy Engineering Research Laboratory
U.S. Environmental Protection Agency
Research Triangle Park, North Carolina 27711

Anne McQueen
Sierra Environmental Engineering, Inc.
3505 Cadillac Avenue K-1
Costa Mesa, California 92626

Frank B. Meserole
Radian Corporation
8501 North MoPac Boulevard
Austin, Texas 78759

Michael Miller
Waste & Water Management Program
Electric Power Research Institute
3412 Hillview Avenue
Palo Alto, California 94303

Stanley J. Miller
Energy and Environmental Research Center
University of North Dakota
Box 9018
Grand Forks, North Dakota 58203

Andrew J. Minchener
British Coal Corporation
Coal Research Establishment
Cheltenham, Glos
GL52 4RZ United Kingdom

George L. Moilanen
Sierra Environmental Engineering, Inc.
3505 Cadillac Avenue K-1
Costa Mesa, California 92626

Paul D. Moskowitz
Biomedical and Environmental Assessment Group
Department of Applied Science
Brookhaven National Laboratory
Upton, New York 11973

Peter K. Mueller
Atmospheric Sciences Program
Electric Power Research Institute
3412 Hillview Avenue
Palo Alto, California 94303

Babu Nott
Electric Power Research Institute
3412 Hillview Avenue
Palo Alto, California 94303

Edward L. Obermiller
CONSOL Inc.
Research and Development
4000 Brownsville Road
Library, Pennsylvania 15129

William Rogers Oliver
Radian Corporation
10389 Old Placerville Road
Sacramento, California 95827

Douglas A. Orr
Radian Corporation
Box 201088
Austin, Texas 78720

Sarah Osborne
Galson Corporation
6601 Kirkville Road
East Syracuse, New York 13215

Ted Palma
Galson Corporation
Raleigh, North Carolina

Stephen Peck
Environmental Sciences Department
Electric Power Research Institute
3412 Hillview Avenue
Palo Alto, California 94303

Warren Peters
U.S. EPA
Office of Air and Radiation
Research Triangle Park, North Carolina 27711

Joseph R. Peterson
Radian Corporation
8501 North MoPac Boulevard
Austin, Texas 78759

Michael W. Poore
Monitoring and Laboratory Division
California Air Resources Board
1309 T Street
Sacramento, California 95814

Frank T. Princiotta
Air and Energy Engineering Research Laboratory
U.S. Environmental Protection Agency
Research Triangle Park, North Carolina 27711

Adrian S. Radziwon
Burns and Roe Services Corporation
P.O. Box 18288
Pittsburgh, Pennsylvania 15236

Richard G. Rhudy
Electric Power Research Institute
3412 Hillview Avenue
Palo Alto, California 94303

David Room
Decision Focus, Inc.
650 Castro Street, Suite 300
Mountain View, California
94041-2055

Edward S. Rubin
Department of Engineering and
Public Policy
Carnegie Mellon University
Pittsburgh, Pennsylvania 15213

Peter W. Sage
Environmental Branch
British Coal Corporation
Coal Research Establishment
Cheltenham, Glos
GL52 4RZ United Kingdom

Pradeep Saxena
Atmospheric Sciences Program
Electric Power Research Institute
3412 Hillview Avenue
Palo Alto, California 94303

Charles E. Schmidt
U.S. DOE/Pittsburgh Energy
Technology Center
P.O. Box 10940
Pittsburgh, Pennsylvania
15236-0940

Christian Seigneur
ENSR Consulting and Engineering
1320 Harbor Bay Parkway,
Suite 210
Alameda, California 94501

Michael Shapiro
Deputy Assistant Administrator
Air and Radiation
U.S. Environmental Protection
Agency
401 M Street, S.W.
Washington, D.C. 20460

Laurence Slivon
Battelle Memorial Institute
505 King Avenue
Columbus, Ohio 43201-2693

Blakeman S. Smith
Electric Power Research Institute
3412 Hillview Avenue
Palo Alto, California 94303

Carolyn B. Suer
California Air Resources Board
2020 L Street
Sacramento, California 95814

George M. Sverdrup
Battelle Memorial Institute
505 King Avenue
Columbus, Ohio 43201-2693

Laszlo Takacs
Occidental Chemical Corporation
Niagara Falls, New York

George E. Taylor, Jr.
Desert Research Institute; and
Department of Environmental and
Resource Sciences
University of Nevado-Reno
Reno, Nevada 95506

Carol May Thompson
Radian Corporation
8501 North MoPac Boulevard
Austin, Texas 78759

Stephen J. Thorndyke
Emission Testing and Control
Section
Environmental Assessment
Technologies Centre
Ortech International
Mississauga, Ontario, Canada

David F. Todd
Monitoring and Laboratory Division
California Air Resources Board
2020 L Street
Sacramento, California 95814

Wangteng Tsai
Systech Engineering, Inc.
3744 Mt. Diablo Blvd., Suite 101
Lafayette, California 94549

Prasad R. Tumati
CONSOL Inc.
Research & Development
4000 Brownsville Road
Library, Pennsylvania 15129

Ralph R. Turner
Environmental Sciences Division
Oak Ridge National Laboratory
Oak Ridge, Tennessee 37831

Peter D. Venturini
Stationary Source Division
California Air Resources Board
2020 L Street
Sacramento, California 95814

Robert G. Vranka
Harding Lawson Associates
7655 Redwood Boulevard
P.O. Box 578
Novato, California 94948

Gayle Edmisten Watkin
Harding Lawson Associates
7655 Redwood Boulevard
P.O. Box 578
Novato, California 94948

Bernard Weiss
University of Rochester
School of Medicine and Dentistry
Rochester, New York 14642

Robert Wells
Galson Corporation
6601 Kirkville Road
East Syracuse, New York 13215

Robert G. Wetherold
Radian Corporation
Box 201088
Austin, Texas 78720

Fred O. Weyman
Radian Corporation
10389 Old Placerville Road
Sacramento, California 95827

Peter Wiederkehr
OECD Environment Directorate
Pollution Control Division
2, rue André-Pascal
75775 Paris Cedex 16, France

Kevin J. Williams
Radian Corporation
Box 201088
Austin, Texas 78720

Ronald E. Wyzga
Electric Power Research Institute
3412 Hillview Avenue
Palo Alto, California 94303-0813

Janice W. Yager
Electric Power Research Institute
3412 Hillview Avenue
Palo Alto, California 94303-0813

Kenneth L. Zankel
Versar, Inc.
9200 Rumsey Road
Columbia, Maryland 21045

Lee B. Zeugin
Hunton & Williams
2000 Pennsylvania Avenue NW
Washington, D.C. 20036

Contents

Chapter 1
General Session

Chairs: Winston Chow, EPRI
Irene M. Smith, IEA Coal Research

Conference on Managing Hazardous Air Pollutants: Keynote Address

Stephen Peck

[Editor's note: Dr. Peck heads the Environmental Sciences Department at EPRI. This department is responsible for R&D on global climate change, air quality, electric and magnetic fields, risk assessment of health and environmental hazards, and other environment-related issues facing the electric utility industry.

Dr. Peck's remarks at the conference are summarized and organized into four sections: conference objectives, research motivation, conference agenda, and summary comments.]

CONFERENCE OBJECTIVES

This conference brought together representatives from electric utilities, other companies, regulatory agencies, the research community, from both the U.S. and abroad, with two goals in mind. The first was to stimulate thinking in an integrated manner across specialties, disciplines, and parties with different approaches and interests. The second objective was to cover where we have been and where we are going in terms of our understanding and management of hazardous air pollutants. Under the second objective we wanted answers to the following questions. What is known and unknown? Where should stakeholders in this issue put their effort? What can be learned in the short term and in the long term? What is at stake? What are the potential effects of under control? What are the potential costs of over control? Finally, how can we manage the risks and uncertainties, and how can engineering and technology help?

RESEARCH MOTIVATION

Two forces motivate research on air toxics: regulatory "push" and planning "pull". The immediate push in the U.S. is the Clean Air Act Amendments (CAAAs) that may impose numerous requirements in both the short and long term. In the short term, the amendments require studies of the electric utility industry, estimates of residual risk from air toxics after utilities comply with other provisions of the amendments, and consideration of controls for utility emissions. The Act also includes many long-term needs for information and technology development as the regulatory process develops.

The second motivating force is the planning pull. Concern about air toxics is only one potential environmental problem that electric companies must address. Resource choice, operating decisions, and control options require integrated consideration for dealing with sulfur dioxide, oxides of nitrogen and particulates, as well as air toxics. These considerations also include short- and long-term needs for information and technology development.

EPRI's intention is to provide information for regulation and planning in an iterative and productive manner. This approach yields information for sound decisions today, with the knowledge that better information will be available in the future.

CONFERENCE AGENDA

The conference was organized around a risk paradigm, running from emissions, transport and fate, exposure and dose, health and environmental effects, risk management options, economic impacts, and, finally, total risks and costs.

The risk paradigm emerged as an appropriate model because the CAAAs explicitly call for a risk assessment and risk-based regulation. The formation of the National Academy of Sciences Risk Review Committee and of the Risk Assessment and Management Commission also acknowledges the uncertainty in the methods and interpretation.

Shifts in the practice of risk analysis are evident already. There is an increased emphasis on multimedia and multipathway assessments. There are attempts to recognize and quantify uncertainty, which are important for avoiding the consequences of excessive conservatism in order to compensate for unanalyzed uncertainty.

This risk-based approach will create a framework that should provide the basis for future decisions on regulations, facility planning, and operation. The conference agenda focused on many important aspects of the risk framework. The major topic areas of the conference included

- Overview — Monday morning
  U.S. and International Perspectives on Air Toxics Issues
- Emissions Topics — Monday afternoon
  Sampling methods, analytical techniques, source-testing programs, source models
- Transport and Fate — Tuesday morning
  Chemical transformations, ambient monitoring techniques, deposition studies, regional models
- Health and Ecological Effects — Tuesday afternoon
  Ecosystem models, mechanistic studies, and bioavailability studies
- Risk Assessment Models — Wednesday morning
  Cancer and noncancer effects, multimedia and multipathway analyses, complex exposure patterns, uncertainty models

- Control Strategies and Technologies — Wednesday morning
  New control technology performance, new combustion technologies, characterization of the byproducts of combustion

SUMMARY COMMENTS

EPRI's previous study of acid rain provides some desirable lessons for work on air toxics. First, it is important to start early. Communication and research should begin as early as possible. Second, the process should include a risk analysis to integrate the separate components of the research in order to understand the implications of research and identify the key uncertainties. The risk analysis should begin early to indicate needs for future research. The process should strive to recognize uncertainty, incorporate it into the analysis, and, through ongoing research, reduce that uncertainty. In this manner, early insights from research efforts and analysis can direct later efforts to obtain the best understanding as quickly as possible.

Finally, cooperation among the stakeholders is critical. The research needs are potentially enormous. Only through cooperation will understanding of the toxics issues develop quickly. The common goal should be to develop a management strategy for toxics that protects public health and welfare and that is economical and efficient.

National Trends in Air Toxics Policy

Michael Shapiro

I. INTRODUCTORY COMMENTS

The U.S. Environmental Protection Agency (EPA) is faced with an enormous task in implementing all parts of the Clean Air Act and especially in implementing the provisions dealing with toxic air pollutants under Title III of the amendments. EPA hopes that, through joint activities with all the affected parties, through consensus building, through consultation, and through a willingness to undertake some novel approaches, the air toxics program can achieve environmental objectives in a flexible and economically efficient manner.

II. EPA'S TOXICS PROGRAM

EPA has had a program for regulating toxic air pollutants since the passage of the 1970 Clean Air Act. The original approach dealt with one pollutant at a time and involved the use of extensive risk assessment. The risk assessment became very controversial because of the early stages of development of EPA's risk assessment capabilities and the debate about the appropriate role of such estimates in the regulatory decision process. The result was a very low output for the program. Only seven toxic air pollutants were regulated in a 20-year period.

Title III of the Clean Air Act Amendments (CAAAs) takes a new approach. EPA will set regulations by source categories, and initially these regulations will be based on achieving emission reductions that reflect the capabilities of control technologies. Although risk assessment does become important in later stages of the toxics program implementation, much of the attention over the next 10 years will be directed toward developing emission standards that are largely driven by technology.

III. EPA'S IMPLEMENTATION OF TITLE III

The Act lists 189 potentially toxic substances. A petition process allows for modification of this list. EPA is charged with identifying all major sources of emissions for these 189 pollutants. EPA issued its final list of source categories in July 1992. The list of source categories contains over a hundred entries and is also subject to modification through a petition process.

Following the issuance of the source category list, EPA developed a regulatory agenda. This agenda was proposed September 1992 and is scheduled for promulgation by Fall 1993. This regulatory agenda is a schedule of when EPA intends to issue regulations for each of the source categories on the list. The Act requires the following basic outline for that schedule:

- First 40 source categories regulated by November 1992
- 25% of all source categories regulated by November 1994
- 50% of all source categories regulated by 1997
- All source categories regulated by the end of the decade

The Act requires EPA to issue standards based on the Maximum Achievable Control Technology (MACT). EPA will decide what constitutes the best technologies that can be applied for both existing and new facilities in each source category. The EPA will then issue regulations to require these technologies. In general, these technologies will have to be in place no later than 3 years following the issuance of standards.

The Clean Air Act (CAA), as revised in 1990, includes provisions for periodic review and update of the MACT standards. It also includes provisions for delaying compliance with the standards through various mechanisms for extensions. One of the most important is the "Early Emission Reduction Credit". If a particular facility can achieve a 90% reduction of its total toxic emissions prior to EPA's issuance or proposal of regulations, the Early Emission Reduction Credit entitles the facility to a 6-year deferral from compliance with the MACT standards. This credit is designed to encourage companies to reduce their toxic emissions early by providing a potential reward down the road. This deferral of compliance will allow recovery of the cost of early investment in controls. This is a good deal both for the environment and for industry. The final regulations concerning early emissions reductions were published in December 1992.

The CAA includes provisions for slight delays in implementation based on either the unavailability of controls or the difficulty of installing controls. Other provisions allow for some compliance extensions if a facility has already installed, for other reasons, the best available control technology or a lowest achievable emission rate technology.

The formal regulatory process allows for risk assessment after EPA has proceeded through a round of regulation on each major source category. Once the MACT regulation is in place for an industry, EPA has up to 8 years to determine whether the risks remaining after those controls are in place are still significant enough to warrant further regulation. In situations where EPA believes further regulations are warranted, the agency will develop regulations using a risk-based approach.

The CAA also incorporates an important provision for modifications. The toxics amendments set up a separate program for reviewing modifications, reconstructions, or new facilities that are built under the MACT provisions. These reviews are done through a permitting program, that in most cases will be run by state agencies.

IV. FOCUS OF RECENT EPA ACTIVITIES

EPA has focused much of its early attention in the following areas:

- *Source Category List* — The source category list was a key focus of early EPA efforts. Much work went into the development of appropriate definitions of categories that are both broad enough to include all of the major sources of toxic air pollutants, and at the same time, specific enough to identify major differences across source types.
- *First Set of Toxic Regulations* — The first toxic regulations that will be issued have to do with the dry cleaners and the emissions of perchloryethelene. EPA is under a court order to propose regulations under the toxics program to affect this source category by November 1991. For this category, EPA will address both the MACT standards for major sources and the Generally Available Control Technology (GACT) for smaller sources. Congress specifically provided GACT to allow smaller sources to be regulated through a less stringent control technology.
- *Early Emission Reduction Credit* — As noted earlier, EPA promulgated regulations for this program early to give industries that are subject to the first round regulations an opportunity to take advantage of this provision and reduce their emissions before the regulations go into place.
- *Hazardous Organics National Emission Standards (HON)* — The HON is the MACT regulation covering the organic chemicals industry. It will set the pattern for the MACT controls for all major sources. EPA proposed the HON in December of 1992.
- *Regulations Concerning Modifications* — EPA will be proposing regulations concerning modifications by the end of 1992. These regulations will lay out EPA's definition of facilities' modifications and procedures for compliance. EPA will require case-by-case revisions to permits to modify existing facilities or build new facilities.

V. RESEARCH NEEDS

In order to put this program in to place, EPA will be heavily dependent on an aggressive research program. The numerous regulations that EPA must issue over the next decade, along with the need to address a residual risk condition in order to do the risk-based regulation, will require the agency to have much more information than it currently has. Some of the key topics to address are listed below.

- The list of 189 chemicals is not fully characterized. Much work is required to understand the various toxicities of those chemicals, their potencies, and their various environmental and health effects. This is a major research priority for the agency over the next decade.

- Improved risk assessment procedures are needed in order to determine the risk-based controls. EPA also requires better information on new control technologies, especially pollution prevention technologies.
- Monitoring will be critical, both in order to understand the toxics problem better and to develop enforceable regulations. There is a major focus of effort underway to improve source monitoring capabilities as well as ambient monitoring capabilities. Availability of continuous emission monitoring (CEM) played a key role in the effectiveness of the acid rain provisions; CEM helped facilitate the proposed regulations and build an enforceable accounting framework. This capability does not currently exist in the toxics program. If the toxics program ultimately uses an approach based on trading or emissions averaging, the ability to identify, measure, and account for the emissions coming out of our facilities requires significant improvement.

VI. IMPACT ON UTILITY INDUSTRY

Because the CAA requires EPA to look at the air toxics emissions from utilities that will result after all other provisions of the Act are in place, the utility industry is not included on a schedule together with most other major source categories. EPA must take into account gains achieved from other provisions that control utility emissions of particulates or gases, and the EPA must determine whether a significant health risk exists after those regulations are in place. EPA will report to Congress on the general problem of air toxics from utilities in November 1993. In addition to this study, The CAA requires a separate study on the problem of mercury releases into the environment from utilities and other sources. This study is scheduled for completion in November 1994. The combination of these two studies will give the agency information on whether additional regulation of the utility industry is needed in the toxics area.

Another study that could affect the utility industry is the Great Waters Study. This study is focusing on the Great Lakes, the Chesapeake Bay, and coastal waters. Under congressional mandate, EPA will look at the effects of deposition of toxic air pollutants on these great waters and determine the need for additional regulation beyond the standards in the Clean Air Act. The report to Congress is due on this issue by November 1993, and by 1995 EPA must decide if additional toxic regulation is required.

This study creates a further impetus to improve our understanding of sources of toxics, their properties, and their distribution in the environment. In order to base decisions with respect to the utility industry on the best science available, EPA intends to work very closely with both EPRI and the Department of Energy. EPA does not have the resources to do extensive monitoring, and hopes to work together with the industry and the Department of Energy to develop information that is necessary to make the best judgments concerning the need for further regulation.

VII. CONCLUDING REMARKS

What will be accomplished over the next decade will depend on the information that industry is able to bring to the table, the cooperation from all parties involved, and the kinds of dialogue that can be created over the next several years. It is important to allocate society's resources to reduce the toxics loading in the environment in a wise manner. Regulations must address key public health and environmental problems, as well as take into account the legitimate needs for flexibility in implementation on the part of industry. By combining good science and good regulation with a creative dialogue among the parties, the air toxics program, like the acid rain and reformulated gasoline programs, will result in a win-win situation for EPA, for the affected industries, and for the environmental community.

The California Experience in Toxic Air Pollutants Control

John S. Lagarias
Carolyn B. Suer
Peter D. Venturini
Donald J. Ames

I. INTRODUCTION

During the last decade, the State of California began an aggressive air toxics control program. At that time, the federal National Emission Standards for Hazardous Air Pollutants (NESHAPs) program was bogged down. Scientific evidence was showing the presence of many toxic compounds in the ambient environment. The ubiquitous presence and persistence of DDT found in the polar icecap region, peregrine falcon eggs, and even in human tissue was a classic example. It raised questions as to whether other contaminants were also present and causing serious public health damage. Improvements in analytical air quality monitoring techniques were detecting known toxics in the atmosphere in parts-per-million and parts-per-billion levels. Some, like benzene, were found to have a half-life on the order of a month of so. Other toxics were persisting in the atmosphere for periods varying from minutes to years.

Ongoing regulatory programs for criteria pollutants, particularly for ozone and particulate matter, were producing side benefits by reducing some toxic pollutants. The regulation of criteria pollutants, however, was not adequate or specific enough to control toxic pollutants whose concentrations were typically several magnitudes lower than those of the criteria pollutants.

This combination of awareness of the presence of known toxic air pollutants in the atmosphere and the absence of guidance or leadership by the federal government led the State of California to initiate its own toxic air pollutant program.

II. AIR TOXICS LEGISLATION OVERVIEW

California's air toxics program is built around the Toxic Air Contaminant Identification and Control Act, known more commonly as Assembly Bill 1807 or the Tanner Bill, after Representative Sally Tanner, its principal author. The bill separates the identification and risk assessment of air toxics from management and control actions. Passage of the legislation in 1983 was long and arduous. To the legislation's credit, it claimed as supporters industrial groups, environmentalists, public interest

Table 1. California Air Toxics Legislation

Year of implementation	Title of legislation
1984	Toxic Air Contaminant Program
1986	Proposition 65
	Landfill Gas Testing
1987	Toxic "Hot Spots" Act
	Air Toxics Monitoring
	Advisory Board on Air Quality and Fuels
1988	Motor Vehicle Air Toxics Bill

groups, and the regulatory agencies. The legislation received broad support because it proposed a logical regulatory process using the best available scientific and technical information. The process also required a review and assessment procedure that evaluated the severity of the possible toxic impacts on public health.

Other toxics-related legislation soon followed. In 1986, voters approved Proposition 65, a safe drinking water initiative. The proposition called for the establishment of a list of carcinogens and reproductive toxicants found in the environment, prohibited the discharge of these pollutants into drinking water, and required public exposure warnings to be posted. A landfill gas testing bill was also adopted in 1986 which required that hundreds of landfills, active and inactive, be tested for the presence of toxics and other pollutants. Landfills were perceived as a possible health risk and this action was taken to quantify that risk.

In 1987, another major piece of air toxics legislation established what is now known as the "Hot Spots" Act. This program required facilities to inventory over 700 toxic substances, conduct risk assessments if necessary, and notify the public of any significant risk. Two additional air toxics bills were also passed in 1987. One bill required the Air Resources Board (ARB) to report on the effectiveness of its toxic air monitoring network and make recommendations for supplemental monitoring. The other legislation established the Advisory Board on Air Quality and Fuels that was commissioned to report on the necessity and feasibility of using mandates and incentives for clean-burning fuels in California. The effort to control motor vehicle toxics was further aided in 1988 with the adoption of the Motor Vehicle Air Toxics Bill. This bill required accelerated control of toxic air contaminants associated with motor vehicle emissions. A summary of these legislative actions is shown in Table 1 with the year of implementation of each bill.

III. CRITERIA POLLUTANT CONTROL BENEFITS

Criteria air pollutants — ozone, nitrogen oxides, sulfur oxides, carbon monoxide, and suspended particulate — are present in the atmosphere in concentrations many orders of magnitude higher than air toxics and have had 2 decades of control actions. Typical annual average concentrations are shown in Table 2, which compares criteria- and toxic-pollutant concentrations. The criteria-pollutant controls for both stationary

Table 2. Comparison of Annual Average Concentrations

Criteria pollutants	Concentration (ppm)
Carbon monoxide	1.9
Ozone	0.02
Sulfur dioxide	0.002
Nitrogen dioxide	0.039
Toxic pollutants	**Concentration (ppm)**
Benzene	0.0039
Carbon tetrachloride	0.00014
Perchloroethylene	0.00055
Trichloroethylene	0.00013
1,3-Butadiene	0.00048
Methyl chloroform	0.0033

Note: These values are from a monitoring station in the Los Angeles area. They are provided as an example and should not be construed as representative of values at other sites in California or representative of statewide averages.

sources and motor vehicles often reduced toxic pollutants in the process. Coatings regulations, designed to reduce volatile organic compounds in paints, have in some instances also reduced contaminants such as perchloroethylene, trichloroethylene, methyl ethyl ketone, and methyl isobutyl ketone.

Recently adopted low-emission vehicles and clean-fuels regulations will reduce not only hydrocarbons, carbon monoxide, and nitrogen oxides, but also benzene, 1,3-butadiene, formaldehyde, and acetaldehyde. Diesel fuel specifications, designed primarily to reduce emissions of nitrogen oxides, particulate matter, and sulfur dioxide, will also reduce benzo(a)pyrene and polycyclic aromatic hydrocarbons.

Many of the stationary-source control devices required by local districts to control particulate matter and reactive organic compounds also reduce toxic metals and organics. While these actions are not typically thought of as part of an air toxics program, the toxics benefits are significant.

IV. TOXIC AIR CONTAMINANT PROGRAM

The Toxic Air Contaminant Identification and Control Program is the foundation of the California program to reduce public exposure to air toxics. The legislation established a two-phase process with clear separation of identification and risk assessment from risk management and control. Each process has a clearly defined role for the various involved state and local agencies and timeframes for action. ARB is the state agency charged with the responsibility of carrying out the program.

The first phase of the process calls for the identification of substances as toxic air contaminants. California's Office of Environmental Health Hazard Assessment (OEHHA) conducts a health evaluation of substances under consideration. ARB prepares an exposure assessment and submits both reports to the Scientific Review

Table 3. Estimated Population-Weighted Statewide Risk from Exposure to Average Ambient Concentrations of TACs

Substance	Potential risk (lifetime cancer cases per million)
Benzene	220
Chromium (VI)	140
Dioxins and dibenzofurans	80[a]
Ethylene oxide	20[a]
Carbon tetrachloride	10
Methylene chloride	10
Ethylene dichloride	10
Vinyl chloride	10
Asbestos	10
Cadmium	7
Inorganic arsenic	6
Ethylene dibromide	4
Trichloroethylene	2
Nickel	2
Chloroform	1

[a] Based on data for the South Coast region.

From Venturini, P. D., Denton, J. E., Howard, K., Huscroft, S. V., Shiroma, G. S., and Suer, C. B., California Air Toxics Program: Review of Progress and Future Direction, presented at Air and Waste Management Association, 84th Annual Meeting and Exhibition, Vancouver, BC, June 16–21, 1991.

Panel (SRP) for review. The SRP, a nine-member appointed panel, determines if the report is based on sound science and if the conclusions and assessments are reasonable and adequate. Once approved by the SRP, the ARB, at a public hearing, makes the determination whether the substance under review should be identified as a toxic air contaminant (TAC).

The Toxic Air Contaminant Identification and Control Act defines a toxic as "an air pollutant which may cause or contribute to an increase in mortality or an increase in serious illness, or which may pose a present or potential hazard to human health." In addition, substances that have been identified as hazardous air pollutants (HAPs) pursuant to section 112 of the Federal Clean Air Act shall be identified by ARB as TACs. With the passage of the new Clean Air Act, the federal process has been obscured since the identification of all 189 compounds as hazardous has not gone through a scientific review and risk assessment process.

The California risk assessment phase identifies those substances of concern in California based on the following criteria: risk of harm to public health, amount or potential amount of emissions, manner of usage, atmospheric chemistry of the compound, and ambient concentrations. All of these factors are considered in selecting substances to enter the identification process.

ARB has identified 15 compounds as TACs up to this time (Table 3). The table also shows the estimated potential risks per million persons exposed to average California ambient concentrations. ARB has purposefully focused on those compounds presenting the highest risk to residents of California. Eight additional compounds are currently under active review by the staff of ARB. ARB's list of substances for the toxic air contaminant program review contains another 209 compounds for further evaluation. These compounds include the federal HAPs, both those known to be emitted in California and those for which emissions information is not currently available for this state.

The second phase of the process, risk management and control, follows the identification of a substance as a TAC. The risk management phase of the program is designed to reduce public exposure to toxics by developing and implementing the best available control strategy. This phase clearly defines the roles of ARB and local districts. ARB prepares a control strategy report in consultation with the districts, affected industries, and the public. The report addresses emission levels, atmospheric persistence, sources of the emissions, technological feasibility of controls, costs, and substitute compounds. This process involves numerous public workshops and meetings with affected industries, and initiates needed research. The completed report is presented to the Board and, if approved, the local districts are to adopt the identified control measures or equivalent within 6 months.

Six control measures have been adopted to date by ARB. As with the identification of substances as TACs, the control measures have targeted the highest-risk facilities. Typically, the toxic emissions from these facilities have been controlled by at least 90% and up to 99.9% in some cases.

The control measure for benzene requires vapor recovery at retail service stations. Existing stations with an annual throughput of greater than 480,000 gal, and all new service stations, must install phase I and II vapor-recovery systems. Smaller existing service stations must install vapor recovery when making major modifications to underground piping. This regulation immediately added over 300 more service stations statewide to the thousands of service stations in the state that had already been equipped with vapor recovery systems to comply with local districts' ozone regulations. More stations will be affected as piping modifications are made.

Two control measures were adopted to reduce emissions of hexavalent chromium. One requires small, medium, and large chrome plating and anodizing shops to achieve either 95, 99, or 99.8% reduction of uncontrolled emissions, depending on plant size, or to emit less than 0.15, 0.03, or 0.006 mg of hexavalent chromium per ampere hour. The second control measure prohibits the use of hexavalent chromium in the circulating water of a cooling tower. Together these two control measures affect over 1400 facilities statewide.

The asbestos control measure banned the use of serpentine rock containing greater than 5% asbestos in surfacing of unpaved areas. Potentially, hundreds of miles of road surfaces are affected by this control measure.

Recently adopted control measures primarily affect the medical community using sterilizers and incinerators. The control measure for sterilizers and aerators requires control of ethylene oxide (EtO) emissions, depending on EtO use, by 99 to 99.9% for sterilizers and 95 to 99% for aerators. Small users of EtO are exempt, but must comply with reporting requirements.

The dioxins control measure for medical waste incinerators requires control of dioxins emissions, from units burning over 25 tons a year, by 99% or an emission rate of less than 10 ng/kg of waste. Other requirements in this regulation call for operator training, source tests, and record keeping. These two regulations affect over 600 facilities in the state.

It is really too early to make a definite statement, but comparisons to EPA's preliminary draft maximum achievable control technology (MACT) standards show ARB adopted regulations for TACs to be more stringent. For example, EPA's proposal for EtO sterilizers would require 98.8% overall control, compared to ARB's 99.8% overall requirement for a sterilization cycle. The California legislation calls for controls to reduce emissions to the lowest level achievable using best available control technology (BACT). When the risk from sources warrants more stringent action (for example, the chrome control measure for large plating and anodizing shops), technology-forcing controls are identified.

The Board has recognized that risk management may not always require regulatory action. Three control decisions have fallen into this category: ethylene dibromide, ethylene dichloride, and carbon tetrachloride. The use of ethylene dibromide and ethylene dichloride has been phased out. The primary, nonpesticidal use of both ethylene dibromide and ethylene dichloride was in leaded gasoline. However, with the declining use of leaded fuel during the 1980s and its eventual elimination, there were no other significant sources of these TACs. Therefore, the Board decided no additional action was necessary. Carbon tetrachloride emissions were voluntarily controlled by facilities before control actions were developed, thereby eliminating the need for ARB regulations.

Other actions to reduce TACs include advisories to local districts and schools. The advisories require no legal action. During the evaluation of benzene emissions from service stations, it was found that health benefits could be gained from the use of hold-open latches on gasoline-dispensing nozzles. Personal exposure to benzene while refueling vehicles can be reduced by 60 to 90% when latches are used. However, several local fire marshals within the state were concerned with the potential for fuel spillage and possible fires during use of hold-open latches. ARB issued an advisory to districts on the benefits of latch use and gave the option to each district to encourage or require greater latch availability, depending on local fire marshal's recommendations.

ARB also issued an advisory to school officials warning of the possible health threat to children from play areas covered with asbestos-containing serpentine rock. Suggestions on how to reduce exposure were made in the advisory if serpentine rock was found on school sites.

A significant consequence of the toxic air contaminant program concerns the pollution prevention measures inherent in the actions taken to date. The identification of a substance by ARB as a TAC is an act that, by itself, has proven to be an effective pollution prevention action. Once a compound is formally identified, industry is put on notice that the TAC may soon be regulated. Carbon tetrachloride is a case in point where no regulatory action was needed because controls were voluntarily put in place. Most control measures adopted also have elements of pollution prevention. For example, sources meeting the chrome-plating regulation are encouraged to use process modifications to meet the emissions limitations. The chrome cooling tower and asbestos-containing serpentine rock regulations prohibit the use of those compounds, thereby preventing the emissions in the first place. Other requirements include operator training, record keeping, and source testing, all in the effort to reduce emissions. Over 1500 facilities statewide are affected by the regulations that require some pollution prevention action.

V. "HOT SPOTS" ACT

A second major piece of legislation which has become a fundamental part of the California Air Toxics Program is the "Hot Spots" Act that was adopted in 1987. The goals of the "Hot Spots" Act are (1) to collect toxics emissions information, (2) to identify facilities having localized impacts, (3) to quantify the risk from emissions, and (4) to notify the affected public of significant risks.

The "Hot Spots" Act requires a facility that emits a toxic substance subject to reporting requirements to file an inventory plan with the local district. Following plan approval by the districts, the facility operator implements the plan and submits data to the district. Each district ranks the facilities for risk assessment as high, intermediate, or low priority. Facilities designated as high priority submit risk assessments, which are reviewed by the district and the OEHHA. If a district determines that there is a significant health risk associated with emissions from a facility, public notification is then required.

To date, over 3000 facilities, representing the largest emitters, have submitted emission inventories in California. Currently, over 800 facilities have been categorized as high priority by the districts. Each district has the opportunity to establish its own significant risk levels. Typically, for carcinogens, this risk level has been set between 1 and 10 in a million.

The "Hot Spots" Act offers substantial benefits to California's Air Toxic Program. It identifies and locates sources of toxic emissions, which may have otherwise remained undetected, It further provides exposure and risk analyses, which aid the local districts and ARB in developing regulations for the control of toxic pollutants.

VI. OTHER AIR TOXIC ACTIONS

A. Motor Vehicle Toxics

Several other legislative mandates have become important elements of the Air Toxics Program. ARB staff estimates that about one half of the risk from air toxics are from motor vehicles. The efforts to control motor vehicle emissions, both criteria and toxic pollutants, have resulted in a significant reduction of risk. In 1990, ARB published a "Motor Vehicle Toxics Control Plan and Review of Schedule" in response to Assembly Bill 4392. This report indicates that there are five "high-risk" substances that account for approximately 98% of potential statewide motor vehicle-related cancer cases. These substances are benzene, 1,3-butadiene, diesel particulate, formaldehyde, and acetaldehyde. The report further identifies benzene and 1,3-butadiene as the two substances presenting over 80% of the risk.

Hydrocarbon reductions result in significant toxics reductions (especially benzene and 1,3-butadiene), while not specifically designed to reduce those pollutants. The new 0.25 g/mile hydrocarbon standard for light-duty vehicles is estimated to prevent 2000 to 3000 potential lifetime cancer cases statewide by reducing cancer-causing hydrocarbons.

Motor vehicle regulations adopted by ARB which will reduce toxics are the low-emission vehicle and clean-fuels regulations. These regulations were developed in an effort to maximize emissions reductions from motor vehicle exhaust by establishing new stringent standards. To meet these standards, motor vehicle manufacturers' certification of vehicles may be dependent upon clean-burning fuels. This integrated approach treats a vehicle and fuel as a system. Implementation of these regulations is expected to reduce a variety of toxic pollutants, including four of the top five "high-risk" substances (benzene, 1,3-butadiene, formaldehyde, and acetaldehyde).

B. Landfill Gas Testing

Landfills have long been perceived as posing a health risk. Legislation passed in 1986 required all active and some inactive landfills in California to be tested for the presence of toxic compounds. To date, over 400 landfills out of the required 634 sites have been tested. The results show that hazardous and nonhazardous waste sites appeared to be similar in their ability to produce toxic gases. In 70% of the sites tested, at least 1 of 10 toxic compounds tested were present (Table 4). Of the toxic gases found, benzene and vinyl chloride are the most significant substances with regard to risk based on emissions and relative toxicity.

Local districts have primary authority over emissions from stationary sources, including landfills. ARB recently approved a suggested control measure (SCM) to reduce landfill gas emissions. The SCM will require the installation of landfill gas collection systems at new, active, and inactive landfills having more than 500,000 tons of waste in place. The SCM contains performance standards to ensure the collection systems are properly installed and to control the leakage of landfill gas. Source tests conducted by ARB and local districts have shown that the toxic air contaminants may be reduced by up to 99% by landfill gas disposal techniques.

C. Proposition 65

In 1986, the voters of California passed Proposition 65. The purpose of Proposition 65 is to prohibit the discharge of carcinogens or reproductive toxicants into

Table 4. Concentration of Specified Contaminants in Landfill Gas Samples

Compound	Number of landfills contaminant detected[a]	Median PPBV	Average[b] PPBV	Maximum[b] PPBV
Perchloroethylene	241	38	1,100	45,000
Trichloroethylene	228	30	840	11,000
Methylene chloride	197	37	4,800	160,000
1,1,1-Trichloroethane	180	2U[c]	650	96,000
Benzene180	132U	2,500	480,000	
Vinyl chloride	160	106U	2,200	72,000
Ethylene dichloride	65	5.1U	600	98,000
Chloroform	58	0.8U	360	11,000
Carbon tetrachloride	31	1.2U	11	2,100
Ethylene dibromide	24	0.3U	4	660

[a] Landfill gas sampling at 340 landfills.
[b] Medians and maximums of the average sampling results from individual sites.
[c] U, nondetected; the number shown is the detection limit.

From California Air Pollution Control Officers Association Technical Review Group Landfill Gas Subcommittee and California Air Resources Board Stationary Source Division, The Landfill Testing Program: Data Analysis and Evaluation Guidelines, approved by the Air Resources Board, September 13, 1990.

drinking water and to require that warnings be provided to individuals if they are exposed to those substances. Currently, that list contains over 485 substances.

The discharge prohibition and warning apply only in those instances where the exposure results in a significant risk. Significant risk is currently defined for these purposes by the California Health and Welfare Agency as one potential excess cancer case per 100,000 based upon lifetime exposure. For reproductive toxicants, "no significant risk" is defined as less than 1/1000 of the no-observable-effect level.

Airborne emissions are also subject to Proposition 65. ARB provides guidance to local districts and the public on air exposures. Additionally, as a part of the notification requirements, any government employee who obtains information regarding the illegal discharge of the listed substances during the course of duty must report this information to local officials.

D. Air Toxics Monitoring

Data obtained from various air monitoring activities are used to evaluate population exposure to toxic air contaminants. ARB's toxics monitoring network has 22 stationary sites and a mobile station covering virtually the entire state. Thirty-eight compounds are currently being sampled (Table 5). Table 5 also identifies the analytical detection limit for each TAC. The limits of detection are commonly in the parts-per-billion range.

The two most populous air districts, the Bay Area Air Quality Management District and the South Coast Air Quality Management District, operate their own

Table 5. Compounds Sampled in ARB Network

Toxic compounds	Limit of detection (ppb)	Toxic compounds	Limit of detection (ng/m^3)
Dichloromethane	1.0	Benzo(a)pyrene	0.05
Chloroform	0.02	Benzo(n)fluoranthene	0.05
Ethylene dichloride	0.20	Benzo(k)fluoranthene	0.05
1,1,1-trichloroethane	0.01	Arsenic	0.40
Carbon tetrachloride	0.02	Beryllium	0.02
Trichloroethylene	0.02	Cadmium	0.20
Ethylene dibromide	0.01	Lead	3.00
Perchloroethylene	0.01	Hexavalent chromium	0.20
1,3-Butadiene	0.04	Total chromium	1.00
Acetaldehyde	0.10	Nickel	1.00
Formaldehyde	0.10	Manganese	1.00
Benzene	0.50	Dibenzo(a,h)anthracene	0.05
Toluene	0.20	Benzo(ghi)perylene	0.05
Ethyl benzene	0.60	Indeno(1,2,3-cd)pyrene	0.05
1,4-Xylene	0.50		
1,3-Xylene	0.60		
1,2-Xylene	0.10		
Chlorobenzene	0.10		
Styrene	0.10		
1,3-Dichlorobenzene	0.20		
1,2-Dichlorobenzene	0.10		
1,4-Dichlorobenzene	0.20		
Methyl ethyl ketone	0.10		
Methyl isobutyl ketone	0.10		

From Loscutoff, W. V., Ambient Air Toxics Monitoring Network, presented at the Air Resources Board Meeting, March 15, 1991.

monitoring networks. To put the size of the California network into perspective, over 90% of the data in the federal air toxics database last year were supplied by the California system.

E. Pesticides

ARB has conducted air monitoring for 17 pesticides since 1984 at the request of the Department of Pesticide Regulation (DPR). DPR has responsibility for conducting a pesticide toxic identification and control program. ARB usually monitors during the 1- or 2-month periods of peak applications of pesticides. Twenty-four samples are collected, for 4 days a week, up to 5 sites, 24 samples are collected. Background comparison data are also taken at sites away from the applications.

Most of the measured pesticide concentrations have been low. Monitoring tests taken by ARB in 1990 for 1,3-dichloropropene (Telone) in Merced County found concentrations presenting significant health risks. Based on this information, all permits for users of Telone were canceled in California by DPR.

VII. SUMMARY

The State of California has developed, and continues to maintain, an aggressive air toxics identification and control program. It follows a logical process which has been addressing the most serious issues first. It uses health-based information, exposure data, technological capability, and peer review processes before taking regulatory action. The California air toxics program attempts to balance science, technology, cost, and health benefits in the process. The process is slow, difficult, and costly, but it is working.

REFERENCES

1. Venturini, P. D., Denton, J. E., Howard, K., Huscroft, S. V., Shiroma, G. S., and Suer, C. B., California Air Toxics Program: Review of Progress and Future Direction, presented at Air and Waste Management Association, 84th Annual Meeting and Exhibition, Vancouver, BC, June 16–21, 1991.
2. California Air Pollution Control Officers Association Technical Review Group Landfill Gas Subcommittee and California Air Resources Board Stationary Source Division, The Landfill Testing Program: Data Analysis and Evaluation Guidelines, approved by the Air Resources Board, September 13, 1990.
3. Loscutoff, W. V., Ambient Air Toxics Monitoring Network, presented at the Air Resources Board Meeting, March 15, 1991.

Implications of the 1990 Clean Air Act Amendments for the Utility Industry[1]

Lee B. Zeugin

I. INTRODUCTION

The 1990 Clean Air Act Amendments significantly revise the regulation of hazardous air pollutants under § 112. Whereas the old § 112 required EPA to take a pollutant-by-pollutant approach and impose health-based emission standards for each pollutant, the new § 112 requires EPA to regulate categories of sources using technology-based emission standards. The amended § 112 specifically identifies 189 hazardous air pollutants, directs EPA to promulgate a list of the "major" and "area" source categories that emit those pollutants, and specifies a schedule within which EPA must promulgate emission standards for all source categories.

The electric utility industry has been singled out under § 112 for further study. The Amendments require EPA to study hazardous air pollutant emission from fossil fuel-fired power plants to determine whether regulation of those units under § 112 is necessary. The purpose of this presentation is to provide an overview of the new § 112 and to explain some of the implications of this regulatory scheme for the electric utility industry.

II. OVERVIEW OF § 112

Section 112(b) of the Clean Air Act Amendments establishes a list of 189 hazardous air pollutants that must be addressed by the new air toxics program. The statute directs EPA to issue a list of categories and subcategories of "major" and "area" sources that emit those pollutants. On July 16, 1992, EPA published its initial list of source categories to be regulated under § 112.[2] Because § 112 requires EPA

[1] The comments presented in this paper are those of the author and are not necessarily those of any client of Hunton & Williams. The author wishes to acknowledge Margaret L. Claiborne for her substantial contributions to this paper.

[2] 57 Fed. Reg. 31, 576 (1992). The Act directs EPA to revise the list from time to time, but no less than every 8 years, if appropriate.

to study emissions from electric utilities, the Agency did not include electric utilities on its list of source categories.

Section 112 distinguishes between "major" and "area" sources of emission. A "major source" is defined as "any stationary source or group of stationary sources located within a contiguous area and under common control that emits or has the potential to emit considering controls, in the aggregate 10 tons per year or more of any hazardous air pollutant or 25 tons per year or more of any combination of hazardous air pollutants."[3] An "area source" is any source that is not a major source.[4]

Once a source is on the list, it will be subject to technology-based emission standards. EPA must establish emission standards for each major source category on the list by the year 2000.[5] For major sources, the emission standards will be based on "maximum achievable control technology" (MACT), that is, the standard must require the maximum degree of reduction in air toxic emissions achievable based on the best technology currently available for the source category in question. Existing sources subject to MACT standards will have to achieve the average emissions limitation achieved by the best performing 12% of the existing sources or the average emission limitation achieved by the best performing five existing sources in that category. New sources must meet the emissions limitation achieved by the best performing plant.

EPA must also promulgate air toxic emission standards for area sources. The Act directs EPA to establish standards for area sources based on "generally available control technology" (GACT). In practice, this standard should be less stringent than the MACT standard for major sources. Regulations for area sources must also be published by the year 2000.

Finally, by November 1996 EPA must investigate and report to Congress on the risks to public health remaining, or likely to remain, after application of MACT standards and make recommendations for legislation necessary to control those risks.[6] The Agency must adopt residual risk-based standards if any source in a source category presents a risk to the maximum exposed individual of more than 1 in 1 million (1×10^{-6}). The risk-based standards must protect public health with an ample margin of safety and prevent adverse environmental effects.

3 CAA § 112(a)(1).

4 CAA § 112(a)(2).

5 More specifically, EPA must establish emission standards for no fewer than 40 source categories or subcategories by 1992, for 25% of the source categories by 1994, for an additional 25% by 1997, and for the remaining categories and subcategories by the year 2000. CAA § 112(e).

6 CAA § 112(f)(1). The report must include (1) the significance of the risks, (2) the available methods and costs of reducing the risks, and (3) the "actual health effects with respect to persons living in the vicinity of sources, any available epidemiological or other health studies, risks presented by background concentrations of hazardous air pollutants, any uncertainties in risk assessment methodology or other health assessment techniques, and any negative health consequences to the community of efforts to reduce such risks." CAA § 112(f)(1).

The air toxics regulations will be implemented by the states under the Title V operating permit program. Once the states have developed an EPA-approved operating permit program, all major sources under § 112 will be required to obtain a permit which specifies the applicable emission standards.

III. THE STUDY PROVISIONS

The Amendments also require EPA to conduct several studies concerning hazardous air pollutants and their associated health risks. The following studies will influence EPA's decision whether or not to regulate the utility industry under the air toxics provisions.

A. The Electric Utility Study

Section 112(n) directs EPA to conduct a 3-year study of "the hazards to public health reasonably anticipated to occur as a result of [air toxic] emissions by electric utility steam generating units..."[7] "Electric utility steam generating unit" is defined as "any fossil fuel fired combustion unit or more than 25 megawatts that serves a generator that produces electricity for sale."[8] Thus, the study provision applies only to fossil fuel-fired power plants.

The purpose of the study is to determine whether air toxics regulations are "appropriate and necessary" after implementation of other provisions of the Act which affect electric utilities, such as the acid rain provisions. The statute does not define "appropriate" or "necessary", and therefore it is not clear what results would trigger the need for regulation. The statute also does not specify when the regulations must be promulgated if such a finding is made.

Finally, if the Agency finds that additional regulation is necessary, it must recommend to Congress "alternative control strategies" to limit emissions of air toxics. Because the provision only refers to "alternative control strategies" for limiting emissions, it is not clear that MACT standards would then be applied to fossil fuel-fired power plants.

The effect of this study provision is to exempt fossil fuel-fired units from listing and regulation under § 112 until after EPA completes the study. The Act requires EPA to complete the study and report its findings to Congress by November 15, 1993. The Agency has recently indicated, however, that it is unlikely to meet this deadline and that its final report to Congress may be 2 years behind schedule.

EPA has also indicated that it has limited resources to conduct the study. It is

7 CAA § 112(n)(A).

8 CAA § 112(a)(8). It also includes a "unit that cogenerates steam and electricity and supplies more than one-third of its potential output capacity and more than 25 megawatts electrical output to any utility power distribution system for sale..." CAA § 112(a)(8).

therefore essential for industry to prepare an accurate factual picture of hazardous air emissions from fossil fuel-fired power plants. This effort should include (1) identifying which of the 189 pollutants are likely to be emitted from power plants at significant levels, (2) collecting air toxic emissions data on those pollutants, and (3) performing risk analyses.

One open question about the utility study is whether EPA will examine other emission sources at a power plant such as fugitive dust emissions from coal piles and emissions from water treatment facilities. Although the definition of electric utility steam-generating unit would seem to limit the study to stack emissions, EPA has indicated that it may study all power plant emissions.

B. Mercury Studies

In addition to requiring EPA to study air toxic emissions from electric utilities generally, § 112(n) requires EPA to study specifically mercury emissions from electric utility steam-generating units and certain other sources.[9] The study must consider "the rate and mass of such emissions, the health and environmental effects of such emissions, technologies which are available to control such emissions, and the costs of such technologies."[10] EPA must report its findings to Congress by November 1994. The Act also mandates a companion study by the National Institute of Environmental Health Science "to determine the threshold level for mercury exposure below which adverse human health effects are not expected to occur."[11] The results of this study must be reported to Congress in 1993. Although the Act does not require EPA to promulgate regulations based on the results of these studies, EPA will likely consider the results when promulgating or revising regulations for mercury emissions. EPA has indicated that the information derived from the mercury studies may be used in the electric utility study.

C. The Great Lakes Study

Section 112(m) requires EPA to study atmospheric deposition of hazardous air pollutants to the Great Lakes, the Chesapeake Bay, Lake Champlain, and coastal waters. This provision requires EPA to establish a monitoring network to investigate the sources of atmospheric deposition, assess deposition rates, and evaluate the effects on public health and the environment. The statute requires EPA to report its findings to Congress by November 1993 and to recommend any air toxics regulations that are "necessary and appropriate" in light of the results by 1995. The utility industry should monitor the Agency's activities under this provision and be prepared

[9] CAA § 112(n)(B).

[10] CAA § 112(n)(1)(B).

[11] CAA § 112(n)(1)(C).

to help develop additional emissions data from plants located in those areas. It may also be necessary for the industry to identify nonutility emission sources that contribute to the pollution problem.

D. Other Studies Relevant to Electric Utilities

Finally, the amendments mandate two studies related to risk assessment. Section 303 establishes a Risk Assessment and Management Commission to "make a full investigation of the policy implications and appropriate uses of risk assessment and risk management in [federal] regulatory programs..."[12] The Commission must make any recommendations resulting from the study available for public comment by May 15, 1994. Those recommendations are likely to include (1) which exposure models are most appropriate, (2) whether the use of risk to the maximum exposed individual as a basis for setting standards is appropriate, and (3) how uncertainties should be reflected in risk assessments. In addition to this effort, § 112(o) requires the National Academy of Sciences (NAS) to conduct a review of risk assessment methodology which will be presented to the Risk Assessment and Management Commission for consideration. NAS will address techniques used to estimate and describe human exposure, carcinogenic potential and other adverse health effects of specific chemicals, and the practical application of risk assessment to Title III of the Clean Air Act.

IV. DEFINITION OF "SOURCE" AND SOURCE CATEGORIZATION

If, as a result of these studies, EPA reports to Congress that additional regulations are in fact necessary to control emissions of hazardous air pollutants from fossil fuel-fired power plants, how the Agency defines "source" and categorizes the industry will become very important issues.

A. Source Categorization and Subcategorization

As previously noted, § 112(c) of the Amendments requires EPA to list the categories and subcategories of major and area sources. If EPA concludes that electric utilities must be regulated under § 112 it may have to include them on that list and schedule them for regulation. How EPA chooses to categorize utilities will have a significant impact on the standards that will eventually be imposed on them.

If EPA categorizes the electric utility industry broadly, breaking it down into subcategories such as coal-fired plants, oil-fired plants, and gas-fired plants, the emission standards will be more stringent than if it categorizes the industry more narrowly. If the Agency takes the broad-brushed approach, existing plants in a particular category will have to achieve the average emission limitation achieved by

[12] CAA § 303(a).

the best-performing 12% of the existing sources or the average emission limitation achieved by the best-performing five existing sources in that category, the result being that older plants will have to meet the same standards as much newer ones. If, however, the Agency breaks the industry down into more specific subcategories, such as coal-fired plants with ESPs that meet new source performance standards, most plants in the category should be achieving essentially the same emissions performance.

Subcategorization may also be a way of limiting further regulation of the industry. If the study results indicate that only certain types of plants emit hazardous air pollutants, EPA could simply list those specific subcategories that need additional controls and exclude the rest of the industry from regulation.

B. Definition of "Source"

EPA has not yet defined "source" for purposes of issuing MACT standards. The Agency has indicated that it plans to set MACT standards for individual sources at a facility rather than setting a single MACT standard covering the entire plant. For purposes of determining whether a site is a "major source" (i.e., whether the plant emits more than 10 tpy of any air toxic or 25 tpy of any combination of air toxics), EPA plans to sum the emissions of the individual sources within a facility. If the total emissions exceed the threshold, the entire plant will be considered a "major source".

The definitions of "source" and "major source" are potentially important to the utility industry because EPA has taken the initial position that, if a facility is classified as a "major source", individual sources within that facility will also be considered "major sources" and therefore subject to MACT standards.[13] For example, if the entire power plant is deemed a "major source" because of boiler emissions, EPA could require MACT for all sources of emissions at the site including cooling towers, water treatment facilities, and coal piles.

V. MODIFICATIONS AND RECONSTRUCTION OF "MAJOR SOURCES"

The definition of "source" is also an important issue for purposes of plant modifications and reconstruction. Under § 112(g), once a state operating permit program is in place, if a modification to a "major source" is proposed and that modification would increase actual emissions of any hazardous air pollutant by more than a *de minimis* amount, the modification cannot begin until the facility demon-

[13] If, alternatively, the statute is interpreted as requiring only those sources of emissions which exceed the 10/25 ton thresholds to be classified as "major" ones, the remaining sources of emissions would be "area" sources. Those sources would ultimately have to meet less stringent GACT standards.

strates that it will meet the MACT for existing sources.[14] Because EPA has not yet defined "source" for purposes of MACT standards, it is unclear whether a modification would trigger MACT compliance for the specific emission point being modified or for the plant as a whole.

This modification provision differs from that under the NSPS and PSD programs in that not all modifications will trigger the rule, only those that will cause more than a *de minimis* increase in actual emissions. Section 112(a)(5) defines "modification" as "any physical change in, or change in the method of operation of, a major source which increases the actual emissions of any hazardous air pollutant emitted by such source by more than a *de minimis* amount or which results in the emission of any hazardous air pollutant not previously emitted by more than a *de minimis* amount." EPA has not yet defined "*de minimis*" for purposes of this section.

Even if the modification would cause more than a *de minimis* increase in emissions, sources making modifications can avoid demonstrating compliance with MACT standards by offsetting any increase in hazardous air pollutants by decreasing other hazardous emissions. More specifically, a physical or operational change that would otherwise fit the definition of modification will not be considered a "modification" if the owner or operator of the source shows that the "increase in the quantity of actual emissions of any hazardous air pollutant from such source will be offset by an equal or greater decrease in the quantity of emissions of another hazardous air pollutant (or pollutants) from such source which is deemed more hazardous..."[15]

One final point to note about the modification provision is that, although it does not apply until after operating permit programs are in place, it has the effect of imposing MACT standards on sources even before EPA establishes MACT standards for that particular category or subcategory. In those cases, the emission limitation will be set on a case-by-case basis.[16]

While modifications under this provision may subject sources to MACT standards for existing sources, a major source that is reconstructed will be subject to MACT standards for new sources.[17] The 1990 Amendments do not define "reconstruction", but EPA is likely to use the definition under the NSPS provisions as a starting point for defining the term.

[14] The modification provision does not apply to a source at all until a state operating permit program is in place.

[15] CAA § 112(g)(1)(A). EPA is in the process of developing regulations for the offset provision. The rule should include a ranking of the 189 hazardous air pollutants based on their effects on human health and the environment. The offset provision applies only in the context of modifications and will not protect major sources from MACT standards once EPA has promulgated them for the source category.

[16] CAA § 112(g)(2)(A).

[17] The maximum achievable control technology for new sources under § 112 is "the emission control that is achieved in practice by the best-controlled similar source, as determined by the Administrator." CAA § 112(d)(3).

VI. THE GENERAL DUTY CLAUSE

While the application of many § 112 provisions turns on the results of the electric utility study, at least one regulatory provision of § 112 creates immediate obligations for electric utilities as well as other industries. Congress enacted § 112(r) to prevent and detect accidental releases of hazardous air pollutants. This section applies to all owners and operators of stationary sources that produce, process, handle, or store certain "extremely hazardous" substances and imposes a "general duty" to prevent and respond to accidental releases of those substances.

EPA must promulgate a list of not less than 100 substances "which, in the case of an accidental release, are known to cause or may reasonably by anticipated to cause death, injury, or serious adverse effects to human health or the environment."[18] Section 112(r) imposes an affirmative duty on owners and operators "to identify hazards which may result from [accidental] releases using appropriate hazard assessment techniques, to design and maintain a safe facility taking such steps as are necessary to prevent releases, and to minimize the consequences of accidental releases which do occur.[19]

Although the statute requires EPA to promulgate regulations under this provision by November 1993, this "general duty" appears to be self-implementing, that is, it can be used as a basis for enforcement actions by EPA, even in the absence of any implementing regulations defining what steps are needed to comply with this "general duty".[20]

VII. CONCLUSION

EPA's study of fossil fuel-fired power plants is well underway. The utility industry needs to continue to work with the Agency and provide the data it needs to complete the study. The industry must also track and participate in the other rulemakings under § 112, keeping in mind that those regulations could have a substantial effect on electric utilities down the road.

[18] CAA § 112(r)(3). EPA proposed a list of 100 toxic substances on January 19, 1993. 58 Fed. Reg. 5102.

[19] CAA § 112(r)(1). The statute defines "accidental release" as "an unanticipated emission of a regulated substance or other extremely hazardous substance into the ambient air from a stationary source." CAA § 112(r)(2)(A).

[20] The statute specifies, however, that the general duty clause cannot be used as a basis for citizen suits or for suits seeking to recover for personal injury or property damage.

Control of Hazardous Air Pollutants in OECD Countries: A Comparative Policy Analysis*

Peter Wiederkehr

I. INTRODUCTION

The highly industrialized countries of the OECD are increasingly affected by a growing number of pollutants that occur in the atmosphere in much smaller concentrations (trace amounts) than the "traditional" air pollutants like sulfur oxides, nitrogen oxides, and ozone. These atmospheric contaminants are often referred to as toxic or hazardous air pollutants. The majority of OECD countries are expressing concern about pollution by heavy metals (e.g., cadmium, mercury), respirable mineral fibers (e.g., asbestos), toxic organic pollutants (e.g., benzene), polycyclic aromatic compounds (including PAHs) and halogenated organic compounds (e.g., vinyl chloride, dioxins). Many of these hazardous air pollutants are recognized to contribute significantly to the cumulative exposure and the total risk posed by air pollutants in urban and metropolitan areas.

In several OECD countries, the concern about the growing list of hazardous air pollutants has reinforced efforts to establish control programs. Some countries are reassessing their original approaches to this complex problem. The OECD project on hazardous air pollutants has proceeded in collecting information on different regulatory and other policy activities in OECD member countries. Detailed case studies have been conducted on the control of hazardous air pollutants of seven countries (France, Germany, Japan, The Netherlands, Sweden, Switzerland, and the U.S.), and a comparative policy analysis has been made. The full report on these country studies will be published in 1993.[1]

In this paper some of the major findings of this policy analysis are being summarized with emphasis on the priorities set for assessment and control, and the results achieved in terms of emission reductions of hazardous air pollutants. Future priorities in national and international action programs relating to monitoring, assessment, and control of hazardous air pollutants are discussed.

* This paper represents views of the author and not necessarily those of the OECD or its member countries.

II. POLICY FRAMEWORK AND REGULATORY APPROACHES

A. Overview of Control Policies

Although the policy principles vary considerably from country to country, generally the control of hazardous air pollutants is integrated into the overall air pollution control policy on stationary and mobile sources. The policy principles can either be described as being effect oriented (using, e.g., quantitative risk assessment) or as source oriented (using technology-based emission controls). These approaches are often used in combination; the difference between the countries' policies lies in the sequence of application or the starting point of the regulatory procedure. For instance, quantitative risk assessment is used primarily in the U.S. and The Netherlands, whereas a technology-driven approach for setting emission standards is used in Germany, Sweden, and Switzerland, complemented by environmental standards or ambient air-quality standards. In France and Japan individual regulations for industrial processes and specific-source categories are being applied.

The following regulatory instruments are used by governments to reduce emissions of hazardous air pollutants:[2-10]

- Setting of *emission standards* for individual pollutants and source categories
- Establishment of *ambient air-quality standards* for individual hazardous air pollutants
- Overall *reduction goals* for compound classes, e.g., VOCs and particulate matter
- *Restrictions* on production, handling, and product use (including pesticides)
- *Action programs* for individual pollutants or classes of pollutants

In most countries several of these regulatory instruments are combined to form the basis for the control policy (see Table 1). Comprehensive lists of technology-based emission standards have been established in Germany and Switzerland, and are being elaborated for the implementation of the Clean Air Act Amendments in the U.S. The procedure of setting ambient air-quality standards for a series of hazardous air pollutants has only been used to a large extent in The Netherlands. General reduction strategies addressing compound classes, e.g., VOCs and particulate matter, have been formulated usually to complement other regulations. Such programs have been applied in France, The Netherlands, Sweden, Switzerland, and the U.S. More specifically, action programs addressing individual substances (e.g., cadmium, mercury, dioxins) have been elaborated and used to a large extent, e.g., in Sweden.

In addition to controls of process emissions, restrictions on the use of certain products containing hazardous pollutants have also been applied, e.g., products containing cadmium and mercury compounds, asbestos, halogenated biphenyls (PCBs), and terphenyls. The regulations differ very much among the countries, both in the number of substances regulated as well as in the degree of restriction. While many substances are banned or restricted in Germany, The Netherlands, Sweden, and Switzerland, only a few restrictions have been applied in France, Japan, and the U.S. The control of halogenated organic compounds including DDT, HCH, HCB, and

Table 1. Approaches to Control of Hazardous Air Pollutants[2-10]

Country	Emission standards	Ambient air quality standards (guidelines)	Reduction goals for VOC
France	Asbestos, Cr^{VI}, cyanide, total heavy metals, Hg + Cd, As	—	30% by 2000 compared to 1980
Germany	160 pollutants, and specific source categories	(concentration of: Pb, Cd, HCl, Cl_2; deposition of: Pb, Cd, Tl, HF, HCl, F)	—
Japan	Cd, Pb, Cl_2, HCl, HF, F, SiF_4, asbestos	—	—
Netherlands	Emissions guidelines	(20 pollutants: VOCs, PAH, halogenated organics, fluorides, H_2S)	50% by 2000 compared to 1980
Sweden	Emission guidelines	Hg, Cd, As, Pb, Zn	50% by 2005 compared to 1988 + action programs
Switzerland	160 pollutants, and specific source categories	concentration of Pb, Cd, Zn; deposition of: Pb, Cd, Zn, Tl	50% by year 1995 compared to 1987
U.S.	Asbestos, inorganic As, Be, Hg, benzene, vinyl chloride; coke oven emissions (planned for 189 air toxics)	Pb	VOC and PM emission-reduction programs

Note: *Germany*: air quality limit values (so-called immission values) have been set in connection with the licensing procedure, and cannot be considered as general ambient air quality standards. The *Netherlands*: ambient air quality values are set as guidelines and are not legally binding standards. In addition, France, Germany, The Netherlands, Sweden, Switzerland, and the United States are signatories to the VOC Protocol under the UN ECE Convention on Long-Range Transboundary Air Pollution requiring a 30% reduction of total VOC emissions by 1999 compared to base year 1988 (except for Switzerland and the U.S. with base year 1989).

PCP that are used as pesticides in agriculture and forestry received particular attention as these substances are used in large quantities and are found worldwide in the environment.

B. Setting Priorities for Assessment and Control

As hazardous air pollutants comprise a large number of compounds with very different environmental impact properties, methods have to be elaborated in order to decide where to focus necessary control. For this purpose several countries, e.g., the U.S. and The Netherlands, use priority substance lists, identifying toxic or hazardous substances for the environment including hazardous air pollutants. The criteria to "list" a pollutant include toxicity, carcinogenic potential, persistence, dispersion properties, and bioaccumulation as well as emission quantities and levels found in the environment. These lists contain single pollutants, (e.g., benzene), classes of pollutants, (e.g., polycyclic aromatic hydrocarbons, chlorinated methanes), and mixtures of pollutants which originated from a specific source, process or activity, (e.g., coke

Table 2. Priority Pollutants of Some OECD Countries[5–8,10–12] and the WHO/Europe[13]

Substance	Canada	The Netherlands	Norway	Sweden	U.S.	WHO/ Europe
Metals and metalloids	6	7	8	7	8	8
Respirable mineral fibers						
Asbestos	1	1			1	1
Particulate matter						
Coarse and fine particles	1	1			1	
Inorganic gases	1	3	1		1	1
Organic compounds						
Nonhalogenated	3	5		1	6	3
Aromatics	7	4		1	8	3
PAH	1	1	1	1	1	1
Halogenated organics	9	10	5	6	11	6
Halogenated aromatics	8	5	5	9	5	
Mixtures of pollutants						
Waste crankcase oils	1					
Mineral oil VOC	1	1				
Gasoline vapors				1	1	
Coke oven emissions					1	

Note: The priority lists of Canada, The Netherlands, Norway, and Sweden contain pollutants being of concern for all environmental media. The lists of the U.S. and the WHO/Europe include priority hazardous air pollutants only.

oven emissions, used crankcase oil, and gasoline vapor). A compilation of the priority pollutant lists of several OECD[5–8,10–12] countries is given in Table 2, including the priority list of the World Health Organization (WHO) issued in 1987 in their Air Quality Guidelines for Europe.[13]

These lists are established by a prioritization method scoring a few hundred pollutants for effects and exposure to the total environment, i.e., using health screening and ecotoxicological evaluation procedures. In order to focus control, an assessment of each pollutant or class of pollutants in the list is made including a series of steps, such as the identification of principal emission sources, the determination of exposure levels, the assessment of health and environmental effects, the setting of control limits for all environmental media, and the identification of candidates for control measures.

It should be mentioned that in some countries, like Germany and Switzerland, a simplified procedure has been used to establish extensive lists of pollutants for setting emission standards.[3,9] About 160 inorganic and organic air pollutants (see Table 3), including carcinogens and toxic metals, have been classified according to their toxicity and ecological impact properties (e.g., photolytic and biological degradability, persistence, and bioaccumulation). In addition, the 1990 Amendments

Table 3. Technology-Based Emission Standards for Hazardous Air Pollutants in Germany and Switzerland[3,9]

Substance	Emission rate[a]	Limiting value
Inorganic Substances in Dust Particles (Total of 20)		
Category I (Cd, Hg, Tl, and compounds)	≥1 g/h	0.2 mg/m^3
Category II (As, Co, Ni, Se, Te, and compounds)	≥5 g/h	1.0 mg/m^3
Category III (Sb, Pb, Cr, Cu, Mn, Pt, Pd, Sn, V, and compounds, CN, F)	≥25 g/h	5.0 mg/m^3
Volatile Inorganic Substances (Total of 10)		
AsH_3, ClCN, $CoCl_2$, PH_3	≥10 g/h	1 mg/m^3
HBr, Cl_2, HCN, F_2, HF, H_2S	≥50 g/h	5 mg/m^3
Other Cl compounds	≥300 g/h	30 mg/m^3
Organic Substances (Total of 103)		
Category I	≥0.1 kg/h	20 mg/m^3
Category II	≥2.0 kg/h	100 mg/m^3
Category III	≥3.0 kg/h	150 mg/m^3
Carcinogenic Substances (Total of 21)		
Category I (asbestos, benzopyrene, other)	≥0.5 g/h	0.1 mg/m^3
Category II (As, Co, Cr, Ni comp., other)	≥5.0 g/h	1.0 mg/m^3
Category III (epichlorohydrine, dibromo-methane, methane, hydrazine)	≥25.0 g/h	5.0 mg/m^3

[a] For emission rates as shown the standard which must be met is indicated in the last column. No standard has been set for lower emission rates.

to the Clean Air Act of the U.S. established a list of 189 hazardous air pollutants or air toxics that will be subject to stringent emission controls.

The pollutants of most concern have been identified in several countries and include carcinogens (e.g., acrylonitrile, benzene, benzidine), toxic metals (e.g., mercury, cadmium, alkylated lead compounds), hazardous organics (e.g., PAH), and persistent halogenated organics (e.g., HCB, PCBs, TCDD/TCDF). Some countries, e.g., The Netherlands and Sweden, have established particular lists for substances that present a serious hazard for human health and the environment. In Germany and Switzerland, the most hazardous air pollutants are listed as the highest toxicity category of the emission standards list. Maximum emission reductions or minimum emission rates are required for these hazardous air pollutants.

III. DISCUSSION AND EVALUATION OF CONTROL POLICIES

A. National Control Programs

In the following section, the national control programs of France, Germany, Japan, Netherlands, Sweden, Switzerland, and the U.S. will be summarized briefly, with emphasis on the control of emissions from stationary sources. The experiences made with these programs will be evaluated and future policy trends highlighted.

In countries where *individual regulations for specific process emissions* have been used, there is little information available on the performance of the control approach. In *France* and *Japan,* the countries using this approach, it is most likely that regulations will continue to be made case by case and pollutant by pollutant.[2,4] Most recent regulatory proposals in France, however, show that a more comprehensive approach is being envisaged in the future. At present, in both countries, no assessment and evaluation of emission reductions achieved for hazardous air pollutants have been made, except for a few heavy metals. It has been clearly stated that more detailed data are needed on emissions and ambient air quality monitoring in order to set regulatory priorities. To date, results of heavy-metals measurements in France show decreasing trends of lead and cadmium of between 20 and 50% in the 1980s. In Japan, long-term ambient air-quality monitoring of heavy metals over the last 20 years showed that concentrations of iron, lead, vanadium, and zinc have decreased by a factor of four to eight. Ambient air concentrations of asbestos, mercury PAH, formaldehyde, and dioxin are currently below health and environmental risk levels.

Future control efforts in *France* will concentrate on specific source categories emitting hazardous air pollutants like waste incineration, solvent-use operations, and mobile sources. There is considerable concern that the beneficial impact of control measures might be offset or even overwhelmed within a few years by growth in these sectors. As technical abatement measures alone will not be sufficient to ensure that environmental quality has improved, particularly in urban areas, supplementary measures affecting growth, structural changes, and economic instruments are being studied. To date, specific measures have not been proposed. The National Plan for the Environment issued in 1990 has identified further priorities and areas of concern for the coming years. In the field of air pollution control, these include the reduction of halogenated organic compounds by 1995, the assessment of pollution and environmental effects from the use of new materials and substances (polymers, ceramics, carbon fibers, memory alloys, selenium, and beryllium), and from the use of new energies (synthetic fuels from coal, use of bituminous shale, biomass fuels, hydrogen technology, solar technology, and nuclear fusion).

In *Japan,* the only four groups of hazardous air pollutants regulated under the provision for harmful substances are[4] cadmium and its compounds, lead and its compounds, chlorine and hydrogen chloride, fluorine, hydrogen fluoride, and silicon fluoride. Asbestos-emitting facilities have been regulated under the provisions for specific particles. These pollutants are continued to be controlled by the implementation of emission standards set for a series of specific sources. In addition, agreements with many industry branches and residents participation have been used frequently in the licensing procedure for a plant. An efficient administrative measure to assure compliance with standards has been the use of public pressure by publishing the names of companies violating emission standards or other administrative regulations. On the other hand, additional emission reductions — even beyond the required limit values — are sometimes used by industry and business establishments as an advertising argument. Future control of hazardous air pollutants from stationary and diffuse sources will focus on the reduction of the use of asbestos and halogenated organic compounds like trichloroethylene, tetrachloroethylene, and other organochlorine solvents.

Countries like *The Netherlands* and the *U.S.*, using mainly *quantitative risk assessment* for setting environmental standards and establishing control measures, recognize that too many substances have not been controlled, and this has been due to the fact the regulatory procedure has been slowed down by this complicated and time-consuming process for control. For instance, in The Netherlands only emission guidelines — no national emission standards — have been issued for hazardous air pollutants, and in the U.S. emission standards for only seven hazardous air pollutants or pollutant groups, and for certain source categories emitting air toxics have been set by the year 1990. In these countries there is a clear tendency to apply a simplified procedure and to introduce technology-based emission standards for numerous air pollutants. Under the 1990 Clean Air Act amendments the U.S. Environmental Protection Agency (EPA) has established a list of 189 hazardous air pollutants for automatic statutory listing, and stationary sources will be subject to technology-based standards for numerous air pollutants. The authorities in The Netherlands will continue the source-oriented control approach, and plan to adopt with some minor changes the set of emission standards set down in the German Technical Instructions on Air Quality Control "(TA Luft)" in order to achieve an overall consistent control approach at national level.

As risk assessment and risk management are integral parts of the policy in *The Netherlands,* this approach in terms of multimedia assessment has been well developed, and a list of 48 priority substances was established, 38 of which are important hazardous air pollutants.[6] Risk limits (maximum acceptable risk levels and target risk levels) have been set for priority substances including carcinogens and noncarcinogens. These risk limits have been used to establish ambient air quality guideline values for 20 hazardous air pollutants (see Table 4).[5] The Dutch authorities stated that the effect-oriented risk approach is important to supplement the source-oriented approach, even more so when a multimedia approach is built in and criteria taking account of effects on human health and the environment are applied. The setting of ambient air quality guidelines allows assessment of pollution levels found and the survey of the effectiveness of emission control measures. Monitoring of hazardous air pollutants, particularly heavy metals, revealed that improvements have been achieved in air quality. Concentrations of cadmium, lead, zinc, and arsenic have decreased by 20 to 60% from 1982/83 to 1988/89. Other organic hazardous air pollutants, e.g., benzene, toluene, and PAH, have not decreased, and concentrations are high particularly in urban areas with high traffic density. In order to focus control measures, the assessment of all the priority substances will be completed in the coming years. The technology-based emission control approach will be pursued with emphasis on VOC control. A VOC reduction target of 50% by the year 2000

Table 4. Ambient Air Quality Guidelines of The Netherlands for Selected Hazardous Air Pollutants (μg/m³)[6]

Compounds	Limit value	Guide value	Target value	Measuring period
Acroleine	20			99.99-P/h
	8			98-P/24-h
	6			95-P/24-h
Acrylonitrile	1		0.1	Year
Benzene	10		1	Year
1,2-Dichloroethane			1	Year
Ethylene	300	120		99.99-P/h
Ethylene	30	12		99.7-P/24-h
Ethylene oxide			0.03	Year
Fluorides	2.8			Day
	0.8			Month
	0.4			Growing season
Formaldehyde	100			99.99-P/h
	40			98-P/24-h
	30			95-P/24-h
Hydrogen sulfide	2.5			99.5-P/h
Methylbromide			1	Year
			100	Hour
Methylenechloride			20	Year
PAH	5	0.5		Year (ng BaP/m³)
Phenol			1	Year
Propylene oxide			1	Year
Styrene			8	Year
Trichloroethylene	50		50	Year
	300			98-P/h
Trichloromethane (chloroform)			1	Year
Tetrachloroethylene	2000	1000	25	Year
	8300			98-P/h
		8300		99.5-P/h
Tetrachloromethane (carbon tetrachloride)			1	Year
Vinylchloride			1	Year

compared to 1981 levels has been decided and will be achieved through a number of agreements with industry in many sectors. The VOC emission reduction strategy will have a great effect in reducing health and environmental risks, as many of the VOCs are hazardous air pollutants.

In the *U.S.* emission reductions of hazardous air pollutants have been achieved through the setting of emission standards under the Clean Air Act, and its ambient air-quality standards program for lead, particulate matter (PM), and ozone. Current estimates are that lead and PM emissions from nontransportation sources have been reduced by 88 and 68%, respectively, since 1970. In addition, as a result of the ozone standard, VOC emissions have been reduced by 15 per cent since 1970. Future work will concentrate on the implementation of the 1990 amendments to the Clean Air Act. A two-phased approach has been decided to regulate 189 hazardous air pollutants:[10] (1) maximum available control technology (MACT) standards are set for all major sources and (2) risk assessment will be used to determine residual risk after MACT has been installed, and health risk-based standards will be set accordingly. EPA must publish by 1992 a list of source categories emitting 10 tons annually of any 1 toxic or 25 tons annually of a combination of toxic pollutants. The agency must then issue MACT standards based on the best-demonstrated control technology or practices of the regulated industry. Within the next years, EPA is required to issue MACT standards for coke oven emissions and for 40 source categories (25% of the source categories listed will have MACT standards set by November 1994, 50% by November 1997, and 100% by 2000); 8 years after the first-phase MACT standards, but 9 years after MACT standard setting for coke ovens and the first part of the 40 source categories standards, the second-phase health risk-based standards are to take effect, if a facility's emissions exceed the calculated cancer risk of 1 to 1 million.

In *Sweden,* the combined approaches using emission guidelines based on best-available control technology together with environmental quality criteria have proven to achieve substantial reductions in emissions of hazardous air pollutants, in particular, heavy metals (cadmium, mercury), PAH, and dioxins. The emission reductions achieved for heavy metals are between 30 and 80%. All these reductions have been gained by a combination of improved abatement technology, changes in production processes, restriction, and substitution of products containing hazardous substances. Future efforts will concentrate on tightening emission standards for carcinogens, persistent organic compounds, and heavy metals. Several action programs have been elaborated and will be implemented to drastically reduce emissions of hazardous air pollutants. The implementation and regular updating of these action programs is considered to be an important policy instrument to follow up emission reductions and improvements of environmental quality. Nevertheless, environmental quality for urban areas is still not satisfactory, as air quality guidelines are exceeded and overall risk from carcinogens is too high compared with acceptable levels. The action program on air pollution[7] has set the objective to meet air quality standards by the year 2000 through the implementation of stringent controls, promotion of clean technology, and improved structural measures such as traffic management, public transport, and economic instruments to promote cleaner motor vehicles.

Countries like *Germany* and *Switzerland* introduced their comprehensive *emission standard programs* in 1986. It is expected that substantial emission reductions of hazardous air pollutants will be achieved. However, it is premature to draw definite conclusions concerning the effectiveness of these control programs. The main advantage of this technology-based approach is the uniformity of regulations addressing numerous substances and various source categories all over the country. The policy in *Germany* for a standard-setting program has been strongly focused on emission standards. For approximately 160 individual hazardous air pollutants, emission standards have been set in the Technical Instructions on Air Quality Control (TA Luft) taking into account their toxicity classification.[3] The pollutants are grouped into inorganic dust, inorganic gaseous compounds, volatile organics, and carcinogens. The emission standards are defined in the form of waste stream limitations for a specified mass flow of that particular pollutant class (see Table 3). Best available control technology, i.e., present state-of-the-art techniques, have to be applied for new installations. As a rule, quantitative risk assessment and cost-benefit analysis are not used in making regulatory decisions. The determination of the appropriate levels for emission standards is made through combining the information on emissions, removal technology, exposure, hazard, and costs. Currently about 40,000 facilities in Western Germany are affected by this program. The impact of the control program is expected to result in total VOC emission reductions of the order of 40% between 1986 and 1995. Furthermore, considerable reductions of hazardous air pollutants are expected, in particular, for heavy metal emissions (about 40% reduction) and emissions from solvent-use operations.

The emission standards approach is supplemented by the application of a set of ambient air quality limit values (so-called immission values, see Table 1) in connection with the licensing procedure. However, air-quality assessment for hazardous air pollutants is not well developed, as these immission values are not proper ambient air-quality standards. In addition, the German governmental authorities feel that several sectors are not controlled well enough and need further improvement (e.g., evaporative emission controls of toxic and halogenated organic compounds). The ongoing procedure of periodically reviewing and tightening the emission standards will be pursued, and it is possible that more immission values will be set in the future. Potential future actions may be taken to restrict the use of tetrachloroethylene — a dry cleaning and degreasing solvent — under the Toxic Chemicals Law. Other future measures are regulations affecting benzene emissions from gasoline stations, and the ban of halogenated organic scavengers in leaded gasoline. Substances of high priority for future measures are dioxins, asbestos, chromium VI, mercury, formaldehyde, benzene, dichloroethane, dichloromethane, trichloroethylene, and perchloroethylene.

The policy in *Switzerland* is both source oriented, using pollution abatement at the source, and effect oriented, relying on ambient air-quality standards.[9] The regulations set forth emission limits for various stationary sources and for about 160 individual inorganic and organic pollutants grouped into classes according to their toxicity and ecological impact properties (see Table 3). These lists correspond to a large extent to those of the German Technical Instructions on Air Quality Control. Overall emission reductions of VOC of at least 40% will be expected from the stringent emission standard program. Generally, hazardous air pollutant emission reductions have not yet been evaluated, except for a few heavy metals like lead, cadmium, and zinc. Emission reductions of more than 60% have been achieved by the combined impact of emission standards and fuel quality requirements. The assessment of air pollution levels in Switzerland is based on the comparison with ambient air-quality standards. These standards are set as impact thresholds, taking into account solely man's and environment's needs for protection. The list of air-quality standards includes hazardous air pollutants, mainly heavy metals such as lead and cadmium in total suspended particles, as well as lead, cadmium, thallium, and zinc in total dust deposition (see Table 1). So far, the effect-oriented parts of the control approach, in particular the setting of ambient air and environmental quality standards, have proven to be essential in defining additional measures to further reduce emissions of hazardous air pollutants including compound classes like VOCs. Pollutants of high priority remain heavy metals, asbestos, and halogenated organic compounds used as industrial solvents and for dry cleaning purposes.

B. The Use of Economic Instruments

As there is a strong tradition in most countries to use "command-and-control" systems for air pollution abatement policy, economic instruments play a minor part and mostly appear as supplementing direct regulations. With regard to environmental effectiveness, an evaluation is difficult to make bearing in mind the results of the limited experience. Only a few systems have been used in the field of hazardous air pollutant controls; subsidies for developing control technology and operating measurement equipment, and tax differentiation with respect to cleaner cars and relating to lead-free gasoline, have been applied in several countries, but their direct effect on emission reductions of hazardous air pollutants remains difficult to evaluate. Even in the U.S., where economic instruments, mainly emissions trading, have been more widely applied in the field of air pollution, there appears to be more of a consensus regarding the economic impact than with respect to environmental effectiveness. Final conclusions on the impact of economic instruments cannot be drawn at present, as more evaluation on environmental benefits of these tools are needed.[15]

C. International Action Programs

Emission reduction programs for hazardous air pollutants have been agreed within the regional sea conventions (North Sea and Baltic Conferences) where priority pollutants have been established aiming at the long-term protection of the marine environment. The Ministerial Declaration of the International Conference on the Protection of the North Sea, March 1990, included nine European countries (Belgium, Denmark, France, Germany, The Netherlands, Norway, Sweden, Switzerland, the U.K.) and the Commission of the European Communities.[14] Giving great emphasis on persistent compounds (metals and organochlorines), it was decided to reduce air emissions of 17 priority pollutants by at least 50% by 1999, and emissions of mercury, cadmium, lead, and dioxins by 70% by the year 1999, taking 1985 emission levels as a basis for the calculation (see Table 5). Furthermore, the countries have agreed on a list of priority activities and source sectors that will be subject to emission regulations based on current state-of-the-art control technology. In principle, best available technology will be applied to all sectors listed, starting, however, with major emission sources that cause the highest pollution. A reference list of several hundred compounds including pesticides has also been proposed from which future priority pollutants will be selected for international action.

Table 5. List of Priority Substances under the North Sea Convention[14]

Substance	Water	Air	CAS Number
1. Mercury	•	•	7439976
2. Cadmium	•	•	7440439
3. Copper	•	•	7440508
4. Zinc	•	•	n.a.
5. Lead	•	•	7439921
6. Arsenic	•	•	7440382
7. Chromium	•	•	n.a.
8. Nickel	•	•	7440020
9. Drins	•		—
10. HCH	•	•	608731
11. DDT	•		50293
12. Pentachlorophenol	•	•	87865
13. Hexachlorobenzene	•	•	118741
14. Hexachlorobutadiene	•		87683
15. Carbon tetrachloride	•	•	56235
16. Chloroform	•		67663
17. Trifluralin	•		1582098
18. Endosulfan	•		115297
19. Simazine	•		122349
20. Atrazine	•		1912249
21. Tributyltin compounds	•		—
22. Triphenyltin compounds	•		—
23. Ethyl-azinphos	•		2642719
24. Methyl-azinphos	•		86500
25. Fenitrothion	•		122145
26. Fenthion	•		55389
27. Malathion	•		121755
28. Parathion	•		56382
29. Methyl-parathion	•		298000
30. Dichlorvos	•		62737
31. Trichloroethylene	•	•	79016
32. Tetrachloroethylene	•	•	127184
33. Trichlorobenzene	•	•	—
34. 1,2-Dichloroethane	•		107062
35. Trichloroethane	•	•	71556
36. Dioxins	•	•	n.a.

Note: n.a. = not applicable; — = not defined.

IV. CONCLUSIONS AND OUTLOOK

In facing the complex problem of controlling hazardous air pollutants, almost every one of the seven OECD countries reviewed has set control priorities based on an estimate of the environmental risk posed by potentially hazardous compounds in the atmosphere. In some countries a priority substance list exists, but there is no internationally agreed priority action list apart from the one taken up by the regional conferences on the protection of the Baltic Sea and the North Sea. This policy analysis shows that in several countries many hazardous air pollutants have not received regulatory attention, in particular the large group of organic compounds comprising numerous hazardous air pollutants. Generally, individual regulations addressing specific process emissions or the use of products predominate, and decisions are being taken on a pollutant-by-pollutant basis. In only three countries (Germany, Switzerland, and the U.S.) have extensive lists of hazardous air pollutants been established that allow a comprehensive control approach of emission sources addressing simultaneously several pollutants and pollutant groups. There is considerable need for more systematic assessment of risks and control options addressing most of the hazardous air pollutants. On the other hand, the efforts made for hazardous air pollutants control showed in several countries (e.g., Sweden, Japan, France) that the *application of best available control technology* reduced emissions substantially. In many cases these control technologies are the same for hazardous air pollutants as for traditional air pollutants. Furthermore, the general application of best available control technology to new emission sources has the potential of considerably reducing emissions of hazardous air pollutants in the medium term. Nonattainment of environmental and air quality standards for traditional pollutants like ozone and particulate matter has led authorities in some countries (e.g., The Netherlands, Switzerland, and the U.S.) to set tighter emission standards for compound classes like VOCs and particulate matter. Therefore, it has been recognized that effect-oriented measures, in particular the *setting of environmental and air quality standards,* contribute to emission reductions of hazardous air pollutants, although these have not been quantified in all cases.

The stringent control measures in various economic sectors have, however, not prevented pollution levels from increasing, as *growth* in these sectors has offset emission reductions achieved. This is particularly the case for the motor vehicle sector where overall traffic growth has overwhelmed the improvements made in emission controls for individual vehicles. In addition, studies in Sweden and the U.S. on outdoor exposure to carcinogens have shown that motor vehicles are one of the major emission sources.[1] Several OECD countries recognized that additional measures have to be implemented to improve air quality, in particular in urban areas, to meet acceptable environmental levels for many hazardous air pollutants.

New *priorities for further action* have been set in several OECD member countries aiming at substantial emission reductions or restrictions on product use. The pollutants in these priority lists include many hazardous air pollutants with emphasis on heavy metals, organic compounds (benzene, polycyclic aromatics), halogenated solvents, and dioxins. In Table 6 a compilation of national priority pollutant lists for immediate action is shown. As hazardous air pollutants may have an impact on humans and several environmental media, detailed emission, ambient air monitoring, and exposure data of most of these priority hazardous air pollutants are needed in order to improve their health and environmental impact assessment. Nevertheless, these lists address only a relatively small number of pollutants compared to the large number of toxic trace pollutants found in ambient air. International cooperation is required as the atmosphere is an important pathway for the transport and dispersion of hazardous air pollutants, and as the many transboundary and global aspects of the problem are becoming increasingly evident.

Table 6. Priority Hazardous Air Pollutants for Further Action[1]

Pollutant	Country
Metals	
Arsenic	S,N
Cadmium	S, CH, US,N
Chromium VI	D, US,NL,CH,N
Lead	S, (US),N
Mercury	D, CH, US,N
Nickel	N
Zinc	N
Organo-tin	S
Toxic fibers	
Asbestos	D, Jap, CH,NL
Fine particulate matter (PM_{10})	US, Jap
Organic compounds	
Nonhalogenated	
Acrylonitrile	US
Formaldehyde	D, US
Benzene	D, CH, US
Polycyclic organics (includes PAH)	US
Phthalates	S
Nonylphenoloxylates	S
Halogenated organics	
Butadiene (chloroprene)	US
Chlorinated paraffins	S,N
Organochlorinated solvents	F, D, Jap, NL, S, CH
Chloroform	US
Methylchloroform	
Brominated flame retardants	S
Ethylenedichloride	D, US
Dioxins	F, D, NL, S, CH, US,N
Carbon tetrachloride	S, US,N
Trichloroethylene	D, Jap, S, CH,N
Tetrachloroethylene	D, F, Jap, S, CH,N
Chlorobenzenes	N
Mixtures of pollutants	
Gasoline vapors	D, NL, US, S, CH
Coke oven emissions	US
Organochlorinated solvents	F,D,Jap,NL,S,CH
Pesticides	
Cresote	S
Hexachlorocyclohexane	D, NL, S, CH
Hexachlorobenzene	D, NL, S, US, CH,N
Pentachlorophenol	D, NL, S, CH,N

Note: National action programs in OECD member countries: F: France; D: Germany; NL: The Netherlands; N: Norway; Jap: Japan; S: Sweden; CH: Switzerland; US: United States.

REFERENCES

1. OECD, *Control of Hazardous Air Pollutants in OECD Countries,* Paris, 1993.
2. Ministry of Environment, Air Quality Agency, *Air Pollution Prevention and Measurement,* Paris,-La Défense, 1988.
3. Federal Ministry of Environment, Nature Conservation and Reactor Safety, *Technical Instructions on Air Quality Control,* Bonn, 1986.
4. Government of Japan, Environment Agency, *Quality of the Environment in Japan,* Tokyo, 1988.
5. Ministry of Housing, Physical Planning and Environment, Air Directorate, *Air Pollution Control Policy in The Netherlands,* The Hague, July 1987.
6. Central Department for Information and International Relations, *Environmental Program of The Netherlands from 1987 to 1991,* The Hague, December 1986.
7. Swedish Environmental Protection Agency (SNV), *Air Pollution '90,* enviro No.10, Solna, Sweden, November 1990.
8. Swedish National Chemicals Inspectorate, *Environmentally Hazardous Chemicals,* Report 10/89, (ISSN 0284-1185), Solna, Sweden, 1989.
9. Swiss Federal Government, *Ordinance on Air Pollution Control,* Bern, December 16, 1985.
10. United States Environmental Protection Agency, Office of Air and Radiation, *Clean Air Act Amendments,* CQ 3934, Washington, DC, November 1990.
11. Under the Canadian Environmental Protection Act, *Report of the Ministers' Priority Substances Advisory Panel,* (ISBN 0-662-16373-7), Ottawa, 1988.
12. The Norwegian State Pollution Control Authority, *Micropollutants in Norway,* SFT-Report No. 80, Oslo, 1987.
13. World Health Organization, Regional Office for Europe, European Series No. 23, *Air Quality Guidelines for Europe,* Copenhagen, 1987.
14. Third International Conference on the Protection of the North Sea, *Ministerial Declaration,* The Hague, March 8, 1990.
15. OECD, *Economic Instruments for Environmental Protection*, Paris, 1989.

The Impact of Recent Legislation on the U.K. Generation Industry

H. E. Evans
W. S. Kyte

I. INTRODUCTION

The generation of electricity, as with so many aspects of life, in the U.K. is increasingly affected by national and international legislation and by international agreements. In the case of electricity generation, these regulations relate to emissions to air, water, and land. It should be recognized at the outset that such regulations are not a new phenomenon as air pollution and water pollution have been of concern since the Middle Ages, if not before. This paper, however, will have a statute of limitations and will only consider legislation which has been passed in the last 12 years and will give predominate attention to the last 4 years as the Water Act of 1989, The Environmental Protection Act (EPA) of 1990, the Third International Conference on the Protection of the North Sea (1990), together with the EC Large Combustion Plant Directive of 1988, provide the major basis for the legislation affecting the generating industry in the U.K. More detail on the progression on legislation during the past 12 years is provided in the following list.

1979	November	UN/ECE Convention on long-range transboundary air pollution
1982	June	Stockholm Conference on acidification of the environment
1982	December	Adoption of EC Directive on air quality standards for lead
1983	August	Implementation of the Health and Safety (emissions into the atmosphere) Regulations, 1983
1984	June	Adoption of EC Industrial Air Pollution "Framework" Directive
1984	October	First North Sea Conference (Bremen)
1985	March	Adoption of EC Directive on Air Standards for NO_2
1985	July	"30% Club" protocol signed by 21 countries at third meeting of the executive body for the UN/ECE convention on long-range transboundary air pollution in Helsinki
1987	November	Second North Sea Conference (London)
1988	January	EC resolution agreed on action program on cadmium

1988	November	U.K. signed Sofia protocol to freeze the level of emissions of nitrogen oxides at 1987 levels by 1994 and by 1996 to agree on further reductions based upon the critical loads approach; EC Directive (88/609/EEC) on emissions from large combustion plants
1989	June	EC Directive (89/427/EEC) on air quality limit values and guide values for sulfur dioxide and suspended particulates
1989	July	Water Act of 1989 set up water and sewerage business in England and Wales and created National Rivers Authority
1989	November	U.K. signed International Declaration at Noordwijk; this recognizes the view of many industrialized nations of the need for stabilization of CO_2 emissions by the year 2000
1990	March	Third North Sea Conference (The Hague); further reductions in the amount of hazardous substances allowed to enter the sea via rivers and the atmosphere agreed
1990	March	U.K. to cease the dumping of sewage sludge in the North Sea by 1998
1990	August	EC Directive (90/415/EEC) on dangerous substances in water adopted
1990	October	U.K. ratified the NOx protocol signed in Sofia in 1988; this commits the U.K. to reduce overall emissions of oxides of nitrogen from all sources to 1987 levels by 1994 and to develop programs for further reductions based on the critical loads approach
1990	October	U.K. stated target of returning carbon dioxide emissions to 1990 levels by the year 2005
1990	October	EC Joint Council Energy/Environmental target for stabilization of CO_2 emissions in general by the year 2000 at 1990 levels
1990	November	Environmental Protection Act received Royal Assent
1991	July	U.K. announcement of an intent to create a new environmental agency

Examination of this list shows that legislation covers solid, liquid, and gaseous environments and, as will be seen later, one of the challenges facing the industry is to identify a mode of operation which will meet the requirements of all the various pieces of legislation in a cost-effective manner.

It is clear that the U.K. has recognized that it is not feasible to consider that pollution takes note of national boundaries, and the Large Combustion Plants Directive on emission of SO_2 and NOx and the North Sea Conference agreements, particularly on emission of toxic metals, are prime examples of this. It is to be expected that this trend to international cooperation on pollution control will be extended in future years. Revisions of the large Combustion Plants Directive in 1994 and other agreements to introduce the critical loads approach for identifying the maximum tolerable emission levels (see later) are examples of this trend.

The aim of this paper is to summarize the way in which the U.K.- and EC-derived legislation impacts on the generating industry in the U.K. Subsequent sections will cover the U.K. electricity industry, the key European legislation which affects its operation, the prime U.K. legislation (the Water Act, 1989, and the Environmental Protection Act, 1990), the way in which this legislation sets targets for the industry and the possible responses, the Best Practical Environmental Option concept, and finally the potential for future legislation which will affect the industry.

II. THE U.K. ELECTRICITY INDUSTRY

As a result of privatization the nature of the U.K. electricity-generating industry has been significantly changed. In England and Wales the precursor company, the Central Electricity Generating Board (CEGB), was effectively a monopoly supplier. The consequence of the privatization of the industry is that in England and Wales there are now a number of competing companies. The nuclear stations have been maintained in government ownership as Nuclear Electric public limited company (plc) and the nonnuclear part of CEGB has been divided into two large companies, PowerGen plc and National Power plc with 40 and 60% of the nonnuclear assets of CEGB, respectively. In addition, the declared aim of the legislation was to encourage third parties to enter the electricity-generating scene. For a number of reasons, predominantly commercial, new generation capacity will be largely based on a gas-fired combined-cycle gas turbine (CCGT) plant. New plants will have to meet source-specific, tight regulations for emissions.

It should also be noticed in passing that there can be conflicts between various environmentally desirable activities. A good example is the fact that reduction in SO_2 emission by fitting flue gas desulfurization (FGD) plant will reduce the overall efficiency of a station and will thus enhance the emissions of carbon dioxide (CO_2) as a result. It will also result in a new discharge to water and there may be additional solid wastes.

III. KEY EUROPEAN LEGISLATION

A. EEC Large Combustion Plant Directive (88/609/EEC)

Legislation adopted as a directive is legally binding on the member states who must introduce in into their national legislation within a specified period. Directive 88/609/EEC (which is known as the Large Combustion Plants Directive) was adopted in November 1988 for plants greater than 50 MW thermal. The directive covers solid, liquid, and gaseous fuel and provides emission standards for new plants for particulates, SO_2, and NOx emission reductions for SO_2 and NOx for existing plants.

For existing plants (pre July 1, 1987) there are targets for reduction of total SO_2 and NOx from a 1980 reference level. Following very detailed negotiations, individual member states have set limits for SO_2 and NOx emission for 1993 and 1998, and values for SO_2 have also been set for 2003.

It should be noted that these emissions refer to all large combustion plants, within each country, and not only to those involved in electricity generation. However, electricity generation is the major activity for plants of this size. The U.K. limits, on a sector basis, for SO_2 and NOx are given in a government publication.[1]

B. North Sea Directives

There have been three International Conferences on the Protection of the North Sea. The first of these took place at Bremen in 1984, the second in the U.K. in 1987, and the third at The Hague in 1990. The driving force for these conferences is the concern felt by those countries bordering on the North Sea that pollution by a wide range of potentially harmful species could reach the North Sea by a combination of river and atmospheric transport in addition to the releases from shipping and offshore platforms based in the North Sea. Taking the most recent two of these conferences, the key conclusions of relevance to the electricity-generating industry are as follows.

1. London, November 1987

There is a need to adopt a precautionary approach to inputs of the most dangerous substances (those that are persistent, toxic, and liable to bioaccumulate) involving both the use of best available technology (BAT) for point sources as well as controls on the manufacture, marketing, use, and disposal of products containing such substances from diffuse sources.

At the same time, the importance of the complementary use of environmental quality objectives (EQOs) and strict emission standards is recognized.

2. The Hague, March 1990

There is also a need to achieve a significant reduction (defined as 50% or more) of a number of substances including cadmium, mercury, zinc, and lead which are released in various ways during the combustion process. They also arise from the use of limestone in the FGD process. The reductions required are of inputs via rivers and estuaries between 1985 and 1995. Reductions are also required for atmospheric emissions by 1995, or by 1999 at the latest, provided that the application of BAT, including the use of strict emissions standards, enables such a reduction.

Table 1. List of Priority Hazardous Substances Identified during the Third International Conference on the Protection of the North Sea

Substance	Water	Air	Substance	Water	Air
1. Mercury	•	•	19. Simazine	•	
2. Cadmium	•	•	20. Atrazine	•	
3. Copper	•	•	21. Tributyltin compounds	•	
4. Zinc	•	•	22. Triphenyltin compounds	•	
5. Lead	•	•	23. Azinphos-ethyl	•	
6. Arsenic	•	•	24. Azinphos-methyl	•	
7. Chrmoium	•	•	25. Fenitrothion	•	
8. Nickel	•	•	26. Fenthion	•	
9. Drins	•		27. Malathion	•	
10. HCH	•	•	28. Parathion	•	
11. DDT	•		29. Parathion-methyl	•	
12. Pentachlorophenol	•	•	30. Dichlorvos	•	
13. Hexachlorobenzene	•	•	31. Trichloroethylene	•	•
14. Hexachlorobutadiene	•		32. Tetrachloroethylene	•	•
15. Carbon tetrachloride	•	•	33. Trichlorobenzene	•	•
16. Chloroform	•		34. Dichloroethane 1,2-	•	
17. Trifluralin	•		35. Trichloroethane	•	•
18. Endosulfan	•		36. Dioxins	•	•

Another need is for dioxins, mercury, cadmium, and lead to achieve reductions between 1985 and 1995 of total inputs (via all pathways) of the order of 70%, provided that the use of BAT or other low-waste technology measures enables such reductions.

There is a need to phase out and destroy, in an environmentally safe manner, all identifiable PCBs by 1995, and by the end of 1999 at the latest.

In addition, international efforts will be made to elaborate by 1991 the order of magnitude of emissions of the hazardous substances (including mercury, cadmium, copper, zinc, lead, arsenic, etc.), reaching the North Sea via the atmosphere as a basis for setting priorities for reduction measures for such emissions.

A list of priority hazardous substances was produced at the third conference.[2] This is shown in Table 1.

Again it should be realized that the conference considered the emissions from all industries, and not specifically from the electricity-generating sector. Indeed, in certain instances significant reduction in emissions, e.g., heavy metal emissions, has already been achieved by action taken by the U.K. government. As a consequence, the extent of the actions to be required of the electricity sector may well be less onerous than it appears at first sight.

The move to increased use of gas, improvement in generation efficiency, the introduction of FDG, and improvement in ESP performance will contribute to the required reductions.

IV. RECENT U.K. LEGISLATION

As mentioned in the introduction, there have been two major new acts in the U.K. in recent years — the Water Act (1989) and the Environmental Pollution Act (1990).

In practice, as far as pollution control is concerned, these two acts are now largely intertwined but for simplicity it is worth dealing with them in chronological order.

A. Water Act of 1989

As a result of the Water Act (1989), the ten Rivers Units of the precursor regional water authorities were combined to form a National Rivers Authority (NRA). This body is responsible for a wide range of topics including water quality, water resources, flood defense, fisheries, recreational use, etc. For electric utilities which use river water as their predominant source of cooling water, the first of these is the most important. Where appropriate, they inherit responsibility for certain EC directives.

The statutory responsibilities upon the NRA for water are as follows:

- To keep deposited maps of controlled waters for public inspection
- To conserve and enhance the amenities of inland and coastal water, and of land associated with such waters
- To achieve water-quality objectives in all controlled waters
- To monitor the extent of pollution in controlled waters
- To maintain registers of water quality objectives, applications for consents, certificates and sampling data, and to make them available to the public.
- To advise and assist the Department of the Environment (DoE) on water pollution matters
- To exchange information with water undertakers on pollution matters
- To determine and issue consents for discharge of wastes into controlled waters (a power to charge for such work exists)

The responsibilities which relate to water quality of most interest to the generators are

- Determination and issuing of consents for discharges into controlled waters
- The monitoring of the extent of pollution in such waters
- Achievement of water-quality objectives

Before any process which results in release of substances on the EEC dangerous-substances directive (76/464/EEC) can be discharged to a river, the NRA must be satisfied that the relevant environmental quality standard is not exceeded. The species of predominant interest are cadmium and mercury (list I substances) and the elements arsenic, chromium, copper, lead, nickel, zinc, boron, iron, and vanadium (list II substances). For each of these substances there is an agreed standard for three reasons: (1) abstraction for potable supply, (2) protection of sensitive aquatic life, and (3) protection of other aquatic life.

It should also be noted that under the control of Pollution Act (1974) some power stations had consent conditions applied to take account of trace metal released from pulverized fuel ash (pfa) disposal schemes. In addition, discharge conditions have in certain instances been applied to temperature rise, suspended solids release, pH, oil, and sometimes chemical cleaning solutions.

The introduction of flue gas desulfurization based on the limestone gypsum process at certain U.K. power stations has introduced a potential new complex effluent from the plant.

B. Environmental Protection Act (1990)

The introduction of this Act provides a basis for wide-ranging control of environmental issues by the U.K. government. The Act is essentially an enabling structure which allows changes to be made quickly via Regulations and Guidance Notes. This gives wide powers to the Secretary of State for the Environment. The Act is being implemented over a period of time for difference processes. It came into force for power stations on April 1, 1991. The Act is made up of nine parts. The two that are particularly relevant to the electricity-generating industry are

- Part 1 — integrated pollution control (IPC) which is administered by Her Majesty's Inspectorate of Pollution (HMIP) and Air Pollution Control (APC) administered by local authorities.
- Part 2 — waste on land.

Regulations under the Act have set out a number of scheduled processes and prescribed substances.

In essence, HMIP issues authorization to operate which will exert control via conditions in the authorization to ensure

- The use of best available techniques not entailing excessive cost (BATNEEC) to prevent/minimize to render harmless prescribed substances
- Compliance with any quality standards or objectives
- Compliance with any national plan
- Compliance with EC obligations or international law

In carrying out their responsibilities under the Act, the enforcing authorities have

- A requirement to consider the best practical environmental option (BPEO) for processes likely to release material to several environments
- To adopt such conditions as NRAs considered appropriate if there is release to water. Effectively the water quality objectives of the Water Act (1989) must be met.
- A requirement to consult certain statutory bodies.
- A requirement to review at least once every 4 years
- A requirement to notify the Waste Regulation Authority that the process produces waste

A number of very important points arise from the introduction of the Environmental Protection Act (1990). A full discussion of the Act is outside the scope of this

Table 2. EPA Schedule of Substances Prescribed for Air

Oxides of sulfur or other sulfur compounds
Oxides of nitrogen or other nitrogen compounds
Oxides of carbon
Organic compounds and hydrocarbons, including partial oxidation products
Heavy metals and other compounds
Smoke, grit, dust, and fumes
Asbestos (suspended particulate matter and fibers), glass fiber, and mineral fibers
Halogens or their compounds
Phosphorus or its compounds

paper but a number of points that impact on the electricity generators are worth elaborating.

1. Access to Information

As part of the new process the generators must apply for authorization to operate both their existing plant and any new plant. The HMIP must keep detailed records related to the authorizations. There is public access to information on these authorizations and their applications as well as the data on emissions of SO_2 and NO_x which operators must report routinely to HMIP.

2. Enforcement

The new act gives the HMIP a range of powers. They are able to issue enforcement, variation, or prohibition notices if the operator is in breach of his duties. It is possible for the HMIP to revoke the authorization of a plant and thus the Inspectorate has wide powers. Penalties for noncompliance include criminal sanctions of unlimited fines and up to 2 years imprisonment.

3. Limits to Emissions

The act provides the Inspectorate with the ability to set specific limits on emissions. These may include such factors as limits on concentration, on the amount of any species which may be emitted in a set period, or requirements related to the characteristics of the substance emitted. These may vary from case to case.

Schedules of substances prescribed for air, water, and land have been produced. These are shown in Tables 2, 3, and 4.

4. National Plans

As mentioned earlier, there is a duty on the Inspectorate to ensure compliance with any international laws. This is a wide-ranging requirement which may relate to any substance from any process being operated. In the case of gaseous emissions, the National Emissions of SO_2 and NO_x are determined by HMIP and then reported by Her Majesty's Government to the EC at Brussels.

Table 3. EPA Schedule of Substances Prescribed for Water

Mercury and its compounds	1,2-Dichloroethane
Cadmium and its compounds	Trichlorobenzene
Hexachlorocyclohexane (all isomers)	Atrazine
DDT (all isomers)	Simazine
Pentachlorophenol and its compounds	Tributyl tin compounds
Hexachlorobenzene	Triphenyl tin compounds
Hexachlorobutadiene	Trifluralin
Aldrin	Fenitrothion
Dieldrin	Azinphos-methyl
Endrin	Malathion
Polychlorinated biphenyls	Endosulfan
Dichlorvos	

Table 4. EPA Schedule of Substances Prescribed for Land

Organic solvents	Pesticides — any chemical substances or preparation prepared or used for destroying any pest, including those used for protecting plants or wood or other plant products from harmful organisms; regulating the growth of plants; giving protection against harmful creatures; rendering such creatures harmless; controlling organisms with harmful or unwanted effects on water systems, buildings, or other structures or on manufactured products; or protecting animals against ectoparasites
Azides	
Halogens and their covalent compounds	
Metal carbonyls	
Organometallic compounds	
Oxidizing agents	
Polychlorinated dibenzofuran and any congener thereof	Alkali metals and their oxides
Polychlorinated dibenzo-p-dioxin and any other congener thereof	Alkaline earth metals and their oxides
Polyhalogenated biphenyls, terphenyls, and naphthalenes	
Phosphorus	

5. Adoption of Best-Available Techniques Not Entailing Excessive Costs (BATNEEC)

Earlier U.K. legislation incorporated the concept of best practical means (BPM). Clearly, this concept was open to difficulty of interpretation and more recently the BATNEEC concept has been introduced. The aim is to ensure than an operator uses the most appropriate techniques and takes note of all other appropriate factors. Thus, the BATNEEC concept includes technology, use of personnel, and design, layout, and maintenance.

The HMIP Chief Inspector's Guidance to Inspectors includes the following:

> The operator should provide full information in the application on the selection of primary process, particularly for a new plant, having regard to the need to use the best available techniques not entailing excessive cost (BATNEEC) to reduce to a minimum the genera-

tion of pollution at source. The operator should select the combination of primary process, pollution abatement techniques and waste treatment and disposal which constitute the best practicable environmental option (BPEO) for minimising the pollution which may be caused to the environment as a whole. The operator should provide a justification for all the likely pollutants produced by the operation of the process.

6. New Plant Standards

On of the objectives of the Environmental Protection Act is to ensure that the existing plant is brought up to new plant standards. It is recognized that upgrading should take note of various criteria in the Air Framework Directive (84/360/EEC), including the provision of not entailing excessive cost and the rate of utilization and length of remaining plant life.

V. INTERACTION OF THE ENVIRONMENTAL PROTECTION ACT (1990) AND THE WATER ACT (1989)

Large power stations come under the control of HMIP, who has a duty to consult a number of other bodies including the National Rivers Authority and Sewage Undertakers.

The NRA's powers under the Water Act remain unchanged. They may review the conditions for scheduled discharges at not less than 2-yearly intervals. However, the NRA's duty to consent the effluents from certain prescribed processes will be subsumed into the HMIP authorization.

The NRA will be closely involved in all discussion with regard to discharges to water, and will be able to place their own limits on what can be discharged in order that Environmental Quality Standards are met for the water course under consideration. It should be noted that HMIP may set more stringent limits under BATNEEC than NRA but cannot set less stringent limits. The task of ensuring that such discharges comply with the consents will fall to HMIP.

The second part of the Environmental Protection Act deals with waste on land. There will be a reorganization of regulation, disposal, and collection authorities through Great Britain. New provisions relating to licensing are introduced as well as a new provision creating offenses of disposing of, or otherwise dealing with, waste without a license.

In addition there is a creation of a new duty of care applicable to waste producers, importers, holders, carriers, and those who treat or dispose of waste, to ensure that the waste is not disposed of, released to the environment, or transferred to another personal illegally.

Offenses under this part of the Act include

- Depositing, treating, or disposing of controlled waste without a waste management license or otherwise than in accordance with the license.

- Treating, keeping, or disposing of controlled waste in a manner likely to cause pollution to the environment or harm to human health.

Conviction for offenses under this part of the Act can lead to unlimited fine or imprisonment for the more serious offenses which are treated under indictment. Lesser penalties are specified for summary convictions.

VI. ADMINISTRATION OF THE LARGE COMBUSTION PLANT DIRECTIVE

It has already been explained that the Large Combustion Plants Directive (88/609/EEC) has identified target reductions for SO_2 and NO_x for the U.K. overall. In practice, the contributions required from the power generators (PowerGen and National Power) have been separately identified by the U.K. government (and are given in Reference 1). It should be recognized that the targets of the Large Combustion Plant Directive will be reviewed in 1994 and there is a possibility that the target emissions for the later years will be revised downward. This is discussed in Section VIII.

A. Company Bubble

In order that the U.K. government can control the emissions of SO_2 and NO_x from the power stations, the "Company Bubble" concept has been introduced for both SO_2 and NO_x on the basis of its expected use during the year. The allocations to each station owned by a particular company (PowerGen or National Power) will be added up to that company's limit or "bubble" specified in the Plan. The companies will be responsible for monitoring the emissions of these species throughout the year by a procedure agreed with HMIP.

When emissions at a station reach a specified "reporting level", which will be at a level of 95% of the plant allocation, the operator will be required by a condition of his authorization to make immediate notification to the relevant enforcing authority.

The plant can be operated after the allocation has been exceeded provided that, on or before its being exceeded, application has been made for an increase in the allocation for the plant together with equivalent offsetting decreases from other plants under the operator's control unless the enforcing authority has notified that it is unwilling to agree to the requested changes.

It should be noted that there is no tradeable permit option within the U.K. at present. (The "Company Bubble" procedure could be considered to be a tradeable permit procedure *within* a single company.) However, discussion within the United Nations Economic Commission for Europe are currently covering such options.

VII. IMPLICATIONS OF THE EXISTING LEGISLATION

In the U.K. there is a combination of new legislation and a new competitive situation. Inevitably, therefore, detailed consideration is being driven to the ways in which the emission targets can be achieved without jeopardizing the commercial position of the generating companies. The companies must take note not only of the government-imposed emission targets, but also of the future requirements for upgrading to new plant standards, and of the possibility that more onerous targets will be set as a result of international agreements.

For sulfur control the available options include (1) the use of natural gas either in a combined cycle gas turbine (CCGT) plant or in conventional boilers, (2) the use of low sulfur coal, (3) fitting flue gas desulfurization plant, and (4) coal purification.

At present, the first three of these options are all being seriously considered by U.K. power producers.

For reduction of NO_x emissions, the major options are (1) fitting low NO_x burners, (2) selective noncatalytic reduction processes (SNCR), and (3) selective catalytic reduction (SCR).

It appears at present that the U.K. targets can be met by a combination of introduction of low NO_x burners and retirement of aging plants.

VIII. POTENTIAL FUTURE LEGISLATION

It is quite clear that the trend is for increasing requirements to minimize emissions from power stations and other industrial plants. In certain areas it is clear that review of the current legislation is scheduled (e.g., the review of the Large Combustion Plant Directive in 1994), and it is probable that additional requirements will emerge as a result of EC legislation and international agreements.

That the Environmental Protection Act (1990) has been written in such a way that additional requirements can be relatively easily introduced into legislation means that the pace of change is likely to increase rather than decrease. It is particularly relevant that HMIP must review the conditions of authorizations at least once every 4 years. In view of this it is particularly important that the industry is well aware of its emissions and their environmental impact in order to make a sensible contribution to the debate and ensure that no unnecessary requirements are imposed. There is a real need for cost-benefit analysis before any additional control requirements are introduced.

For pollutants that are released by a number of industries, it is essential that legislators continue to take an industry sector, rather than a specific industry, approach.

A. Critical Loads

As the understanding of the way in which acid rain affects the environment has progressed, it has been realized that an emission target for a power station or group of power stations is not necessarily the optimum approach to legislation. In view of the varying susceptibilities of different parts of the world to acidification, as a result of differing regional geological properties, it is now considered necessary to identify "critical loads" for each area. With this information, together with an understanding of the dispersion of pollutants from the various industrial processes, it is possible to attempt to identify the extent of emission control required from any particular area.

B. CO_2 Emissions

If targets for CO_2 emission are set in due course, this may provide a considerable constraint on generation. The current trend to substitute aging power stations with high-efficiency gas-fired combined-cycle gas turbine stations means that there is inevitably a reduction in CO_2 emission per unit generated. In this context, the use of flue gas desulfurization is not beneficial as overall CO_2 release is enhanced due to the decreased system efficiency and, to a lesser extent, as a result of CO_2 released from the limestone used.

IX. BEST PRACTICAL ENVIRONMENTAL OPTION (BPEO)

An inevitable consequence of Integrated Pollution Control (IPC) is that attention is focused, quite properly, not only on the amount of pollution emitted from a process but also on the way in which that pollution is handled. While this is clearly a sensible route to follow, it is less clear how the decision on the most favorable option is to be arrived at. This is an area which clearly warrants further consideration; it is necessary to identify not only the options which are available, but also to determine both the short- and long-term consequences which may arise. This means that a very detailed understanding of both the technical options for treating the pollutant and also the potential for harm which may arise over the short and long term for each of the options is required.

It should also be recognized that each change to an established process may alter the way in which pollutants are released. An example is the fitting of flue gas desulfurization. It is quite clear that the removal of SO_2 is beneficial in terms of gaseous discharge but it immediately produces a problem of product gypsum which has to be either sold or disposed of as waste. Further consideration identifies that there will be second-order issues. For example, the scrubbing process will remove certain material which was originally discharged through the stack, in addition to the SO_2 itself. The question then arises of whether the other impurities removed from the gas phase reach the environment with the waste water or are removed as sludge from a water-treatment plant and how each of these routes impacts on the environment overall.

A further complication arises in that, if IPC and BPEO are properly applied, conflicts can occur with media-specific regulations. Regulatory bodies will need to address this conundrum if the full benefits of IPC are to be realized.

REFERENCES

1. The United Kingdom's Programme and National Plan for Reducing Emissions of Sulphur Dioxide (SO_2) and Oxides of Nitrogen (NOx) from Existing Large Combustion Plants, December 20, 1990.
2. Final Declaration of the Third International Conference on the Protection of the North Sea, The Hague, March 8, 1990.

Air Toxics: The Research Challenge of the 1990 Clean Air Act Amendments

Frank T. Princiotta
Douglas G. McKinney

I. INTRODUCTION

Toxic emissions from stationary air pollution sources have been recognized as a national health problem for several decades. Unfortunately, a large percentage of these hazardous air pollutant (HAP) emissions have not been controlled because, between 1977 and 1990, the risk-based Federal regulatory program to reduce them only addressed a few pollutants. Disputes between industry, the U.S. Environmental Protection Agency (U.S. EPA), and environmental groups over health-effects data, exposure levels, and what constitutes an *ample margin of safety to protect the public health* were the primary impediments to progress and limited the number of National Emission Standards for Hazardous Air Pollutants (NESHAPs) issued to only ten. The lack of progress toward reducing HAP emissions raised public concern and convinced Congress to design a new regulatory approach which would avoid the past impediments and accelerate emission reductions. The purpose of this paper is to describe the major provisions of Title III of the 1990 Clean Air Act Amendments (CAAA) which formally established this modified approach and the research which the U.S. EPA has initiated to support state and Federal regulatory officials implement the extensive new provisions.

II. MAJOR TITLE III PROVISIONS

Prior to the passage of the 1990 CAAA, all Federal regulations to reduce HAP emissions were chemical specific, with emission levels set based on a specific risk level. This strategy was substantially changed when Congress modified section 112 (now Title III) of the Clean Air Act to establish a two-phased approach for regulating HAPs. The first phase requires national *technology-based* standards for new and existing major stationary sources which emit more than 10 tons of one or 25 tons of a combination of the 189 HAPs listed under section 112 (b)(1). Emission levels promulgated for each specific category or subcategory of major sources will be based on the emission-reduction capability of the Maximum Achievable Control Technology (MACT) available (Section 112(d) of the Amendments defines MACT for existing and new sources). Once all the technology-based standards are completed (10 years), the second phase of the program will be initiated. This latter phase will determine if additional standards are necessary for a given source category or subcategory to protect human health and the environment from *residual risks* caused by any remaining emissions. The two-phased approach should accelerate HAP emission reductions because the initial phase will reduce the majority of the risks without the need for contentious risk assessments.

In addition to the two-phased regulatory strategy, Title III includes a variety of special studies which will generate data to support future national air toxics policies. These studies require the U.S. EPA to either further characterize air toxic problems or improve the methods and techniques used to define air toxic risks. Several of the most significant required studies are described below.

- **Area Source Program**: The purpose of this program is to reduce 75% of the cancer incidence resulting from area sources (defined as stationary sources which emit less than 10 tons of a single HAP or 25 tons of a combination of HAPs) of hazardous air pollutants in urban areas. The study requires the U.S. EPA to initiate an urban toxics research program and to produce a Comprehensive National Strategy which must identify the 30 pollutants that pose the greatest threat to public health in urban areas and ensure 90% of the area sources which emit these pollutants are subject to standards. The strategy developed must be provided to Congress by November 1995.
- **Great Waters**: The purpose of this study is to identify and assess the extent of atmospheric deposition of hazardous air pollutants to the Great Lakes, the Chesapeake Bay, Lake Champlain, and coastal waters. Within 5 years, the Administrator must determine whether further emission standards (beyond those required under other portions of section 112) are needed to prevent serious adverse effects to any of the Great Lakes or coastal waters.
- **Coke Oven Production Technology**: Requires DOE and EPA to jointly undertake a 6-year study to assess coke oven production emission control technologies and assist in the development and commercialization of technically practical and economically viable control technologies which have the potential to significantly reduce emissions of hazardous air pollutants from coke oven production facilities.
- **Electric Utilities:** This study requires the Administrator to report to Congress by November 1993 on hazards to public health due to emissions of hazardous air pollutants from utilities and to develop standards for utilities if he concludes they are necessary based on the results of the study. In addition, a special assessment of mercury emissions from utilities including possible control options is due by November 1994.
- **NAS Risk Assessment**: EPA is required to make arrangements with the National Academy of Sciences to conduct a review of the risk assessment methodology EPA uses to determine carcinogenic risks associated with exposure to hazardous

air pollutants and to discuss possible improvements in the methodology. A report is due by May 1993.

Based on the results of these studies, the U.S. EPA may promulgate additional emission standards to protect specific geographic areas such as the Great Lakes or to include sources such as utility boilers which are not currently subject to any Title III regulations.

III. RESEARCH STRATEGY

The data needed to support this new two-phased regulatory approach are extensive because of the large number of pollutants and source categories subject to regulation. The U.S. EPA's Office of Research and Development (ORD) has recognized the challenge of this new regulatory strategy and has modified its existing research program to provide the critical data and methods needed. Development of detailed chemical-specific risk assessments emphasized since the late 1970s has been replaced by a broader research program which focuses on providing the data needed to regulate specific source categories. The new strategy will initially focus on developing information to support the phase-one MACT standards and special studies. In future years, emphasis will shift toward developing advanced risk assessment methods and other data needed to support residual risk determinations. Based on this fundamental change in research emphasis, the new goals of the Federal HAP research program are to

- Support the technology-based standards by developing source test methods which will be used to determine compliance with MACT standards and by evaluating existing and innovative control technologies and pollution prevention approaches applicable to source categories subject to MACT standards
- Characterize sources of air toxics and determine exposure levels and risks associated with those air toxic sources where little or no information is available; much of this assessment work is currently aimed at supporting the large number of special studies in Title III (Area Sources; Great Waters)
- Provide data to support residual risk determinations

Research in the following six major areas is now being conducted to support these broad goals: source test methods, emissions data and emission reduction approaches, ambient monitoring, exposure assessment, health studies and assessments, and technology transfer. These research areas were chosen for emphasis based on near-term data needs identified by the U.S. EPA's Office of Air Quality Planning and Standards (OAQPS) and an independent evaluation by ORD scientific staff of the research needed to respond to future HAP problems such as residual risk. Table 1 provides a summary of how each research area supports the new Title III provisions.

Table 1. Relation of Research Areas to Key Title III Provisions

Research areas	Key Title III provisions		
	MACT standards	Special studies	Residual risk
Source test methods	X	X	X
Emissions data and emission reduction approaches	X	X	X
Ambient monitoring		X	X
Exposure assessment		X	X
Health studies and assessments		X	X
Technology transfer	X	X	X

IV. RESEARCH DESCRIPTIONS BY AREA

A. Source Test Methods

Currently there are no established methods for measuring emissions of specific HAPs from many of the source categories for which the U.S. EPA will develop MACT standards over the next 10 years. ORD is responding to this need by developing measurement methods for the list of 189 air toxics to support compliance with MACT and any future standards. The program is designed to have stationary source test methods available for a relatively large number of pollutants in the shortest time and at the least cost through development of screening methods and generic methods which identify and quantify a number of different pollutants. However, for those substances that have unique chemical properties, specific test methods will need to be developed.

B. Emissions Data and Emission Reduction Approaches

This research area includes efforts to develop and evaluate data on HAP emissions, control technologies, and pollution prevention approaches. The emissions data research is focused on developing and improving area source emission estimation methods and providing emission inventories for three to five urban areas to support the Area Source Program. Emissions data are also being generated to support the periodic assessments of deposition to the Great Lakes and coastal waters. The emission reduction research includes (1) developing innovative control processes and pollution prevention approaches for source categories where existing technologies are either not available or not cost effective; (2) providing guidance on proper engineering practices to ensure that existing technologies which will be used as the basis for many MACT standards perform to their expected design efficiencies in the field; and (3) conducting engineering studies to demonstrate how existing control and prevention approaches which have only been used in a limited number of applications can be applied to other sources to reduce HAP emissions. Initial emphasis will be placed on approaches that could be used to comply with the phase one technology-based standards.

C. Ambient Monitoring

The goal of this research is to develop and evaluate, both in the laboratory and under field conditions, methods for sampling and analysis of ambient air pollutants. There are no satisfactory ambient methods for more than half of the HAPs listed under Title III. These methods are needed to identify and quantify HAPs which pose risks in urban air and to major lakes and coastal waters and to monitor the progress of the overall air toxics regulatory program. In addition, the ambient data will be used in the future to support residual risk determinations by providing information on how the phase-one MACT standards have impacted ambient air toxic concentrations. Initially efforts are focused on monitoring a wide variety of air toxics in several urban areas. These monitoring data directly support the Area Source Program which requires monitoring in a representative number of urban locations to characterize risks from area sources.

D. Exposure Assessment

The ORD program is developing data on the atmospheric lifetime and fate of air toxic chemicals. Such information will also be generated for selected nontoxic substances that react to form potentially toxic products. The Agency requires data on air toxics to estimate human and ecological exposures to those compounds and to help devise strategies to eliminate or limit exposures. Research is also underway to identify and quantify the contributions of specific sources of air toxics to ambient concentrations. Source signature data will be gathered to reconcile emissions and ambient concentration data. The exposure assessment work directly supports the Area Source Program which requires consideration of atmospheric transformation and other factors which can elevate public health risks from such pollutants. In addition, the data on exposure will be critical to making residual risk decisions.

E. Health Studies/Assessments

In the health effects area, ORD is evaluating the nature and magnitude of noncancer health effects associated with exposure to toxic air pollutants including developing methods to quantify developmental and reproductive toxicity of chemicals emitted to the air by point and area sources. Other health research is also planned to assess the neurotoxicity of the 189 HAPs in animals and humans, to identify which air toxics are liver toxicants, and to improve cancer risk extrapolation for air toxics through biologically based dose-response modeling. The assessments portion of this research area will provide noncancer and cancer health risk assessments on air toxics in support of regulatory activities. The major activities are to develop chronic Reference Concentrations known as RfCs (defined as an air concentration for a chemical which is anticipated to pose little or no public health risks) for Title III compounds and to develop acute noncancer assessments of the 189 listed air toxics. The health research will support many of the special studies required under Title III including the area source program which requires EPA to identify the 30 (or more) air toxics that pose the greatest threat to public health and will provide the foundation for future residual risk assessments after MACT has been applied to point sources.

F. Technology Transfer

The technology transfer program will provide technical assistance on a variety of HAP issues ranging from control technology performance to health risk assessments. The information will be provided to users through three technical centers currently in full operation. The type of assistance each center provides is described below.

- The Control Technology Center (CTC) provides technical assistance to state and local air pollution control agencies on matters relating to air pollution control technology. A hotline is used to provide quick responses to telephone requests for assistance. In addition, more in-depth studies are initiated to respond to specific state or local requests for assistance which cannot be answered without initiating a formal study. Results from these longer-term assistance projects are issued as technical guidance through publication of technical reports, development of personal computer software, and workshops on control technology matters.
- The Air Risk Information Support Center (Air RISC) provides scientific assessment assistance to states, regions, and local agencies on air toxics health risk issues. The types of guidance provided include (1) rapid response to health risk questions; (2) training in use of risk assessment guidelines; (3) development of cancer and noncancer assessments; and (4) review/consultation on state regulatory actions.
- The Emission Measurement Technical Information Center (EMTIC) promotes consistent and accurate application of emission test methods for development and enforcement of national, state, and local emission prevention and air pollution control programs. The center has gathered information on testing needs, held workshops, and sent test methods and quality assurance information to regional, state, and local agencies.

V. CONCLUSION

Title III of the 1990 CAAA has substantially changed the process used to regulate HAPs and is expected to substantially decrease emissions over the next 10 years as MACT standards are promulgated for hundreds of source categories and subcategories. However, the effectiveness of the program will depend heavily on the quality of scientific data used to develop and enforce each emission standard issued. The U.S. EPA's ORD has redesigned its air toxics research program to ensure this scientific data is generated in time to meet mandated regulatory deadlines. The combination of near-term research to support MACT standards and special studies with longer-term research to support future residual risk determinations will ensure the challenging new regulatory approach is implemented with the best scientific information available.

Chapter 2
Emissions Sources

Chair William P. Peel
Canadian Electrical Association/Alberta Power
Co-Chair: Charles E. Schmidt
U.S. Department of Energy —
Pittsburgh Energy Technology Center

An Overview of CARB-Adopted Source Test Methods for Toxic Compounds and Results of Testing Natural Gas-Fired Utility Boilers

David F. Todd
William V. Loscutoff

I. INTRODUCTION

Since 1983, the California Air Resources Board (CARB) has been developing and adopting, for regulatory purposes, nonvehicular source test methods for toxic air contaminants. Those methods have been used extensively to develop data for California's two major air toxic programs: (1) the Toxic Air Contaminant Identification and Control Act, also known as AB1807, which was signed into law in 1983 and (2) the Air Toxics "Hot Spots" Information and Assessment Act, also known as AB2588, which was passed in 1987.[1] Specifically, toxic source test methods are used for both generating data and enforcing toxic air control measures or regulations as part of the Toxic Air Contaminant program and for generating source-specific toxic air emissions inventory to satisfy the "Hot Spots" Act requirements. The methods are also used for engineering purposes such as determining toxic air emissions from specific sources or evaluating control devices for the control of toxic air emissions. Presently, CARB has 25 Toxic Air Contaminant source test methods[2] available.

This paper presents a brief overview of the CARB toxic air contaminant source test methods and the results of toxic air emissions tests conducted on three natural gas-fired utility boilers.[3]

II. SOURCE TEST METHODS FOR TOXIC AIR CONTAMINANTS

In order to respond to programmatic needs, the CARB has developed and adopted 25 toxic air contaminant source test methods. Eight of these methods apply to metals as shown in Table 1. Metals methods are presently being reviewed for possible inclusion in an overall, multiple-metals method. Table 2 lists methods adopted for inorganic compounds other than the metals listed in Table 1. In some cases, like bulk asbestos testing, methods have been developed to respond to a specific rule-development need. In this case, the method is designed to provide data in order to determine compliance with rule limits rather than data for engineering purposes.

Source test methods for specific volatile organic compounds like vinyl chloride, benzene, formaldehyde, ethylene oxide, and volatile organic compounds in general are shown in Table 3. The vinyl chloride method was adopted before 1986 because

Table 1. Air Toxics Test Methods: Metals

Number	Compound	Adopted or amended
12	Lead	March 1986
101, 101A	Mercury	March 1986
104	Beryllium	March 1986
423	Arsenic	January 1987
424	Cadmium	January 1987
425	Chromium	September 1990
433	Nickel	September 1989
Draft 436	Multimetals	June 1992

Table 2. Air Toxics Test Methods: Other Inorganics

Number	Compound	Adopted or amended
13A and B	Fluoride	March 1986
421	HCl	January 1987
426	Cyanides	January 1987
427	Asbestos	March 1988
434	Chlorine	September 1990
435	Asbestos — bulk	June 1991

Table 3. Air Toxics Test Methods: Volatile Organics (VOCs)

Number	Compound	Adopted or amended
106	Vinyl chloride	June 1983
410A and B	Benzene	March 1986
422	Volatile organics	September 1990
430	Formaldehyde	September 1989
431	Ethylene oxide	September 1989

Table 4. Air Toxics Test Methods: Semivolatile Organics

Number	Compound	Adopted or amended
428	Dioxins/furans/PCBs	September 1990
429	PAH	September 1989
	Miscellaneous	
401	% VOC in waste	March 1986
432	Dichloromethane and TCE in coatings	September 1989

Table 5. Tested Natural Gas-Fired Utility Boilers and Selected Process Information

Company	Power plant	Boiler	Test date	Fuel, MCFH[a]	Electricity, (MWatts)[b]
LA DWP	Haynes	#4	6/11/91	950	120
SDG&E	Encinas	#1	6/13/91	259	23.7
PG&E	Morro Bay	#4	6/19/91	2936	332

[a] MCFH — thousand cubic feet per hour.
[b] MWatts — megawatts.

California had adopted an ambient air quality standard for vinyl chloride prior to the implementation of the Toxic Air Contaminant program. The volatile organic compounds method, Method 422, was recently used to determine volatile organic emissions from natural gas-fired boilers. This will be discussed later.

Table 4 shows the methods adopted for semivolatile organics and two miscellaneous test methods, also adopted in response to specific rule development needs. The semivolatile methods are often used at combustion sources such as incinerators and are generally the most expensive to conduct.

III. SOURCE TESTS OF UTILITY BOILERS

Emissions tests of three natural gas-fired utility boilers for selected toxic or potentially toxic volatile organic compounds were conducted by the CARB staff using CARB Test Method 422. The three tested power plants included the Los Angeles Department of Water and Power (DWP) Haynes facility (Long Beach, CA), San Diego Gas and Electric (SDG&E) Encinas facility (Carlsbad, CA), and Pacific Gas and Electric (PG&E) Morro Bay facility (Morro Bay, CA). Only one natural gas fired-utility boiler at each of the facilities was tested. The tested boilers, test dates, and selected process information are shown in Table 5.

The purpose of this test was to determine the presence of selected aromatic and halogenated volatile organic compounds in the natural gas and stack emissions. Although all of the tested utility boilers are designed to burn both natural gas and fuel oil, the test was conducted only when the boilers were fired with natural gas.

IV. TEST METHODS

CARB Method 422 was followed to determine the amount of selected aromatic and halogenated volatile organic compounds. Grab samples of the stack gas and natural gas fuel were collected in Tedlar® bags in accordance with CARB Method 422. No liquid knockouts were used for bag sampling (a Method 422 option) and no moisture condensed in the sample bags.

Within 72 h, the samples were analyzed by gas chromatography (GC) as specified in Method 422. Stack samples were analyzed by a gas chromatograph (GC) with a photoionization detector (PID) followed by an electron capture detector (ECD). (With the DB-624 column in the GC/PID/ECD instrument, o-xylene and styrene co-eluted and are reported together in this report.) Natural gas fuel samples were analyzed by GC/mass spectrometry (GC/MS). GC/MS was also used to confirm the results of the GC/PID/ECD for three selected stack samples.

V. QUALITY ASSURANCE PROCEDURES

Quality assurance procedures were used by the CARB staff to ensure the reliability of the results. These procedures included checking the sampling bags for contamination prior to sampling and collecting blank and spike bag samples in the field.

Before going into the field, the stack sample bags were checked for contamination. Out of 12 sample bags, 9 were reported to contain toluene ranging in concentrations from 0.4 to 1.5 ppbv, indicating toluene contamination is likely in any of the sample bags.

At each power plant, a blank bag was collected by running 99.9% pure nitrogen through the bag-sampling train into the blank bag. A check of the nitrogen collected through the sampling train prior to testing indicated the presence of trichloromethane, toluene, and xylene in the nitrogen, but only toluene shows up in a majority of the blank bags.

Summa-polished canisters containing low- and high-concentration spikes were taken to each power plant. The low-concentration spikes contained about 5 ppb of all of the compounds of interest except toluene and m,p-xylene. The high-concentration spikes contained about 100 ppb of all of the compounds of interest except toluene and benzene. At each power plant, part of the contents of the high- and low-spike canisters were transferred to empty sample bags. The low concentration spike canisters and bags traveled with the stack samples for analysis. The high-concentration spikes traveled with the fuel samples for analysis. The results generally indicated low recovery.

None of the results of the pretest bag checks, sample blanks, or sample spikes was used to "correct" or in any way change the results of the fuel and stack-sample analyses that are presented in this report.

VI. TEST RESULTS AND DISCUSSION

The analytical results by GC/MS for volatile organic compounds in the natural gas fuel samples are summarized in Table 6. From those results, natural gas contains four of the volatile organic compounds of interest — benzene, toluene (methyl benzene), ethylbenzene, and o-xylene (dimethyl benzene). The concentrations of these four compounds varied from power plant to power plant. Natural gas to the

Table 6. Summary of Natural Gas Analysis, ppbv

Compound	Minimum reporting Limits[a]	Haynes average	Encinas average	Morro Bay average
Dichloromethane (methylene chloride)	380			
Trichloromethane (chloroform)	140			
1,1,1-Trichloroethane (methyl chloroform)	NA[b]			
Tetrachloromethane (carbon tetrachloride)	470			
1,2-Dichloroethane (ethylene dichloride)	320			
Trichloroethylene (TCE)	140			
Tetrachloroethylene (PERC)	400			
1,2-Dibromoethane (ethylene dibromide)	540			
Benzene		8,325	6,403	3,060
Toluene (methyl benzene)		11,150	8,980	3,950
Ethylbenzene		2,600	578	172
Chlorobenzene	240			
m,p-Xylene (dimethyl benzene)	NA			
o-Xylene		675	762	213

[a] GC/MS minimum reporting limits for natural gas analysis are elevated because low injection volumes were used to offset the relatively high concentrations.
[b] NA — not applicable.

Table 7. Summary of Stack Sampling Results (and Results at 12% CO_2), ppbv

Compound	Minimum reporting Limits	LA DWP Haynes average	SDG&E Encinas average	PG&E Morro Bay average
Carbon dioxide (12%)		14% (12%)	6.5% (12%)	9.5%
Dichloromethane	3.3			
Trichloromethane	0.25			
1,1,1-Trichloroethane	1.3	0.8 (0.7)[a]	1.2 (2.1)	
Tetrachloromethane	0.18			
1,2-Dichloroethane	3.3			
Trichloroethylene	3.3			
Tetrachloroethylene	0.33	0.9 (1.6)[a]		
1,2-Dibromoethane	0.25			
Benzene	0.8	1.3 (1.1)	1.2 (2.3)	0.7 (0.8)
Toluene	0.25	1.4 (1.2)	1.8 (3.2)	2.2 (2.7)
Ethylbenzene	0.18	0.1 (0.1)[a]		
Chlorobenzene	0.43			
m,p-Xylene	1.2	0.3 (0.3)[a]		0.3 (0.4)[a]
o-Xylene and styrene[b]	1.8			

Note: For averaging, concentrations below the minimum reporting limits were given a value of zero. For risk assessment, CARB policy requires those compounds be given a value equal to half the minium reporting limit. Numbers in parentheses are concentrations to 12% CO_2.

[a] Compound was detected in only one of four stack-sample bags collected at this facility.
[b] o-Xylene and styrene co-eluted together and are reported together.

Table 8. Destruction Efficiencies of Selected Volatile Organic Compounds (Stack Flow Based on Fuel Flow and U.S. EPA F-Factors)

Compound	Haynes #4	Encinas #1	Morro Bay #4
Benzene			
In, lb/day	38.5	8.1	43.7
Out, lb/day	0.07	0.03	0.13
Destroyed, %	99.8	99.7	99.7
Toluene			
In, lb/day	60.8	13.4	66.6
Out, lb/day[a]	0.09	0.05	0.47
Destroyed, %	99.9	99.7	99.3
Ethylbenzene			
In, lb/day	16.34	0.99	3.34
Out, lb/day	<0.03	<0.01	<0.11
Destroyed, %	>99.8	>98.7	>96.8
o-Xylene			
In, lb/day	4.24	1.31	4.14
Out, lb/day	<0.13	<0.05	<0.45
Destroyed, %	>96.9	>96.0	>89.2

[a] In a pretest bag check of all of the sample bags, toluene contamination ranged from less than 0.25 (minimum reporting limit) to 1.5 ppb for an average of 0.7 ppb (concentrations below the minimum reporting limits were given a value of zero). These results were not corrected to account for contamination, blank bag results, or spike bag results.

[b] Compound was below the minimum reporting limit in all stack-sample bags. "Out" calculations based on minimum reporting limit. For risk assessment, CARB policy would require "Out" calculations be based on half the minimum reporting limit.

Haynes power plant had the highest concentrations for three of the four compounds — benzene, toluene, and ethylbenzene. Natural gas to the Morro Bay power plant had the lowest concentrations of the four compounds.

The analytical results by GC/PID/ECD for selected aromatic and halogenated volatile organic compounds in the stack samples are summarized in Table 7. Benzene and toluene were found in almost all of the stack samples. Other compounds such as 1,1,1-trichloroethane and tetrachloroethylene were also reported, but only in one or two stack-sample bags.

Based on the natural gas fuel analysis (Table 6), the number of volatile organic compounds of interest could be reduced to the four compounds that were reported in the fuel — benzene, toluene, ethylbenzene, and o-xylene. Table 8 presents the destruction efficiencies of the three utility boilers for those four compounds. From Table 8, over 99.6% of the benzene was destroyed in the three utility boilers. However, GC/MS confirmation on three of the stack-sample bags failed to confirm the presence of benzene in the stack samples, so the percent of benzene destroyed could be higher.

Over 99% of the toluene in the natural gas was destroyed in the three utility boilers. The toluene contamination reported in the stack-sample bags was relatively high ranging from less than 0.25 to 1.5 ppb with an average of 0.7 ppb. The toluene in the stack samples ranged from 1.0 to 3.5 ppb including contamination. Therefore, some of the toluene reported in the stack samples may be the result of toluene in the bags before sampling began. GC/MS confirmed the presence of toluene in the three confirmation bags.

The percent destroyed shown in Table 8 for ethylbenzene and o-xylene is also low. Neither compound was reported in the stack samples and this was confirmed by GC/MS analysis on the three confirmation bags. The percent destruction for ethylbenzene and o-xylene is based upon the minimum reporting limits for those compounds in the stack samples.

VII. CONCLUSIONS

In conclusion, it can be reported that volatile organics exist in natural gas going into the tested utility boilers. It can also be reported that emissions of these volatile organics from utility boilers is low, even though the stack-sample results are clouded by low concentrations of VOCs, high toluene contamination, a lack of benzene confirmation by GC/MS, and low recoveries for the spike samples. If there is concern about the amount of volatile organic compounds emitted, a more sensitive test method with a lower reporting limit is needed.

REFERENCES

1. Lagarias, J. et al., "The California Experience in Toxic Air Pollutants Control", presented at EPRI Conference, Managing Hazardous Air Pollutants, Washington, DC, November 4–6, 1991.
2. California Air Resources Board, "Stationary Source Test Methods Volume III, Methods for Determining Emissions of Toxic Air Contaminants from Stationary Sources", California Air Resources Board, Sacramento, 1990.
3. Todd, D. F., Ouchida, P. K., and Lew, G., "Emissions Tests on Natural Gas-Fired Utility Boilers at Three Power Plants", California Air Resources Board, Sacramento, 1991.

Trace Element Contents of Commercial Coals

E. L. Obermiller
V. B. Conrad
J. Lengyel, Jr.

I. INTRODUCTION

Because of an increased emphasis on coal quality in the utility industry, CONSOL began a systematic comprehensive analysis program for all of its product coals in 1981. Monthly composites of shipped coals are analyzed twice per year. The database includes analyses of most of the trace elements listed in the Air Toxics title of the 1990 Clean Air Act Amendments for product coals (both raw and washed). This database is particularly useful for addressing utility environmental concerns, because it represents actual commercial coal shipments. Most published data on the trace elements contents of coal are limited to a compilation of core or channel sample analyses, or hypothetical washabilities.

In this paper the CONSOL comprehensive database is used to evaluate trace element variability within a single mine over time and variability within a single seam as it relates to commercial shipments. A comparison of the trace element contents of coals in different regions of the U.S. and the effect of coal cleaning on trace elements will be discussed. All of these will be related to the in-lab analytical variability of the trace element analyses and to variability in ash, sulfur, and Btu contents. Finally, a brief discussion will be made of a trace element analytical variability study involving five commercial laboratories.

II. COAL ANALYSIS DATABASE

The CONSOL coal analyses database has been generated since 1981 and includes trace element analyses made in 1982 through 1988, 1990, and 1991. The program is ongoing. The samples of each coal shipped during the months of March and September are composited by the CONSOL regional laboratories and sent to R&D for a comprehensive suite of all of the commercially important coal analyses. The database provides an excellent evaluation of the variations of coal characteristics over time in a given mine as they relate to commercial coal shipments. The database also can provide information on in-seam variations and the effect of coal preparation, at least indirectly. Because not all product coals are shipped every month, the number of coal

Table 1. Repeatability of Trace Element Analyses in CONSOL R&D Analytical Laboratory

	Percent relative standard deviation	
Trace element	**ICP-MS**	**ICP-AES**
As	3.4	3.8
Be	13.4	2.8
Cd	4.9	3.2
Cr	3.9	2.8
Li	5.6	2.6
Ni	6.2	8.1
Mn	3.3	4.6
Mo	3.9	—
Pb	4.8	3.2
Sb	3.4	7.1
Se	10.9	6.0
Sn	2.7	6.8
Tl	5.7	13.7
V	5.3	2.8
Zn	4.7	2.4
Average	6.3	5.5
F[a]	5.5	5.5
Hg[b]	17.6	17.6

[a] By hydropyrolytic-ion chromatographic method.
[b] By gold-film method.

samples analyzed varies from 35 to 50 each time they are collected. The database now contains analyses of about 800 composite samples, and about 3100 individual trace elemental analyses on over 200 coal samples. The coals used for this paper are all product coals, and, generally, those for which a number of trace element analyses exist in the database. All of the results are reported on a dry coal basis and as the element, not the oxide as is common in reporting the major ash constituents such as SiO_2, Al_2O_3, etc.

III. METHODS OF ANALYSIS

The trace elements included in the program are listed in Table 1. In addition, thorium and uranium were analyzed for some samples. These were chosen based on a literature survey related to their frequency of occurrence in coals and potential environmental impact. Before 1987, most trace element analyses were done using a combination of inductively coupled plasma-atomic emission spectroscopy (ICP-AES) and atomic absorption spectroscopy (AAS) including hydride and graphite furnace techniques.[1] Beginning in 1987, analyses were done using inductively coupled plasma-mass spectroscopy (ICP-MS)[2] and ICP-AES for all elements with the exception of Hg and F. Mercury is determined using the gold film method[1] and fluorine using a hydropyrolysis-ion chromatographic method.[1,3]

To assess the variability of trace elements in a mine or seam, the precision of the analytical procedures must be determined. Table 1 shows the elements and the associated analytical repeatabilities as determined in the CONSOL R&D laboratory. These data were generated using National Institute (NIST) Standard Reference Material 1633a. Most percent relative standard deviations (PRSDs) are below 10% with Be, Hg, and Se higher than 10%. All of the trace element determinations in our laboratory are done in duplicate and our instruments are calibrated with NIST or other certified reference materials. Based on the results in Table 1, the analytical variability in the trace element analyses reported here is small compared to the natural variability of the samples. Ash, sulfur, and Btu content were determined by ASTM methods. The analytical variabilities (repeatabilities) of these parameters (from ASTM) are 2, 5, and 0.5%, respectively, when analyzing a coal sample with <10% ash, 2.5% sulfur, and 13,000 Btu.

IV. COALS

The database consists predominantly of analyses of eastern and midcontinent coals ranging in rank from hi-vol to mid-vol bituminous. Both steam and metallurgical coals are represented. Several lignite samples are also included. In most cases, the coals are representative of output from a single mine during the compositing period. However, in some cases they represent preparation plant products and could include coals from more than one source. The important fact is that all of the coals represent commercial products as shipped to the customer. Therefore, they can be used to evaluate the real-world variability in trace element contents likely to be encountered by the coal customer. Because these analyses were performed on monthly composite samples, the trace element contents represent averages over that period of time. It is possible that more instantaneous samples would show greater variability.

A. Comments on Statistical Presentation of Data

Throughout this paper, variability is reported as a percent relative standard deviation (PRSD). This is a useful and convenient means of representing the data. However, in a sense, it may underestimate variability, particularly if it becomes important to control trace element contents of coal within particular limits. Assuming that the data are normally distributed, the area encompassed by one standard deviation on either side of the mean value is 68.3% of the total distribution. Therefore, 31.7% of the samples analyzed would fall outside the range indicated by the PRSD. Two standard deviations encompass 95.4% of the normal distribution. This is arguably a more reasonable measure of natural variability in a regulatory situation.

V. RESULTS AND DISCUSSION

There are many technical specifications which are important in coal purchasing. Among the most common are ash and sulfur contents and heating value. Coal mining plans are devised and coal preparation plants are operated to control these parameters to meet coal-sale specifications. Therefore, the variability in these parameters can be used as a gauge against which to compare the variability in trace element contents. If a trace element is more variable than ash or sulfur, this implies a level of natural variability that is not controlled by those measures which are taken in coal production to guarantee product constancy. On the other hand, the degree to which trace element content is correlated with ash or sulfur can be used to evaluate the potential success of controlling trace element content by conventional coal-cleaning practices.

A. Correlation of Trace Element Contents with Ash and Sulfur

The concentrations of the individual trace elements are plotted against ash content in Figures 1 to 3. These figures indicate the range of trace element contents in the database. Average trace element concentrations range from <0.1 (mercury and cadmium) to 76 ppm (fluorine); the median value is 1.91 ppm (selenium). Most of the trace elements are variable over about an order of magnitude. The ash contents are somewhat less variable, most falling in a band from about 6 to 15%.

As might be expected, the trace element concentrations are correlated with ash content, but not strongly. The correlation coefficients (r^2) for linear regressions of trace elements with ash and with sulfur are plotted in Figures 4A and B. The lack of a stronger correlation reflects the variety of different coals in the database, as well as the natural variability for a given coal. As noted above, the analytical precisions for the trace element analyses in this database are typically in the range of 5 to 10% relative standard deviation, so analytical error is a minor component of the overall variability.

Figures 5A and B present the x-coefficients from the correlations expressed as a percentage of the average concentration of the element. These represent the average relative decrease (or increase) in a trace element for each 1% (absolute) increase in ash or sulfur. Most of the elements show a positive correlation with ash. The correlation with sulfur content (Figures 4B and 5B) is poor, and the x-coefficients show no consistent trend in relating trace element content to coal sulfur. There is no particular reason to expect a correlation of trace element content with sulfur, except to the extent that sulfur is itself correlated to ash content. For the coals in this database, the sulfur content is only poorly correlated with ash ($r^2 = 0.09$).

B. Trace Element Variability in a Single Mine

The correlations described above were based on an analysis of data for all the various coals in the database. Figure 6 presents the trace element variability for clean coal produced from a single Pittsburgh seam mine over the period 1982 to 1990. The variability data are shown as a PRSD, defined as the standard deviation in the data expressed as a percentage of the average value. The numbers above the bars are the average concentrations of ash and sulfur (wt% dry coal), and the trace elements (ppm whole coal). The coal averaged 7.8% ash and 2.5% sulfur. The relative variations in ash and sulfur were 7 and 9%, respectively. The PRSDs for the trace elements range from 5 to over 50%; a typical value is about 25%, or three times the PRSDs of the ash or sulfur.

C. Trace Element Variability in Adjacent Mines

Figure 7 is a plot of the PRSDs for data on coals from two adjacent Pittsburgh seam mines over the period 1986 to 1990. These two mines produce similar unwashed coals (15% ash, 4% sulfur). As the figure shows, the relative ash and sulfur variabilities are both less than 10%, similar to those in the previous case. Of course, on an absolute basis the variabilities are higher. The trace elements show PRSDs of 5 to 40%, with a typical value around 20%. Therefore, although the data include products from two different mines, the relative variabilities of the trace element contents are comparable to those of a product of a single mine. This suggests some consistency in the distribution of trace elements within a seam.

D. Trace Element Variability from a Preparation Plant

It is not uncommon for a coal preparation plant to process feeds from different sources. The data plotted in Figure 8 were obtained for coal samples from a single prep plant which processed different Ohio deep (Ohio 8 seam) and surface (Ohio 9 seam) coals. The ash and sulfur PRSDs are around 10%, and the trace element PRSDs range from 15 to 50%, with a typical value of 35%. This is approximately twice the trace element variability as that found for the single-source coals described above. The ash and sulfur variabilities are also somewhat greater, indicative of a wider range of product coals in this data set.

E. Trace Element Variability in Other Seams

The results presented above were obtained mostly for coals from the Pittsburgh seam (Ohio 8 is the Pittsburgh seam in Ohio). For comparison, Figures 9A and B show variability data for coals from the Pocahontas 3 and 4 seams, and the Illinois 6 seam. The results are generally similar to those for the Pittsburgh seam coals. For the metallurgical coals, the PRSDs range from 15 to 50%, with a typical value of around 30%. This larger PRSD for the metallurgical coals is a result of including coals from two seams, and three mines, in the calculation. For the Illinois 6 coal, the variabilities range from 15 to 40%, with a typical value of 25%. The Illinois 6 data include analyses of coals from three mines.

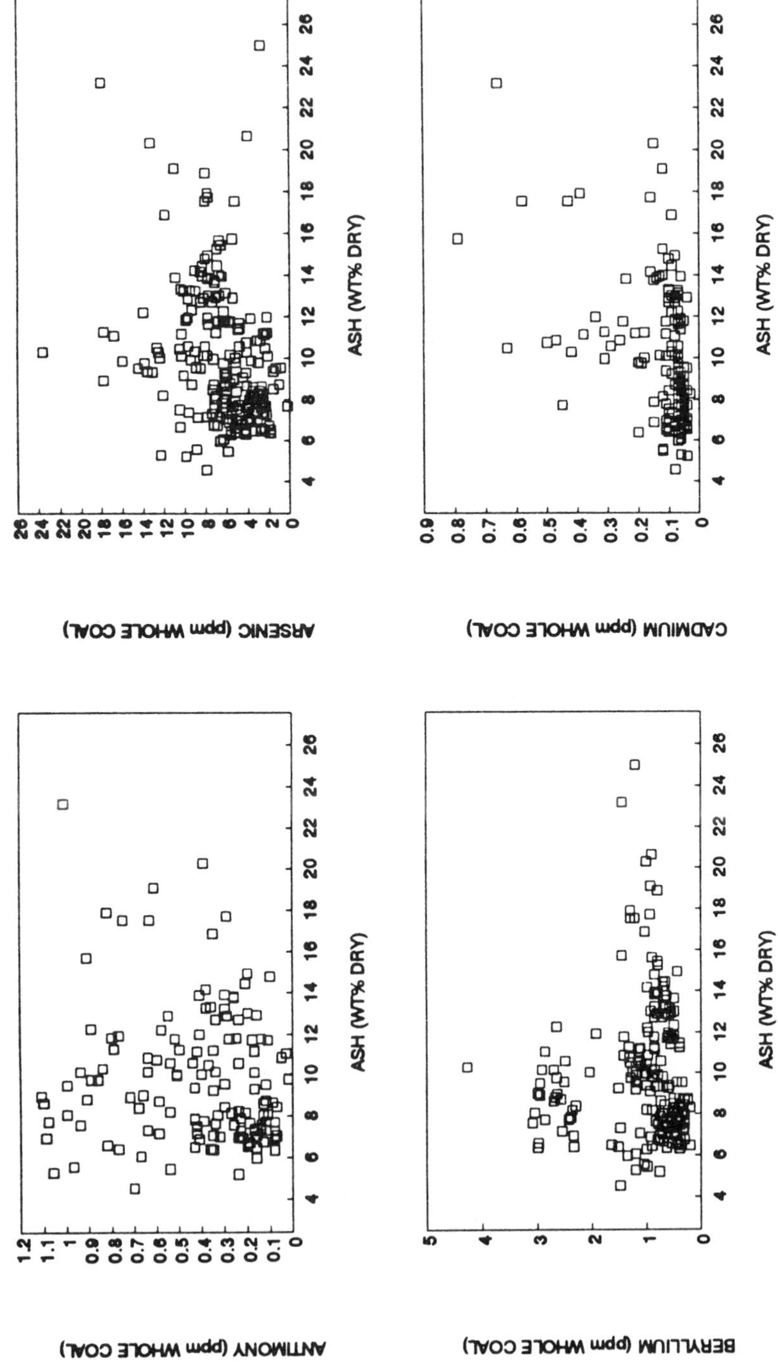
ARSENIC (ppm WHOLE COAL)
ASH (WT% DRY)
CADMIUM (ppm WHOLE COAL)
ASH (WT% DRY)
ANTIMONY (ppm WHOLE COAL)
ASH (WT% DRY)
BERYLLIUM (ppm WHOLE COAL)
ASH (WT% DRY)

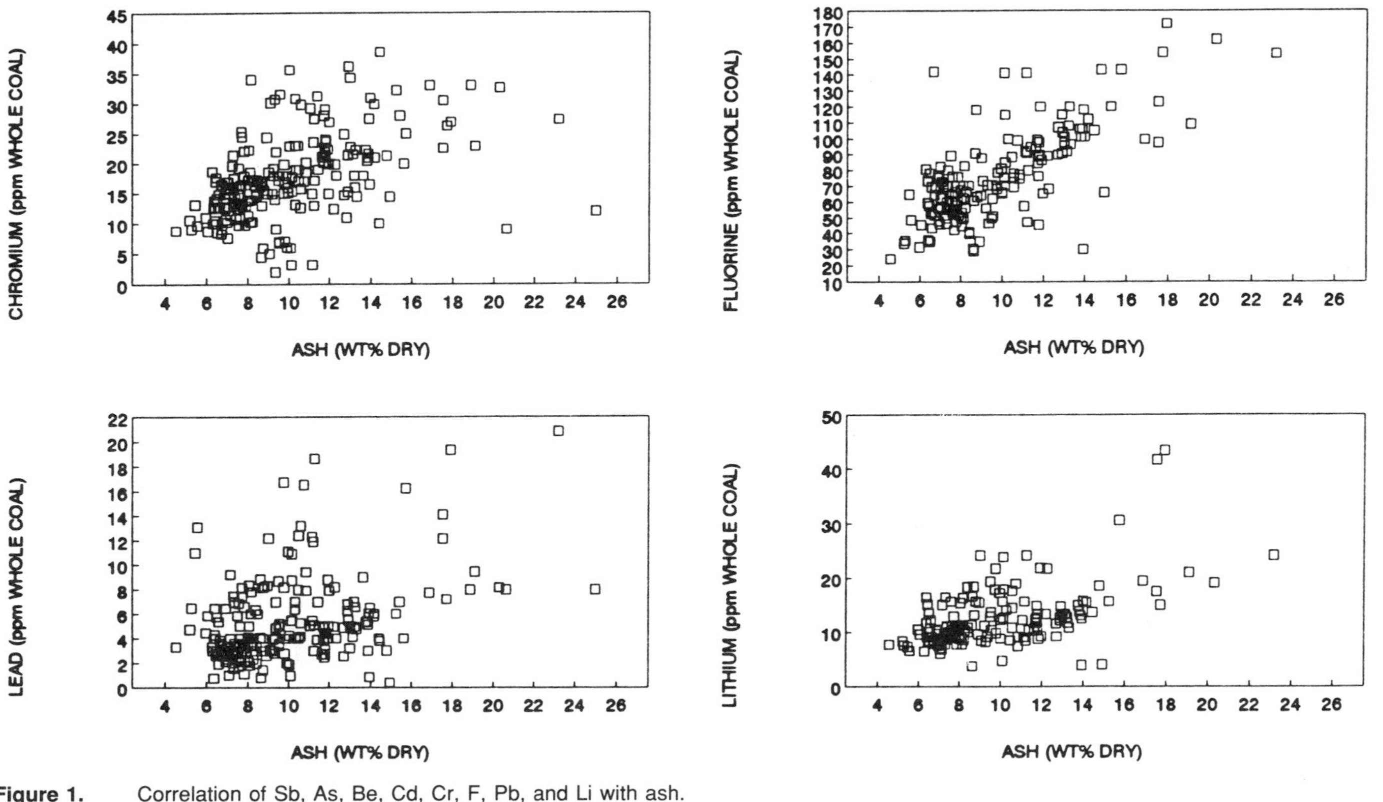

Figure 1. Correlation of Sb, As, Be, Cd, Cr, F, Pb, and Li with ash.

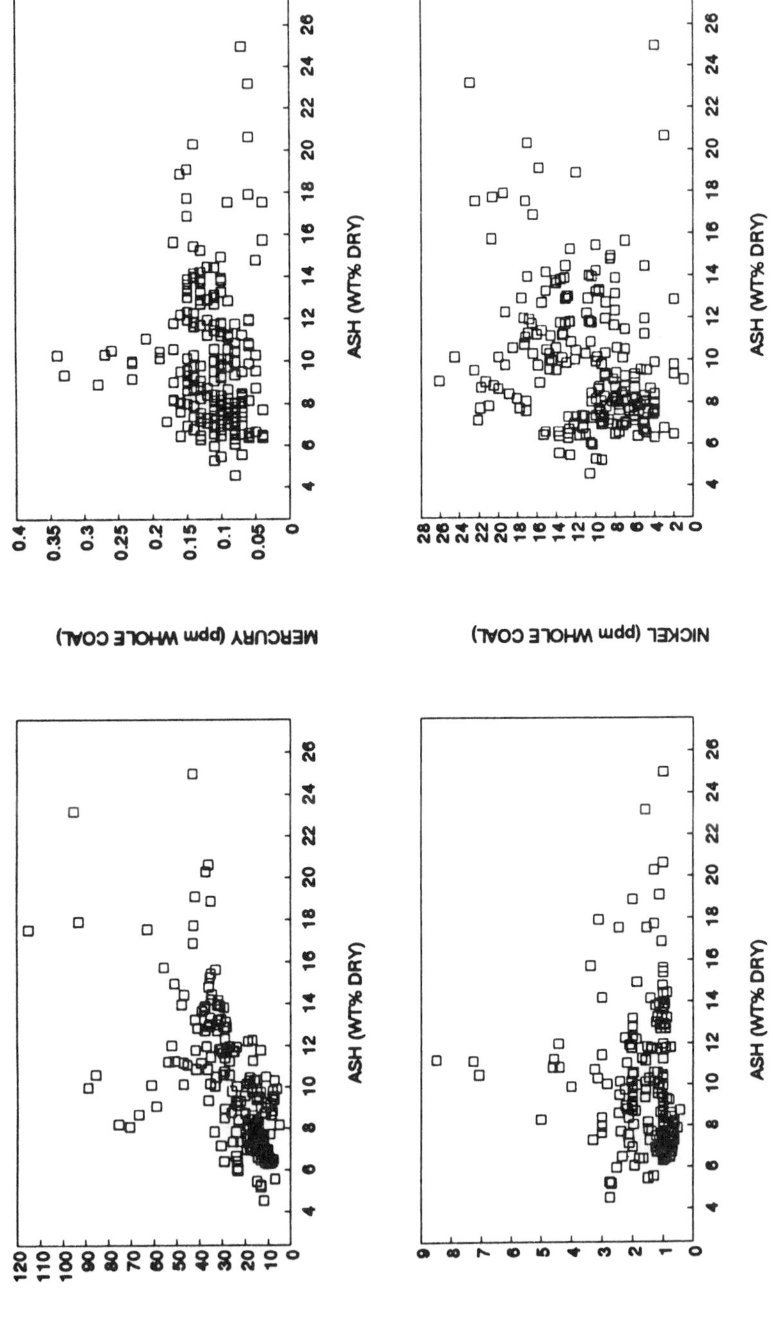
MERCURY (ppm WHOLE COAL)
ASH (WT% DRY)
NICKEL (ppm WHOLE COAL)
ASH (WT% DRY)
MANGANESE (ppm WHOLE COAL)
ASH (WT% DRY)
MOLYBDENUM (ppm WHOLE COAL)
ASH (WT% DRY)

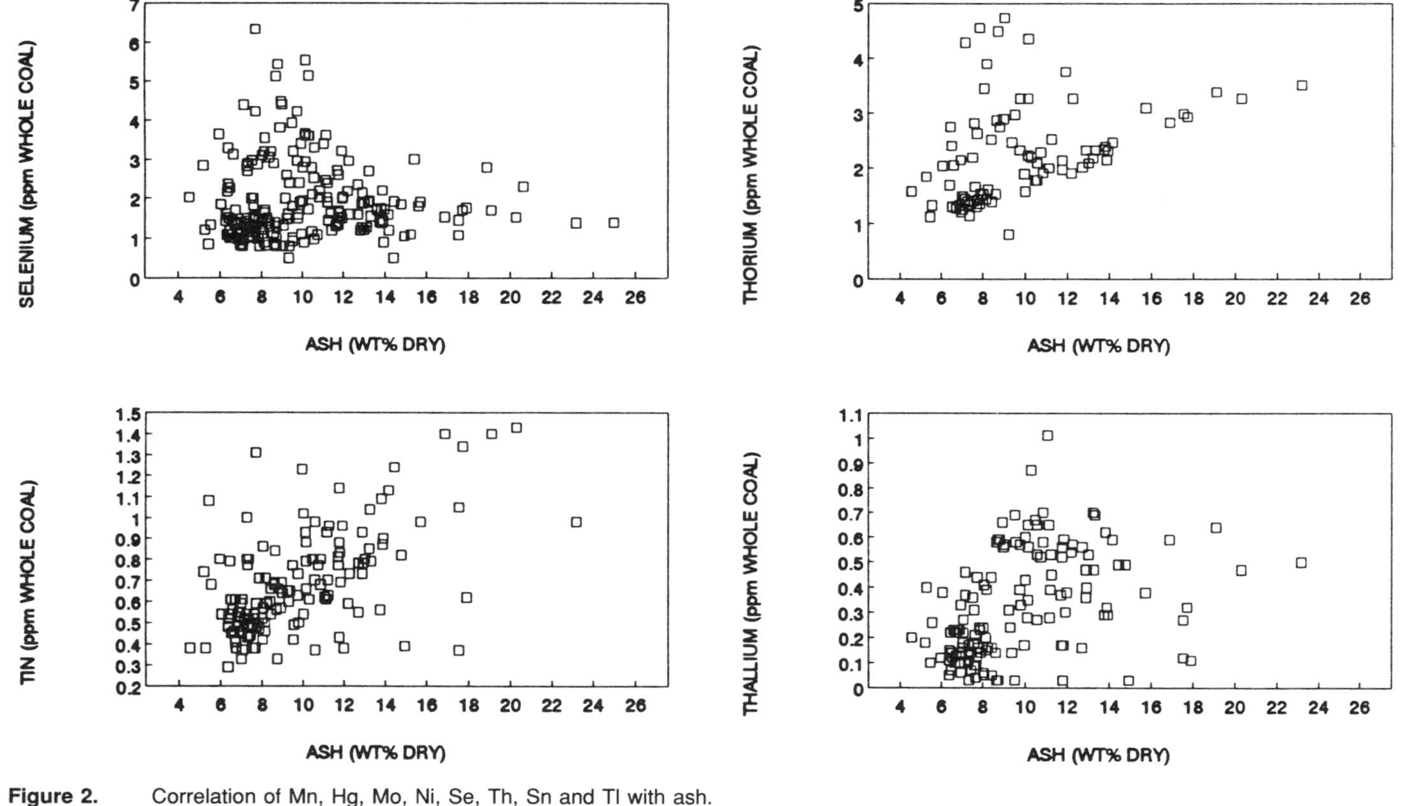

Figure 2. Correlation of Mn, Hg, Mo, Ni, Se, Th, Sn and Tl with ash.

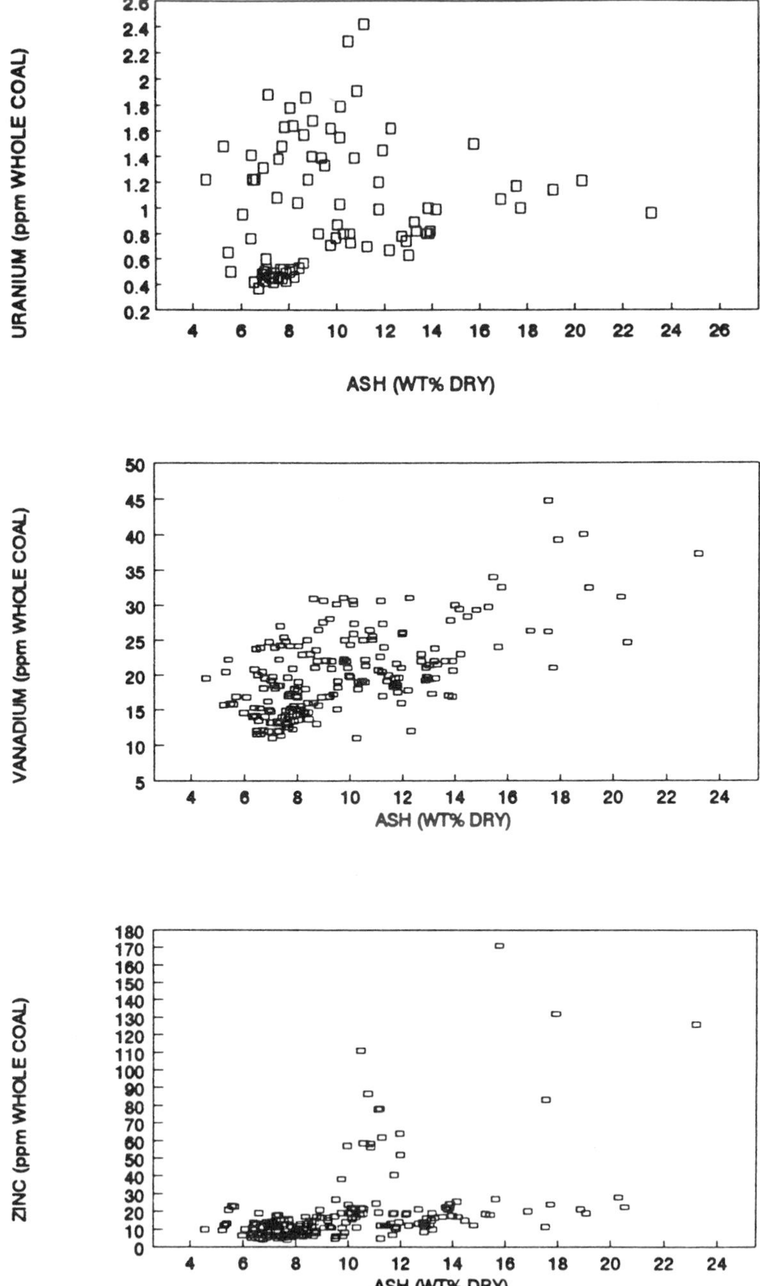

Figure 3. Correlation of U, V and Zn with ash.

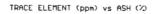

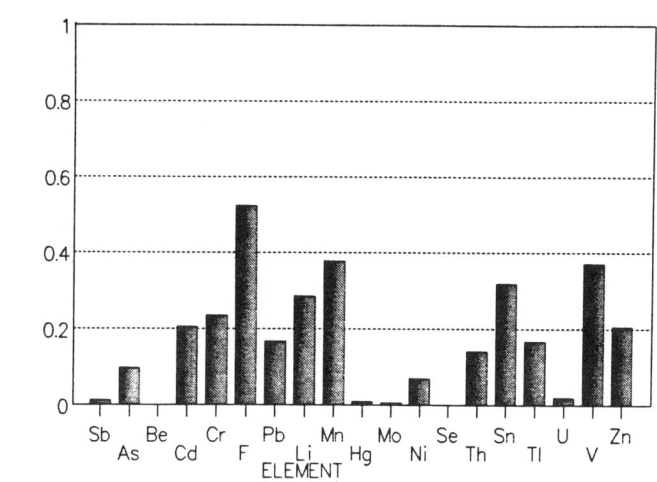

TRACE ELEMENT (ppm) vs SULFUR (%)

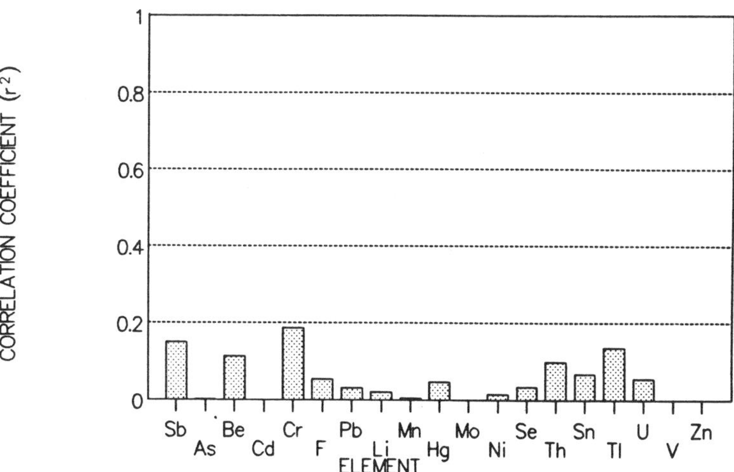

Figure 4. Correlation coefficients (r^2) for regression of trace elements with ash and sulfur.

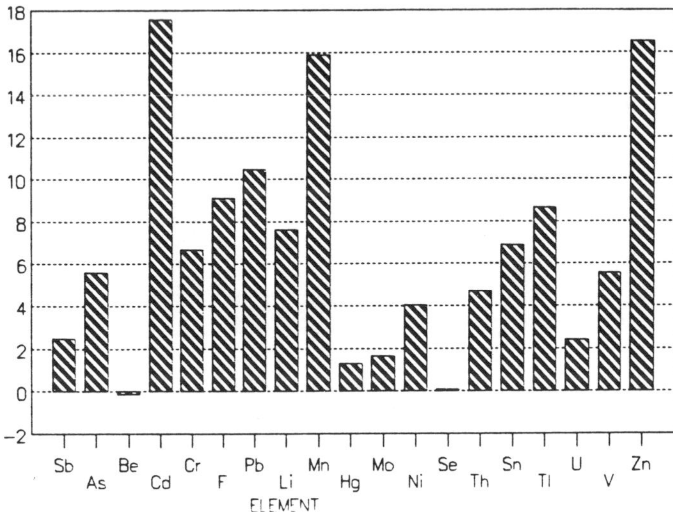

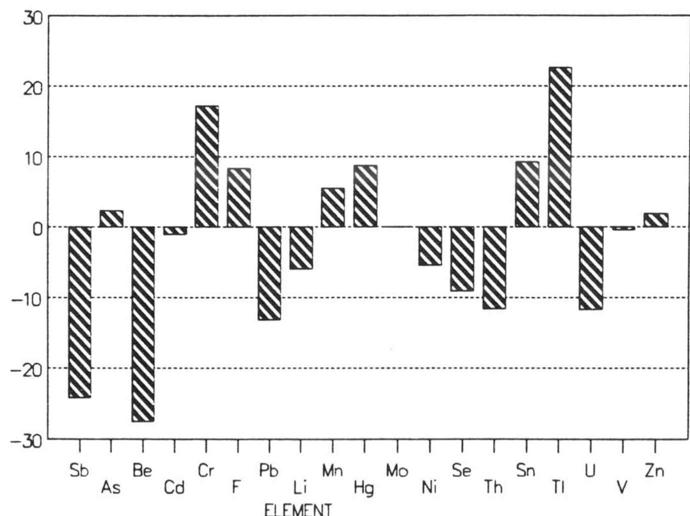

Figure 5. X-coefficient in regression of trace elements with ash or sulfur expressed as a percentage of the average trace element concentration. This number represents the average reduction in a trace element per percent ash or sulfur for all coals in the database.

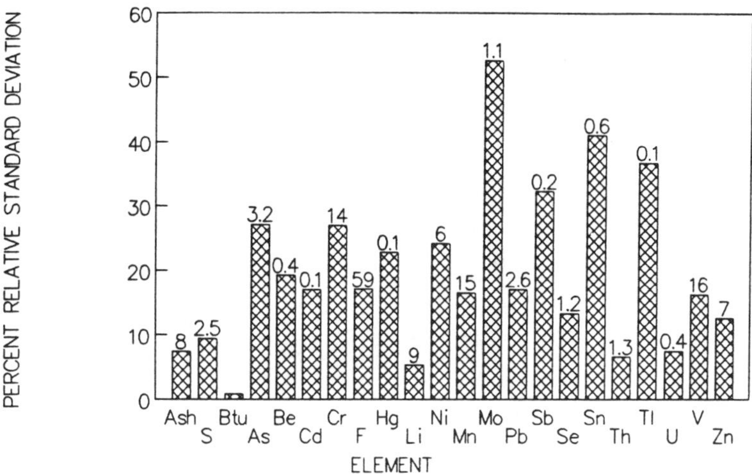

Figure 6. Trace element variability in clean coal from a single Pittsburgh seam mine (1982 to 1990). Number over the bar is the average value of the parameter.

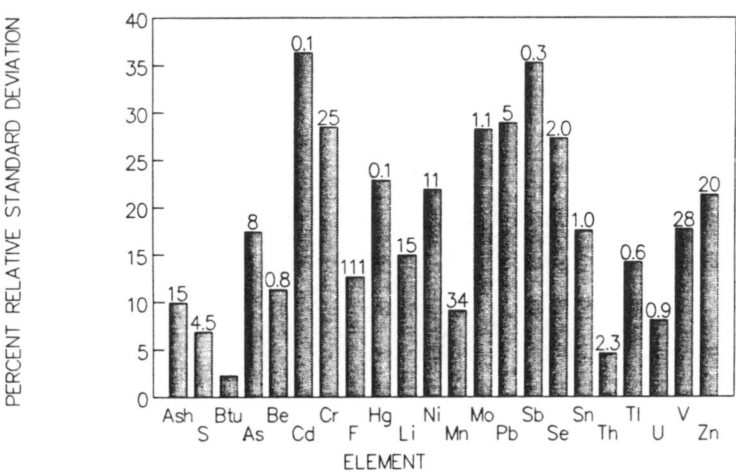

Figure 7. Trace element variability for coals from two adjacent Pittsburgh seam mines (1982 to 1990).

F. Comparison of Clean and Raw Coals

As discussed above, there is a weak correlation of individual trace elements with ash content for the database as a whole. However, when comparing the trace element contents of clean and raw coals from a single source, there is a much clearer

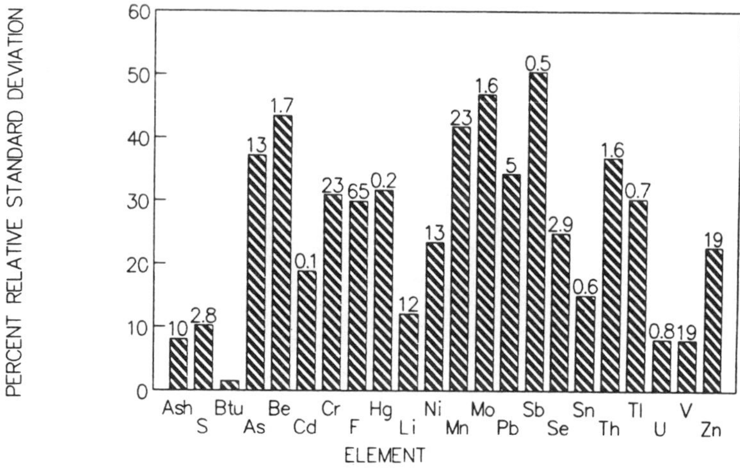

Figure 8. Trace element variability for Ohio preparation plant products (1982 to 1990).

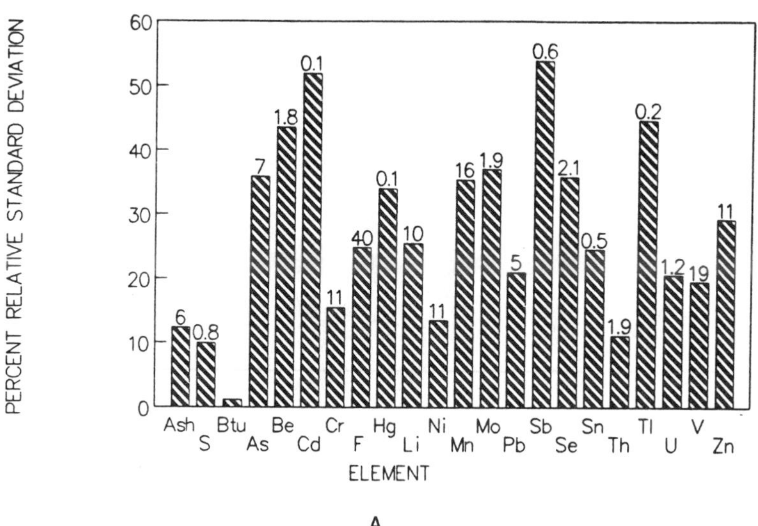

Figure 9. (A) Trace element variability for Central Appalachia metallurgical coals from three mines (1986 to 1990). (B) Trace element variability, Illinois, 6 seam coals from three mines (1985 to 1990).

relationship between ash and trace element concentrations. Figures 10A and B show the reduction in ash, sulfur, and trace elements based on a comparison of clean and raw coals from two different Pittsburgh seam mines. In both cases, the ash reduction is about 40%. All of the trace elements are also reduced in concentration; most of the

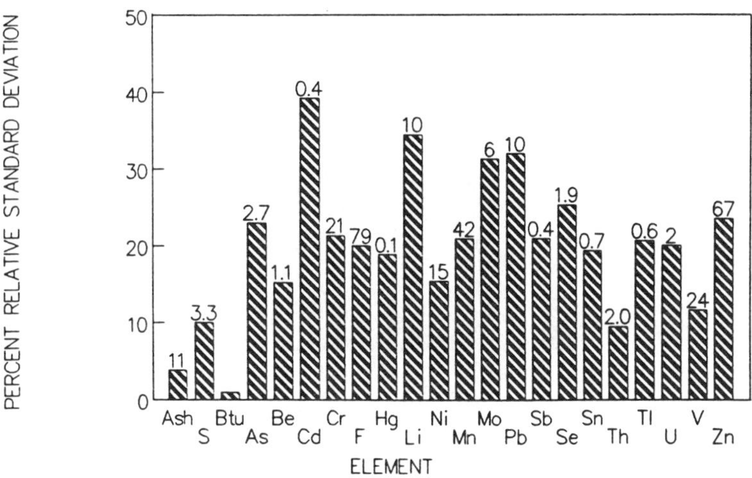

Figure 9B.

reductions are in the 30 to 50% range, a level very comparable to the ash reduction. The two consistent exceptions are selenium and mercury, which show smaller reductions. It would be questionable to draw any general conclusions about the behavior of these two elements from these limited data. Several authors have reported washability data for trace elements. Gluskoter et al. classified elements as having organic or inorganic affinity.[4] However, the results shown in Figures 10A and B appear to be inconsistent with his ranking. On the other hand, Ford and Price,[5] studying the washability of a Pittsburgh seam coal, also show low removals of mercury and selenium, consistent with the results reported here. This may suggest that the affinity of a given trace element for the organic matrix or mineral matter may be coal specific.

Figure 11 compares the trace element variabilities for clean and raw coals from a single mine in the Pittsburgh seam over the period 1982 through 1990. Although the ash content is considerably less variable for the clean coal, the trace element variabilities do not show a consistent result. For 8 of the 17 elements, the PRSD is greater for the clean coal than for the raw coal. It would seem safe to conclude that the relative variability of the trace elements is roughly the same for clean and raw coals. However, because coal cleaning removes about 40% of most trace elements, the variability on an absolute basis is less for the clean coal than for the raw coal from the same source.

G. Average Variabilities by Element

Figure 12 is a plot of the average PRSDs by element for all of the bituminous coal product coal groups. The data are plotted as the mean PRSD for each element ±1

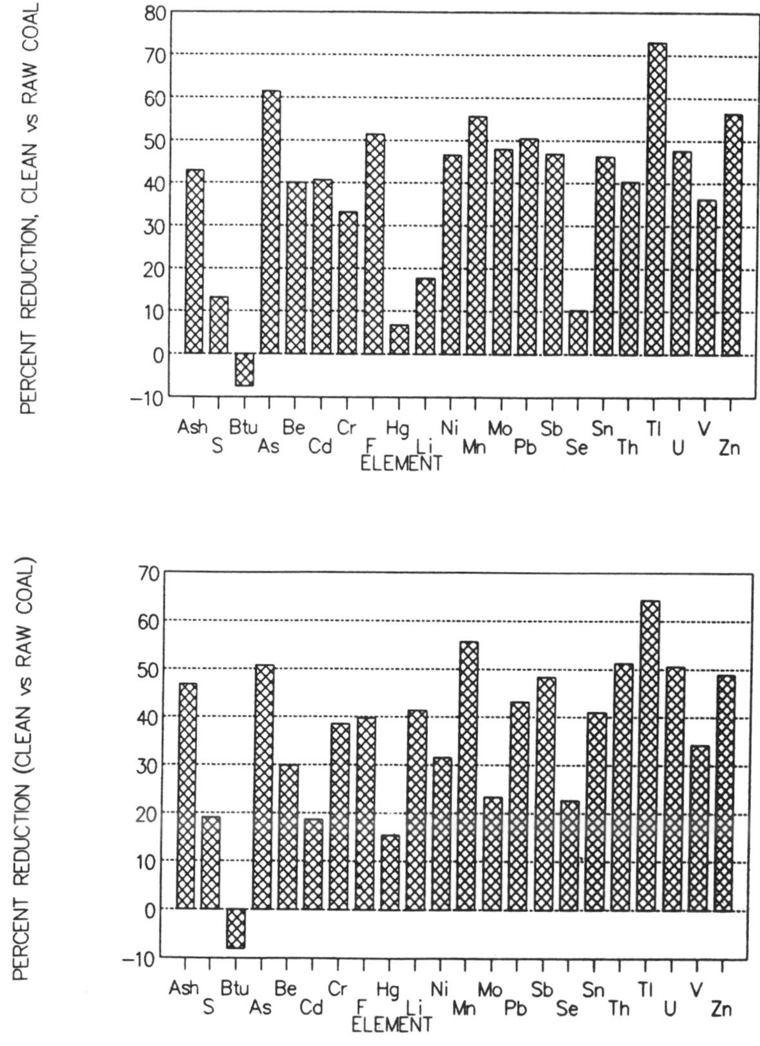

Figure 10. Trace element reduction, raw vs. clean coals, Pittsburgh seam mines (1982 to 1990).

standard deviation. In general, the least abundant elements (Cd, Sb, Tl) show the highest variability, perhaps reflecting some analytical imprecision, although the variability for mercury, which is low in abundance, is relatively small. The average variabilities for thallium and uranium are also small, but this may be an artifact. Thallium and uranium were analyzed for only about 40% of the samples. Their low PRSDs may reflect this more limited, and therefore more consistent, sample base. The other elements have average variabilities in the 20 to 50% range. A PRSD of

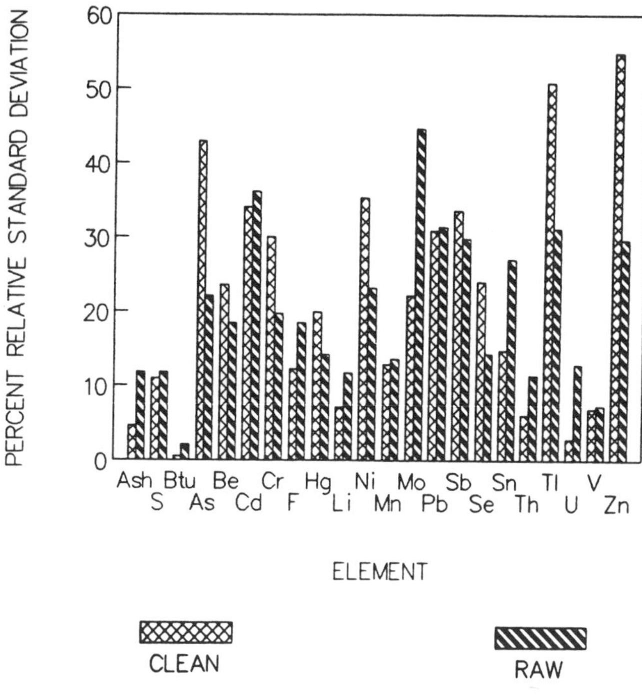

Figure 11. Trace element variabilities, raw and clean coals from a single Pittsburgh seam mine (1982 to 1990).

60% would appear to be a conservatively high estimate for most trace elements in this database.

H. Analytical Precision

As described earlier, the precision and accuracy of the trace element analysis has an obvious impact on any assessment of natural variability. For the work reported here, consistent attention has been paid to ensuring the quality of the analytical data. This has included methods development specific to coals, replicate analyses to guarantee precision, and frequent analyses of standard samples to ensure accuracy. The result is a level of analytical accuracy (Table 1) which is acceptable for determining the variabilities reported here. We are concerned that this level of analytical

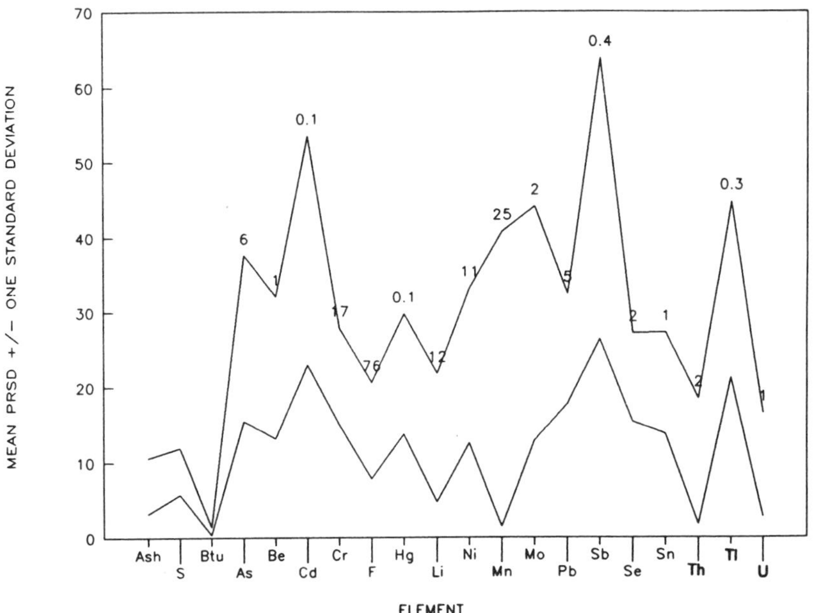

Figure 12. Average trace element variabilities for all bituminous coals in database. Area between lines includes mean ±1 standard deviation.

accuracy is not common in commercial labs. To assess this, we conducted a blind round robin in which identical samples were sent to five reputable commercial laboratories once a year for 3 years. The results of this work, reported in detail elsewhere,[1] indicate that intralaboratory reproducibility and interlaboratory repeatability may be relatively poor compared to natural variability. These data are summarized in Table 2. The intralaboratory reproducibilities are of the same magnitude as the natural variabilities reported above. The interlaboratory differences are generally about twice as great.

VI. CONCLUSIONS

The concentrations of 19 trace elements were measured for commercial coals shipped between 1982 and the present. The following conclusions can be drawn from analysis of these data.

- Trace element contents varied over an order of magnitude for coals in this database. No coals of commerce appear to be predictably low in trace elements.
- Individual trace element concentrations are weakly to moderately correlated with

Table 2. Interlaboratory Repeatability and Intralaboratory Reproducibility of Trace Element Analyses at Five Commercial Laboratories

Element	Intralaboratory reproducibility, %[a]	Interlaboratory repeatability, %[a]	Interlaboratory accuracy as % bias[b]
Antimony	44	91	+22
Arsenic	35	61	+141
Beryllium	32	65	—[c]
Cadmium	24	21	−33
Chromium	29	32	+27
Copper	26	40	+6
Fluorine	28	38	—[c]
Lead	39	66	−18
Lithium	39	107	+163
Manganese	18	31	−16
Mercury	28	56	+25
Nickel	43	117	+81
Selenium	41	106	+16
Thallium	63	—	—[c]
Tin	54	87	—[c]
Vanadium	26	40	−26
Zinc	40	87	+39
Average	36	65	+33

[a] Based on coefficient of variation.
[b] Based on NIST SRM 1635 (only one analysis per lab).
[c] No certified or literature value available.

coal ash content over the range of coals examined. Trace element contents are not correlated with sulfur content of the coal.

- Trace element variabilities, expressed as a percent relative standard deviation, typically ranged from 20 to 40% for coals from a single mine, from mines within the same seam, and even for geographically proximate coals from different seams. A PRSD of 60% would appear to be a reasonably conservative estimate of the variability of the trace element contents for the coals examined.
- Coal cleaning to remove ash is effective in removing a portion of many trace elements. In general, the degree of trace element removal is approximately the same as the ash removal. A few elements do not appear to be as effectively reduced by coal cleaning, and even complete ash removal would not guarantee complete trace element removal. No technology for trace element removal other than coal washing is currently available.
- Relative variabilities are comparable for clean and raw coals from the same source. However, because cleaning removes about 40% of the ash and trace elements, variability is reduced on an absolute basis.
- Analytical accuracy and precision were carefully controlled in obtaining the data reported here. Given adequate care, it is possible to reduce analytical error to a level that is small compared to the natural variability of the trace elements studied. However, as a blind round robin of commercial labs shows, without adequate analytical procedures, analysis errors can be as great or greater than natural variability.

ACKNOWLEDGMENTS

The authors acknowledge R. M. Meyer and B. D. Singh of CONSOL Inc. for the organization and management of the comprehensive analysis program, and all in CONSOL who contribute to sample acquisition and analysis.

REFERENCES

1. Obermiller, E. L. and Lengyel, J., Jr., Interlaboratory variability and accuracy in trace element analyses of coal, in *Proc. 4th Annual Pittsburgh Coal Conference*, September 28, 1987, pp. 148-159.
2. Conrad, V. B., The Quest for the Universal Trace Element Analyzer: Inductively Coupled Plasma-Mass Spectrometry Applied to Coal and Coal Derived Materials, Indiana University of Pennsylvania, Indiana, PA, October 1988.
3. Conrad, V. B. and Brownlee, W. D., Hydropyrolytic-ion chromatographic determination of fluoride in coal and geological materials, *Anal. Chem.*, 60, 365, 1988.
4. Gluskoter, H. J., Ruch, R. R., Miller, W. G., Cahill, R. A., Dreher, G. B., and Kuhn, J. K., Trace elements in coal: occurrence and distribution, *Illinois State Geological Survey Circular 499*, 1977.
5. Ford, C. T. and Price, A. A., Evaluation of the Effect of Coal Cleaning on Fugitive Elements, Final Report, Phase III, U.S. Department of Energy Report DOE/EV/04427-62, July 1982.

Assessment of Air Pollutants from Coal Utilization Plant

P. W. Sage
R. J. Gemmill

ABSTRACT

Increasing public awareness and corresponding governmental response means that environmental issues continue to grow in social and political priority. Concerns have developed from local issues such as urban smoke emissions and smogs to aspects of regional and global scale such as acidification and anthropogenic greenhouse gas contribution. There has also been a growing need to account for source emissions of toxic trace pollutants such as heavy metal elements, polycyclic aromatic hydrocarbons, and dioxins. Most recently there has been a requirement to examine precursor emissions which are subsequently responsible for secondary pollutant formation, e.g., volatile organic compounds and nitrogen oxides and their roles in the formation of photochemical oxidants including ozone.

This paper discusses stack sampling and analytical approaches used by the British Coal Corporation's (BCC's) Coal Research Establishment (CRE) to assess a comprehensive range of air pollutant emissions from coal use. Such studies form an integral part of BCC's research effort to develop and adopt clean coal technology. The benefits of continuous gas monitoring are considered together with comparison of the British and U.S. approach to dust sampling. Specialist techniques currently used to monitor trace elements and trace hydrocarbon compounds are also described. Pertinent results are presented.

I. INTRODUCTION

The development of new environmental concerns has usually been in advance of full understanding of the associated science, methods for reliable quantification, and/or technology for abatement processes. Progressively more stringent legislative requirements and the economic advantages of a greener image have started to encourage a growing number of industrial companies to take a more serious look at the environmental consequences of their processes and products. BCC continues to respond positively to environmental challenges and this paper reviews the methods currently used by CRE to assess atmospheric emissions from coal use.

II. METHOD SELECTION FOR ASSESSING STACK EMISSIONS

Selection of sampling and analytical techniques depends on a number of criteria. Often, the first approach is to identify standard methods suitable for the less-than-benign composition of fossil fuel stack gases. The U.K. is most influenced by standards published by its own British Standards Institute (BSI) and the Comité Européen de Normalisation (CEN — the European Committee for Standardization). Other standards issuing organizations that are usually checked include the International Standards Organization (ISO), the American National Standards Institute (ANSI)/American Society of Testing and Materials (ASTM), the U.S. Environmental Protection Agency (EPA), the Association Française de Normalisation (AFNOR), and Verein Deutscher Ingenieure (VDI). In addition, regulatory organizations such as Her Majesty's Inspectorate of Pollution (HMIP), the U.K. Health and Safety Executive (HSE), and the EEC (in their Directives) sometimes specify reference methods for use in conjunction with the legislative limits they advocate. Where standard methods are not available, it is usual to identify the methods in most common use elsewhere so that at least data obtained can be compared with information already in existence.

It is convenient to categorize stack sampling into manual and automated (continuous) methods, and also into particulate and gaseous discharges. Reference test (or standard) methods are usually manual because they are generally considered more specific and more accurate. The specific methods used by CRE are referred to below.

III. PARTICULATE EMISSIONS

Most particulate emission tests in the U.K., including those undertaken by CRE, are carried out using BCURA (British Coal Utilisation Research Association) equipment and procedures[1] conforming to BS3405.[2]

The BCURA train is shown schematically in Figure 1. Materials are mostly stainless steel. A sharp-edged interchangeable nozzle is positioned into the oncoming gas stream at representative points in the duct.[1] Isokinetic sampling of the flue gas ensures collection of particulates in their correct quantity and size distribution. The coarse fraction of the sampled particulates (nominally greater than 8 μm) is removed by the system's miniature cyclone and hopper; the remainder is caught on the back-up glass/quartz wool filter. Typically, 80% of particulates in coal-fired stacks are retained in the cyclone hopper.

Duct flow is characterized using pitot static tubes of British Standard specification or ones calibrated against British Standard tubes. The isokinetic sampling rate is achieved by controlling the pressure drop cross the BCURA cyclone by means of an infinitely variable valve fixed to the sampling pump. The weights of solids collected are determined by difference to enable calculation of the overall flue emission rate which is then compared with the Clean Air Act limits.[3] Additionally,

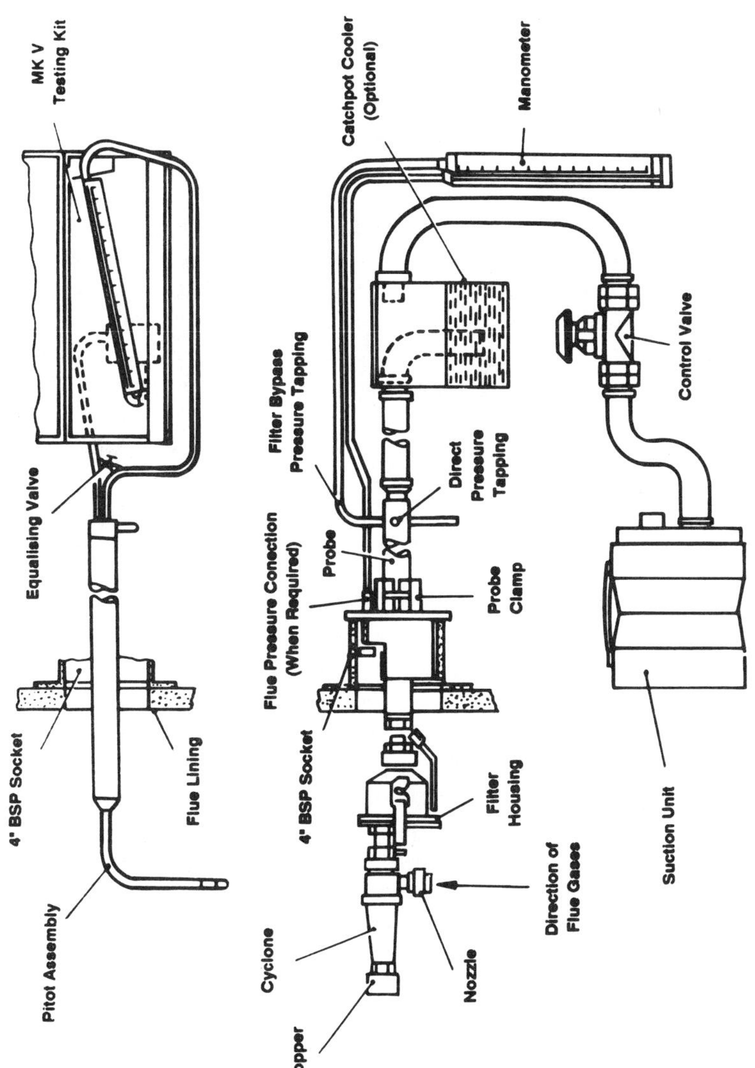

Figure 1. Schematic arrangement of the BCURA sampling train.

Table 1. Comparison of Simultaneous Dust Sampling using BCURA and Mark IIIA Apparatus

	Solids flowrate		
Test	**BCURA (kg/h)**	**Mk3a (kg/h)**	**Mk3a (% of BCURA)**
1	14.5	13.8	95
2	11.6	12.3	106
3	15.9	18.1	114
4	18.2	15.0	82
5	20.7	15.8	76
Mean	16.2	15.0	93

the cyclone solids are sized to establish the relative proportions of "nuisance", inhalable, and respirable dust. Sizing down to 63 μm is by sieving analysis, below 63 μm is by Coulter Count. In certain cases size analysis data are applied to mathematical models to enable prediction of the atmospheric dispersion of the particulate emissions. Chemical analysis of collected particulates can yield data on emissions of other species for example trace elements. The accuracy of BCURA measurements is mainly affected by the number of points surveyed in the duct. For a survey involving 8 points (as normally used by CRE) the accuracy is estimated to be within ±8.5%.[1]

Another U.K. method which is recognized by BS3405 is the Mark IIIA filter apparatus. This consists of a stainless steel probe with an elongated mild steel filter housing and swan-necked sampling tube of variable inlet area (again by means of interchangeable nozzles). The filter housing holds a woven glass fiber bag to collect sampled particulates. Isokinetic sampling is undertaken at representative points[1] by measuring the pressure drop across an external orifice plate controlled by an in-line infinitely variable valve.

CRE has undertaken comparison trials on the two different sets of apparatus using a bituminous coal-fired boiler fired by twin-traveling grate stokers and rated at 180,000 lb/h steam (approximately 50 MW). The overall results of the trials are shown in Table 1. It is interesting to note reasonable agreement between the measurements, the mean difference being 7% (reduced collection for the Mark IIIA). The Mark IIIA method is less expensive than the BCURA apparatus (£3000 compared to £4500) and it is more portable. The BCURA apparatus benefits from use of a smooth-walled cyclone and hopper allowing retrieval of collected solids for size distribution analysis. Also, its small collecting hoppers and individual filters are far more conducive to incremental point sampling. Both techniques are straightforward to operate with practice although the BCURA technique requires fewer simultaneous adjustments during sampling.

On occasions, CRE uses a multistage cascade impactor system for stack gas particulate sampling. This is advantageous for measuring low dust loadings and obtaining *in situ* size analysis. It can also be used at high temperatures.

The main dust sampling apparatus used in the U.S. is the EPA Method V Sampler[4] shown schematically in Figure 2. This apparatus is versatile and can be

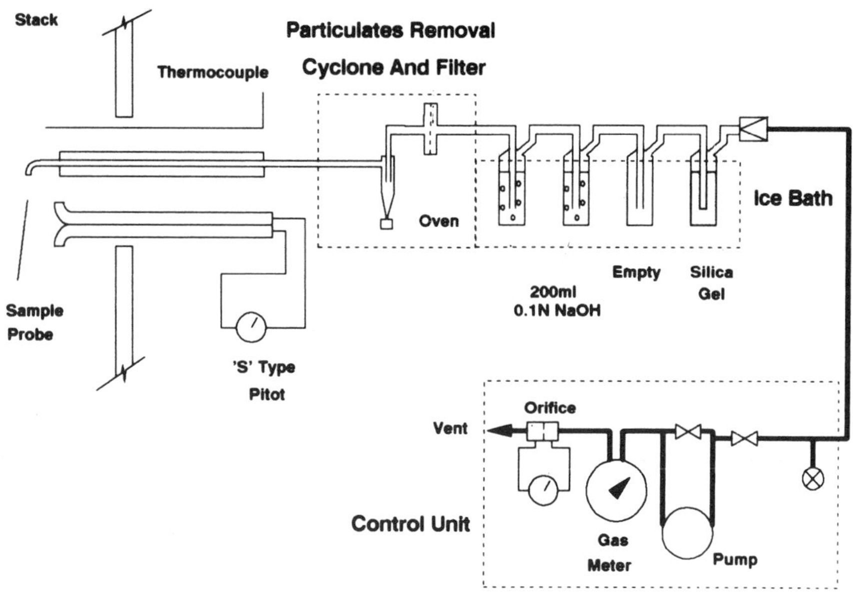

Figure 2. The EPA Mark V sampler.

adapted to monitor most pollutant species. However, it is heavy and cumbersome, and often difficult to use in ducts without the aid of an overhead support rail. The long length of the probe makes it prone to dust deposition which can result in significant sample losses. At £15,000, the EPA apparatus is more expensive than the BCURA or the Mark IIIA systems.

IV. MONITORING OF PRINCIPAL GASEOUS EMISSIONS

Alternatives for assessing gaseous emissions include spot sampling followed by chemical or instrumental analysis, proprietary gas analyzer tubes, continuous monitoring using permanently installed equipment, and periodic monitoring using mobile or portable instrumentation. While constituting elements of the operator's portfolio, neither gas analyzer tubes or spot sampling are sufficient to meet new legislative requirements. It is continuous and periodic instrumental monitoring which is of main importance.

Flue gas monitoring requires a high level of specialist knowledge, and expensive, complex equipment. Capital expenditure of approximately £25,000 is required to measure oxides of nitrogen using a chemiluminescent analysis system. Setting up analyzers for principal combustion gases costs up to £200,000. Ongoing costs including skilled analyst time may amount to £100,000 per year. This level of investment in monitoring can only be afforded by the biggest operators such as power companies and large-scale industry.

A cheaper option for smaller scale industry is to measure flue gas concentrations over a period of days, under representative boiler-operating conditions. Emissions from boiler plants are relatively constant so long as operating parameters and fuel quality are consistent. Periodic monitoring is therefore adequate for a wide range of industrial boilers and substantially cheaper. CRE operates two mobile laboratories for flue gas monitoring. The typical cost of assessing emissions from several boilers on a single site is less than £10,000. This option is often more attractive to an operator than the installation of continuous monitors with associated running costs and technical back-up.

The CRE emissions monitoring laboratories are based on 6.5 tonne integral vans (Figure 3). Gas analyzers are housed inside (Figure 4) together with their associated sampling equipment, air conditioning and air exchange units, and storage for compressed gas cylinders and other miscellaneous equipment. The flue gas is sampled using a 20-mm od steel sample probe which is packed with glass wool to act as a coarse particulate filter. The probe is connected to a heated line through which the sample passes to the prefilter unit in the mobile laboratory. This unit removes all fine particulates by means of a microfiber filter and divides the sample gas into four separate substreams. Three streams run via separate heated sample lines to hydrocarbon, nitrogen oxides, and water vapor analyzers, while a fourth heated line carries gases to the inlet of a permapure dryer. The latter stream is further divided at the inlet of the dryer with one half passing through a refrigeration unit, and then onto carbon monoxide, carbon dioxide, and oxygen analyzers; the other half passes through the permapure dryer and then to a sulfur dioxide analyzer. All the analyzers have a 4- to 20-mA signal which is relayed to a chart recorder and data-logging facility. The analyzers are regularly calibrated (at least every 4 h) using standard concentration and zero gases.

Nitrogen oxide analysis is based on chemiluminescence and the gas-phase reaction between ozone and nitric oxide. About 10% of the nitrogen dioxide produced is in an excited state. Subsequent transition to the normal state gives rise to a light emission which is measured by a photomultiplier tube. Total oxides of nitrogen are determined by first passing the flue gas through a converter to reduce any nitrogen dioxide to nitric oxide. Thus, nitrogen dioxide is not measured directly but deduced by difference.

Water vapor is determined by an infrared technique whereby the signal from a reference cell and the flue gas sample are compared. Knowledge of the water-vapor content is used to convert the other gas analysis data to a dry basis. Sulfur dioxide, carbon monoxide, and carbon dioxide are also determined by an infrared technique, although for these constituents there is no reference cell.

Oxygen concentration is determined by paramagnetism. Oxygen molecules are paramagnetic and therefore attracted toward a magnetic field. This property is used to displace a light beam sensed by photocells. The degree of displacement is proportional to the oxygen content. Knowledge of the flue gas concentration of oxygen allows standardization of the other flue gas constituents to a reference value (normally 6% O_2 as prescribed by U.K. and European legislation).

Hydrocarbon concentration is determined by flame ionization whereby the sample is burned in a hydrogen flame and the carbonium ions produced migrate to a collection electrode and thereby generate an electrical signal. This analysis does not differentiate between different hydrocarbons and results are given as "methane equivalents".

Although CRE has the experience of sampling a variety of gaseous species by batch methods, only two species are routinely assessed by such an approach, HCl and N_2O. The technique for HCl is based on two British Standard Methods.[5,6] Flue gas is drawn through an inert silica sampling probe into two Dreschel bottles containing 3% w/v hydrogen peroxide. A flow rate of 1 L/min is typically used and a total volume of 20 L normally sampled. The resulting solutions and washings of all tubing and glassware are then analyzed by dionex chromatography. Additional analyses are obtained for blank samples of the reagents to provide a background chloride concentration. Consistent results are obtained by the method so long as the system is properly passivated (usually by the first sample which is then discarded). Good correlation is found between emission concentration and chlorine content of coal.

Nitrous oxide sampling involves extraction of flue gas through a silica wool particulates filter, followed by a caustic scrubbing agent (200 ml of 0.1 *M* KOH) to remove SO_2 and NO_2, and finally a desiccating agent [$CaCl_2$ or $Mg(ClO_4)_2$]. A sample of this conditioned flue gas is then sealed in an in-line gas bottle for analysis by gas chromatography utilizing an electron capture detector. Flow rate through the system must be maintained at space velocities greater than 500/h with a minimum purge time of about 10 min to allow complete clearing of the sampling train, and to allow the scrubbing solution to come to equilibrium with other flue gas constituents. CRE is currently commissioning a photoacoustic semicontinuous analyzer for N_2O determination.

V. ASSESSMENT OF TRACE EMISSIONS

Trace emissions are species present in very low concentrations, low parts per million or less. CRE's work has mainly been concerned with trace elements, polycyclic aromatic hydrocarbons (PAH), and dioxins (polychlorinated dibenzo-para-dioxins and polychlorinated dibenzofurans). Our assessments have shown that coal utilization is not a significant source of these emissions. However, the onus is on industry to continue to demonstrate its environmental acceptability, particularly for unassessed processes and new technological developments.

Trace emissions are inherently difficult to measure accurately because concentrations are so low. CRE has been careful to choose sampling and analytical methods believed to be the most suitable for evaluating coal utilization and, to a lesser extent,

Figure 3. CRE mobile emissions monitoring laboratory.

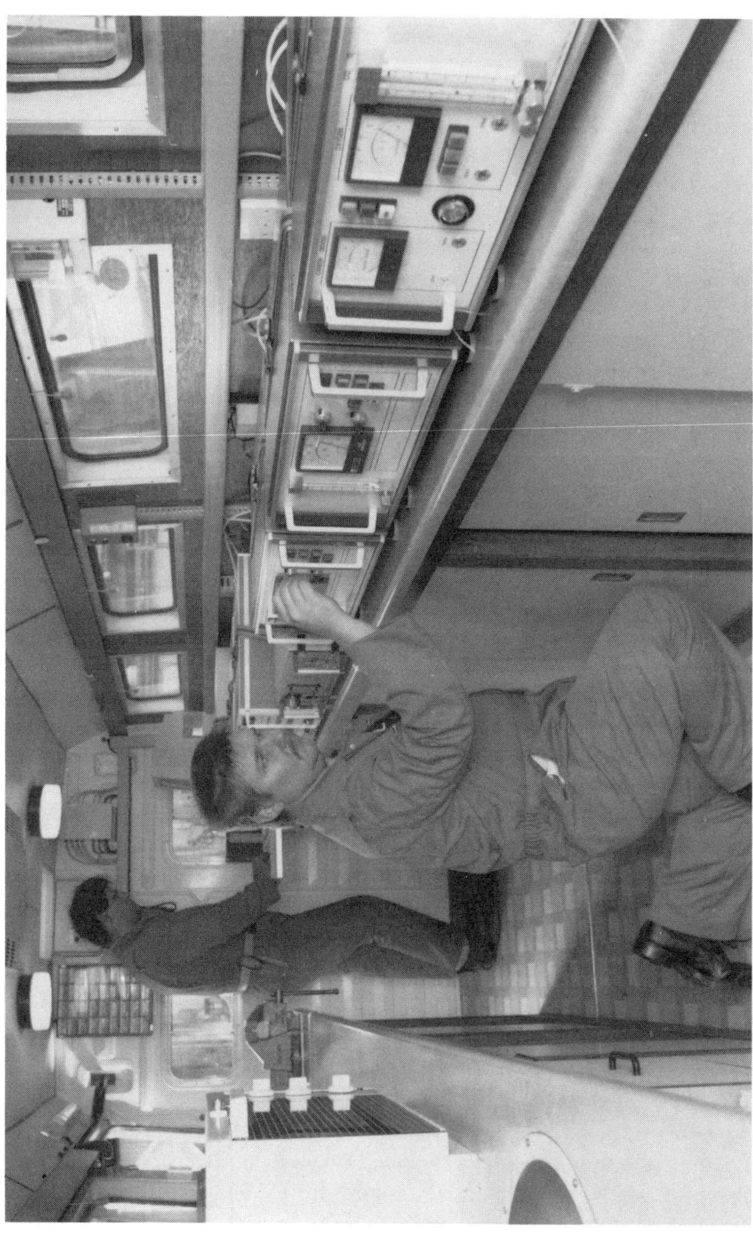

Figure 4. Interior of the mobile emissions monitoring laboratory.

Table 2. Trace Element Emissions from a Typical Industrial Scale Coal Combustion Boiler[8]

	Concentration, mg/m³			
Element	**Emission**	**Estimated glc**	**Recommended limit, 1/30 OEL**	**Recommended limit, estimated glc**
As	3.4×10^{-3}	5.0×10^{-8}	3.3×10^{-3}	6.6×10^{4}
Cd	3.6×10^{-4}	5.3×10^{-9}	1.7×10^{-3}	3.2×10^{5}
Cr	1.7×10^{-1}	2.5×10^{-6}	1.7×10^{-2}	6.8×10^{3}
Co	8.5×10^{-2}	1.3×10^{-6}	3.3×10^{-3}	2.5×10^{3}
Cu	2.4×10^{-1}	3.5×10^{-6}	6.7×10^{-3}	1.9×10^{3}
Pb	1.6×10^{-1}	2.4×10^{-6}	5.0×10^{-3}	2.1×10^{3}
Mn	3.3×10^{-1}	4.8×10^{-6}	1.7×10^{-1}	3.5×10^{4}
Hg	5.5×10^{-4}	8.1×10^{-9}	1.7×10^{-3}	2.1×10^{3}
Ni	2.9×10^{-1}	4.3×10^{-6}	3.3×10^{-3}	7.7×10^{2}
V	2.4×10^{-1}	3.5×10^{-6}	1.7×10^{-2}	4.9×10^{3}
Zn	5.5×10^{-1}	8.1×10^{-6}	1.7×10^{-1}	2.1×10^{4}

Note: glc — ground level concentration, OEL — occupational exposure limits.[11]

most comparable with work elsewhere. The large numbers of individual trace elements, PAH and dioxins, makes it necessary to select representative cogeners from each group for practical study. Selection is based on recommended practice, species of greatest toxicity, legislative requirement, and special interest to coal.

Trace element sampling utilizes a Universal Stack Sampler based on the EPA Method V standard apparatus for dust sampling. Stack gas is drawn isokinetically through a heated glass or quartz probe and particulates filter. It then passes on to a cooled impinger train consisting of two traps of 0.5 *M* nitric acid, a third containing water, and a fourth containing drying agent to protect the flow control unit. Both the particulates fraction and the vapor phase component, trapped by the impinger solutions, are analyzed for trace element content. Analysis is mainly undertaken by atomic absorption and inductively coupled plasma spectrometry.

Measurements from a range of coal combustion and coal gasification plants have consistently identified similar findings. First, the trace elements are predominantly associated with the ash and particulate streams rather than the vapor phase. Second, the trace element concentration of solid residues increases with decreasing particle size, indicating a volatilization/condensation mechanism. Third, discharges to atmosphere are well below levels of concern.

Legislation has not yet been set in the U.K. for trace element emissions from coal-fired processes. However, the European Directive[7] for new municipal waste incineration plant sets limits of 5 mg/m³ for the combined total of Pb, Cr, Cu, Mn; 1 mg/m³ for the combined total of Ni and As; and 0.2 mg/m³ for Cd and Hg. Table 2 shows typical emissions of 11 elements measured from an industrial scale (10 MW) coal combustion boiler.[8] Table 3 shows typical concentrations of 14 elements measured in coal gasification industrial fuel gas (IFG).[9] Levels are well below the incineration plant limits. The tables also show corresponding estimates of ground-level concen-

Table 3. Trace Element Concentrations in Coal Gasifier Industrial Fuel Gas (IFG)[9] and Environmental Fate Assuming Total Discharge on Combustion

	Concentration, mg/m³			
Element	IFG concentration	Estimated glc	Recommended limit, 1/30 OEL	Recommended limit, estimated glc
As	5.1×10^{-1}	7.5×10^{-6}	3.3×10^{-3}	4.4×10^{2}
B	1.4×10^{0}	2.1×10^{-5}	3.3×10^{-1}	1.6×10^{4}
Ba	7.7×10^{0}	1.1×10^{-4}	3.3×10^{-1}	2.9×10^{3}
Cd	$<1.4 \times 10^{-2}$	2.1×10^{-7}	1.7×10^{-3}	$>8.3 \times 10^{3}$
Cr	2.4×10^{-1}	3.5×10^{-6}	1.7×10^{-2}	4.8×10^{3}
Cu	3.0×10^{-1}	4.4×10^{-6}	6.7×10^{-3}	1.5×10^{3}
Hg	$<1.0 \times 10^{-2}$	1.5×10^{-7}	1.7×10^{-3}	$>1.1 \times 10^{4}$
Mn	1.3×10^{0}	1.9×10^{-5}	1.7×10^{-1}	8.7×10^{3}
Mo	7.1×10^{-2}	1.0×10^{-6}	1.7×10^{-1}	1.6×10^{5}
Ni	1.6×10^{-1}	2.4×10^{-6}	1.7×10^{-2}	7.1×10^{3}
Pb	7.5×10^{-1}	1.1×10^{-5}	5.0×10^{-3}	4.5×10^{2}
Se	3.6×10^{-2}	5.3×10^{-7}	3.3×10^{-3}	6.3×10^{3}
V	4.7×10^{-1}	6.9×10^{-6}	1.7×10^{-2}	2.4×10^{3}
Zn	5.1×10^{-1}	7.5×10^{-6}	1.7×10^{-1}	2.2×10^{4}

Note: glc — ground level concentration, OEL — occupational exposure limits.[11]

trations and compares them with 1/30 of the values of occupational exposure limits.[10] Ratios of 440:32,000 demonstrate a comfortable margin of safety regarding potential hazard to health.

Sampling of PAH and dioxins uses a CRE version of the Universal Stack Sampler (based on EPA Method 23). This differs from the trace element apparatus by incorporating a condenser, a water fall-out pot, and an XAD-2 sorbent trap. The nitric acid and water bubblers are replaced with two of toluene and a third cardice cooled trap. The toluene acts as a back-up vapor phase hydrocarbon extractant to the XAD-2 resin. The cardice trap ensures no sample escapes the system in aerosol form. PAH and dioxins are desorbed from the particulates and XAD-2 resin by soxhlet extraction. Analysis makes use of high-resolution gas chromatography mass spectrometry. This technique is imperative for accurate detection of very low concentrations of dioxins. High-pressure liquid chromatography provides a good alternative method for PAH. Standard (labeled) PAH and dioxin spikes are applied to the sampling apparatus before use and during subsequent sample clean up so that sampling and analysis efficiency can be checked against the spike recovery rates. Overall, CRE follows the strict acceptance criteria for dioxin analysis laid down by Ambidge et al.[11]

Initial PAH and dioxin measurements have been undertaken on a domestic open fire operated to BS 3841.[12] NATO/CCMS 2,3,7,8 Tetrachloro dibenzo-*p*-dioxin toxic equivalent factors of 0.03, 0.008, and 0.01 ng/Nm³ have been found for bituminous coal, anthracite, and a manufactured smokeless fuel, respectively. The most stringent legislation set in Europe is 0.1 ng/m³ TCDD TEQ for a new incineration plant. Only low concentrations of carcinogenic PAH have been detected.

The most recent area of method development that CRE has become involved in is sampling and analysis of speciated volatile organic compounds. It is planned to build up a coal-fired inventory for different plants and to use corresponding compound ozone-producing potentials to calculate coal's overall contribution to tropospheric ozone levels. However, this work is in its infancy and the first step is to identify the most suitable measurement techniques and validate them. It seems most likely that they will be based on grab samples using Tedlar bags or adsorbent resin tubes followed by analysis by gas chromatography or GC-MS techniques.

ACKNOWLEDGMENTS

The authors acknowledge the contribution by colleagues at CRE to some of the practical data referred to in this paper. The views expressed are those of the authors and not necessarily those of British Coal Corporation.

REFERENCES

1. Hawksley, P. G. W., Badzioch, S., and Blackett, J. H., *Measurement of Solids in Flue Gases*, 2nd ed., The Institute of Fuel, London, 1977.
2. British Standard Method for Measurement of Particulate Emission Including Grit and Dust (Simplified Method), BSI, BS 3405, 1983.
3. The Clean Air (Emission of Grit and Dust from Furnaces) Regulations 1971, SI No. 162, HMSO 1971.
4. EPA Stationary Source Sampling Methods Notebook (NPS and NESHAPS Methods), The McIlvaine Company, Illinois.
5. British Standard Methods for the Sampling and Analysis of Flue Gases, Part 4, Miscellaneous Analyses, BSI, BS 1756:Part 4:1965.
6. British Standard Methods for Analysis and Testing of Coal and Coke, Part 8, Chlorine in Coal and Coke, BSI, BS 1016:Part 8:1977.
7. Council Directive of 8 June 1989 on the Prevention of Air Pollution from New Municipal Waste Incineration Plant (89/369/EEC), Off. J. Eur. Comm. No. L 163, 32, 14.6.89.
8. Hughes, I.S.C. and Littlejohn, R.F., Trace Element Emissions from AFBC, in *Proc.Int. Conf. Am. Soc. Mech. Eng. Fluidised Bed Combustion,* Boston, 1987.
9. Industrial Fuel Gas (IFG) from Coal: The Fate of Trace Elements and Hydrocarbons, European Coal and Steel Community (ECSC) Project No. 7220-EC/842, 1988–1991.
10. Occupational Exposure Limits, U.K. Health and Safety Guidance Note EH40/91, HMSO, 1991.
11. Ambidge, P.F. et al., Acceptance criteria for analytical data on polychlorinated dibenzo-p-dioxins and polychlorinated dibenzofurans, *Chemosphere*, 21, 8, 999, 1990.
12. British Standard Method for the Measurement of Smoke from Manufactured Solid Fuels for Domestic Open Fires, BSI, BS 3841:1972 (1990).

Low Level Emissions of Hazardous Organic Pollutants from Coal-Fired Power Plants — Fact or Artifact?

K. E. Curtis
N. Krishnamurthy
S. J. Thorndyke

I. INTRODUCTION

Ontario Hydro is a large publicly owned power utility that supplies the electricity requirements for the province of Ontario in Canada. To meet this demand there is at present an installed generating capacity of about 30,500 MW which is derived from 21% hydroelectric, 41% nuclear, and 38% fossil fuel-fired generation.

This last category is made up of six major generating stations, as shown in Table 1. Note that all but the oil-fired Lennox station are fueled with various types of coals.

The release of air contaminants from these plants has been governed by various regulations under the Ontario Environmental Protection Act, the most recent being Ontario Regulation 308. None of these sets emission limits at the source but instead regulate their point of impingement concentrations at receptor sites. This requires the use of dispersion modeling to calculate these concentrations from physical stack parameters and measurements of contaminant emission rates in the stack.

The Ontario government is planning major revisions to Regulation 308 under the Environmental Protection Act with the proposed Clean Air Program (CAP). If implemented, this would change entirely the way air pollutants are regulated in Ontario.

II. THE ONTARIO CLEAN AIR PROGRAM

A. Highlights of Proposed Changes under CAP

- Control of emissions at the source by installation of appropriate control equipment to meet specific emission limits
- Level of control will be tied to the degree of hazard of the contaminant
- Source registration of all emissions to air, water, and soil
- Two-part Certificates of Approval, one for construction and one for operation. The second will require periodic renewal and demonstration of compliance with emission limits and new ambient air quality standards.

Table 1. Ontario Hydro's Fossil Fuel-Fired Generating Stations

Station	Peak capacity (MW)	Fuel
Atikokan	215	Western Canadian lignite
Lakeview	2180	Medium sulfur U.S. bituminous coal
Lambton	2040	Regular sulfur and low sulfur U.S. bituminous coals
Lennox	2230	Low sulfur residual or crude oils
Nanticoke	4340	Blend of western Canadian and regular sulfur U.S. bituminous coals
Thunderbay	320	Western Canadian lignite, western Canadian and U.S. bituminous coals

Table 2. CAP Source Registration — Levels of Concern

Level 1	Highest hazard; requires LAER control technology; contaminants listed — 37
Level 1/2	Under review; contaminants listed — 33
Level 2	Moderate hazard; requires BACT; contaminants listed — 76
Level 3	Low hazard; requires reductions using demonstrated and generally accepted technology; contaminants listed — 10
Level 0	Unclassified at this time; contaminants listed — 33

Note: The total number of chemical substances that are listed for source registration is 535, of which 189 have been classified as shown above.

B. Source Registration

A vast range of contaminants will need to be considered during source registration. The draft CAP documents list 535 chemical substances. Of these, 189 are already included in an interim list of classifications based on their degree of hazard, or level of concern, as shown in Table 2.

C. The *De Minimis* Concept

The proposed legislation makes provisions to limit the control requirements for very small sources or very low emission rates from complex sources. Limits are in kilograms per year and have or will be set for some 548 substances, with a maximum limit of 1000 kg/yr for any contaminant. Some examples of these that are topical for this discussion are shown in Table 3. Note that for a large source such as a multiunit power plant, this upper limit of 1000 kg/yr translates into an in-stack concentration of about 20 mg/m^3.

III. SUMMARY OF ONTARIO HYDRO'S AIR EMISSION PROGRAMS

Regular, routine measurements are carried out for particulate, SO_2, NO_x, CO_2,

Table 3. CAP Small Source Designation Limits

Most restrictive	
Total dioxins (PCDDs) (level 1)	1×10^{-7} kg/year
Asbestos (level 1)	8×10^{-7} kg/year
Benzo-a-pyrene (BaP) (level 1)	7×10^{-4} kg/year
46 other organic compounds	0.002 kg/year.
Others of interest	
Polychlorinated biphenyls (level 1)	0.04 kg/year
Formaldehyde (level 1)	2.0 kg/year
Biphenyl (level 1?)	2.0 kg/year
Naphthalene (level 2?)	8.0 kg/year
Upper limit	
Any substance	1000 kg/year

CO, and O_2 concentrations in the flue gases. Continuous emission monitors (CEMs) will be required under CAP. Continuous opacity monitors are installed on each coal-fired unit. An extensive trace element emission characterization study was carried out in the 1970s.

Prior to 1986, only minor studies had been attempted for measurement of any organic emissions. Organics make up most of the list of substances to be registered. For example, 28 of the 37 Level 1 contaminants and 30 of the 33 Level 1/2 contaminants are organic substances.

In 1986, selected organic emissions were measured at the coal-fired Lambton generating station (GS). These included dioxins and furans, chlorobenzenes, chlorophenols, polychlorinated biphenyls (PCBs), and a number of polyaromatic hydrocarbons (PAHs).

A similar program was carried out at a nonradioactive incinerator at the Bruce Nuclear Power Development (BNPD) site in 1988 followed by a very comprehensive study at the coal-fired Lakeview GS in 1989.

Other large programs are presently under way at the coal-fired Nanticoke GS and the oil-fired Lennox GS. Plans are in place to extend this work to include an oil-fired auxiliary steam plant and a radioactive waste incinerator at the BNPD site.

These programs are very costly and basically are all being carried out in anticipation of CAP and the requirement for source registration. It is considered preferable to have the information available before CAP and avoid the need for costly control technologies and continued compliance testing for air toxics that may not even be detectable from these sources.

IV. SOME CONCERNS REGARDING CAP

A. Source Registration

The number of contaminants to be considered for source registration is very large, presently at 535. Smaller, source-specific lists should be developed in consultation with industry.

B. Inadequacy of the *De Minimis* Concept

For many of these contaminants, the *de minimis* concept is inadequate for a large source with a very high volume of flue gas emitted per year. For example, the annual emission limit for polychlorinated dibenzo-p-dioxins (PCDDs) is 0.1 mg. For a large source such as the Lakeview GS, this is equivalent to an in-stack concentration of about 0.002 pg/m^3! With sample sizes limited to about 10 m^3, this would require an analytical detection limit of 0.02 pg per sample before the "absence" of this contaminant could be proven. This is at least three to four orders of magnitude below that achievable with present technology.

C. Validity of Existing Data for Low Level Organic Emissions

Finally, there is even some doubt about the validity of reported low level organic emission data that appear to be above both the *de minimis* emission limit and the detection limits of the analytical methods. It is this concern that is addressed in the remainder of this paper.

V. DETAILED REVIEW OF SOME OF THE RESULTS FROM THE LAKEVIEW GENERATING STATION STUDY

A. Plant Description

Lakeview GS is an eight-unit station, each with a nominal capacity of 300 MW. Coal is the predominant fuel although oil is also used during cold start-up. Each unit is equipped with an electrostatic precipitator and the flue gases are discharged to the atmosphere through four stacks, each one servicing two units.

Two different types of boilers are used. Units 1 and 2 and units 5 to 8 are wall-fired Babcock and Wilcox boilers while units 3 and 4 are Combustion Engineering corner-fired boilers. Both types were tested during this program.

B. The Test Program

This program was carried out by Ortech International under contract to Ontario Hydro.

Triplicate tests were performed at nominal steady loads of 100, 200, and 300 MW. In addition, the organic emissions were measured during unit start-up (cold boiler start), a period when it was suspected that combustion conditions would be at their worst.

Measured parameters during steady load tests were particulate matter, combustion gases (including nitrous oxide and total hydrocarbons), 32 metals and acid gas anions, a wide range of chlorinated aromatics, selected aldehydes and ketones, and

Table 4. Selected Emission Rates from Lakeview GS

Contaminant	SSDL	Cold start	Load of 100 MW	Load of 200 MW	Load of 300 MW
Dioxins and furans[a]	3.2 pg/s	0.37 μg/s	0.14 μg/s	0.56 μg/s	0.22 μg/s
Benzo-a-pyrene (BaP)	22 ng/s	0.19 μg/s	0.06 μg/s	1.39 μg/s	0 μg/s
Polychlorinated biphenyls (PCBs)	1.3 μg/s	27 μg/s	32 μg/s	83 μg/s	66 μg/s
Naphthalene	0.25 mg/s	56 mg/s	0.80 mg/s	0.79 mg/s	1.14 mg/s

[a] Expressed as total dioxins, with furans equal to 2% of dioxins.

38 PAH compounds. Fly-ash samples were also analyzed for these parameters. The metals and acid gas anions were not measured during unit start-up.

C. Review of Selected Emission Data

1. Rationale for the Review

The review of some of the emission data for a few of the more important organics from the study was carried out to determine their significance when compared to the CAP Small Source Designation Limits (SSDLs).

Those chosen for closer scrutiny were the two with the most restrictive SSDLs, dioxins plus furans and benzo-a-pyrene (BaP). Also reviewed were the ever topical PCBs and, probably the most commonly found PAH, naphthalene. The emission rates for these contaminants are shown in Table 4 and have been taken directly from the Executive Summary of the final report from Ortech International to Ontario Hydro. Unfortunately, this is only too often all that is seen by the various interested parties.

Note that, except for BaP at 300 MW, all these rates are above the corresponding SSDLs. Most of them, being Level 1 contaminants, would therefore require Lowest Achievable Emission Rate (LAER) controls. The question is, are these results fact or artifact?

2. General Aspects of the Reviews

There are two aspects to the generation of emission rates from a stack-testing program. The first is the sampling itself, in which a representative sample of the stack gas is obtained and the volumetric flowrate of the exhaust gases is measured. These methods are now reasonably well established and should not cause any major difficulties for an experienced, competent stack-testing team. The most likely problem would be contamination of the sample and this is monitored by the use of field blanks.

Table 5. Types of Samples Analyzed for Organic Contaminants

Extracts from the probe/filter of test trains	23
Extracts from the adsorbent/impingers of test trains	23
Extracts from complete field blank trains	11
Blank solvents	6
Total samples analyzed	63

The other component of a test is the analysis of the collected samples. It is here, during attempts to quantitate very small amounts of these analytes, that uncertainties can occur. This is not to demean the current technologies in use or the efforts of those that practice this profession. It is more that the matrices of these samples are extremely complex and that measurements at or near the detection limits of the methods are becoming more the norm rather than the exception.

Consequently, the actual analytical results that were used to generate the various emission rates have been examined. In all, there were 63 samples analyzed for each of the organics sampled, as shown in Table 5. Of these, 46 were from 23 actual stack tests (1 probe/filter sample and 1 adsorbent/impinger sample per test train), 11 were field blank train samples, and 6 were blank solvents used during sample recovery.

3. Dioxin and Furan Review

Considering first dioxin, there were 315 data points in the raw analytical results. This number was derived from 63 samples, each with a result for 5 congener groups (total tetrachloro through to total octochloro).

The first observation of significance was that 243 of these results were reported as "not detected", giving a frequency of detection of only about 20%. The detection limits quoted ranged from 0.07 to 3.0 ng, with the average being 0.28 ng. This is four orders of magnitude above the 0.02 pg that would be required to achieve SSDL exemption.

Although this demonstrated the failure of the SSDL concept, it was perhaps unfair in that one would not expect to detect dioxins in the blanks and perhaps the probe/filter samples. The latter is because these emissions are generally expected to be found in the gaseous state at stack conditions.

Consequently, the data were reexamined by combining corresponding probe/filter and adsorbent/impinger samples to give 23 test train results. This resulted in a near reversal of the frequency of detection with about 80% of the trains showing at least one result out of the ten congener group measurements above its detection limit. It is interesting to note that this same frequency of detection of about 80% was exhibited by the field blank train samples. Also, at least one congener group was detected in four of the six blank solvent samples as well, giving a 70% detection rate. Given the similarity of these detection frequencies, one must suspect the presence of some contamination or analytical artifacts.

Table 6. Examples of Positive and Questionable Detections of Dioxin

	Unit 4 (300 MW)		Unit 4 (cold start)	
	Test (ng)	Blank (ng)	Test (ng)	Blank (ng)
Probe/filter				
Tetra-CDD	ND	ND	0.24	ND
Penta-CDD	ND	ND	ND	ND
Hexa-CDD	ND	ND	ND	ND
Hepta-CDD	ND	ND	ND	ND
Octa -CDD	0.25	0.7	0.26	0.6
Adsorbent/impingers				
Tetra-CDD	ND	ND	ND	ND
Penta-CDD	ND	ND	0.67	ND
Hexa -CDD	ND	ND	2.7	ND
Hepta-CDD	ND	ND	1.9	ND
Octa -CDD	1.3	ND	2.0	ND

Note: ND = not determined.

Those familiar with the analysis of dioxins, especially for samples derived from combustion, might recognize that the detection of only a low level of one congener group in a sample is not usually considered very strong evidence of their presence. Examples of this versus a sample exhibiting strong evidence of detection are given in Table 6. Only one of the eight samples shown, the adsorbent/impingers extract from unit 4 start-up, clearly indicated the presence of measurable amounts of dioxins.

Based on these examples, an arbitrary assumption was made that the detection of only one congener group per sample, or two per test train, did not clearly confirm the presence of dioxins. The analytical data were then reexamined on a per-train basis and the frequency of detection fell from the previously quoted 80% to only 30% for the test trains, 10% (1 in 11) for the blank trains, and 0% for the blank solvents.

A summary of all these observations gave the following table of detection frequencies for the actual test train samples:

Load	Unit 7	Unit 3/4	Projected for 8 units	Presence of dioxins confirmed?
Cold start	1/3	2/2	10/22	Yes
100 MW	0/3	1/3	2/24	No
200 MW	0/3	3/3	6/24	?
300 MW	0/3	0/3	0/24	No

These observations are in no way considered to be proof of the presence or absence of dioxin emissions but are presented in this way only to illustrate the dilemma faced by both industry and regulating bodies when faced with test results of this type. However, it would appear that these emissions at conditions other than cold start-up are, at best, questionable.

The review of the furan analytical data gave a similar pattern of results, as shown below:

All samples:

Detection frequency for all analyses was 56/315 = 18%
Range of detection = 0.03 to 5.0 ng
Average detection limit = 0.19 ng

Test trains and blanks:

Test train detection frequency was 15/23 = 65%
Blank train detection frequency was 0/11 = 0%
Blank solvent detection frequency was 0/ 6 = 0%

Greater than two congener groups:

Positive test train detection frequency was 11/23 = 48%

Again, a summary of these detection frequencies for the actual test train samples gave

Load	Unit 7	Unit 3/4	Projected for 8 units	Presence of furans confirmed?
Cold start	3/3	2/2	22/22	Yes
100 MW	1/3	1/3	8/24	?
200 MW	0/3	2/3	4/24	No
300 MW	1/3	1/3	8/24	?

As with the dioxin results, only the emissions during cold start-up were clearly confirmed, with the others being questionable.

4. Benzo-a-Pyrene Review

Since there are no congener groups in this analysis, the total number of results reported for review was 63. Of these, eight results were above the detection limit, giving a detection frequency of only 13%. Grouping these into train samples and solvent blanks showed the following:

Test train detection frequency was 4/23 = 17%
Blank train detection frequency was 1/11 = 9%
Blank solvent detection frequency was 2/ 6 = 33%

Two of the test train results and one blank solvent result were at or below the average detection limit of 2.6 ng (range of 0.3 to 10 ng), giving realistic detection frequencies of only about 10% for all three categories of samples.

A summary of this information for the actual test train samples gave

Load	Unit 7	Unit 3/4	Projected for 8 units	Presence of BaP confirmed?
Cold start	0/3	1/2	2/22	No
100 MW	0/3	0/3	0/24	No
200 MW	0/3	1/3	2/24	No?
300 MW	0/3	0/3	0/24	No

The question is, do these results justify the reported emission rates of 0.19, 0.06, and 1.4 mg/s for station loadings of cold start, 100, and 200 MW, respectively? All are above that required for SSDL exemption, which would require an analytical detection limit of 0.12 ng. It would appear not, with only the zero-emission rate at 300 MW being confirmed.

5. Naphthalene Review

This is a very different example of questionable emission data. Again, this analysis produced 63 results but in this case all were above the detection limit of the method. The problem here was that these positive results also included all the blank trains and all the blank solvent samples, often with amounts at or near that found in the corresponding test trains. Since there is no provision made in these test methods to allow for field blank corrections, the results from those samples were ignored during calculation of the emission rates for naphthalene.

A summary of the analytical results for naphthalene is shown below, expressed as the average quantity of the analyte found in samples of the same type:

Load	Average quantity of naphthalene found in samples (μg)		Average quantity in blanks (μg)	Presence of naphthalene confirmed?
	Unit 7	Unit 3/4		
Cold start	125	2450	1.3	Yes
100 MW	3.3	5.1	1.1	?
200 MW	2.4	3.3	1.6	No
300 MW	3.3	2.9	3.6	No
		Overall average for blanks =	1.7	
		Standard deviation =	1.5	

Relatively large amounts of naphthalene were found in each of the five cold start-up tests. Only one other test, unit 4 at 100 MW, gave a result that was significant when compared to the field blank samples. Even then, the average for all six tests at 100 MW was less than three standard deviations of the blank values and is most likely not significant.

6. Polychlorinated Biphenyl Review

There were 9 congener groups (dichloro through to decachloro) reported for the PCB analysis of the 63 samples, giving a total of 547 results for review.

All samples:

Detection frequency for all analyses was 106/567 = 19%
Range of detection = 0.4 to 20 ng
Average detection limit = 2.8 ng

In contrast to most of the previous results, this detection limit is below that of the 7 ng from a 10 m^3 sample that is required for SSDL evaluation.

Test trains and blanks:

Test train detection frequency was 21/23 = 91%
Blank train detection frequency was 4/11 = 36%
Blank solvent detection frequency was 2/ 6 = 33%

Again using the assumption that the detection of only one congener group per sample, or two per test train, did not clearly confirm the presence of the analyte, then the detection frequency for PCBs was

Greater than two congener groups:

Positive test train detection frequency was 16/23 = 70%
Positive blank train detection frequency was 1/11 = 9%

A summary of this information for the actual test train samples gave

Load	Unit 7	Unit 3/4	Projected for 8 units	Presence of PCBs confirmed?
Cold start	3/3	2/2	22/22	Yes
100 MW	0/3	3/3	6/24	?
200 MW	0/3	3/3	6/24	?
300 MW	2/3	3/3	18/24	Yes

This showed that the results reported for PCBs were probably a fair indication of the actual emissions of this contaminant. The differences in the results between the two types of units are still of some concern but there were no obvious reasons for rejecting any of this test data.

VI. SUMMARY AND CONCLUSIONS

The emission rates of a large number of organic air toxics were recently measured at Ontario Hydro's coal-fired Lakeview GS. The final report from Ortech International indicated measurable emissions for most of these species over the full range of operating loads for the units tested, from cold start-up to 300 MW.

Following the current Ontario air emission regulations, dispersion modeling was used to determine compliance with point of impingement concentration standards and guidelines. No organic was found to yield a result within an order of magnitude of the allowed concentrations and most were much lower than that.

Changes to this regulation proposed under CAP would cause a drastic reversal of this with most organics exceeding their SSDL exemption and requiring LAER control technology.

This realization lead to a closer examination of some of the actual measurements that were made and subsequently used to calculate the reported emission rates. The raw analytical data for the analyses of dioxins and furans, BaP, naphthalene, and PCBs were reviewed.

Although somewhat cursory in nature, this process has indicated that most of these results are questionable at best. Only the cold start-up emissions were found to be significant for dioxins, furans, and naphthalene. None of the BaP results was significant. At loads other than start-up, only the reported PCBs emissions appeared to be justified.

The nature of this review and the assumptions made during it by no means represent a definitive analysis of the the Lakeview GS emission data. However, it does raise some important questions regarding the low level emission rates that are often reported for these contaminants. Results of this type warrant more scientific and mathematical evaluation before they can be considered real and before decisions are made on the need for costly control technologies.

Comprehensive Assessment of Toxic Emissions from Coal-Fired Power Plants

Thomas D. Brown
Charles E. Schmidt
Adrian S. Radziwon

ABSTRACT

This paper briefly describes both recent and ongoing studies being conducted to assess hazardous and toxic substances from a variety of coal-fired electric utility power generation and environmental control subsystems. Also, current and future U.S. Department of Energy plans to augment these assessments will be addressed.

I. INTRODUCTION

The trace elements associated with the mineral matter in coal and the various compounds formed during coal combustion have the potential to produce air toxic emissions from coal-fired electric utilities. The recently enacted Clean Air Act Amendments (CAAA) contain provisions that will set standards for the allowable emissions of 190 hazardous air pollutants (HAPS). These 190 air toxics indicated in Table 1 can be associated with any number of source categories that emit pollutants to the environment. Many of these HAPS could possibly be emitted from coal-fired electric generating stations. Coal-fired electric utility boilers will be studied by the U.S. Environmental Protection Agency (EPA) to determine if regulation is appropriate and necessary.

Title III, the Hazardous Air Pollutants section of the CAAA, requires the EPA to determine stationary source categories that have the potential to emit any of the 190 HAPS listed in the act. Coal-fired electric utilities are contained in a draft list of 750 sources that the EPA has already developed. The EPA will designate as major sources those stationary sources that could emit 10 tons/year of any single HAP or 25 tons/year of a combination of HAPS. After November 15, 1991, all major sources must be regulated over a 10-year period, according to a schedule provided in the CAAA. Determination of whether or not coal-fired power plants need alternative control strategies for any HAPS emissions that may warrant regulation will be made before November 15, 1993.

Table 1. The 190 Hazardous Air Pollutants Listed in Clean Air Act Amendments of 1990

Acetaldehyde
Acetamide
Acetonitrile
Acetophenone
2-Acetylaminofluorene
Acrolein
Acrylamide
Acrylic acid
Acrylonitrile
Allyl chloride
4-Aminobiphenyl
Aniline
o-Anisidine
Asbestos
Benzene (including benzene from gasoline)
Benzidene
Benzotrichloride
Benzyl chloride
Biphenyl
Bis(2-ethylhexyl)phthalate (DEHP)
Bis(chloromethyl)ether
Bromoform
1,3-Butadiene
Calcium cyanamide
Caprolactam
Captan
Carbaryl
Carbon disulfide
Carbon tetrachloride
Carbonyl sulfide
Catechol
Chloramben
Chlordane
Chlorine
Chloracetic acid
2-Chloroacetophenone
Chlorobenzene
Chlorobenzilate
Chloroform
Chloromethyl methyl ether
Chloroprene
Cresols/cresylic acid (isomers and mixture)
o-Cresol
m-Cresol
p-Cresol
Cumene
2,4-D, salts, and esters
DDE
Diazomethane
Dibenzofurans
1,2-Dibromo-3-chloropropane
Dibutylphthalate
1,4-Dichlorobenzene(p)
3,3-Dichlorobenzidene
Dichloroethyl ether [bis(2-chloroethyl)ether]
1,3-Dichloropropene
Dichlorvos
Diethanolamine
N,N-Diethyl aniline (*N,N*-dimethylaniline)
Diethyl sulfate
3,3-Dimethyoxybenzidine
Dimethyl aminoazobenzene
3,3′-Dimethyl benzidine
Dimethyl carbamoyl chloride
Dimethyl formamide
1,1-Dimethyl hydrazine
Dimethyl phthalate
Dimethyl sulfate
4,6-Dinitro-o-cresol and salts
2,4-Dinitrophenol
2,4-Dinitrotoluene
1,4-Dioxane (1,4-diethyleneoxide)
1,2-Diphenylhydrazine
Epichlorohydrin (1-chloro-2,3 epoxypropane)
1,2-Epoxybutane
Ethyl acrylate
Ethyl benzene
Ethyl carbamate (urethane)
Ethyl chloride (chloroethane)
Ethylene dibromide (dibromoethane)
Ethylene dichloride (1,2-dichloroethane)
Ethylene glycol
Ethylene imine (aziridine)
Ethylene oxide
Ethylene thiourea
Ethylidene dichloride (1,1-dichloroethane)
Formaldehyde
Heptachlor
Hexachlorobenzene
Hexachlorobutadiene
Hexachlorocyclopentadiene
Hexachloroethane
Hexamethylene-1,6-diisocyanate
Hexamethylphosphoramide
Hexane
Hyzadrine
Hydrochloric acid
Hydrogen fluoride (hydrofluoric acid)
Hydrogen sulfide
Hydroquinine
Isophorone
Lindane (all isomers)
Maleic anhydride
Methanol
Methoxychlor
Methyl bromide (bromomethane)
Methyl chloride (chloromethane)
Methyl chloroform (1,1,1-trichloroethane)
Methyl ethyl ketone (2-butanone)
Methyl hydrazine
Methyl iodide (iodomethane)
Methyl isobutyl ketone (hexone

Table 1. The 190 Hazardous Air Pollutants Listed in Clean Air Act Amendments of 1990 (continued)

Methyl isocyanate	Titanium tetrachloride
Methyl methylacrylate	Toluene
Methyl-tert-butyl ether	2,4-Toluene diamine
4,4-Methylene bis (2-chloroaniline)	2,4-Toluene diisocyanate
Methylene chloride (dichloromethane)	*o*-Toluidine
Methylene diphenyl diisocyanate (MDI)	Toxaphene (chlorinated camphene)
4,4′-Methylenedianiline	1,2,4-Trichlorobenzene
Naphthalene	1,1,2-Trichloroethane
Nitrobenzene	Trichloroethylene
4-Nitrobiphenyl	2,4,5-Trichlorophenol
4-Nitrophenol	2,4,6-Trichlorophenol
2-Nitropropane	Triethylamine
N-Nitroso-*N*-methylurea	Trifluralin
N-Nitrosodimethylamine	2,2,4-Trimethylpentane
N-Nitrosomorpholine	Vinyl acetate
Parathion	Vinyl bromide
Pentachloronitrobenzene (quintobenzene)	Vinyl chloride
Pentacholorphenol	Vinylidene chloride (1,1-dichloroethylene)
Phenol	Xylene (isomers and mixture)
p-Phenylenediamine	o-Xylenes
Phosgene	m-Xylenes
Phosphine	p-Xylenes
Phosphorus	Antimony compounds
Phthalic anhydride	Arsenic compounds (inorganic Incl. arsine)
Polychlorinated biphenyls (aroclors)	Beryllium compounds
1,3-Propane sultone	Cadmium compounds
b-Propiolactone	Chromium compounds
Propionaldehyde	Cobalt compounds
Propoxur (bargon)	Coke oven compounds
Propylene dichloride (1,2-dichloropropane)	Cyanide compounds[a]
Propylene oxide	Glycol ethers[b]
1,2-Propylenimine (2-methyl aziridine)	Lead compounds
Quinoline	Manganese compounds
Quinone	Mercury compounds
Styrene	Fine mineral fibers[c]
Styrene oxide	Nickel compounds
2,3,7,8-Tetrachlorodibenzo-*p*-dioxin	Polycyclic organic matter[d]
1,1,2,2-Tetrachloroethane	Radionuclides (including radon)[e]
Tetrachloroethylene (perchloroethylene)	Selenium compounds

Note: For all listings above that contain the word "compounds" and for glycol ethers, the following applies: unless otherwise specified, these listings are defined as including any unique chemical substance that contains the named chemicals (i.e., antimony, arsenic, etc.) as part of that chemical's infrastructure.

[a] X′ CN where X = H′ or any other group where a formal dissociation may occur. For example, KCN or $Ca(CN)_2$.

[b] Includes mono- and diethers of ethylene glycol, diethylene glycol, and triethylene glycol R-(OCH_2CH_2)n-OR′ where n = 1, 2, or 3, R = alkyl or aryl groups, and R′ = R, H, or groups that, when removed, yield glycol ethers with the structure: R-(OCN_2CH)n-OH. Polymers are excluded from the glycol category.

[c] Includes mineral fiber emissions from facilities manufacturing or processing glass, rock, or slag fibers (or other mineral-derived fibers) of average diameter (1 μm or less).

[d] Includes organic compounds with more than one benzene ring and boiling points greater than or equal to 100°C.

[e] A type of atom that spontaneously undergoes radioactive decay.

Considerable actual HAPS emission data already exist for many of the stationary sources that will be designated as major sources under Title III. In these cases, the EPA will be able to use sound scientific data to prepare regulations. In contrast, a limited database exists for coal-fired utility boilers. Much of the technical literature concerning toxic emissions from coal combustors consists of calculated values based on test burns under controlled conditions or incomplete material balance studies that related the emissions of trace metals to the inorganic composition of the input coal. Also, there is a high degree of uncertainty concerning much of the data on some of the more volatile components contained in the generated flue gas. For example, results from a literature survey indicated that wet scrubbers may remove anywhere from 20 to 80% of the mercury from utility boiler flue gas.

Conventional air pollution control subsystems have the potential to remove many of the air toxic emissions from flue gas generated from the combustion of coal. There is a lack of precise analytical data on the removal of toxics across environmental control devices, such as electrostatic precipitators, baghouses, and wet limestone scrubbers. However, the relative concentrations of some of the toxic materials could also be increased as a result of using these technologies, or toxics could be formed when chemicals are added to the flue gas stream to increase particulate collection efficiency. Further, some of the more advanced SO_2 and NO_x mitigation technologies involve furnace injection of a sorbent and combustion modification, respectively. To date, little information exists on the effects these advanced technologies have on the amounts of toxic substances formed in the combustion zone.

Efforts are under way to develop a more complete database on potential HAPS emissions from electric utilities. During the first phase of a two-phase program, the Canadian Electric Association conducted a study to examine air, water, and ash pathways for trace constituents released to the environment from four Canadian coal-fired generating stations. The second phase of the program dealt with the environmental dispersion and biological implications of the release.

The Electric Power Research Institute (EPRI) has begun to assess the emissions from power plants under the PISCES (Power Plant Integrated Systems: Chemical Emission Studies) program. This activity involves the use of a consistent and comprehensive analytical protocol that evaluates all inputs and outputs concerning pollution control and process streams at the utility. To date, the EPRI study has gathered analytical information at 6 utility sites for 24 of the 190 hazardous pollutants listed in Title III of the CAAA.

The Pittsburgh Energy Technology Center (PETC) of the U.S. Department of Energy (DOE) has two current investigations, initiated before passage of the CAAA, that will determine the air toxic emissions from coal-fired electric utilities. DOE has contracted with Battelle Memorial Institute and Radian Corporation to conduct studies focusing on the potential air toxics, both organic and inorganic, associated with different size fractions of fine particulate matter emitted from power plant stacks. Table 2 indicates the selected analytes to be investigated during these studies. PETC is also developing guidance on the monitoring of HAPS to be incorporated in

Table 2. Elements and Compounds for the Battelle and Radian Air Toxics Studies

Element	Compound
Arsenic	Ammonia
Barium	Radionuclides (Ra, Po, U, etc.)
Beryllium	Sulfates
Cadmium	
Chromium	Benzene
Chlorine (as Cl^-)	Toluene
Cobalt	Formaldehyde
Copper	Polycyclic aromatic hydrocarbons
Cyanide	
Fluorine (as F^-)	
Lead	
Manganese	
Mercury	
Molybdenum	
Nickel	
Phosphorus (as PO_4^{3-})	
Selenium	
Vanadium	
Other elements associated with instrumental neutron activation analysis (INAA)	

the Environmental Monitoring plans for the demonstration projects in its Clean Coal Technology Program.

These ongoing DOE air toxic emissions studies, which were initiated before passage of the CAAA, are somewhat limited in scope and therefore cannot provide all the information necessary for Title III considerations and requirements. Consequently, there is a need to expand and broaden these studies so as to increase the technology base on toxic emissions from coal-fired utilities.

II. RECENT/CURRENT TOXIC EMISSIONS STUDIES

A. Canadian Electric Association Study

A major toxic-emissions study was performed by the Battelle Pacific Northwest Laboratories for the Canadian Electric Association (CEA). The objectives of the CEA-sponsored study were the following:

- Identify release pathways for trace elements in coal-fired generating stations and to quantify the releases.
- Document the accuracy and reliability of various analytical and sampling procedures.
- Identify the effect of operational parameters on the release of trace elements to the environment.
- Determine the effects of trace releases on living organisms.

- Compare the quantity of trace constituent releases from power plants to those from other man-made sources.
- Provide a basis for determining the effectiveness of controls.

The CEA program did not study emissions associated with acid rain. This program dealt strictly with coal-fired power generating stations. The four stations studied and the types of coal used were the following:

- Battle River — subbituminous C (low sulfur)
- Poplar River — lignite (medium sulfur)
- Nanticoke — bituminous (low sulfur)
- Lingan — bituminous (high sulfur)

The CEA study was divided into two phases. Phase I work dealt with obtaining and analyzing data pertaining to emissions to the environment and quantifying those emissions and identifying the pathways to the environment. This phase was completed in early 1985. Phase II consisted of work on dispersion in and effects on the environment. Our discussion of the CEA work will be limited to Phase I since this paper is chiefly concerned with identifying and quantifying trace toxic emissions to the environment.

To determine the pathways to the environment a number of streams were sampled and analyzed at each plant:

- Feed coal
- Bottom ash
- ESP hopper ash
- Inlet and outlet ash sluice water
- Ash lagoon water
- Stack flue gas
- Flue ash (emitted particulates)
- Miscellaneous, site-specific samples

These materials were sampled for up to 45 elements in addition to polycyclic aromatic hydrocarbons (PAHs). Material balances were made based on the averages of several runs. Closure to within 20% was found for 37 elements. Elements for which closure was not obtained were fluorine, silicon, phosphorus, cadmium, boron, and mercury.

Some elements become enriched in the flyash as it passes through the system from the furnace to the stack. The elements included boron, zinc, gallium, arsenic, selenium, molybdenum, cadmium, antimony, and lead. Of these, only antimony, arsenic, cadmium, lead, and selenium are on the EPA's HAPS list. Other elements have patterns of enrichment that varied from plant to plant. These included sodium, vanadium, chromium, manganese, cobalt, nickel, copper, barium, and uranium. Of these, chromium, manganese, and cobalt are found on the EPA's HAPS list. Most

Table 3. Trace Element Releases with the Flue Gas

Element	Percent of total element in coal released with the flue gas
Chlorine	49 — 99.0
Chromium	0.1 — 8.7
Manganese	0.1 — 1.0
Cobalt	0.09 — 1.5
Arsenic	0.74 — 9.3
Selenium	3.5 — 73.0
Antimony	0.2 — 2.5
Mercury	79.0 — 87.0
Lead	0.2 — 1.4

elements were found to be part of the silicate matrix and showed no enrichment patterns. It should also be noted that enrichment was most pronounced in the smallest-size particles.

The volatile and gaseous elements were all found to be depleted from the ash and are assumed to be emitted as vapors. These include fluorine, sulfur, chlorine, bromine, and mercury. Chlorine and mercury are on the EPA's HAPS list. Two elements, selenium and arsenic, are not gases but are released to the atmosphere. From 4 to 73% of the selenium in the coal is released while 1 to 9% of the arsenic is released with the flue gas. It is interesting to note that arsenic emissions tended to be inversely proportional to the calcium content of the coal.

Measurements of PAH emissions were obtained at three of the plants. At all three plants, the emission levels were very small and ranged from 0.15 to 0.66 g/h/MW (≈0.17 to 0.73 lb/h for a 500-MW plant). The largest single-compound emission found was benzo(a)pyrene. Other relatively high PAH emissions included 9,10-dihydroanthracene, 9-methylanthracene, 9,10-dimethylanthracene, and 1,2-benzofluorene.

The CEA study also investigated the release of radionuclides. There was significant evidence of the fractionation of radionuclides in their movement through the system. Enrichment was found for ^{210}Pb and ^{238}U. Essentially all of the ^{222}Rn is released as a gas. While the concentration of radon in the stack gas is 30 to 90% over ambient, it is quickly diluted to near ambient levels. Overall, the release of all radionuclides is of minimal significance.

Table 3 summarizes the emissions of the elements listed on the EPA HAPS list. The data resulted from measurements at several power plants. These plants use the coals described previously.

B. EPRI's Field Chemical Emissions Monitoring (FCEM) Project

The EPRI air toxic-emissions study is being carried out by Radian Corporation. The objectives of the current study are

- Develop material balances around the combustion system and associated air pollution control equipment.

- Provide a preliminary basis for partitioning around control devices.
- Provide an indication of the long-term, uncontrolled variability for species emitted from a conventional power plant.

This project will involve measuring the quantities of select chemical species at key points in a power plant to close the material balance for those selected chemical substances. The FCEM project is a follow-up project to an earlier EPRI project called PISCES. The PISCES project consisted of an exhaustive literature review to obtain as much existing data as possible on the emission of chemical species from power plants. These data were then organized into a database that contains information on both individual power plants and on the chemical characteristics of various streams within those plants. The PISCES project also served to identify gaps in the existing data on power plant emissions. These data gaps are most prevalent in the category of trace chemical or element atmospheric emissions.

The current FCEM project seeks to eliminate those gaps pertaining to the emission of trace elements from power plants. This will quantify the concentration of select materials in streams leaving the plant. The project will also result in probabilistic concentration profiles for select chemical species.

The sampling and analysis portion will take place at select coal-, oil-, and gas-fired conventional power plants (fossil fuel, steam turbine generator). While the end results will be specific to these power plants, they will be selected so as to be representative of a significant number of U.S. power plants. The tests will be made for both organic and inorganic species in both controlled and uncontrolled streams. The specific chemical species, as well as the specific power plants, will be selected to provide a solid baseline for future work. Compounds to be the subject of these analyses are expected to include arsenic, cadmium, chromium, lead, mercury, nickel, selenium, vanadium, zinc, benzene, formaldehyde, chlorides, fluorides, and PAH. Streams to be sampled for a coal-fired unit include feed coal, coal pile runoff, bottom ash, economizer ash, ESP gas inlet and outlet, ESP ash, FGD system inlet, FGD system outlet, stack gas, and FGD waste. Appropriate power generation and environmental control subsystems of oil- and gas-fired units will be sampled for comparison of toxic emissions.

The end result of the FCEM project will be reasonably complete emission profiles on trace toxic species from a number of power plants that are representative of many U.S. power plants. This information will be obtainable from the PISCES database and will serve as the benchmark for further work that needs to be done in quantifying trace emissions from power plants.

C. Battelle Memorial Institute

Battelle Memorial Institute and its subcontractor Keystone/NEA will make a correlation between air toxics produced by a laboratory combustor and two operating coal-fired electric utility boilers. A characterization of air toxics associated with the surfaces of fine particles and vapor phase constituents of the stack flue gas of the selected coal-fired units will be made. Diluted, cooled flyash particles with adsorbed and condensed material on the surfaces and hot gas flyash particles without a majority of these absorbed and condensed materials will be collected in three size fractions from the stacks. These size fractions are <0.6, 0.6 to 2.0, and 2.0 to 5 mm.

An innovative source dilution sampler will be utilized to simulate plume cooling and collect the diluted, cooled particles that may have an increased concentration of certain toxic substances. The hot gas samples, particulate and vapor phase, will be collected by EPA Modified Method 5 procedures. The differences in the two samples will provide information on the characteristics of surface-layer composition of fine particles, particularly materials of air toxic concern.

Laboratory studies can be more useful under certain circumstances than full-scale studies because of the flexibility to examine emissions from developing pollution control technologies (i.e., furnace and duct sorbent injection, flue-gas conditioning, and from various combustion configurations). If possible, the coals used by the two coal-fired electric utilities will be used in the laboratory combustion studies, which will indicate the efficiency of using a well controlled laboratory-scale combustor to simulate emissions from a full-scale unit. Additional results from the Battelle laboratory combustion work will include the further development of more advanced sampling methods for collection of flyash and vapor phase constituents from flue gas. The results will also assist DOE and EPRI in determining which toxic substances to sample in future emissions characterization studies.

D. Radian Corporation

Radian Corporation will collect size-fractionated particles from the stack of a full-scale coal-fired utility boiler and characterize the particles for both bulk- and surface-chemical composition. The sampling will take place over two different time periods ranging from 3 to 4 weeks. This will enable the collection of fine particles during a high-load season (winter) and a lower-load season (spring), and during load swings. Particulate samples will be collected from the stack effluent under both hot stack and dilution-cooled conditions.

A source dilution sampler will be utilized to simulate the cooling and dilution that the flue gases and particles experience while entering the atmosphere at the stack exit. A relationship will be determined between the chemical materials found and the size of particles. Also, an evaluation and subsequent characterization will be performed on the effects of cooling and dilution upon the surface condensation of volatile species. In addition, the carbon content of the particulate matter will be determined in an attempt to correlate any organic compounds found on the dilution-sample particulate with the amount of carbon.

Other considerations within this project include the differences in the potential health impacts of each fraction as a function of particle size and the leachability of the toxic chemicals from the particles.

III. FUTURE TOXIC-EMISSION STUDIES

A collaborative effort has been initiated by the DOE, the Utility Air Regulatory Group (UARG), EPRI, and the EPA to expand the study of hazardous pollutant emissions from utility boilers. This effort will involve measurements at a number of power plants having different boiler designs, NO_x control methods, particulate control devices, and SO_2 removal systems (wet and dry). From these measurements, it is anticipated that the EPA will be able to predict the potential air toxic emissions from coal-fired boilers in 1995 and in the year 2000 (after controls are installed to meet the requirements of the acid rain title of the CAAA). Measurements from plants firing bituminous or subbituminous coal will be used to evaluate the entire range of existing power plant configurations and will form the basis for this study.

The DOE, through the Office of Project Management at PETC, will issue a solicitation for proposals to assess selected hazardous/toxic pollutants from a number of utilities that utilize different pollution control and process subsystems while burning either bituminous or subbituminous coal. An objective of this solicitation will be to determine the removal efficiencies of pollution control subsystems for these selected pollutants and the concentration of the respective pollutants associated with the particulate fraction of the flue gas stream as a function of particle size. A further objective is to determine mass balances for selected pollutants for a variety of different input and output streams of the power plants and subsequently for the entire power plant.

This solicitation will be announced early this fiscal year. A Commerce Business Daily (CBD) announcement was issued in early October 1991, addressing the solicitation. Attachment 1 is the CBD announcement. The primary goal of this work will be to produce concise, consistent data on the hazardous emissions from a number of coal-fired utilities before December of 1992.

The DOE plans to incorporate monitoring of HAPS in the Environmental Monitoring Plans for projects in the Clean Coal Technology Program. The primary objective is to quantify the mass flow rate of the listed HAPs in stack gases emitted to the atmosphere at Clean Coal demonstration sites. A secondary objective is to quantify the removal of HAPs in gaseous streams across pollution control subsystems. Monitoring would be conducted under both baseline and demonstration operating conditions.

Results from all the DOE studies will provide input to the congressionally mandated study being conducted by the EPA to assess the impacts of the listed HAPs emissions from coal-fired electric utilities, as required in Subtitle III of the CAAA of 1990. In addition, the data will provide a basis for evaluating the potential effects of air toxics regulation on existing pollution control and auxiliary processes being utilized at electric utilities and on the commercialization of technologies demonstrated under the Clean Coal Technology Program.

Determination of the Level of Hazardous Air Pollutants and other Trace Constituents in the Syngas from the Shell Coal Gasification Process

D. C. Baker
W. V. Bush
K. R. Loos

I. INTRODUCTION

The Shell coal gasification process (SCGP) produces a medium-BTU syngas, consisting predominantly of CO and H_2. Determination of trace constituents in the syngas is an integral part of the SCGP development program. This paper reviews the results of environmental characterizations focusing on trace constituents in the SCGP syngas, particularly hazardous air pollutants (HAPs). The results verify that very few HAPs or any other trace constituents are present in the syngas, and, for those few species present, their concentrations are extremely low. These results are used to estimate emissions of HAPs and other trace constituents for a 500-MW SCGP combined-cycle facility. Such a facility would not be defined as a major source of HAPs based on Title III of the 1990 Clean Air Act Amendments (CAAA), since the level of HAPs in the syngas (even prior to combustion) will be well below 10 tons/year for any one HAP and well below 25 tons/year for all HAPs.

II. SHELL COAL GASIFICATION PROCESS

Shell research into high-pressure, slagging coal gasification began in 1972 as an extension of the technology developed for high-pressure oil gasification. A 6-tons-per-day (TPD) process development unit began operation in 1976, a 150-TPD pilot plant was operated between 1978 and 1983, and a 250- to 400-TPD demonstration unit was operated at Deer Park, TX, from 1987 to 1991. This latter plant, designated SCGP-1, began operation with a demonstration phase on Illinois No. 5 coal. The demonstration phase included a 1528-h continuous run which ended in a voluntary shutdown. After this phase, the program focused on gasifying a variety of diverse feedstocks (18 total) and on providing data for both process optimization and confirmation of new features.

During the demonstration on Illinois No. 5 coal, a comprehensive environmental sampling and analytical program was conducted. The program was repeated for periods of 150 to 300 h of continuous operation on other feedstocks as follows: Blacksville No. 2, SUFCo, Drayton, and El Cerrejon coals; Texas lignite; and petroleum coke.

III. SCGP TECHNOLOGY

A block flow diagram of the process is shown in Figure 1. Crushed coal is fed to a mill for pulverization and drying and then is stored under nitrogen prior to entering the gasifier feed system. Nitrogen is used to pressurize and pneumatically transport coal in a dense phase to the burners. Pulverized coal, oxygen, and steam enter the gasifier through horizontally opposed burners and react to produce a medium-BTU syngas. Oxygen and steam consumption are dependent on coal composition and are controlled on the basis of syngas quality.

The gasifier consists of an outer pressure vessel and an inner, water-cooled membrane wall. The membrane wall encloses the gasification zone and has two outlets. One outlet at the top of the gasification zone allows hot syngas to exit where it is quenched with cooled, recycled, solids-free syngas to ensure solidification of entrained ash particles (flyslag) prior to entering the syngas cooler. The other outlet at the bottom of the gasification zone allows molten slag to exit into a water-filled slag bath where solidified, dense, inert granules form.

Flyslag is removed by dry filtration and recycled to the gasifier to increase overall carbon conversion and improve slagging efficiency. Wet slag is depressurized, dewatered, and sent to by-product utilization.

The wet-scrubbing system removes water-soluble constituents in the syngas, such as hydrogen halides and ammonia. Other constituents removed from the syngas include hydrogen cyanide and trace elements.

Sulfur-containing compounds (primarily H_2S) are removed from the syngas in a Sulfinol-based, acid-gas removal system and then routed to a Claus/SCOT sulfur plant to produce saleable, elemental sulfur.

IV. RESULTS OF ENVIRONMENTAL CHARACTERIZATIONS

A. Elemental Balances and Emissions

The syngas cleanup sequence for SCGP involves dry filtration, water scrubbing, cooling, and multistaged solvent contacting. Consequently, the syngas is expected to be essentially free of HAPs and other trace elements inherent in coal, and this expectation was confirmed by the SCGP-1 experience.

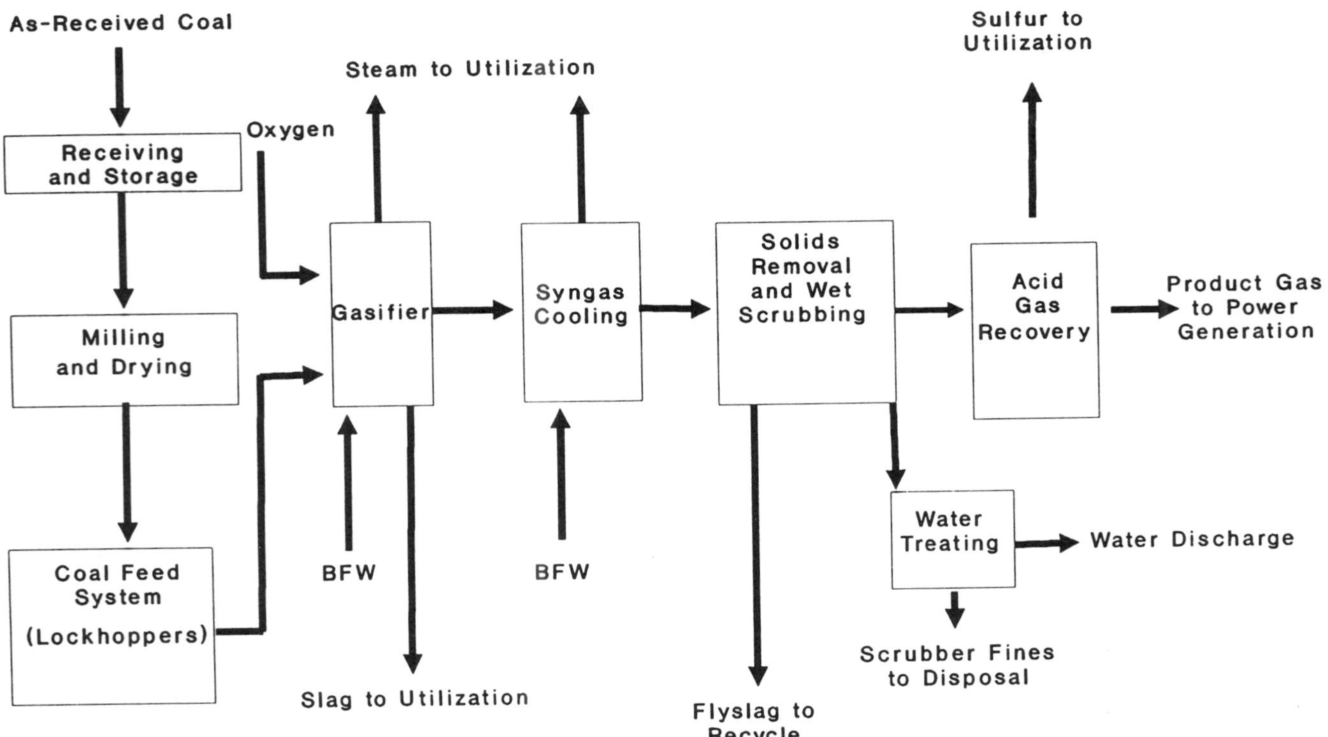

Figure 1. Shell coal gasification process block flow diagram.

Table 1. Elemental Analyses of Composite Samples of Blacksville No. 2 and Drayton Coals

	Blacksville No. 2	Drayton
Al, %	0.95	1.05
Ca, %	0.28	0.76
Fe, %	1.11	0.69
Mg, %	0.04	0.03
P, %	0.01	0.02
K, %	0.10	0.04
Si, %	1.77	2.40
Na, %	0.03	0.01
Ti, %	0.07	0.08
Sb, ppm	0.07	0.10
As, ppm	1.56	4.49
Ba, ppm	77.75	134.50
Be, ppm	0.63	0.08
B, ppm	21.60	55.62
Br, ppm	1	1
Cd, ppm	0.93	0.10
Cl, ppm	1180	500
Cr, ppm	14.18	1.90
Co, ppm	3.05	2.81
Cu, ppm	6.98	7.84
F, ppm	121	152
Pb, ppm	6.08	2.31
Mn, ppm	19	190.70
Hg, ppm	0.16	0.13
Mo, ppm	0.67	0.77
Ni, ppm	11.50	8.59
Se, ppm	0.94	0.31
As, ppm	<1	0.16
Tl, ppm	0.12	0.03
Th, ppm	0.23	0.28
Sn, ppm	0.80	1.21
U, ppm	0.05	0.44
V, ppm	12.80	11.85

Elemental analyses of two coals from SCGP-1 are given in Table 1. (Additional coal analyses from SCGP-1 are given elsewhere.[1-3]) Elemental balances and distributions, determined for these two coals and given in Table 2, confirm that after gasification the majority of trace elemental matter is tightly bound in the glassy matrix of the inert slag. As expected, the halides are concentrated in the aqueous scrubbing water where they are easily neutralized and removed as salts. Only small percentages of trace elemental matter partition into the syngas. The small percentages in the syngas, given in Table 2, largely correspond to those trace elements that occur at the lowest concentrations in the feed coal; thus, their concentrations in the syngas are also extremely low.

Table 2. Typical SCGP Elemental Balances and Distributions

	SCGP-1			
Element	**% in solids**	**% in effluent water**	**% to sulfur plant[a]**	**% in product gas**
Al	102.0	0.0	0.0	0.0
Ca	97.4	0.7	0.0	0.0
Fe	97.8	0.1	0.0	0.0
Mg	121.8	1.7	0.0	0.1
P	98.3	0.6	0.0	0.1
K	117.0	3.2	0.0	0.0
Si	103.6	0.1	0.0	0.0
Na	144.4	3.2	0.0	0.0
Ti	94.3	0.0	0.0	0.0
Sb	89.7	10.8	0.3	2.4
As	214.9	1.8	0.3	0.3
Ba	47.8	0.3	0.0	0.1
Be	212.0	1.9	0.0	0.3
B	128.4	60.0	0.0	0.1
Br	10.1	304.1	0.0	0.0
Cd	50.7	0.7	0.1	1.4
Cl	6.6	111.6	0.0	0.0
Cr	76.2	0.2	0.0	0.0
Co	61.4	0.2	0.1	0.0
Cu	206.3	0.2	0.0	0.4
F	20.4	26.7	0.0	0.0
Pb	176.6	0.1	0.1	2.1
Mn	45.8	1.7	0.0	0.0
Hg	25.1	5.3	1.7	3.4
Mo	480.2	0.5	0.2	3.3
Ni	86.6	1.3	0.1	0.0
Se	189.4	165.0	13.1	0.5
Ag	13.3	18.0	0.8	0.8
Tl	200.7	2.5	0.2	1.4
Th	117.2	0.9	0.0	0.0
Sn	55.3	8.5	0.1	0.0
U	115.0	1.8	0.5	3.4
V	100.7	0.1	0.0	0.0
Zn	130.1	1.6	0.2	0.2

Note: Average of data from Drayton and Blacksville No. coals.

[a] Prior to control technology.

Elemental concentrations in the syngas entering and exiting the acid-gas removal are given in Table 3. Concentrations are given for HAPs and a number of other elements. These data confirm that small amounts of elemental matter (gaseous or enriched on fine particulates) which entered the acid-gas removal system were effectively removed from the syngas. In an effort to identify the disposition of elemental matter entering the acid-gas removal system, analyses were performed on the concentrated acid gas, Shell's Sulfinol solvent (Table 4), particulates in the solvent surge tank, and location-specific deposits. It was determined that the concentrated acid gas (Table 2), and certain deposits from the concentrated acid gas stream,

Table 3. Analyses of Syngas Entering and Exiting the Acid Gas Removal System

Sampling method	Carbon bed	Impinger train	Carbon bed	Impinger train	Impinger train	Impinger train
Coal	El Cerrejon	Drayton	El Cerrejon	Drayton	Blacksville #2	Illinois #5
Concentration factor	~1000	~10	~1000	~10	~10	~5
	Sour syngas		**Purified syngas**			
Element	ppmw	ppmw	ppmw	ppmw	ppmw	ppmw
Al	0.294	0.030	0.835[a]	0.040	<0.020	<0.100
Ca	0.011	1.242	0.425[a]	1.363[b]	0.023	0.390
Fe	2.000	0.100	0.070	0.017	0.016	<0.050
Mg	0.049	0.417	1.650[a]	0.452	<0.020	0.080
P	0.129	0.121	0.250	0.210	<0.020	<0.500
K	2.000	0.250	6.750[a]	0.020	<0.020	<0.060
Si	0.070	0.083	0.790[a]	0.082	<0.020	0.860
Na	1.000	0.500	6.100[a]	0.005	<0.020	N/A
Ti	0.017	<0.001	<0.001	0.005	<0.020	<0.050
Sb[c]	<0.035	0.022	<0.030	<0.001	<0.003	<0.010
As[c]	0.010	0.003	<0.001	<0.001	<0.007	<0.010
Ba	<0.018	<0.001	<0.019	<0.001	0.019	<0.050
Be[c]	N/A	<0.001	N/A	<0.001	<0.003	<0.050
B	N/A	0.017	N/A	0.017	0.010	0.07
Br	<0.001	<0.001	<0.001	<0.001	<0.001	N/A
Cd[c]	<0.022	<0.001	<0.015	0.002	<0.003	<0.050
Cl[c]	0.060	<0.080	0.590[a]	<0.080	7.600[b]	N/A
Cr[c]	<0.001	<0.001	0.003	<0.001	<0.003	<0.050
Co[c]	<0.006	<0.001	<0.002	<0.001	<0.003	<0.050
Cu	0.007	0.008	0.004	0.007	0.024	<0.050
F[c]	N/A	<0.001	N/A	<0.001	<0.001	N/A
I	<0.038	<0.001	<0.085	<0.001	<0.001	N/A
Pb[c]	0.002	0.096	0.002	0.040	0.048	0.010
Mn[c]	0.015	<0.001	0.02	<0.001	<0.001	<0.050
Hg[c]	<0.002	0.004	<0.002	0.002	<0.007	<0.010
Mo	0.003	0.040	0.002	<0.001	<0.001	<0.050
Ni[c]	2.000[b]	0.003	<0.001	<0.001	<0.003	<0.050
Se[c]	0.009	<0.001	<0.001	<0.001	<0.007	<0.010
Ag	<0.017	0.007	<0.015	0.002	<0.001	N/A
Sr	0.002	N/A	0.027	N/A	N/A	<0.010
Tl	<0.003	0.055	<0.002	<0.001	<0.003	<0.010
Th	<0.006	<0.001	<0.005	<0.001	<0.001	N/A
Sn	<0.035	<0.001	<0.028	<0.001	<0.001	<0.010
U	<0.009	<0.001	<0.011	<0.001	<0.003	N/A
V	<0.002	<0.001	<0.003	<0.001	<0.001	<0.010
Zn	<0.002	0.170	0.032	<0.001	0.013	<0.050

[a] Values not used in Table 8 calculations due to Sulfinol entrainment.
[b] Suspect values not used in Table 8 calculations.
[c] HAPs listed in Title III, 1990 CAAA.

Table 4. Analyses of Samples of Sulfinol[a]

Element	Sample 1 (ppmw)	Sample 2 (ppmw)	Sample 3 (ppmw)
Al	N/A	0.031	0.020
Sb	0.003	<0.01	0.049
As	0.026	0.111	0.016
Ba	N/A	<0.01	0.027
Be	<0.001	<0.01	<0.005
B	0.23	1.86	<0.005
Cd	0.013	<0.01	0.025
Ca	0.54	0.383	2.2
Cl	361	126	10
Cr	0.56	0.964	0.096
Co	N/A	<0.01	0.029
Cu	0.015	<0.01	0.160
F	94	10.3	1.1
Fe	N/A	2.25	5.3
Pb	<0.01	<0.01	0.053
Mg	N/A	0.153	0.070
Mn	0.45	1.12	0.420
Hg	<0.002	<0.005	<0.005
Mo	0.027	0.036	0.020
Ni	<0.005	<0.01	1.3[b]
P	N/A	0.7	<0.05
K	6.2	3.54	0.5
Se	<0.01	3.85	0.026
Si	N/A	0.624	<0.05
Ag	N/A	<0.01	0.009
Na	9.3	6.44	0.788
Tl	N/A	<0.01	0.008
Th	N/A	<0.01	<0.005
Sn	0.007	<0.01	0.009
Ti	N/A	<0.01	N/A
U	N/A	<0.01	<0.005
V	N/A	0.016	0.086
Zn	0.4	<0.01	0.199

Note: N/A = not analyzed.

[a] Values preceded by < represent the reporting limit for that element in that sample.
[b] Sampled after petroleum coke run.

contained small amounts of arsenic, mercury, and selenium. Suitable control technology is available for removal, encapsulation, and disposal of these trace elements.

Finally, gasification of petroleum coke provided an opportunity to study the fates of nickel and vanadium at concentrations about two orders of magnitude higher than those usually found in coals. SCGP-1 data confirmed that nickel and vanadium concentrations in the purified syngas were below detection (7 and 2 ppbw, respectively). Consequently, gasification of petroleum coke presents no new environmental issues, particularly with respect to HAPs.

B. Investigations into Trace Organics

For SCGP gasification conditions, thermochemical calculations predict that few

Table 5. Analyses of Syngas for Major Components and Organics[a]

H_2O and N_2 free basis	Conc (% vol)	Method
Carbon monoxide	32.2	Off-line GC-FID
Hydrogen	67.8	Off-line GC-FID
Organic compound	**Conc (ppmw)**	**Method**
Methane	80	On-line GC-FID
Ethane	<0.01	Off-line GC-FID
Benzene	ND	Tenax GC-FID[b]
Anthracene	ND	EPA 625[c]
Benzo(a)-pyrene	ND	EPA 625[c]
Fluorene	ND	EPA 625[c]
Naphthalene	ND	EPA 625[c]
Phenanthrene	ND	EPA 625[c]
Phenol	ND	EPA 625[c]
Pyrene	ND	EPA 625[c]

Note: ND = not detected.

[a] Data from Blacksville No. 2 coal.
[b] Detection limit of 0.005 ppbw.
[c] Data from methylene chloride impingers. Detection limit of 0.025 ppbw.

hydrocarbon molecules heavier than CH_4 would be produced. This prediction is supported by the data generated at SCGP-1.

The environmental characterization program at SCGP-1 included a multimedia search for organics.[4-6] Results of this multimedia search indicated that no organics were detected in the slag or flyslag, and only a few ppbw of formaldehyde were occasionally detected in the scrubbing water. As anticipated from thermochemical predictions, no aromatic or polynuclear aromatic hydrocarbons were detected in the syngas (Table 5).

Due to the occasional presence of formaldehyde in scrubbing waters and the desire for lower detection limits, more extensive and longer-term sampling of the syngas was conducted. Only a few trace constituents were detectable at ppbw levels in the syngas as summarized in Table 6. These trace constituents are combustible in a gas turbine. The distribution of two of these trace constituents (i.e., aldehydes and organosulfur species) can shift as a result of downstream reactions, as given in Table 7; however, their total quantity remains extremely low.

C. Investigations into Radionuclides

Due to excellent capture of fine particulate matter in the SCGP cleanup sequence involving dry filtration, water scrubbing, cooling, and multistage solvent contracting, the syngas should be free of radionuclides. This expectation was confirmed by plant balances on gross a and gross b counts and by direct syngas analyses.

Recoveries in the slag and flyslag of gross a and gross b counts from the feed coal

Table 6. Analyses of Trace Constituents of SCGP-1 Syngas

	Carbon bed	GC-FID	Carbon bed	GC-FID	Carbon cartridge	Tenax cartridge	Resin cartridge	Summa canister	Methylene chloride impinger
Coal	**El Cerrejon**	**Drayton**	**El Cerrejon**	**Drayton**			**El Cerrejon**		
Conc factor	**~1000**	**1**	**~1000**	**1**	**~10**	**~10**	**~5**	**1**	**~5**
	Sour syngas (ppmw)				**Purified syngas (ppmw)**				
Trace constituents									
Methane	NA	90	NA	60	NA	NA	NA	NA	NA
Ethane	NA	<0.1	NA	<0.1	NA	NA	NA	NA	NA
Other hydrocarbons (includes some HAPs)	0.006	NA	0.0005	NA	0.060 0.094 0.069	0.003 0.011 0.005	NA	0.012 <0.004 0.004	<0.025
Aldehydes[a]	NA	NA	NA	NA	NA	NA		NA	NA
Formaldehyde							0.015 0.015 <0.015		
Acetaldehyde							0.140 0.040 0.100		
Sulfur compounds	NA		NA		NA	NA	NA		
Carbonyl sulfide[a]		1110 1300 1160		50 3 3				1.0 1.15 1.43	NA
Carbon disulfide[a]		10 10 30		<1 <1 <1				<0.001 <0.001 <0.001	NA
Methyl mercaptan		40 <5 45		<1 <1 <1				<0.01 <0.01 <0.01	NA
Methyl disulfide		NA		NA				0.205 0.280 0.800	NA
Methyl trisulfide		NA		NA				0.295 0.060 0.330	NA
Tetrahydro-thiophene		NA		NA				0.060 0.440 1.2	<0.025
THT, 1,1-dioxide		NA		NA				<0.001 <0.001 <0.001	1.4
Sulfur dioxide		<1		<1				NA NA NA	NA
Nitrogen compounds	NA		NA		NA	NA	NA	NA	NA
Ammonia		<1 <1 2		1 3 <1					
Hydrogen cyanide[a]		<1 <1 <1		<1 <1 <1					

Note: NA = not analyzed.

[a] HAPs listed in Title III, 1990 CAAA.

Table 7. Possible Reactions of Formaldehyde and Methyl Mercaptan

Reaction	Free energy of Rxn (25°C)
$CH_3SH + CH_2O \rightarrow CH_3CH_2O + H_2S$	–11.1 kcal/mol
$4CH_3SH$ + THT, 1,1-dioxide $\rightarrow 2(CH_3)2S_2$ + THT + $2H_2O$	–9.7 kcal/mol (est.)
$2(CH_3)2S_2 + 2H_2S$ + THT, 1,1-dioxide $\rightarrow 2(CH_3)2S_3$ + THT + $2H_2O$	–5.5 kcal/mol (est.)

Note: THT = tetrahydrothiophene. Shell's Sulfinol solvent is a mixture of THT, 1,1-dioxide, water, and amine.

were 98 to 115%. The radioisotopes radium-226, radium 228, and thorium-228 were lower in syngas-exposed sorbent than in the virgin sorbent. Also, estimated radon concentrations (based on mother isotopes on syngas-exposed sorbent) were less than background levels. These results preclude both the means and the need for assigning radionuclide concentrations above background to the syngas.

V. EXPERIMENTAL METHODS

A. Elemental Species

Gas sampling was conducted by means of ice-chilled, aqueous-impinger trains connected to fast-flowing loops of gas streams of interest. Volatile metals were captured in an impinger train using a solution of hydrogen peroxide in nitric acid, followed by an impinger filled with potassium permanganate in sulfuric acid to ensure capture of mercury. A separate impinger train filled with sodium carbonate solution was used to capture halides, sulfides, and selenides. Analysis was obtained by atomic absorption (AA), inductively coupled plasma mass spectrometry (ICP-MS), or ion chromatography (IC) depending on the element.

For longer-term sampling, a series of carbon beds was placed in fast-flowing loops of syngas entering and exiting the acid-gas removal system, and the streams were processed to concentrate elements by three orders of magnitude. Analysis of virgin carbon and syngas-exposed carbon was obtained by proton-induced X-ray emission spectroscopy (PIXE) from which concentrations of elements in the syngas were calculated. For the syngas stream exiting the acid-gas removal system, the sampling port was upstream of a knock-out vessel for entrained solvent. During 1 h sampling, the chance of an event of solvent entrainment was small. During longer-term sampling, events of solvent entrainment did occur. Values that are biased high due to solvent entrainment are noted.

B. Organic and other Inorganic Species

Process-derived solids were extracted with methylene chloride and analyzed for semivolatile organics and formaldehyde. The scrubbing water was analyzed directly for the same compounds and volatile organics. Syngas, both entering and exiting the acid-gas removal system, was passed for 1 h through ice-chilled impinger trains filled with methylene chloride which was analyzed directly for semivolatile organics. Grab samples were also obtained by passing through cartridges filled with Tenax, and then thermally desorbing the material into a gas chromatograph with a flame ionization detector (GC-FID) for analyses of volatile organics. Additional grab samples of syngas were taken directly into high-pressure cylinders; attachment of valving equipped with septums allowed for subsequent laboratory sampling by high-pressure syringes and analyses by GC-FID. Also, the purified syngas was sampled directly into a SUMMA® canister and shipped overnight for analysis by gas chromatography/mass spectrometry (GC/MS). Aldehydes in the purified syngas were sampled by means of dinitrophenylhydrazine-coated, absorbent-filled cartridges followed with detection by high-performance liquid chromatography (HPLC).

For longer-term sampling, a series of carbon beds, both upstream and downstream of the acid-gas removal system, were used to concentrate traces of adsorbable material by about three orders of magnitude. During this time, smaller samples of syngas were also taken through carbon-filled cartridges. Both sets of carbon were then extracted with carbon disulfide and analyzed by GC-FID.

C. Radionuclides

Analyses for radionuclides involved using the long-term, syngas-exposed carbon beds which were analyzed for gross a count, gross b count, radium-226, radium-228, and thorium 228, and these measurements were compared to the virgin carbon.

VI. COMMERCIAL APPLICATIONS

The first commercial application of SCGP will be in an integrated coal gasification combined cycle (ICGCC) power plant which is currently under construction in The Netherlands and scheduled to start up in 1993. The plant is deigned to gasify 2000 TPD of coal in a single gasification train and to produce 253 MW (net) at a thermal efficiency of 41.4%, based on coal higher heating value (HHV).

In the U.S., Shell has participated with EPRI and others in engineering and feasibility studies of coal gasification combined cycle (CGCC) plants for a number of electric utilities. Shell, General Electric, and Air Products recently completed a joint study of SCGP-based CGCC integration schemes assuming two GE 7F gas turbines and two Shell gasifiers. One case from the study resulted in an estimated net power output of 516 MW at a thermal efficiency of 43.6% (HHV basis). For such a plant, the total amount of HAPs in the clean syngas is estimated to be 4.2 tons/year as given in Table 8. Since the majority of this material is carbonyl sulfide which will be combusted in the gas turbine, the actual total emissions of HAPs from the nominal 500-MW, SCGP-based CGCC plant are estimated to be less than 1 ton/year. Similarly, emissions of other trace constituents are estimated to be extremely small based on data given in Table 8.

Table 8. Estimated Amounts of Trace Constituents in the Syngas for a 4000-TPD Shell Coal Gasification Plant[a]

Metals and compounds; halides and compounds	Other inorganics	Organics
Sb, As, Be, Cd, Cr, Co, Pb, Mn, Hg, Ni, Se, HCl, HF	COS, CS_2, HCN	Aldehydes and hydrocarbons
0.7 TPY	3.3 TPY[b]	0.2 TPY[b]
	Non-HAPs	
Aluminum through Zinc	NH_3	Hydrocarbons methyl mercaptan, disulfide, trisulfide,
2.3 TPY	2.9 TPY[b]	1.3 TPY[b]

Note: TPY = tons per year.

[a] 516 MW (net) at 43.6% thermal efficiency (HHV) on Pike County coal.
[b] These species will be converted in the combustion turbine to SO_x, NO_x, N_2, CO_2, and H_2O.
[c] HAPs = hazardous air pollutants, Title III, 1990 CAAA.

VII. CONCLUSIONS

Extensive environmental testing has clearly demonstrated that the syngas from the SCGP contains extremely low levels of HAPs and other trace constituents. Analyses of elemental matter in the process solids, along with monitoring of the clean syngas, show that the majority of the trace elemental matter is tightly bound in the glassy matrix of the inert slag and not present in the syngas. Also, other analyses of the SCGP syngas verify that, even at parts-per-billion levels, only a few organic compounds heavier than methane are detectable. Consequently, emissions of HAPs from a nominal 500-MW, SCGP-based CGCC power plant are expected to be extremely small. Such a facility would not be defined as a major source of HAPs based on Title III of the Clean Air Act Amendments.

REFERENCES

1. Mahagoakar, U. and Krewinghaus, A. B., Shell Coal Gasification Project: Gasification of SUFCo Coal at SCGP-1, EPRI GS-6824, May 1990.
2. Mahagaokar, U. et al., Shell Coal Gasification Project: Gasification of Six Diverse Coals, EPRI GS-7051, November 1990.
3. Phillips, J. N. et al., Shell Coal Gasification Project: Gasification of Eleven Diverse Feeds, EPRI GS-7531, May 1992.
4. Bush, W. V. et al., Environmental Characterization of SCGP-1. I. Gaseous Effluent Streams, 15th Biennial Low-Rank Fuels Symposium, May 22–25, 1989.
5. Baker, D. C. et al., Environmental Characterization of SCGP-1. II. Aqueous Effluent, 6th Annu. Pittsburgh Coal Conference, September 25–29, 1989.
6. Perry, R. T. et al., Environmental Characterization of SCGP-1. III. Solids By-Products, 7th Annu. Pittsburgh Coal Conference, September 10–14, 1990.

Air Toxics Monitoring Issues

Robert M. Mann
Winston Chow

I. INTRODUCTION

Control of toxic substance emissions from industrial sources is required by the Clean Air Act amendments of 1990 (CAAA).[1] To respond to this legislation, the electric utility industry must be able to answer several questions. (1) What is or is not being emitted from a facility? (2) Are methods available to accurately measure the emissions? (3) What risks do specific emissions pose? (4) Can "critical" emissions be controlled? (5) What are the costs for controls? The focus of the Field Chemical Emission Monitoring (FCEM) project is to help answer questions 1, 2, and 4.

Currently the Electric Power Research Institute (EPRI) is working with the electric utility industry to assess the extent of emissions from fossil fuel-fired power plants equipped with both conventional and innovative controls. EPRI's Power Plant Integrated Systems Chemical Emission Studies program (PISCES) has compiled data from literature and private sources to document previous test results for air toxics. Previous monitoring efforts for most power plants have generally focused on collection of data to meet regulatory requirements for criteria pollutants such as particulate matter, SO_2, and NO_x since most of facility controls were designed specifically with those constituents in mind. Control effectiveness for each of 189 hazardous air pollutants (HAPs) listed in the CAAA was not available to utilities when they installed these units. However, information about control effectiveness is crucial should a utility desire to add an emission control system or to modify or replace an existing system.

Information presented in this paper includes (1) data requirements, (2) EPRI's strategy for monitoring HAPs in the utility industry, and (3) data collection and validation methods to ensure accurate emissions and control performance data are obtained. Also, EPRI must determine whether these HAP emissions data are internally consistent for the utility configurations tested and, if so, whether the data are adequate to define the degree of environmental health risk posed by emissions from fossil fuel utilities. Are additional utility configurations required and what additional monitoring data are required for that purpose?

II. DATA REQUIREMENTS

The CAAA requires monitoring of hazardous air pollutants in emissions of select industries and defines actions required if specific target levels are exceeded. Many of the CAAA components can be eliminated from utility consideration since their presence in either the feed to the process or in the outlet streams is not consistent with basic chemical principles. EPRI established a PISCES Program Advisory Committee to focus the air toxics monitoring effort, and consulted its Utility Air Regulatory Group regarding the impact of the new CAAA regulations. Table 1 presents a subset of components selected by EPRI as target hazardous air pollutants.

III. CURRENT TEST PROGRAM

The PISCES program was expanded in 1990 to add a FCEM effort to monitor the levels of selected air toxics in plant streams for fossil fuel-fired power plants. This new information will update and enhance the PISCES database. These components have now been measured in the discharge streams of several power plant facility configurations to assemble data on their presence and distribution in power plant streams and to evaluate the effectiveness of control technologies.

To date, seven facilities have been tested. The objectives of the testing have been to look at the inlet/outlet streams of the boiler and the inlet/outlet streams of control devices, to define stack emissions, and to define the control device removal efficiency for hazardous air pollutants. The project has included (1) the collection of samples at representative stream locations, (2) collection of multiple sample sets, (3) use of specific collection/analysis methods to generate data, (4) use of alternate collection/analysis techniques where problems are encountered, and (5) use of good statistical procedures to provide proper data interpretation.

Most of these components, though emitted in significant annual amounts [tons per year (TPY)], are found at very low concentrations in the fuel and/or in the discharge streams. For instance, the mercury concentration is generally below 0.1 mg/kg (parts per million) in most power plant solid streams. These low levels can result in difficulties in detecting these components in the inlet fuel, in tracking their distribution through the power plant, and finally in defining how well the component is controlled.

To perform the unit characterization, several conditions must be considered before monitoring. The unit operating conditions must represent normal and/or planned operations. Critical process streams must be identified to determine the unit performance. Stream flow rates must be measured to define the conditions of the unit and allow the development of material balances. Monitoring methods must be selected so that valid, complete, and essential data are collected.

IV. MONITORING METHODS

Several different monitoring methods are required to measure trace elements, trace-level volatile and semivolatile organic compounds, and various acid gas constituents. Each method consists of collecting a composite sample of the gas over an

Table 1. EPRI FCEM Project Hazardous Air Pollutant Target List

	Trace Elements	
Arsenic Cadmium Copper Mercury Selenium	Barium Chromium Lead Molybdenum	Beryllium Cobalt Manganese Nickel Vanadium
	Acid-Forming Anions	
Chloride	Fluoride Sulfate	Total P
	Volatile Organic Compounds	
Benzene	Formaldehyde	Toluene
	Semivolatile Organic Compounds	
Polynuclear aromatic hydrocarbons		
Acenaphthene Benzo(a)anthracene Benzo(g,h,i)perylene Dibenzo(a,h)anthracene Indeno(1,2,3-cd)pyrene Pyrene	Acenaphthylene Benzo(a)pyrene Benzo(k)fluoranthene Fluoranthene Naphthalene 2-Methylnaphthalene 7,12-Dimethylbenz(a)anthracene	Anthracene Benzo(b)fluoranthene Chrysene Fluorene Phenanthrene 3-Methylcholanthrene
Additional polycyclic organic matter		
1,2-Diphenylhydrazine 2-Chloronaphthalene 4-Aminobiphenyl Benzidine Dibenz(a,j)acridine	1-Chloronaphthalene 2-Naphthylamine 4-Bromophenylphenyl ether Butylbenzylphthalate Diphenylamine	1-Naphthylamine 3,3-Dichlorobenzidine 4-Chlorophenylphenyl ether Dibenzofuran C-Nitrosodiphenylamine

extended period of time, performing any special on-site tasks to preserve the sample before analysis, returning the sample to the lab, and performing the off-site chemical analyses. Several of the methods have been prepared to use in characterizing solid and hazardous waste and are documented in a reference document generally referred to as SW-846.[2]

The Multiple Metals Train (SW-846 Draft Method 0012) procedure is used to collect and determine particulate and vaporous metallic constituents in gas streams. The digested filter material, probe wash, and impinger solutions are analyzed by ICP/AAS methods. This gas sample collection method also is used to determine particulate loading.

VOST system (SW-846 Method 0030) is used to collect benzene, toluene, and other volatile organic compounds in gas streams. Analysis is conducted on the thermally desorbed resin columns by GC/MS (SW-846 Method 8240).

An impinger train (SW-846 Method 0011) is used to collect formaldehyde and other aldehyde species in gas streams. The method, as applied by the project, is for nonisokinetic sample collection, since aldehydes are assumed to be primarily associated with the vaporous phase of the stack gas. Chemical analysis is by HPLC to determine the dinitrophenylhydrazine derivative of the aldehyde.

Modified Method 5 (SW-846 Method 0010) is used to collect semivolatile principal organic compounds. The particulate filter, gas stream condensate, and vaporous-sorbed organics are extracted with methylene chloride, the solvent concentrated by vacuum distillation, and the concentrate analyzed by GC/MS (SW-846 Method 8270).

Anion-containing acid gases (HCl, HF, SO_2, and H_3PO_4) are collected with an impinger train (variation of EPA Method 5) by nonisokinetic collection of flue gas samples. A buffered impinger solution was selected to allow easy analysis of the sample for chloride and sulfate by ion chromatography (IC). Fluoride and phosphate are analyzed by ion-selective electrode and spectrophotometric methods, respectively, since their concentrations are too low to measure by IC in the presence of significant chloride and sulfate concentrations.

V. METHOD/DATA VALIDATION

Each method used to collect data must be able to provide a valid means of measurement. For the FCEM project this means either providing a stream flow rate, a solids mass flow rate, a chemical concentration of a constituent in the stream or in a single phase of a multiphase stream.

When possible, the FCEM project uses material balances (around the boiler, air pollution control device, the full facility, etc.) to validate data consistency. Other data validation checks include a comparison of ash metal content of inlet and outlet process solids, determination of spike recovery of blank samples, spike recovery of stream samples, and collection and blind testing of standard reference materials (SRM). For example, to validate the chemical results for particulate emissions, a sample of NBS (NIST) fly ash was collected with the multiple metals train and shipped to the lab for analysis. This collection, preparation, and analysis of a SRM of known composition provides a validation of the complete sampling and analysis system. Figure 1 presents the measured results of that check sample analysis for comparison with certified results. This procedural check allows the project to evaluate not only the field or lab collection/analysis procedure, but the preparation, analysis, and reporting steps associated with generating stream results. Although the procedure recovered 97% of the fly ash mass and generally 100 ± 30% of most elements, the results for As, Hg, and Se indicate poor recovery with the existing procedure. Additional work is required to identify the cause of the high recovery of these elements.

Several data quality checks, similar to the one above for elemental determination of the flue gas particulates, can be performed to show whether the application of a method can be validated, the method requires modification, or a completely new method may be required. The following paragraphs present a summary of findings from the current study and recommendations for planning at future sites.

A. Multiple Metals Train

The train has worked well in providing data for a number of elements from a single sample. Limitations observed are as follows:

- Filters used to capture the particulate matter provide a background matrix that can produce interference and bias in the analysis of certain metals. The use of glass-fiber filters results in at least a fivefold increase in certain trace contaminants (As, Ba, Be, Pb, Mn) as opposed to contaminants observed from the use of quartz. This becomes most critical for streams with low particulate loading. Alternate filter materials with low trace contaminants and resulting in minimal background or matrix impact on the sample analysis are desired.
- Mercury performance and capture with the train have not been well documented. Recent methods investigation[3,4] has shown that mercury may not be quantitatively recovered from mercury collection impingers unless specific preparation modifications are made. The FCEM project has identified potential limitations of the current mercury chemical analysis method which, if not addressed by the analyst, will result in incomplete mercury recovery from the vaporous metals collection solution.
- In several instances the level of particulate matter in the gas stream is so high that it produces difficulties when using the normal Method 5 type approach for particulate collection. In those high solids loading cases, either a shorter collection time or a collection device of higher sample capacity is required. The project has used in-stack thimbles in instances of high flue gas particulate loading.

B. VOST Train

Volatile organic compound data for benzene and toluene have been extremely variable and often nonreproducible with the VOST collection/analysis method. In many cases, data have varied by two to three orders of magnitude when collected at a single location. Inlet and outlet data around a control device have not been consistent enough to allow determination of control device performance for either benzene or toluene removal. At one site, accumulation of acidic condensate in the VOST tube prevented the analysis of the samples. A procedural modification was required to neutralize the acidic components; it provided good recovery of surrogate spikes added to samples.

An alternate method is required to identify whether problems come from the sample collection procedure, the analysis, or both. The FCEM project plans to perform a side-by-side performance test of the VOST method and the TO-14 canister method[5] to check the applicability of the VOST method to power plant flue gas.

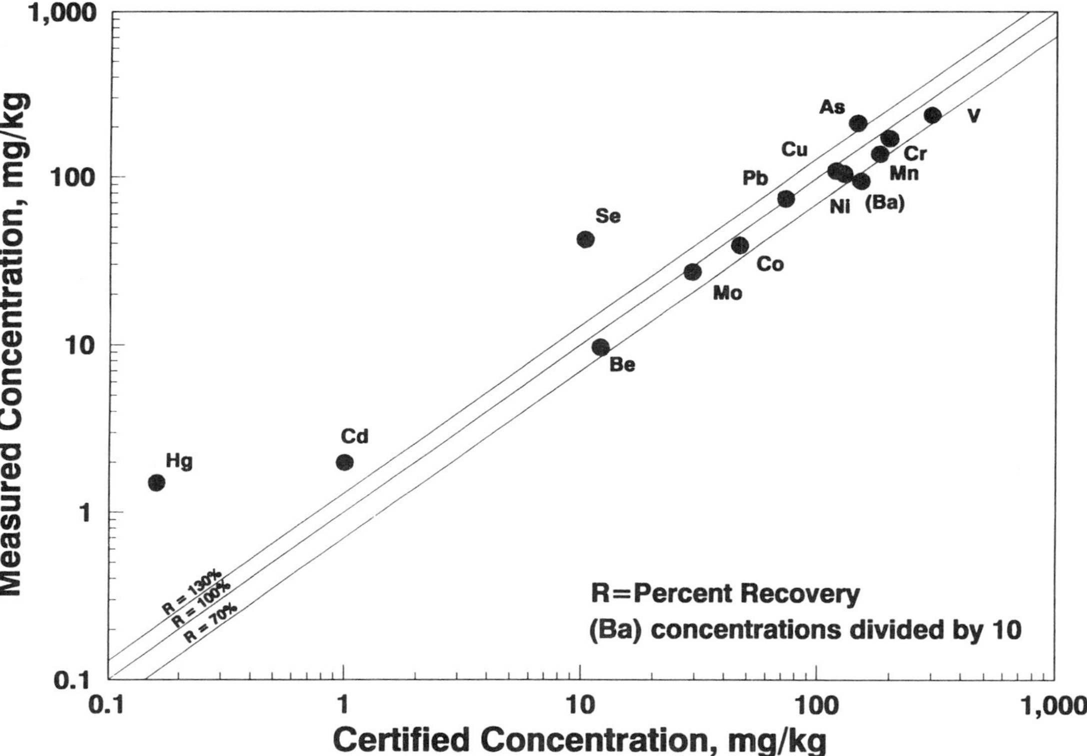

Figure 1. Recovery of trace elements in NBS fly ash collected with the multiple metals train.

C. Aldehyde Train

No problem has been identified in the use of the formaldehyde method (SW-846 Method 0011) for flue gas analysis. High field blank results were identified at one site. The acetone probe rinse must not be performed since it interferes with the method.

D. Semivolatile Organic Train

The Modified Method 5 procedure to determine PNAs and POM has resulted in no significant organic compounds being observed. Although a few semivolatile components were detected in the flue gas, generally all components were at a level near the method detection limit. Although spike addition to field blanks from the collection train have resulted in good surrogate spike recoveries, recovery of precollection spikes to flue gas samples should be determined for method validation.

E. Anions Train

The FCEM project has used an Anions Train to determine the level of acid gases (HCl, HF, SO_2, and H_3PO_4) in the flue gas. Inconsistencies in mass balance results for chlorine and fluorine and the inability to recover phosphate spikes in the samples have forced a reassessment of this train. Although separate and individual gas collection methods exist for all but H_3PO_4, their use is time and labor intensive. At this point, a reassessment of priorities may make it necessary to determine HCl concentrations with a method considered uniquely efficient for its determination. A higher priority for HCl may be warranted because results indicate the potential for high HCl levels in several flue gas streams, and because of the inability of the current method to provide good material balance closures around power plant systems. Although similar difficulties exist for the original method when applied to HF, the level of HF is assumed to be approximately an order of magnitude below the level of HCl based upon the ratio of chlorine to fluorine in the coal.

F. Bulk Solids Collection

Collection and analysis of representative bulk samples from the process inlet and outlet streams are necessary to provide material balance information. This requirement often poses a real challenge to the FCEM program. Most facilities were designed to effectively "handle" the process streams and not necessarily to allow easy subsampling at minimal effort. As an example, it is usually easy to obtain representative coal stream samples that can be referenced in time to the boiler operations; however, similar access to bottom ash and fly ash samples is not always available. Therefore, depending upon stream accessibility, some facilities may be better suited to testing than others.

G. Solids Preparation/Analysis

Metals results for some samples suffer from the high level of detection of existing methods when applied to the complex ash matrix of both the coal and ash samples. In many cases the inability of the analyst to ash the coal sample without losing volatile trace elements limits the results. Although instrumental neutron activation analysis (INAA) has provided concentration data for many elements in coal, the method's limit of detection for several target elements has often been above that required by the project to develop material balances. Additional work is required to validate a coal preparation method such as microwave digestion. Lime, limestone, and FGD solids present matrices different from that of coal ash and may present analytical problems for some elements.

VI. RISK ASSESSMENT REQUIREMENTS

Current monitoring efforts, although unique in providing information, do not necessarily provide it in the format required for risk assessment models. For instance, the current program collects emissions data for arsenic, chromium, and mercury. However, the data do not define whether these elements are present in the most or least toxic forms, or whether the streams will release freely these elements into the environment. Collecting this kind of information requires a different approach and monitoring method. A method must be found that is specific to collection and segregation of the specific toxic component, while making sure that it does not change form during the collection, handling, or analysis steps of its measurement. For example, the analysis of total chromium is frequently performed by either ICP or AAS methods, while chromium (VI) analysis has been performed by reaction with diphenylcarbazide and spectrophotometric analysis of the complex.

Both the State of California's Air Resources Board (CARB) and the U.S. EPA independently have researched and established methods[6,7] to measure chromium (VI) in process emissions since it is assumed to be the chromium form of environmental concern. Separate methods were drafted, were researched by each group, and currently are required for data collection for their specific regulatory compliance reporting efforts. The CARB method collects the particulate matter on a filter while sorbing vaporous metals in an impinger solution train. The EPA method scrubs both particulate and vaporous emissions with a recirculating impinger solution. Although these methods are required for specific applications, their use must be validated prior to acceptance of the collected data. Radian has found that each method has required modifications to provide valid data for the utility industry. The filtration of particulates appears to result in fair to good recovery of pretest spikes when applied to flue gas particulate filters.

Radian testing has shown that SO_2, present in the flue gas, interferes with the alkaline solution collection of chromium (VI) by producing a sulfite or bisulfite ion which reacts with available chromium (VI) during gas-sample collection. When a flue gas sample was obtained for analysis, spiking studies did not recover chromium (VI) when added at levels 100 and 1000 times the EPA method detection limit. Therefore, it will be impossible to measure chromium (VI) in SO_2-laden flue gas using the approved EPA method. However, if properly researched and validated, alternate methods should be available to provide valid data for the utility industry. Also, it is possible that vaporous phase chromium (VI) does not exist in measurable concentrations in the presence of high SO_2 concentrations.

The FCEM project has experienced a similar situation in attempting to determine the arsenic species in flue gas emissions. Although methods exist to determine arsenic species in water,[8,9] validated collection techniques have not been identified that allow arsenic speciation in particulate or vapor phase samples collected from flue gas. Research is required to define and validate such techniques.

Alternate collection procedures have been used to try to determine mercury speciation of flue gas samples. A full understanding of the chemistry of how and where individual mercury species are collected and an understanding of their recovery efficiency is not yet available.

VII. FUTURE UTILITY STUDIES

Other innovative technologies that may require testing are fluidized bed combustion, integrated gasification combined cycle, and NO_x emission reduction systems. Alternate combustion conditions employed in these technologies may change the chemistry of specific constituents in the streams and produce interferences in air toxics measurement.

Finally, EPRI's FCEM project, although not specifically focused to develop test methods, has demonstrated that validation of measurement methods is critical to the present air toxics study. Specific compound spiking in blank and stream samples is required to define whether the collection method and analysis method in combination are functioning properly on each stream to which they are applied.

As utilities in the U.S., Canada, Western Europe, and throughout the world prepare to examine air toxics issues and legislation, additional studies will be performed similar to the one which we describe. EPRI, as a participant in these studies, can provide guidance in study design, study review, and finally in prescribing approaches to monitoring to fill data gaps.

REFERENCES

1. Clean Air Act Amendments of 1990. PL 101-549; November 15, 1990.
2. Test Methods for Evaluating Solid Waste — U.S EPA/OSW, SW-846.
3. Amos, C. K. et al., "Developmental Testing for New Mercury Stack-Gas Sampling and Analytical Methods for Municipal Waste Combustors." Presented at AWMA Interna-

tional Symposium on Measurement of Toxic and Related Air Pollutants, Durham, NC, May 1991.
4. Communication with Frank Wilshire of U.S. EPA RTP, North Carolina, concerning new results in mercury method validation.
5. Winberry, W. T., et al., "Compendium Methods for the Determination of Toxic Organic Compounds in Ambient Air." U.S. EPA/AREAL, RTP, North Carolina, EPA/600/4-89/017, June 1988.
6. CARB Cr(VI) Draft Method 426, from correspondence with Bill Vance, California DHS, Sacramento, CA, 1990.
7. Methods Manual for Compliance with the BIF Regulations, U.S. EPA/OSW, Washington, DC, EPA/530-SW-91-010, December 1990.
8. Fieldman, C., *Anal. Chem.,* 51(6), 664, 1979.
9. Glaubig, R. A. and Goldberg, S., *Soil Sci. Soc. Am. J.,* 52, 536, 1988.

Mercury Speciation in Flue Gases: Overcoming the Analytical Difficulties

Nicolas S. Bloom

ABSTRACT

Several methods for the separation and analysis of the individual chemical species of mercury [Hg^o, Hg(II), CH_3Hg] have been evaluated in the laboratory, and the promising ones subjected to limited field testing. Iodated activated carbon traps were found to be >99% efficient for the collection of Hg^o, and total Hg in stack gases, over a wide range of temperature, Hg concentration, and matrix gases. These traps are simpler and more economical than the complex bubbling systems currently employed for stack-gas monitoring. Bubbling of stack gases through various solutions as a way to separate the oxidized species [Hg(II), CH_3Hg] was investigated as a first stage in tandem with iodated carbon for Hg^o collection. These experiments were less successful, with some solutions (0.1 *N* HCl and 0.1 *N* KCl in the presence of SO_2) showing in-bubbler oxidation of Hg^o, thus falsely inflating the Hg(II) levels observed. Other solutions, such as deionized water, and 0.1 *N* K_2CO_3 were found to promote the in-bubbler reduction of Hg(II) to Hg^o, thus underestimating the true Hg(II) content of the gas stream. A solid sorbant of KCl/soda-lime granules appeared to work well, although it remains possible that on-column reactions occur under actual field-sampling conditions. Methylmercury was effectively trapped and retained with all of these collection options. Methylmercury and Hg(II) were then determined by aqueous phase ethylation, cryogenic GC separation, and cold-vapor, atomic fluorescence detection. High ratios of Hg(II) to CH_3Hg required an extraction step to separate the CH_3Hg, prior to analysis. Limited data for coal and municipal waste burners indicate dramatic short-term changes in the Hg^o/Hg(II) ratio. Little, if any, methylmercury or particulate-phase mercury was detected at these sites.

I. INTRODUCTION

Recent evidence showing elevated fish-tissue mercury levels in remote lakes has stimulated a renewed interest in the magnitude and form of mercury emissions from combustion sources.[1-3] Due to its low boiling point, mercury is volatilized during combustion, and then may leave the stack either in the elemental (Hg^o) or oxidized ($HgCl_2$, HgO, or CH_3HgCl, etc.) form. Previous studies have indicated that, unlike other metals, very little mercury is emitted in the particulate phase.[4-6]

Hg^o has a relatively long atmospheric half-life (≈1 year)[7] and so is relatively well mixed in the global atmosphere before being deposited as a function of atmospheric oxidation-reduction reactions and precipitation.[8,9] By contrast, ionic mercury should have a much shorter atmospheric residence time and so will more effectively be scavenged and deposited locally, near the source.[10] Thus, in assessing the impact of emissions and emission-reduction strategies on the mercury cycle, it is vital that high-quality speciation information be obtained.

Currently, however, no reliable methods exist to measure the chemical speciation of mercury as it leaves the stack. In fact, due to the inability to achieve a closed mass balance for combustion sources, controversy continues even as to the accuracy of current standard methods for total mercury analyses in these complex matrices. Over the past 8 years, several groups have made initial measurements of mercury speciation in stack gases, using operationally defined selective trapping techniques.[11-13] The results of these studies have shown a surprisingly large and variable fraction of the total mercury emissions (30 to 80%) in the ionic form, with some suggestion of a significant level of organomercury species.[13] At this point it remains unclear whether these levels represent actual stack conditions or are the result of species conversion within the sampling train itself.[12] The high reported variability suggests analytical artifact, although the observed sensitivity of gas-phase mercury speciation to temperature and trace flue gas constituents (HCl, SO_2, etc.) may result in large-scale variations in actual species distributions.[11,12]

In this paper, we will present preliminary results from laboratory and field experiments attempting to sort out the actual from the experimentally induced variability in mercury speciation. Although much work still remains to adequately validate any methodology, we are now beginning to see which approaches clearly will not work, as well as those which show future promise.

II. EXPERIMENTAL

A. Reagents and Apparatus

Iodated Carbon Traps. Available from MSA (#459003 Pittsburgh, PA).[14] Each trap contains 0.2 g iodated carbon granules sealed into a 6 mm O.D. × 5-cm long glass tube. Prior to sampling, the sealed ends are broken off, and the tube is connected to the sampling train with Teflon friction-fit connectors. After sampling, tubes are sealed with Teflon end plugs until analysis.

Potassium Chloride/Soda-Lime Traps. 90 g reagent grade 8- to 12-mesh soda-lime [$NaOH + Ca(OH)_2$] is mixed with 10 g KCl and 50 mL H_2O. The mixture is thoroughly stirred until all KCl is dissolved and soaked into the soda-lime granules. After saturation, the granules are dried at 200°C overnight, and then cooled in a desiccator. They are then crushed and sieved to provide a 12- to 24-mesh fraction. The fines are discarded. Traps were made from 10 cm lengths of 6.4-mm I.D. thin-walled Teflon tubing. Each trap contains 1.0 g of KCl/soda-lime held in place with silanized glass wool plugs. The ends of the traps are closed with Teflon plugs before and after sample collection.

Potassium Chloride Solution (0.1 *M*). Dissolve 7.6 g reagent grade KCl to make 1.0 L with deionized water.

Hydrochloric Acid Solution (0.1 *M*). Dissolve 7.9 mL concentrated low mercury HCl to make 1.0 L with deionized water.

Potassium Carbonate Solution (0.1 *M*). Dissolve 14.0 g reagent grade K_2CO_3 to 1.0 L with deionized water.

Gases. Pure gases were high-purity laboratory grade (99.997%). Air was hospital-grade breathing air. Synthetic stack gas was a custom mixture containing 5% O_2, 15% CO_2, 80% N_2, and 1340 ppmv SO_2. All gases were prepurified for Hg by passing through gold-coated sand traps.

Water. Water for dilutions and equipment cleaning was 18 MΩ double deionized.

Bubblers. 150 mL all-Teflon digestion vessel with cap containing two 3.2-mm tubing ports (Savillex, Minnetonka, MN).

Filters. Particulate mercury was collected either on 5-cm long quartz wool plugs in 5-mm I.D. quartz tubes or on 47-mm quartz fiber filters (0.7 mm nominal retention size) in Teflon-coated stainless steel filter holders.

Hg^o Diffusion Cell. Known concentrations of Hg^o in gas streams were provided using a thermostated mercury diffusion cell. The cell consisted of a Teflon bubbler vessel into which was placed a small vial containing 1.0 g of elemental mercury. The vial was fitted with a 10-mm silicon-rubber septum. The entire vessel was then thermostated at 40.0°C, with a constant flow-through of gas. The output of this cell was found to be 1.48 ± 0.05 ng $Hg^o \cdot min^{-1}$ after several weeks equilibriation. Generally, laboratory experiments were run at 0.165 $L \cdot min^{-1}$ flow rates, resulting in gas phase Hg^o concentrations of 9.0 $mg \cdot m^{-3}$.

Stack Sampling Probe. A 2-m section of 12-mm O.D. heavy-wall quartz tubing (now stainless steel) was wound with silicon heat tape over a 60-cm section at one end (Figure 1). This was wrapped with 3 cm of fiberglass insulation, held in place with metal foil tape. The probe temperature was maintained using a proportional controller with type-J thermocouple feedback.

Stack gases were collected into the sampling train with a 6.5-mm O.D. silanized quartz tube inserted through the sampling probe. The probe was put into the stack up to the heated handle (1.4 m). Gases were pulled into the sampling train with a small vacuum pump at 0.75 atm pressure and a flow rate of 0.5 ± 0.1 $L \cdot min^{-1}$. Flow rate and volume sampled were quantified with a digital mass flowmeter/totalizer (Kurz, Monterey, CA). When bubblers were used, they were situated to provide a minimum length of Teflon tubing (<20 cm) between the heated end of the probe and the bubbler. Any condensation in this tube was collected into the first bubbler. Prior to the pump/flow monitor, an iodated carbon trap was placed to eliminate back diffusion of Hg from the pumping system.

Connections and Fittings. Connections were made using either Teflon friction-fit tubing or Teflon compression fittings. All tubing was either 3.2-mm O.D. Teflon FEP or 6.4-m m O.D. silanized quartz.

B. Methods

Analysis. Samples for total mercury were analyzed following wet oxidation, using $SnCl_2$ reduction, dual gold amalgamation, and cold vapor atomic fluorescence (CVAFS) detection.[15] The CVAFS system has an instrumental detection limit of about 0.1 pg Hg. In the case of iodated carbon traps, the entire contents were emptied into a 18.2-mL Teflon vial (Savillex, Minnetonka, MN) and 3.0 mL of 7:3 HNO_3/H_2SO_4 added. The contents were digested in sealed vials for 3 h at 80°C and then diluted to 18.2 mL with 0.02 *M* BrCl.[16]

The KCl/soda lime trap contents (including glass wool plugs) were added to 125 mL of 10% HCl in Teflon bottles and mechanically shaken until the trapping material was dissolved (1 to 2 h). Total mercury in both the dissolved KCl/soda lime solution and in bubbler solutions was determined on 18.2-mL aliquots, oxidized with 0.5 mL 0.2 *N* BrCl. Typically, 0.200 mL of these digestates was analyzed for total mercury, giving a detection limit of about 0.04 mg $\cdot$ m^{-3} Hg in the gas stream — assuming 100-L samples were collected.

Inorganic ionic mercury [also called "acid labile", "reactive", "easily reducible", or Hg(IIa)] in the bubblers and dissolved KCl/soda-lime traps was determined by direct $SnCl_2$ reduction, dual gold amalgamation, and CVAFS detection, with no preoxidation step. Methylmercury was determined on these same samples, following CH_2Cl_2 extraction, using aqueous phase ethylation, cryogenic GC separation, and CVAFS detection.[17]

C. Laboratory Experiments

Laboratory experiments were carried out to assess the potential for species conversion in bubblers and on traps, as well as to observe the useful temperature range of iodated carbon traps. Two bubblers, containing either 100 mL of 0.1 *M* KCl, HCl, or K_2CO_3 solution, were deployed in tandem, and then bubbled overnight (12 to 18 h at 0.165 L $\cdot$ min^{-1}) with Hg^o in both air and argon. This was used to assess the potential of the medium to inadvertently oxidize Hg^o. To assess the potential for the errant reduction of Hg(II), or volatilization/decomposition of CH_3Hg, the first of the bubblers was spiked with a known quantity of these species (typically 5 mg Hg as $HgCl_2$or 0.2 mg CH_3HgCl), and then the sample purged overnight with Hg-free air. Mercury species and concentrations were measured in both bubblers, as well as Hg^o on an iodated carbon backup trap.

For the KCl/soda-lime traps, a similar protocol was employed to observe the oxidation of Hg^o passing through the traps. However, observation of the reduction of

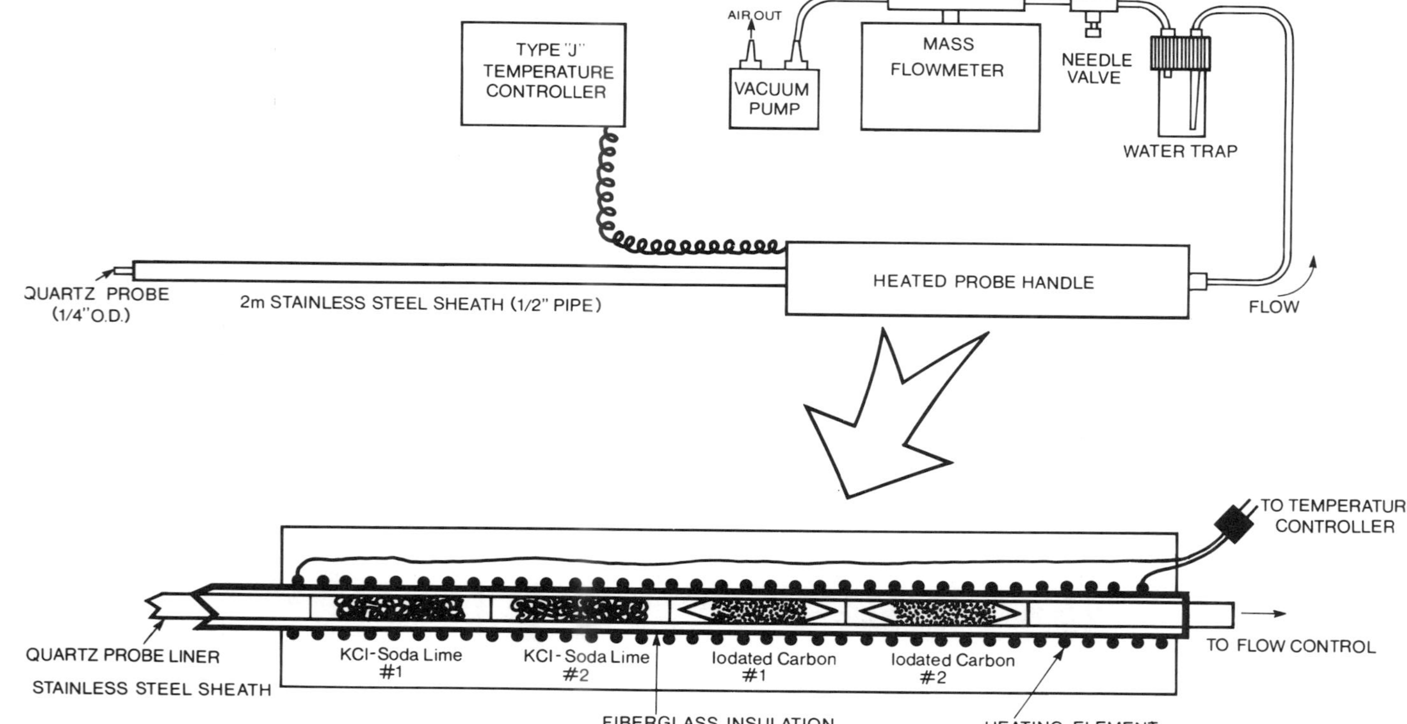

Figure 1. Schematic diagram of heated sampling probe, showing the placement of solid sorbant traps.

ionic Hg on the traps was impossible due to lack of a quantifiable gas phase Hg(II) source at this time. Retention of methylmercury was assessed by the addition of CH_3HgCl in propanol to the glass-wool plug, then evaporating the solvent with carrier gas.

For both KCl/soda-lime traps and the iodated carbon traps trapping effectiveness was investigated at temperatures up to 145°C to simulate the heated sampling probe or direct insertion of traps into the stack.

D. Field Tests

The equipment and methods found to be promising in the laboratory were tested in the field at three sites. Two were coal burning power plants, and one was a municipal waste incinerator. In the first power plant, ($Hg_{tot} \approx 2\ \mu g \cdot m^{-3}$) samples were collected without heated probe, directly into dual-tandem bubblers containing 0.1 M KCl solution. In the second power plant, ($Hg_{tot} \approx 6\ \mu g \cdot m^{-3}$) the heated probe was used with tandem KCl/soda-lime traps. At the waste incinerator ($Hg_{tot} \approx 150\ \mu g \cdot m^{-3}$) both methods were tried at different times. In all cases, iodated carbon traps were utilized to collect the Hg^{o} passing through the oxidized mercury traps. At the waste combustor, tandem iodated carbon traps were also inserted directly into the stack (140°C) to obtain total mercury values to compare with the sum of speciation. The second trap in each tandem set was compared with the trap blank values to indicate the degree of breakthrough from the primary sample trap.

III. RESULTS AND DISCUSSION

Iodated Carbon Traps. Iodated carbon traps were evaluated for mercury retention under a range of temperatures, matrices, and sample volumes. As can be seen from Table 1, under the conditions encountered in our work, no significant breakthrough of mercury (<1%) was observed. One case of breakthrough was observed at the incinerator sampling where water vapor was deliberately allowed to condense in cooled traps. In this case, 8.4% breakthrough was observed, indicating the importance of keeping the traps heated above the dew point.

Even in the case of a 7-h collection in artificial flue gas containing 1340 ppm SO_2, no degradation of the traps' mercury retention capacity was observed. Also, the traps' capacity seem unaffected by temperature, up to at least 145°C. This allows the traps to be used in heated probes, or directly inserted into the hot flue gas, thus minimizing the potential loss of low volatility species [e.g., Hg(II)] on tubing walls.

The trap blanks are very low and consistent between lots. In two sets of blank tests, we observed 1.13 ± 0.28 ng Hg $\cdot$ $trap^{-1}$ (n = 12) for MSA lot #21 and, 14 months later, 0.85 ± 0.76 ng Hg $\cdot$ $trap^{-1}$ (n = 10) for MSA lot #22. The values are probably even closer, given that in more recent analyses we no longer digest the copper retaining screen from the trap together with the granules of carbon. Most of the blank with the iodated carbon trap analysis (>80%) comes from the traps themselves, with the remainder a result of the residual mercury in the digestion reagents.

Bubbler Experiments. Table 2 summarizes the effect of various bubbler solutions on the speciation of mercury. Generally, the oxidation of Hg° to Hg(II) was small (<1%), especially in comparison to observed levels of Hg(II) at field sites (Table 3). However, a single laboratory experiment using artificial flue gas containing 1340 ppm SO_2 did show a considerably greater rate of Hg° oxidation — 7.56%. This is of concern, given the SO_2 content of coal plant flue gases. It also seems somewhat surprising, since aqueous SO_2 is known to reduce Hg(II) to Hg°.[8] Generally, the presence of acid (HCl, SO_2) seems to promote the oxidation of Hg° in solution.

Reduction of added Hg(II) to Hg° shows a characteristic relationship to pH, with greater reduction rates at higher pH. A similar phenomenon has been observed in natural lakes of varying pH.[7] Neutral pH water showed the most dramatic volatilization of Hg(II), where, under argon bubbling, more than 70% was lost after 24 h. This is clearly an indication of the importance of complexing ligands in retaining Hg in solution. Only a single field case with a spiked solution has thus far been evaluated. In this case, a bubbler containing 0.1 *M* KCl was spiked with 500 ng Hg(II) and 43 ng CH_3HgCl, and bubbled with 20 L of gas from a waste incinerator which had first been passed through iodated carbon traps to remove Hg. Of the 543 ng Hg present, 538 ng remained in the bubbler, and a further 45 ng Hg was found on the back-up iodated carbon. Speciation of this sample was unsuccessful so it is unknown if Hg(II) or methylmercury was selectively lost.

The combined potential for either oxidation of Hg° or reduction of Hg(II) ultimately casts doubt upon all of the solutions investigated, although 0.1 *M* HCl and 0.1 *M* KCl are worth further field investigation with spiked and blank bubblers. A point of particular importance is whether the effects seen after 12 to 24 h of bubbling in a laboratory context are still important in the 1- to 2-h field sampling situation.

In all cases, the bubblers maintained their spiked CH_3HgCl content quite well. In the worst case, 5.5% of the CH_3HgCl was volatilized from the KCl solution after bubbling for 24 h. Given both the extremely small fraction of methylmercury found in field samples (Table 3) and that sampling times are much shorter than 24 h, the losses of CH_3HgCl would be unobservable compared to the analytical noise. No investigation has yet been undertaken to observe the spurious production of methylmercury in the bubblers, but again, given the field results, this would certainly be small.

Analytically, the methylene chloride extraction of methylmercury from both 0.1 *M* KCl field bubblers and dissolved KCl/soda-lime traps gave very good recoveries. The mean extraction efficiency for matrix spikes was 85.2 ± 11.1% (n = 7), indistinguishable from the reported value for deionized water of 82 ± 3%.[17] Blanks due to actual methylmercury were quite low (0.21 ± 0.005 ng $\cdot$ m^{-3} (n = 3) for 100-L sample sizes), but the following artifact related to high Hg(II) levels was noted.

The field bubbler and dissolved KCl/soda-lime trap solutions contained orders of magnitude more Hg(II) than methylmercury — a situation never before faced with the aqueous phase ethylation technique. In these samples, we found that a

Table 1. Percent Retention of Hg° by the First of Tandem Iodated Carbon Traps

Temp (°C)	Matrix	Flow rate ($L \cdot min^{-1}$)	Volume (m^3)	(Hg°) $\mu g \cdot m^{-3}$	% Retained
27	Coal gas	0.45	0.10	0.94 ± 0.85	99.2 ± 0.4 n = 4
45	Air	0.40	0.55	3.89	99.8
85	Air	0.40	0.53	3.72	99.9
105	Air	0.40	0.64	11.8	99.9
114	Synthetic gas[a]	0.40	0.22	5.08	100.0
115	Coal gas	0.50	0.04	1.72 ± 1.11	97.9 ± 3.2 n = 8
140	Air	0.40	0.39	11.3	100.0
143	Waste incinerator	0.54	0.02	133 ± 37[b]	99.7 ± 0.3 n = 4
145	Air	0.40	0.50	3.60	100.0

[a] Synthetic "stack gas": 5% O_2, 15% CO_2, 80% N_2, 1340 ppmv SO_2.
[b] Total Hg (approximately 70% Hg(II) + 30% Hg°).

Table 2. Percent of Hg Species Conversion per 24 h Bubbling

	0.1 *M* HCl	0.1 *M* KCl	DDW	0.1 *M* K_3CO_3	Na lime
Air bubbling					
Hg° oxidized	1.34	0.26 ± 0.38	0.20	0.47	0.07 ± 0.08
Hg(II) volatilized	<0.2	0.97	3.60	3.35	—
CH_3Hg volatilized	0.02	5.5	2.0	0.29	—
Argon bubbling					
Hg° oxidized	0.72	0.30	0.40	0.37	—
Hg(II) volatilized	<0.2	0.40	73.4	7.47	—
CH_3Hg volatilized	—	—	—	—	—
Synthetic "stack gas"[a]					
Hg° oxidized	—	7.57	—	—	1.2
Hg(II) volatilized	—	0.14	—	—	—
CH_3Hg volatilized	—	<0.5	—	—	<0.5

Note: Blank values indicate that no data were collected.

[a] 5% O_2 15% CO_2 30% N_2 + 1340 ppmv SO_2.

small fraction of the Hg(II), rather than being ethylated to diethylmercury, was actually methylethylated, by an impurity in the reagent, to methylethyl mercury — the analyte for the ethylation of methylmercury. This resulted in spurious methylmercury observations which were later corrected once the effect was verified. The level of the methylating impurity varies with the batch of sodium tetraethylborate. In some, it is low enough to present no significant peaks at [Hg(II)] levels a 1000-fold higher (CH_3HgCl), but in others, even 10^2 more [Hg(II)] will yield significantly inflated methylmercury concentrations. The use of the methylene chloride preextraction step helps minimize the magnitude of the effect, but does not eliminate it, as up to 10% of the Hg(II) present may be carried over in the extraction. Currently, we are developing a new distillation method,[18] which will effectively separate >99.9% of the Hg(II) from the CH_3HgCl, allowing a very clean ethylation of the distillate.

Table 3. Mercury Speciation at Various Combustion Sites

		Mean mercury concentrations ($\mu g \cdot m^{-3}$ as Hg)[a]				
Site	n	$Hg_{(tot)}$[b]	$Hg_{(part)}$	Hg^o	Hg(II)	CH_3Hg
Coal plant #1	4	1.48	0.015	0.94	0.41	0.0075
S.D.		0.77	0.008	0.85	0.40	0.0006
Waste incinerator	6	146.0	0.10	42.0	104.0	<0.2
S.D.		37.7	0.09	36.5	68.8	—
Total (direct)	4	132.6	—	—	—	—
S.D.		36.5	—	—	—	—
Coal plant #2	5	6.41	0.13	1.26	5.01	0.0012
S.D.		1.80	0.16[c]	0.53	1.74	0.0008

[a] These data represent the results of ongoing analytical research and are presented here for illustrative purposes only. Their absolute accuracy has not yet been verified.
[b] Sum of species, except direct total for waste incinerator.
[c] Hot side of ESP.

KCl/Soda-Lime Traps. Solid phase traps are much more convenient than bubblers to use in the field, and may be inserted into the heated probe, thus minimizing potential losses on the tubing walls leading to the bubblers. Several laboratory results are presented in Table 2. Most significant is the observation that little Hg^o is oxidized to Hg(II) on these traps. In fact, the mean value of 0.07% oxidized in air is for four experiments, conducted at trap temperatures ranging from 27 to 145°C. At 115°C, in synthetic stack gas, only 1% of Hg^o was oxidized and less than 0.5% of CH_3HgCl spiked onto the column was lost in 7 h.

It is difficult to assess the potential for trapped Hg(II) to be volatilized, due to our inability, as yet, to add known quantities in the gas phase. In three experiments where the $HgCl_2$ was added as an aqueous solution and then dried with carrier gas, losses were significant (7 to 55%). However, losses from aqueous solution, especially upon drying, are well known, and so may not be representative of what would happen to adsorbed gaseous species.

The results from tandem trapping at a coal-fired power plant are illustrative. At this site, six tandem sets were collected using a heated (115°C) probe. The results of first and second trap analyses indicate that $81.2 \pm 6.3\%$ of the ionic mercury was found on the first trap. In addition, analyses of the dissolved traps for labile Hg(II) and total mercury indicated $100.2 \pm 3.1\%$ of the mercury present was as Hg(II) (only 0.02% of the ionic mercury was methylmercury). This picture is consistent either with the KCl/soda-lime traps having a 60% efficiency in oxidizing Hg^o (not consistent with lab studies) or a trapping efficiency of at least 82% for Hg(II) with little revolatilization. Further field studies with loaded traps placed behind iodated carbon traps to measure revolatilization rates will be performed to resolve this question.

Precision, Accuracy, and Detection Limits. Mean results for field blanks, as well as the associated detection limits (3 S.D.) for five stack gas species, are presented in Table 4. These values, based upon a 0.10-m^3 sample volume, are orders of magnitude lower than the total Hg concentrations found in even the cleanest power plants, and similar to those attainable for total Hg with EPA standard methods (101-A; multi-metals) using a 2.0-m^3 sample volume. These results, of course, represent only the analytical evaluation of what is found on the traps, and does not consider inaccuracies which may be imposed by the sampling system itself. The trapping efficiencies are based upon a comparison of tandem traps from actual field samples. More work must be done to accurately quantify and optimize the efficiencies for the oxidized species as well as to verify that only trapping, and not species interconversion, is taking place.

Field Results. Mean results from two coal-fired power plants and a municipal waste incinerator used in this study are illustrated in Table 3. These data are reported to illustrate the concentrations used to generate the previous analytical results and to provide general insights into the potentially important and unimportant species in combustion flue gases. Because they all represent active methods development sites, their absolute accuracy is not verified. The plants are listed in the order sampled, over a period of 8 months, and most likely represent increasingly accurate results. This is largely due to refinements in the sampling equipment, as the analytical methods and trapping materials have remained constant throughout this study. Unfortunately, few intercomparison data are yet available for the field work at these sites.

At power plant #1, the results of three runs on the previous day, using the EPA multimetals train, gave significantly higher results: $Hg_{tot} = 6.97 \pm 0.59$ mg · m^{-3}, with 0.91 ± 0.35 mg · m^{-3} in the back-up permanganate traps. In addition, mass balance considerations at this site lead researchers to expect on the order of 12 mg · m^{-3} Hg_{tot}. These results point to losses in the sampling system which, in retrospect, are quite likely at that site given the high levels of oxidized Hg. Due to time constraints in reporting to the site, a heated probe was not used (in fact, a 1-m length of Teflon tubing at ambient temperature was used to connect the probe to the bubblers). The likelihood that ionic Hg was lost in the sampling train is bolstered by the fact that our reported values for Hg^o, which is not surface reactive, (1.03 ± 0.94 ng · m^{-3}) are very similar to those reported from the back-up permanganate bubblers of the EPA method (0.91 ± 0.35 ng · m^{-3}).

At the second coal plant, as well as at the waste incinerator, heated probes were used and rinsings of the probe linings were analyzed, giving better recovery rates. In the incinerator case, this is verified by the measurement of total mercury by direct insertion of iodated carbon traps (no tubing between trap and gas stream) on alternate runs with speciation analysis. Although the Hg emissions were variable over time at this site, the total mercury (133 ± 37 ng · m^{-3}) and the sum at speciation (146 ± 38 ng · m^{-3}) are quite in concordance.

IV. CONCLUSIONS

The results presented here represent preliminary data in a continuing project to develop and verify methods to accurately quantify the mercury speciation in combustion flue gases. The need for such development is clearly evidenced by the thus-far

Table 4. Field Blanks, Detection Limits, and Trapping Efficiencies for Mercury Species

	Mercury concentrations ($\mu g \cdot m^{-3}$ as Hg)				
	$Hg_{(tot)}$	$Hg_{(part)}$	Hg^o	Hg(II)	CH_3Hg
Blanks	0.012[a]	0.001	0.009	0.008	0.00021
S.D.	0.008	0.0004	0.008	0.001	0.00005
N	—	3	10	3	5
D.L.[b]	0.04	0.003	0.04	0.03	0.001
Efficiency	99.9	—	99.9	81.2	>90[c]
S.D.	0.2	—	0.1	6.3	—
N	4	—	21	6	4

[a] Composite of Hg^o and Hg(II). Directly measured $Hg_{(tot)}$, detection limit equals that for Hg^o.
[b] 3 S.D. field blanks, 100-L sample size.
[c] Based on four points where CH_3Hg was barely detected.

poor concordance of mass balance calculations for total mercury emissions with field measurements. The results so far, however, clearly point to the sampling train as a likely source of losses. The unanticipatedly large fraction of surface-reactive oxidized mercury (30 to 100%) found in these stack gases means that greatest attention must be paid to minimizing the contact of sample gas with surfaces before the trapping media. Foremost in this process should be attempts to place the sampling medium (i.e., solid sorbant traps) directly into the stack, at *in situ* temperature, and with no lead-in tubing. To the degree that this is not possible, the following factors should improve collection efficiency of volatile species: increased sampling flow, shortened tubing, heated probes, and silanized glass surfaces.

Once the ability to retain all of the mercury present in the sampling traps is attained, testing must verify that *in situ* mercury speciation is maintained during sample collection, transport, and analysis. This is difficult to do, as realistic means of spiking traps with known quantities of species such as Hg(II) must be developed, and then the requisite studies be conducted either in the field or with a laboratory combustion simulator.

From this work, several results are apparent. First, it is possible, with compact, easily operated equipment, to collect samples for mercury in combustion gases. Iodated carbon is efficient, economical, and easy to use in this respect. Second, no major analytical difficulties exist in adapting current mercury speciation techniques to the problem of flue gas analysis. Finally, while continued investigation of all speciation fractions at a basic level is warranted, it is clear that the most important species on which to concentrate efforts are Hg^o and Hg(II). Contrary to earlier reports,[13] little non-Hg(II) oxidized mercury seems to be emitted by combustion processes.

ACKNOWLEDGMENTS

This work was supported largely through the generous support of the Electric Power Research Institute (Palo Alto, CA) under RPM 3177-03. In addition, I would like to thank Eric Vondergeest for his work in collecting field samples and Sharon K. Goldblatt for her editorial expertise.

REFERENCES

1. Bjorklund, I., Borg, H., and Johansson, K., Mercury in Swedish lakes—its regional distribution and causes, *Ambio*, 13, 118, 1984.
2. Grieb, T. M., Driscoll, C. T., Gloss, S. P., Schofield, C. L., Bowie, G. L., and Porcella, D. B., Factors affecting mercury accumulation in fish in the upper Michigan peninsula, *Environ. Toxicol. Chem.*, 9, 19, 1990.
3. Boutacoff, D., New focus on air toxics, *EPRI J.*, March, 5, 1991.
4. Germani, M. S. and Zoller, W. H., Vapor phase concentrations of arsenic, selenium, bromine, iodine, and mercury in the stack of a coal-fired power plant, *Environ. Sci. Technol.*, 22, 1079, 1988.
5. Lindberg, S. E., Mercury partitioning in a power plant plume and its influence on atmospheric removal mechanisms, *Atmos. Environ.*, 14, 227, 1980.
6. Anderson, W. L. and Smith, K. E., Dynamics of mercury at coal fired power plant and adjacent cooling lake, *Env. Sci. Technol.*, 11, 75, 1977.
7. Fitzgerald, W. F., Cycling of mercury between the atmosphere and oceans, in *The Role of Air-Sea Exchange in Geochemical Cycling*, Brent-Menard, P., Ed., D. Reidel, Dordrecht, The Netherlands, 1986, 363.
8. Munthe, J., The aqueous oxidation of elemental mercury by ozone, *Atmos. Environ.*, 26A, 1461, 1992.
9. Iverfeldt, Å. and Lindqvist, O., Atmospheric oxidation of elemental mercury by ozone in the aqueous phase, *Atmos. Environ.*, 20, 1145, 1986.
10. Lindqvist, O., Jernelöv, A., Johansson, K., and Rhode, H., Mercury in the Swedish Environment, Global and Local Sources — SNV PM 1816, Swedish Environment Protection Agency, S-171-85 Solna, Sweden, 1984.
11. Hall, B., Lindqvist, O., and Ljungström, E., Mercury chemistry in simulated flue gases related to waste incineration conditions, *Environ. Sci. Technol.*, 24, 108, 1990.
12. Lindqvist, O., Fluxes of mercury in the Swedish environment: contributions from waste incineration, *Waste Mgt. Res.*, 4, 35, 1986.
13. Brosset, C., Transport of airborne mercury emitted by coal burning into aquatic systems, *Water Sci. Technol.*, 15, 59, 1983.
14. Moffitt, A. E. and Kupel, R. E., A rapid method employing impregnated charcoal and atomic absorbtion spectrometry for the determination of mercury, *Am. Ind. Hyg. Assoc. J.*, 32, 614, 1971.
15. Bloom, N. S. and Fitzgerald, W. F., Determination of volatile mercury species at the picogram level by low-temperature gas chromatography with cold-vapor atomic fluorescence detection, *Anal. Chim. Acta*, 208, 151, 1988.
16. Bloom, N. S. and Crecelius, E. A., Determination of mercury in seawater at sub-nanogram per liter levels, *Mar. Chem.*, 144, 49, 1983.

17. Bloom, N. S., Determination of picogram levels of methylmercury by aqueous phase ethylation, followed by cryogenic gas chromatography with cold vapor atomic fluorescence detection, *Can. J. Fish Aquat. Sci.*, 46, 1131, 1989.
18. Horvat, M. and Bloom, N. S., Comparison of different isolation procedures for separation of methylmercury from natural water and sediment samples, *Anal. Chim. Acta,* in press, 1993

Chapter 3
Atmospheric Chemistry, Measurements, and Models

Chair: Peter W. Sage, British Coal
Co-Chair: Peter K. Mueller, EPRI

Ambient Air Monitoring for Mercury Around an Industrial Complex

Ralph R. Turner
Mary Anna Bogle

I. INTRODUCTION

Public and scientific interest in mercury in the environment has experienced an upsurge in the past few years, due in part to disclosures that fish in certain waters, which have apparently received no direct industrial discharges (e.g., Everglades National Park), were contaminated with mercury.[1] Atmospheric releases of mercury from fossil fuel energy generators, waste incinerators, and other industrial sources are suspected to be contributing to this problem. Such releases can be evaluated in a variety of ways, including stack sampling, material balance studies, soil/vegetation sampling, and ambient air monitoring.

Ambient air monitoring of mercury presents significant challenges because of the typically low concentrations (nanograms per cubic meter) encountered and numerous opportunities for sample contamination or analyte loss.[2] There are presently no EPA-approved protocols for such sampling and analysis. Our approach was developed with simplicity as a main consideration. Experience[3] showed that commercially available iodated charcoal sampling media could be successfully applied to outdoor ambient air monitoring for Hg^o. High capacity, low blanks, and the ability to trap all important species of mercury with negligible interferences[4] were noteworthy features of iodated charcoal leading to its initial selection.

A. Study Area and Objectives

Elemental mercury was used in large quantities at a nuclear weapons plant in Oak Ridge, TN, between 1950 and 1963 in a process similar to chloralkali production. Soil and water contamination with mercury were known to be present at the facility; however, outdoor ambient air contamination had not been investigated prior to the present study. In addition, one large building (9201-4) still contained original process equipment with mercury residuals. The objectives of this study were to establish a monitoring network for mercury which could be used (1) to demonstrate whether or not human health and the environment was being protected and (2) to establish a baseline against which to compare the effects and effectiveness of future decontamination and decommissioning activities at the facility.

II. METHODS

A. Monitoring Sites

Four ambient air sampling stations for mercury were established in 1986 along a line roughly parallel to the predominant up- and down-valley wind directions (Figure 1). Additional sites (Stations 5 and 6) were operated temporarily further east and further west of the facility during parts of 1987, 1988, and 1989 (Figure 1). Current sources of airborne Hg^o were suspected to be (1) fugitive emissions from the large building (9201-4) still containing process equipment, (2) Hg-contaminated soil throughout the facility, and (3) a coal-fired steam plant adjacent to Building 9201-4. Stations 2 and 3 were located immediately southwest and southeast of Building 9201-4 and the steam plant. Stations 1 and 4 were located at greater distances from these sources and near the perimeter of the facility area.

B. Air Sampling

Mercury was collected by pulling ambient air sequentially through a Teflon filter, through a flow-limiting orifice, and finally through an iodated charcoal sampling tube (Mine Safety Appliances, Part No. 459003). Figure 2 illustrates the mercury collection assembly. The flow-limiting orifice was used to restrict air flow through the collection system to approximately 1 L/min or less. Particulate mercury was collected for 28 days on the Teflon filter, while gaseous mercury was collected for 7 days in the charcoal absorber prior to replacement.

The use of iodated charcoal to collect mercury vapor was first described by Moffitt and Kupel[5] and has been evaluated by others.[4,6-8] These authors reported that iodated, activated charcoal efficiently absorbs elemental mercury vapor and most other volatile species of mercury. The capacity of the 150-mg iodated, activated charcoal tubes used in this study is 0.5 to 1.0% of the charcoal weight (75 to 150 μg of mercury). Lindberg[8] observed insignificant breakthrough at up to 20 μg total mercury loading per tube.

C. Analysis

Mercury collected on the filters and charcoal was analyzed by cold vapor atomic absorption (CVAA) spectrophotometry after digestion in nitric-perchloric acid.[9] The separate charcoal sections in the tubes were combined for chemical analysis. Typically, the charcoal digests were diluted to 50 mL but only up to 8 mL could be used per analysis due to matrix suppression, detected by spiking, at larger analyte volumes. The analytical detection limit by this method as applied in the Oak Ridge National Laboratory (ORNL) Environmental Analysis Laboratory, which performed all the analyses, was 0.5 ng. Tube blanks were measured with each new lot number of tubes

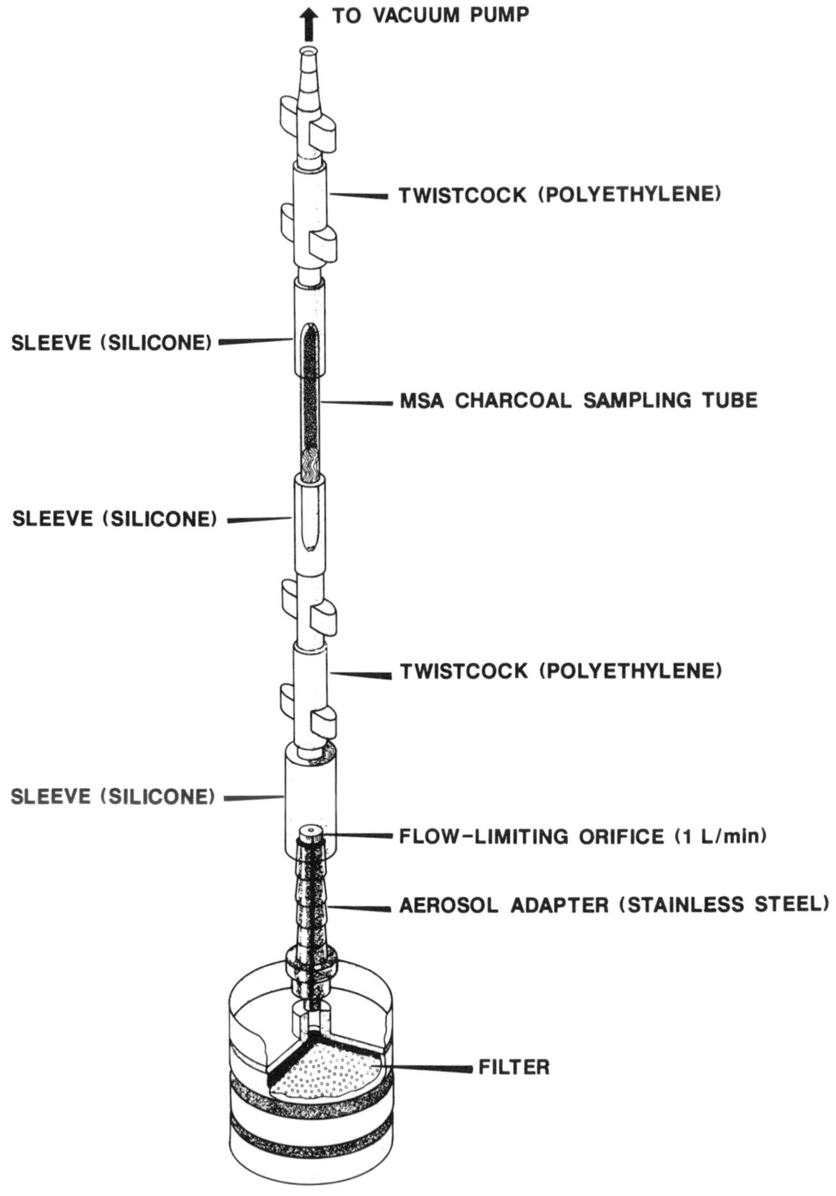

Figure 1. Locations of ambient air monitoring sites for mercury and the 1988 annual wind rose.

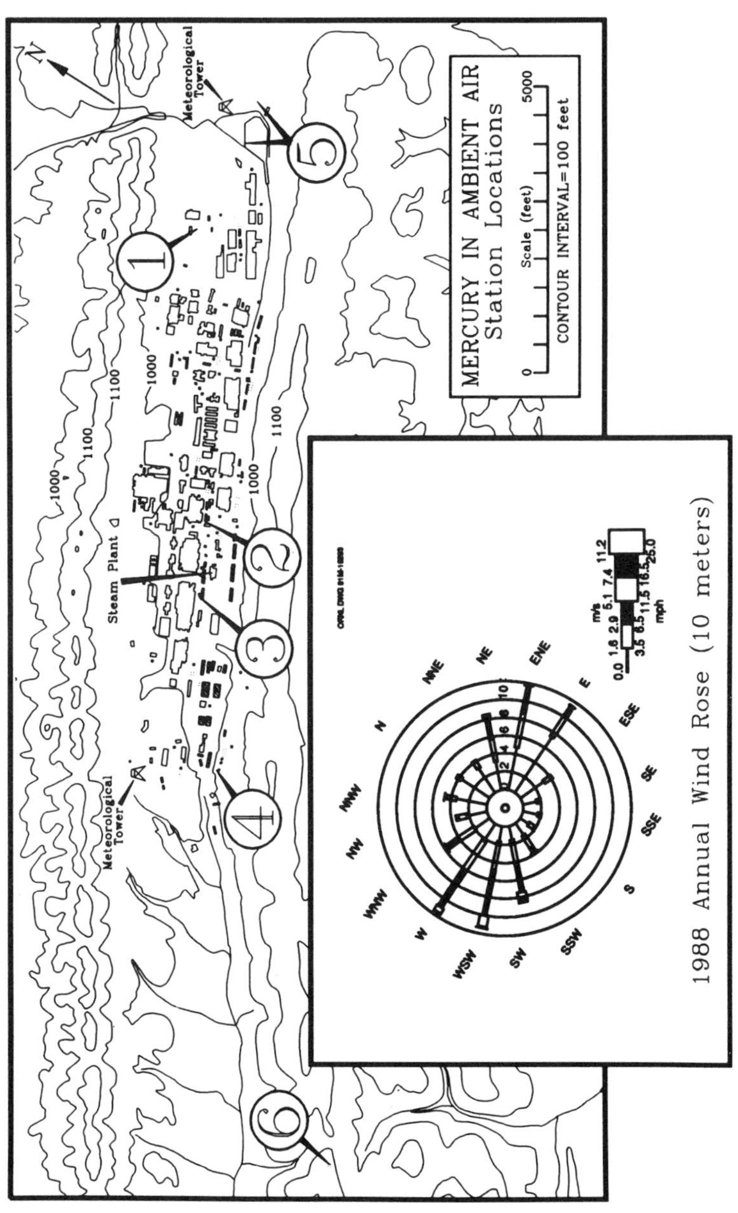

Figure 2. Assembly for ambient air mercury sampling.

and with every other batch of 24 to 30 tubes analyzed. Charcoal tube blanks were typically 1 to 2 ng, while filter blanks were always reported as <1 ng per filter. A private laboratory (Brooks Rand, Ltd., Seattle, WA) verified our charcoal tube blanks by digesting ten tubes in sulfuric acid with analysis by two-stage gold amalgamation and atomic fluorescence spectrophotometry.[10] Blanks varied from 1.11 to 3.70 ng per tube and averaged 1.82 ng per tube. The same laboratory loaded one tube with 18 μg of mercury without observing breakthrough.

D. Data Reduction and Quality Assurance

Average air concentration during the sample collection period (7 days for vapor mercury, 28 days for particulate mercury) was calculated by dividing the total quantity of mercury collected on the charcoal and filters by the total volume (uncorrected to standard temperature and pressure) of air sampled during the sampling period. The volume of air sampled was calculated as the average of flow measurements, determined with a Gilmont Size No. 2 rotameter, taken at the beginning and end of each 7-day period. Rotameter readings varied by 3% or less from week to week except when leaks developed (rarely), an orifice clogged, or a pump failed (rarely). Experience has shown that problems (rare) maintaining constant air flow were almost always traceable to a partially clogged critical orifice. Ordinarily, the upstream Teflon filter protects the orifice against dust invasion and the charcoal media causes very little pressure drop. We also learned that attempting to place the charcoal tube upstream of the critical orifice without protecting the orifice with an in-line filter can lead to flow reduction if any fine particulate charcoal escapes from the tube. Beginning in February 1988, hourmeters were installed in line with the air sampling pumps to record the number of hours of actual pump operation. This allowed for a more accurate determination of the volume of air sampled in case of power outages (rare). Whenever doubt existed as to pump running time or air flow rate, no average mercury concentration was recorded for the week. Although we have not employed mass flow-controlled air pumps, this modification would represent an improvement and be required if concentrations corrected to STP are desired. No replicate air samples were collected using the charcoal tubes during this study. However, Lindberg et al.[11] found the average measurement precision to be ±6% at concentrations of 4 to 15 ng/m^3 using the same method at Station 6 in 1990.

Performance of the charcoal tubes compared favorably with a modified version of EPA Method 101 (40 CFR 61, App. B) which employs impingers filled with iodine monochloride. In this comparison charcoal tubes and impingers were run simultaneously in a building (9201-4) with air concentrations approaching the OSHA threshold limit value of 50 μg/m^3. Additional comparisons between charcoal tubes and gold-plated quartz absorbers[12] analyzed by atomic fluorescence spectrophotometry[10] were also favorable.

Full descriptions of all methods and quality assurance results are given in Turner et al.[13]

Table 1. Annual Results of Weekly Ambient Air Monitoring for Gaseous Mercury (1986 through 1990)

Site	Year	N	Gaseous mercury concentration (μg/m^3)		
			Minimum	Maximum	Average
Station 1	1986	34	0.003	0.058	0.011
(east end of plant)	1987	52	0.001	0.033	0.009
	1988	52	0.003	0.036	0.010
	1989	52	0.003	0.012	0.006
	1990	51	<0.001	0.018	0.006
	1986–1990	241	<0.001	0.058	0.008
Station 4	1986	27	<0.001	0.034	0.017
(west end of plant)	1987	52	0.007	0.067	0.032
	1988	52	0.007	0.407	0.041
	1989	52	0.006	1.187	0.143
	1990	50	0.002	0.025	0.011
	1986–1990	233	<0.001	1.187	0.053
Station 3	1986	31	0.033	0.197	0.108
(SW of	1987	52	0.044	0.465	0.174
Building 9201-4)	1988	51	0.028	0.340	0.137
	1989	52	0.024	0.250	0.101
	1990	51	0.001	0.277	0.068
	1986–1990	237	0.001	0.465	0.119
Station 2	1986	15	0.026	0.137	0.070
(SE of	1987	52	0.036	0.226	0.109
Building 9201-4)	1988	52	0.017	0.384	0.097
	1989	51	0.017	0.206	0.072
	1990	51	0.018	0.162	0.071
	1986–1990	221	0.017	0.384	0.086
Station 5[a]	1987	20	0.006	0.039	0.016
	1988	52	0.004	0.412	0.046
	1989	37	0.002	0.009	0.004
	1987–1989	109	0.002	0.412	0.026
Station 6[b]	1988	47	0.002	0.016	0.006
(Control site)	1989	47	<0.001	0.015	0.005
Total	1988–1989	94	<0.001	0.016	0.006

[a] Site discontinued September 19, 1989.
[b] Site discontinued October 31, 1989.

III. RESULTS AND DISCUSSION

The annual maximum, minimum, and average concentrations of gaseous mercury in air at the six sampling locations are summarized in Table 1. Figure 3 displays the time trends in gaseous mercury concentration in air at each site. The data indicate that on-site gaseous mercury concentrations have been, with one exception (Station 4, October 10, 1989 to October 17, 1989, 1.18 mg/m^3), well below the value derived from the U.S. Environmental Protection Agency's (EPA) National Emission Standard for Hazardous Air Pollutants (NESHAP) for mercury in ambient air (1.0 mg/m^3,

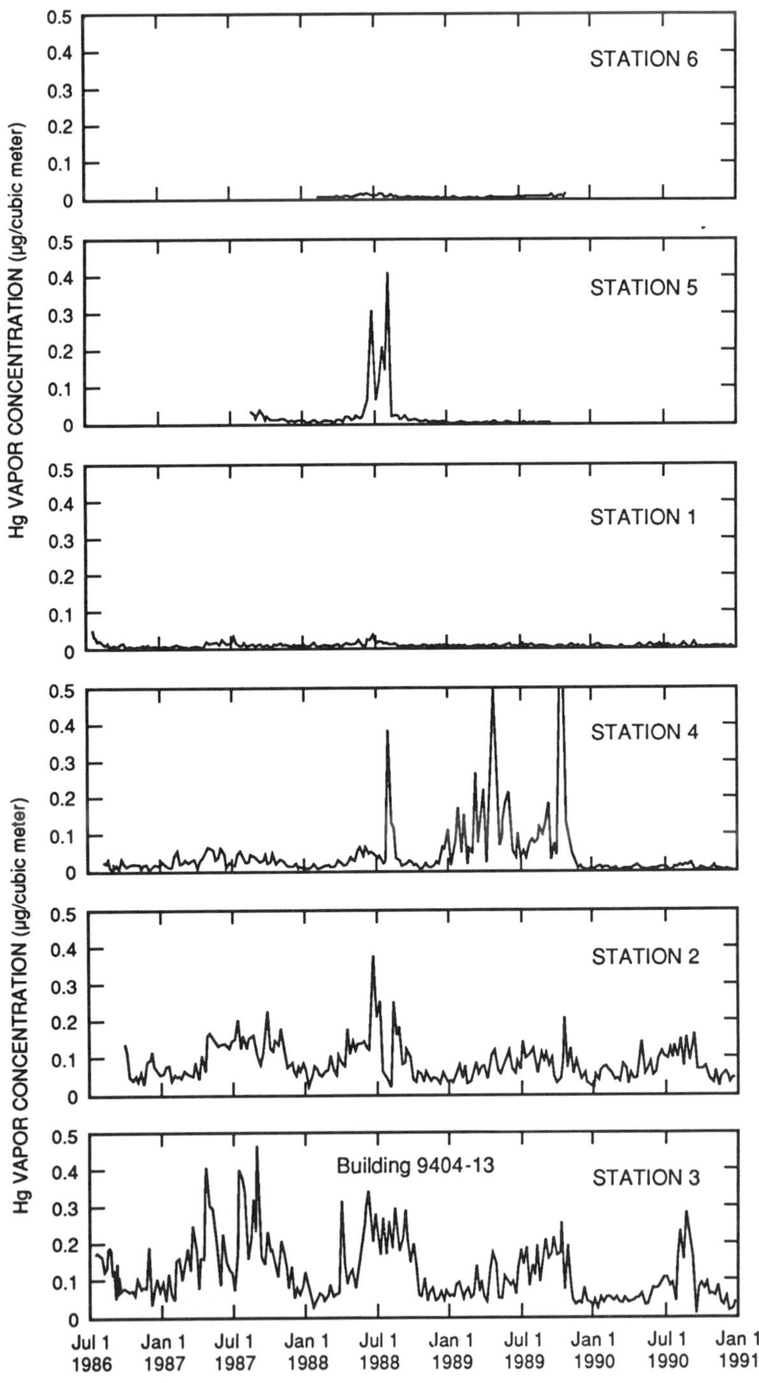

Figure 3. Time trends in gaseous mercury concentrations (7-day averages).

Table 2. Results of Monthly Ambient Air Monitoring for Particulate Mercury (July 1986 through April 1989)

Site	Sampling period	N[a]	Particulate Hg concentration (μg/m3)		
			Maximum	Minimum	Average
Station 1 (east end of plant)	7/18/86–4/18/89	34	0.00079	0.00001	0.00006
Station 4 (west end of plant)	8/12/86–4/18/89	33	0.00068	0.00002	0.00007
Station 3 (SW of Building 9201-4)	7/15/86–4/18/89	34	0.00084	0.00002	0.00018
Station 2 (SE of Building 9201-4)	9/23/86–4/18/89	33	0.00039	0.00002	0.00011
Station 5	8/19/87–4/18/89	22	0.00082	0.00001	0.00024
Station 6 (control site)	2/9/88–4/18/89	15	0.00007	<0.00001	0.00003

[a] N = number of observations.

30-day average) and the Office of Health and Safety Administration (OSHA) industrial hygiene workplace standard of 50 mg/m^3 (8-h average). The monitoring sites (Stations 2 and 3) located southeast and southwest of the steam plant and Building 9201-4 have usually shown the highest concentrations of gaseous mercury among the five sites located in or on the perimeter of the facility, although there are notable exceptions. As would be expected, the lowest concentrations of gaseous mercury were measured for the control site (Station 6) located about 2 miles west of the facility. Concentrations of particulate mercury (Table 2) have consistently been <1 ng/m^3 at all sites. Although collection of particulate mercury samples was continued after April 1989, no analyses have been performed.

Except for the site (Station 4) at the west end of the facility, results (Table 1) indicate significant decreases (Student's *t*-test at the 1% level) in annual means for ambient mercury vapor measured at the plant sites during 1989 and 1990 when compared with the means for 1986 through 1988. There are several possible explanations for this trend. The decrease in ambient mercury recorded at these sites in 1989 and 1990 may be related to the 80% reduction in the tonnage of coal burned at the steam plant beginning in 1989. Prior to 1989, the completion of several major engineering projects, involving extensive earth moving and storm sewer cleaning, may have temporarily elevated mercury air concentrations because of disturbance of mercury-contaminated soil and sediment. Lastly, the period from 1986 through 1988 was characterized by drought conditions which may have increased emission rates of mercury from contaminated soils in the plant area. However, unlike the three other Y-12 Plant sites, the west-end site (Station 4) showed a significant increase in annual mercury vapor concentration from 0.030 mg/m^3 from 1986 through 1988 to 0.143 mg/m^3 in 1989. This increase is thought to be related to the widespread construction and accompanying earth-moving activities undertaken at the west end in 1989, resulting in increased mercury vaporization from newly exposed contaminated soils.

In 1990, mercury vapor levels at this site decreased to an annual average of 0.011 mg/m^3, the lowest level recorded for this site since the monitoring program was established. As might be expected, average ambient mercury levels (0.005 mg/m^3) at the control site (Station 6) showed no change between 1988 and 1989.

Mercury concentrations in air at the control site (Station 6) have ranged from 0.001 to 0.016 mg/m^3, with the highest values occurring during the summer months. Background ambient air mercury concentrations for nonmineralized and nonurbanized areas of the continental U.S. are reported to range from 0.003 to 0.009 mg/m^3.[14] Thus, the average value for Station 6 (0.006 mg/m^3) is quite typical of "background" continental air and demonstrates that the site was a good choice as a control site. Lindberg et al.[11] have performed a more detailed analysis of the data for this site and could not conclusively establish any effects of mercury emissions from the weapons plant although measured concentrations were higher during periods of northeasterly wind flow.

The control site (Station 6) was established in part to take advantage of other measurements, such as sulfur dioxide, which are routinely monitored by others [National Oceanic and Atmospheric Administration/Atmospheric Turbulence and Diffusion Laboratory (NOAA/ATDL)]. The site has experienced occasional fumigations from the steam plant, as evidenced by elevated sulfur dioxide (SO_2) concentrations during meteorological conditions favorable to plume transport from the steam plant stack located 2 miles northeast of Station 6. However, an analysis of the data did not show the expected correlation between SO_2 and mercury at the site. The long integration time (7 days) for mercury sampling probably precluded detecting a short-term event such as a fumigation (generally an overnight phenomenon of a few hours' duration).

Mercury concentrations in air at Stations 2 and 3 exhibited the highest average concentrations (0.119 and 0.086 mg/m^3, respectively) among all sites. Concentrations were typically higher during spring and summer than during the cooler periods (Figure 3). As noted earlier, these monitoring sites are located, respectively, southwest and southeast of Building 9201-4. The sites represent the nearest outdoor monitoring points for any fugitive emissions from this building and were deliberately located east and west of this building to be within the plume of this building whenever the wind was blowing up or down valley, the most common directions (Figure 1). Air monitoring data (unpublished) collected inside this building have shown a strong temperature dependence, with the highest concentrations of mercury occurring in July and August. Because of the very complex terrain created by the buildings in the weapons plant, it is not clear whether the outdoor monitoring points intercepted any mercury plume from this building.

Data interpretation for Stations 2 and 3 is further complicated by the fact that the steam plant is located immediately south of Building 9201-4 and emits approximately 35 kg/year of mercury vapor (assuming 70,000 kg of coal are burned per year with 0.5 mg Hg/kg coal) through stacks that are only 100 ft tall. Beginning in December 1988, the steam plant began using natural gas for 80% of its steam production. This reduction in coal usage would certainly reduce the emission of mercury vapor from the steam plant because natural gas contains much less mercury than coal on a unit heat value basis. Fugitive and stack emissions from the steam plant may account for some of the elevated mercury concentrations in ambient air at the monitoring points at Stations 2 and 3 or elsewhere as suggested by the significant decreases in annual averages after coal usage was reduced. However, because much more coal is burned in the winter, the higher summer concentrations at the ambient air monitoring stations are inconsistent with an exclusive coal combustion source for the mercury and suggest the importance of volatilization of Hg^0 from contaminated soils or buildings.

IV. CONCLUSIONS

The data for mercury in ambient air at the weapons plant suggest that the environment and human health have been protected from releases of mercury to the atmosphere during the monitoring period. With only one exception, the highest observed concentrations (7-day average) at any monitored site have been <50% of the NESHAP criterion (1.0 mg/m^3) and <1% of the OSHA industrial hygiene standard of 50 mg/m^3. These criteria (1.0 and 50 mg/m^3) are intended to protect both off-site human populations that are exposed continuously and on-site workers exposed for 8-h work shifts 40 h/week, respectively. The data do show that ambient air mercury concentrations in the plant area are elevated well above natural background and may have reached greatly elevated concentrations for short periods in very localized areas. Although the present monitoring program is sensitive to fluctuations in levels of ambient mercury caused by local disturbances and operational changes at the facility, it is inadequate to pinpoint areas within the facility where mercury concentrations may be greatly elevated or to determine the magnitude and duration of these excursions. Additional monitoring sites operated on a weekly integration period would not improve this situation. Rather, a program aimed at obtaining air samples at short integration times (e.g., 8 h day/night) at a large number of sites during hot weather would more likely pinpoint sources and yield better estimates of the magnitude and duration of excursions. If available, the use of a real-time monitor with sufficient sensitivity to quantify Hg^0 concentrations of <<1 mg/m^3 would provide an even better evaluation of sources, magnitudes, and durations. At present, no such instrument is known to be commercially available.

ACKNOWLEDGMENT

This work was funded by the Environmental Management Department, Health, Safety, Environment, and Accountability Division of the Oak Ridge Y-12 Plant which is managed for the U.S. Department of Energy by Martin Marietta Energy Systems under contract DE-AC05-84OR21400. Publication No. 4092, Environmental Sciences Division, Oak Ridge National Laboratory.

REFERENCES

1. Raloff, J., Mercurial risks from acid's reign, *Science News*, 139, 152, 1991.
2. Lindquist, O., Johansson, K., Aastrup, M., Andersson, A., Bringmark, L., Hovsenius, G., Hakanson, L., Iverfeldt, A., Meili, M., and Timm, B., Mercury in the Swedish environment — recent research on causes, consequences and corrective methods, *Water, Air, Soil Pollut.*, 55, 1, 1991.
3. Lindberg, S. E. and Turner, R. R., Mercury emissions from chlorine production solid waste deposits, *Nature*, 268, 133, 1977.
4. Braman, R. S. and Johnson, D. L., Selective absorption tubes and emission techniques for determination of ambient forms of Hg in air, *Environ. Sci. Technol.*, 8, 996, 1974.
5. Moffitt, A. E., Jr. and Kupel, R. E., A rapid method employing impregnated charcoal and atomic absorption spectrophotometry for the determination of mercury, *Am. Ind. Hyg. Assoc. J.*, 32, 614, 1971.
6. Schroeder, W. H. and Jackson, R., An instrumental analytical technique for vapor-phase mercury species in air, *Chemosphere*, 13, 1041, 1984.
7. Lindberg, S. E., Mercury partitioning in a power plant plume and its influence on atmospheric removal mechanisms, *Atmos. Environ.*,14, 227, 1980.
8. Lindberg, S. E., Author's reply 'Mercury partitioning in a power plant plume and its influence on atmospheric removal mechanisms', *Atmos. Environ.*, 15, 631, 1981.
9. Feldman, C., Perchloric acid procedure for wet-ashing organics for the determination of mercury (and other metals), *Anal. Chem.*, 46, 1606, 1974.
10. Bloom, N. S. and Fitzgerald, W. F., Determination of volatile mercury species at the picogram level by low temperature gas chromatography with cold-vapor atomic fluorescence detection, *Anal. Chim. Acta*, 209, 151, 1988.
11. Lindberg, S. E., Turner, R. R., Meyers, T. P., Taylor, G. E., Jr., and Schroeder, W. H., Atmospheric concentrations and deposition of Hg to a deciduous forest at Walker Branch Watershed, Tennessee, USA, *Water, Air, Soil Pollut.*, 56, 577, 1991.
12. Fitzgerald, W. F. and Gill, G. A., Subnanogram determination of mercury by two-stage gold amalgamation and gas phase detection applied to atmospheric analysis, *Anal. Chem.*, 51, 1714, 1979.
13. Turner, R. R., Bogle, M. A., Heidel, L. L., and McCain, L. M., Mercury in ambient air at the Oak Ridge Y-12 Plant, Y-12 Report Y/TS-574, Oak Ridge, TN, July 1986 through December 1990.
14. McCarthy, J. H., Jr., Meuschke, J. L., Ficklin, W. H., and Learned, R. E., Mercury in the Atmosphere, in *Mercury in the Environment,* U.S. Geological Survey Prof. Paper 713, Washington, DC, 1970, 37.

Sources and Chemistry of Chlorides in the Troposphere: A Review

Pradeep Saxena
Peter K. Mueller
Lynn M. Hildemann

I. BACKGROUND

Inorganic chlorides are ubiquitous in the troposphere. For instance, the oceans emit a large amount of particles containing chloride salts. However, in comparison to ambient sulfate and nitrate concentrations and anthropogenic emission rates of their precursors, chloride concentrations and emission rates in continental environments are much smaller in general. Thus, chlorides have been taken to be of far less importance to air quality issues such as acidic deposition, ambient concentrations of particulate matter and acidity, and visibility impairment. Consequently, the presence of chlorides in the continental troposphere has sparked few scientific enquiries. The U.K. is an exception where much work has been done to characterize chloride emissions and atmospheric chloride levels. Anthropogenic hydrogen chloride (HCl) emissions in the U.K. are the highest in western Europe and, due to the high chloride content of English coal, the HCl/sulfur dioxide (SO_2) emission ratio for the U.K. is much higher than that for the U.S.[1,2]

The recent emergence of the following issues provides impetus for furthering our knowledge of atmospheric chloride compounds:

1. *Air toxics.* This is the most important issue because the Clean Air Act Amendments of 1990 designate HCl as a hazardous air pollutant (HAP or an "air toxic"). To be able to relate emission rates of HCl to ambient concentrations and human health risks requires an understanding of the formation and removal mechanisms in the atmosphere as well as its toxicity. Moreover, because in the atmosphere chloride salts are converted into HCl and vice versa (as discussed below), the health hazards of anthropogenic HCl emissions ought to be assessed within the context of the total HCl and chloride emissions from all sources.
2. *Atmospheric acidity, aerosols, and visibility.* Comparing maximum 24-h concentrations (in moles per cubic meter of air volume) from a 1-year data set for the Los Angeles area, HCl concentrations are 30 to 120% of HNO_3 concentrations.[3] Whereas HNO_3 — the predominant gaseous inorganic acid in the Los Angeles area — is produced from atmospheric transformations of NO_x emissions, no large anthropogenic sources of HCl are known to exist in the area. It is widely

accepted that the reactions of nitric and sulfuric acid with sea-salt aerosols generate HCl in coastal environments.[4-9] Given that chloride salts are present not only in sea salt but also in primary particulate matter emissions from natural (e.g., soil dust) and anthropogenic sources, the importance of HCl displacement reactions to ambient acidity and aerosol properties needs to be assessed. For some areas, an adequate representation of gas- and particle-phase chlorides may be necessary in air quality models when these models are used to address atmospheric acidity and visibility.

3. *Climate and cloud condensation nuclei (CCN).* It has been postulated that sulfate aerosols — formed by the oxidation of oceanic dimethyl sulfide (DMS) and/or anthropogenic SO_2 — are important contributors to CCN, which influence the earth's radiation budget and, hence, climate.[10-13] Because chlorides are ever present in marine environments, their presence and interactions, particularly with sulfur compounds, ought to be considered in delineating the linkage between climate and CCN. For instance, the oxidation of SO_2 — an intermediate product of DMS-to-sulfate conversion[14] — can be catalyzed by chloride ions.[15]

This paper presents an overview of

- Emissions of HCl and other chlorides in the U.S.
- Reported ambient and precipitation concentrations
- Chemistry of chlorides in the troposphere

This information is synthesized to estimate the relative importance of anthropogenic HCl gas and particulate chloride salt emissions and to explore the potential hazards of ambient HCl as an air toxic.

Here "HCl" refers to gaseous hydrogen chloride, and "chloride", to chloride ions or inorganic salts in the solid or aqueous phase.

II. EMISSIONS

The combustion of chlorine- or chloride-containing fossil fuels (e.g., coal and oil) and the incineration of chlorine- or chloride-containing refuse (e.g., paper, plastics such as polyvinyl chloride) are the two major anthropogenic sources of HCl.[1,6,16,17] HCl manufacturing, operating automobiles that run on fuels containing lead antiknock compounds, and launching of solid-fueled spacecraft are other known anthropogenic sources.

Table 1, taken from the 1985 NAPAP inventory,[18] shows that practically all of HCl emissions in the U.S. emanate from the combustion of coal, particularly bituminous coal. In general, high-sulfur bituminous coals from the eastern U.S. have higher chloride content than western subbituminous and lignite coals.[19]

The estimates of HCl emissions from coal combustion in Table 1 do not take into consideration the presence of any emission control such as a flue gas desulfurization system or particulate control device (baghouse or electrostatic precipitators). Avail-

Table 1. 1985 U.S. Anthropogenic HCl Emissions

Source	Emission factor[a]	Emissions (tons/year)
Utility boilers		
Anthracite coal	0.91	822
Bituminous coal	1.90	629,939
Lignite	0.01	128
Industrial boilers		
Anthracite coal	0.91	181
Bituminous coal	1.90	48,447
Lignite	0.01	2
Commercial and industrial boilers		
Anthracite coal	3.07	269
Bituminous coal	1.48	2,693
Lignite	0.35	24
HCl manufacturing	0.20	1,068
Municipal incineration		
Solid waste	5.00	7,683
Industrial Incineration		
Solid waste	5.00	1,294
Liquid waste	5.35	133
Total		692,683

[a] In pounds per ton of fuel or waste burned; for HCl manufacturing in pounds per ton of HCl manufactured.

From Saeger, M., Langstaff, J., Walters, R., et al., The 1985 NAPAP Emissions Inventory (Version 2): Development of the Annual Data and Modelers' Tapes, Alliance Technologies Corporation, Chapel Hill, NC, prepared for the U.S. Environmental Protection Agency (EPA-600/7-89-012a), Research Triangle Park, NC, 1989.

able literature data indicate that these systems together can remove a substantial fraction of HCl — 50 to 80% — from exhaust gas. Therefore, these estimates are likely to be much higher than the actual emissions for some individual sources.

In the U.S., HCl emissions are about 5% of SO_2 emissions on a molar basis, compared to the U.K. where they are about 12%.[1,18] Recognizing that coal combustion is the major source of both SO_2 and HCl, the local and regional air quality impacts of HCl emissions are likely to be small when compared to total sulfur (i.e., SO_2 and sulfate).

Because stronger acids (HNO_3 and H_2SO_4) can react with chloride salts in atmospheric aerosols to produce HCl,[7,9,20-22] assessments of the significance of HCl emissions must consider particulate chloride emissions as well as HCl.

Table 2 shows estimates of primary particulate chloride emissions that we derived by combining total particulate matter emissions and speciation profiles developed by EPA.[18,23] For chloride emissions from fuel combustion (industrial, agricultural, institutional, and transportation) and manufacturing, we selected the profile with the maximum chloride fraction when more than one speciation profile was applicable to a particulate source category. Therefore, for these sources, our estimates represent the upper limit based upon the available data. Estimates of

Table 2. 1985 U.S. Chloride Emissions

Source	Emissions (tons/year)
Utility boilers (all fuels)	500
Industrial boilers (wood)	18,700
Industrial boilers (other fuels)	1,100
Residential wood burning	5,500
Agricultural field burning	2,600
Aluminum production	1,700
Plywood and particle board production	2,900
Light-duty gasoline vehicles (exhaust)	1,900
Municipal solid waste incineration	1,400
Unpaved road travel (soil dust)	15,000 to 213,600
Wind erosion and dust devils (soil dust)	2,800 to 92,000
Other sources	2,800
Total	57,000 to 345,000

chloride emissions due to suspension of soil dust by wind erosion, dust devils, and vehicular traffic on unpaved roads are also included in Table 2; because the magnitude of these emissions potentially can be very large, our estimates are in the form of ranges encompassing the ranges of particulate matter emission rates and speciation profiles reported in EPA data sets. As for reliability, Table 2 represents "first-cut" estimates that are based upon very limited data concerning (1) emission rates of soil dust and (2) speciation profiles for all sources. Therefore, Table 2 represents a scoping characterization of national level chloride emissions. Considering chloride emissions from fuel combustion and manufacturing first, combustion of biomass (wood, agricultural waste, etc.) is the largest source; particulate matter emissions from these sources can contain over 20% chloride. However, chloride emissions from fuel combustion and manufacturing are small in comparison to direct HCl emissions from these sources (e.g., Table 1). Suspension of soil dust, on the other hand, may account for as much as 30% of total HCl and chloride emissions in the U.S. Using the midvalue for total emissions in Table 2, the U.S. total HCl and chloride emissions from human activities involving fuel combustion and manufacturing and suspension of soil dust are approximately 890,000 TPY, with 70% of this flux being emitted by coal-fired electric generating plants.

Although we are unable to compare the relative rates of HCl and chloride emissions in the U.S. from human activities vs. natural phenomena because information is not available on sea-salt aerosol fluxes being carried over the continent, the available information does permit an order-of-magnitude comparison on global scale. Global HCl and chloride emissions in Table 3 are taken from NRC[6] with the following additional assumptions:

1. 90% of sea-salt aerosol, which is produced by breaking of waves in the oceans, is deposited back into the oceans. The remaining 10% is carried over to continental atmospheres.[24-26]
2. Since approximately one fourth of global coal combustion takes place in the U.S.

Table 3. Global HCl and Chloride Fluxes over the Continents

Source	Estimated emissions (tons/year)
Sea salt (10% of total)	77 million to 660 million
Volcanos	9,000 to 8 million
Weathering of crystal rocks	1,500 to 110,000
Forest fires	<1.6 million
Human activites	3.5 million

and since coal combustion accounts for 90% of all HCl and chloride emissions from fuel combustion and manufacturing activities in the U.S., we assumed that global HCl emissions from human activities are four times the U.S. emission.

Oceans are the largest worldwide source which sends 8×10^7 to 7×10^8 TPY of chloride flux over the continents. In comparison, global emissions of HCl and chloride from human activities are only 4×10^6 TPY.

III. AMBIENT AND PRECIPITATION CONCENTRATIONS

Table 4 presents a summary of reported ambient HCl and chloride concentrations from the recent literature. Although the long-term measurements in Table 4 are predominantly from coastal environments, reported HCl measurements for inland, coastal, and marine environments all average within a narrow range of 0.2 to 3.0 mg/m^3.

Examining the contribution of HCl to total gas-phase acidity, annual average HCl concentrations (in equivalents per cubic meter) in the Los Angeles area are much smaller than concentrations of other volatile acids and represent only 3 to 12% (range over site locations) of total gas-phase acids.[3] Formic and acetic acids are the most abundant gas-phase acids in the Los Angeles area; together they account for 70 to 90% of gas-phase acids.[3,27] However, considering reported measurements of strong, inorganic acids (HNO_3 and HCl) alone, annual average HCl concentrations in the Los Angeles area represent 15 to 50% of gas-phase strong acids.

Although concurrent measurements of HCl and HNO_3 are limited, data from the U.K. also indicate that concentrations of HCl are similar to those of HNO_3 and that together these two acids account for most of the gas-phase strong acidity.[28-30]

In explaining these concentrations, two sources of HCl have been proposed: coal-fired power plants (chiefly in England) and the reaction of HNO_3 with coarse-mode sea salt and soil dust.[9,21,22,28] Hitchcock et al.[7] concluded that sulfuric acid produced in coastal environments would also lead to volatilization (i.e., degassing) of HCl. These postulates are supported by the evidence of chloride loss from sea-salt aerosols and the presence of nitrate in the coarse fraction. This reaction will be discussed in the next section.

Long-term continental particle-phase chloride concentrations range from 0.1 to 4.2 mg/m^3 (Table 4). The contribution of chlorides to total particle mass and anions

Table 4. Recently Reported Ambient HCl and Chloride Concentrations

Location	Averaging period	Date	HCl[a] ($\mu g/m^3$)	Chloride[b] ($\mu g/m^3$)	Ref.
Rural eastern U.S.	2300 24-h samples	1978–1979	—	0.05–0.5	55
Los Angeles, CA	1 year	1986	0.8–1.8(8)	—	3
	1 week	1986	1.6	—	27
Denver, CO	3 months	1987–1988	—	0.13–0.15(3)	56
Essex, U.K.	2 3-month periods	1986–1987	0.34–1.13(4)	0.89–3.1(4)	28
Lancaster, U.K.	65 24-h samples	1979–1981	—	4.2	38
Phoenix, AZ	3 weeks	1983	—	0.89	31
Bavaria, Germany	1-3 months	1986–1987	—	0.42–1.2(3)	57
Hampton, VA	3 30-min periods	1984	2.0	—	58
Tucson, AZ	?	?	Up to 3	—	59
Urban areas, Germany	?	?	0.2–3.0	—	60
Marine atmosphere	?	?	1–2	—	61

a Number of sites in parentheses.

b The reported values for Denver represent total particle-phase chlorine measured using X-ray fluorescence (XRF); for the remaining locations, data represent chloride ions measured using ion chromatography or spectrophotometry.

is small but appreciable: chloride represents about 13 and 17% (in equivalents per cubic meter) of major anions (nitrate, sulfate, and chloride) in Phoenix and Essex, respectively.[28,31]

Data from the 1985 to 1987 Utility Acid Precipitation Study Program (UAPSP) show that, in the eastern U.S., measured chloride concentrations in precipitation (Table 5) range from 0.05 to 0.61 mg/L. For most locations, chloride concentrations (in equivalents per liter) are smaller than sulfate and nitrate concentrations by approximately a factor of 5 to 25 and chloride represents 2 to 7% of major anions. However, for Winterport, ME, and Uvalda, GA, chloride represents 35 and 25%, respectively, of major anions. These sites are closer to the coast than other UAPSP sites; however, the marine influence is not uniformly evident (e.g., Underhill, VT), perhaps because of variations in the direction from which precipitation-causing air masses originated.

By comparing Na/Cl ratios in 1982 through 1986 precipitation data from the National Acid Deposition Program (NADP) with the corresponding value for sea salt, Wagner and Steele[32] suggested that precipitation chloride concentrations in the northeastern U.S. — where they found Na/Cl ratios to be smaller than the sea salt value — may be attributable to coal-fired power plants. However, this speculation is not based upon an analysis of actual emission data or chemical transformations, and, in general, our understanding of the sources of chloride content of rain is much weaker than that of sulfate and nitrate.

In California (Table 5), precipitation chloride concentrations (in equivalents per liter) are comparable to sulfate and nitrate concentrations at inland sites and are much higher than sulfate and nitrate concentrations at coastal sites. As California is practically free of coal combustion and precipitation is generally associated with air masses coming from the Pacific Ocean, these concentrations suggest that chloride emanates from sea salt or other sources (e.g., soil dust, agricultural burning). The contribution of chloride to total anions (in equivalents per liter) ranges from 5 to 84% and even at Lake Tahoe, which is about 250 km from the coast, chloride contribution is as much as 25%. Saxena et al.[33] found precipitation chloride concentrations at several inland sites in California to be comparable to nitrate and sulfate concentrations.

IV. ATMOSPHERIC CHEMISTRY

Reactions causing transformations of chlorides can be divided into two categories:

1. Conversion of chloride salts in particle phase to gaseous HCl and vice versa. The former reaction is referred to variously as "degassing", "displacement", or "ion exchange".
2. Conversion of chlorides to chemically active forms of chlorine such as Cl_2, Cl, and $ClNO_3$.

Table 5. Precipitation Concentrations of Chloride and other Anions

Location	Averaging period (years)	Date	Precipitation-weighted mean concentration (mg/l)			Ref.
			Chloride	Sulfate	Nitrate	
Cleofield, KY	1	1987	0.11	2.09	1.49	62
Alamo, TN	1	1987	0.14	2.06	1.49	62
Winterport, ME	1	1987	0.61	0.93	0.75	62
Uvalda, GA	1	1987	0.46	1.09	0.77	62
Underhill, VT	1	1987	0.06	1.46	1.44	62
Yampa, CO	1	1987	0.06	0.64	0.80	62
Round Lake, WI	1	1987	0.05	1.20	1.30	62
Eureka, CA	2	1984–1986	2.50	0.14	0.59	63
San Jose, CA	2	1984–1986	1.51	0.46	0.65	63
Bakersfield, CA	2	1984–1986	0.31	0.98	1.24	63
Pasadena, CA	2	1984–1986	0.81	1.34	0.87	63
S. Lake Tahoe, CA	2	1984–1986	0.15	0.39	0.26	63

In addition, chlorides in aerosols may catalyze the oxidation of SO_2 to sulfate. We discuss all of these reactions in this section.

A. Formation of HCl and NH_4Cl

It is commonly accepted that the following mass balance equations represent HCl formation by reactions between particulate chlorides and gaseous nitric and sulfuric acids:

$$NaCl + HNO_3 \rightarrow NaNO_3 + HCl \quad (1)$$

$$2NaCl + H_2SO_4 \rightarrow Na_2SO_4 + 2HCl \quad (2)$$

These reactions are inferred from the observations that chloride/sodium ratio in sea-salt aerosols is often smaller than that in sea water and that chloride depletion from aerosols is accompanied by enrichment in sulfate and/or nitrate.[7,21,22] In thermodynamic terms, Equations 1 and 2 can be described by the following reactions:

$$NaCl\ (s) + HNO_3\ (g) \rightleftharpoons NaNO_3\ (s) + HCl\ (g) \quad (3)$$

where $K_{298} = 3.96$[36]

$$HCl\ (g) \rightleftharpoons H^+ + Cl^- \quad (4)$$

where $K_{298} = 2.04 \times 10^6\ M^2\ atm^{-1}$.[8]

Equation 3 explains the displacement of HCl from dry, coarse sea-salt or soil-dust particles as proposed by Hildemann et al.[9] As for aqueous particles, production and/or transport of H_2SO_4 (mostly to fine fraction) and HNO_3 (to fine and coarse fractions) would lower aerosol pH, leading to HCl evaporation as predicted by Equation 4.

Experiments conducted by Clegg and Brimblecombe[8] confirmed these mechanisms and suggested that, while acidification by H_2SO_4 would lead to HCl displacement at all stages of acidification (although as pH diminishes, more and more of sulfate is retained as HSO_4^-, reducing the amount of H^+ ions generated by additional H_2SO_4), acidification by HNO_3 could result in equilibrium being reached between gas- and aqueous-phase HNO_3, which may reduce HCl displacement.

The HCl released by reactions described above or emitted directly can react with gaseous ammonia to form NH_4Cl as follows:[29,34,35]

$$NH_3\ (g) + HCl\ (g) \rightleftharpoons NH_4Cl\ (s) \quad (5)$$

where $K_{298} = 9.6 \times 10^{15}\ atm^{-2}$.[36]

$$NH_3\ (g) + HCl\ (g) \rightleftharpoons NH_4^+ + Cl^- \quad (6)$$

where $K_{298} = 2.1 \times 10^{17}\ M^2\ atm^{-2}$.[36]

The presence of ammonium chloride in atmospheric aerosols was confirmed by Harrison and Sturges[37] using an X-ray diffraction technique to identify specific compounds. In addition, ammonium chloride concentrations of the order of 10 mg/m^3 have been inferred in Lancaster, England, and Riverside, CA.[38,39] Wall et al.[21] found a good correlation between chloride loss from coarse sea-salt particles and the sum of HCl and fine-mode chloride. Hara et al.[40] concluded that the observed differences between the size distributions of summer and winter aerosols in Tokyo stem from the temperature dependence of the dissociation constants of ammonium chloride and ammonium nitrate.

Recognizing that ammonium chloride deliquesces at a relative humidity (RH) of 80% and assuming that it exists as an external mixture, as has been suggested by Allen et al.,[29] ammonium chloride would exist as a dry solid below an RH of 80%, provided that the product of partial pressures of ammonia and HCl exceeds the dissociation constant of ammonium chloride. At other RHs, ammonium chloride would exist as an aqueous solution. More data are needed to judge whether atmospheric aerosols exist as internal or external mixtures and whether equilibrium actually prevails in the atmosphere.

Modeling studies indicate that for some conditions — such as cool, marine environments — the time scales for reaching equilibrium may be too large to justify the assumption of equilibrium between the vapor and the aerosol phases.[41] Measurements of NH_4Cl evaporation from aerosols[42] also suggest that the attainment of equilibrium may be constrained by some as-yet-unidentified kinetic considerations. In addition, using thermodynamic principles, Wexler and Seinfeld[43] showed that the relative humidity of deliquescence for a salt in an internal mixture is lowered by the presence of other salts. These findings need to be verified with ambient data.

In summary, the net result of Equations 3 through 6 is the transfer of some of the coarse-mode chloride to gaseous HCl and fine-mode $NH_4Cl(s)$ or chloride ion. However, a complete quantitative characterization of the distribution of chloride between vapor and particle phases would require the use of realistic models. More ambient data are needed to evaluate the available models.

B. Conversion to Chemically Active Chlorine Compounds

Whether HCl or chlorides are converted to other chlorine compounds in the atmosphere has implications for the

- Rate of HCl displacement from aerosols
- Role of chlorides in tropospheric chemistry
- Removal mechanisms of chlorine compounds
- Toxicity of chlorine compounds

Unfortunately, data on this subject are woefully lacking. Some studies (Table 6) indicate that certain reactions do indeed convert chlorides to chemically active forms

Table 6. Tropospheric Reactions of Chlorides

No.	Reaction	k_{298}[a]	Ref.
	Gas-Phase Reactions		
1[b]	$HCl + O_3 \xrightarrow{hv} Cl$	N/A	44
	Heterogeneous Reactions		
2	$O_3(g) + Cl^-(aq) \rightarrow O_2(g) + ClO^-(aq)$	2.0×10^{-4}	45, 48
3	$ClO^-(aq) + 2H^+(aq) + Cl^-(aq) \rightarrow Cl_2(g) + H_2O(l)$	N/A	6, 45
4	$2NO_2(g) + NaCl(s) \rightarrow ClNO(g) + NaNO_3(s)$	N/A	46
5	$N_2O_5(g) + NaCl(s) \rightarrow ClNO_2(g) + NaNO_3(s)$	N/A	47
6	$ClONO_2(g) + NaCl(s) \rightarrow Cl_2(g) + NaNO_3(s)$	N/A	47
	Aqueous-Phase Reactions		
7	$SO_4^- + Cl^- \rightarrow SO_4^{2-} + Cl$	2×10^8	64
8	$Cl^- + OH + H_3O^+ \rightarrow Cl + 2H_2O$	4×10^9	64
9	$NO_3 + Cl^- \rightarrow NO_3^- + Cl$	7.1×10^7	64

[a] In $M^{-1}s^{-1}$ for two-body reactions and $M^{-2}s^{-1}$ for three-body reactions; N/A: not available.
[b] The stoichiometry or other reaction products not reported in the reference.

such as Cl atoms but the overall significance and even the kinetics of these reactions is not known.

Behnke and Zetzsch[44] found, using smog chamber experiments, that molecular chlorine is formed by the reaction of HCl and ozone on glass surfaces (Reaction 1 in Table 6). They also found that the reaction was catalyzed by sunlight, which also photolyzed molecular chlorine to Cl atoms.

Yeatts and Taube[45] studied the reaction between chloride and ozone (Reactions 2 and 3); however, these reactions were judged too slow to be of any importance in converting atmospheric chloride (e.g., in sea salt) to chlorine.[6]

Reactions between NO_2 and NaBr produced BrNO in the laboratory experiments conducted by Finlayson-Pitts and Johnson.[46] The investigators concluded that, in marine environments that are affected by anthropogenic NO_2 emissions, an analogous reaction between NaCl and NO_2 (Reaction 4) may compete with the HCl displacement reaction in removing chloride from sea-salt aerosols. While HCl is relatively unreactive, ClNO can rapidly photolyze to yield atomic chlorine that in turn can initiate photooxidation of hydrocarbons by abstracting hydrogen from hydrocarbons in a manner hydroxyl radicals do.

Another experimental study conducted by Finlayson-Pitts and co-workers[47] suggests that reactions of NaCl with N_2O_5 and $ClONO_2$ (Reactions 5 and 6) may convert chemically inert NaCl to photochemically active chlorine compounds in both the troposphere and the stratosphere. The reaction with N_2O_5, in particular, may be important in cold, polluted marine environments.

Reactions of chloride ions with SO_4^-, OH, or NO_3 (Reactions 7 through 9 in Table 6) in the aqueous phase can lead to the formation of atomic chlorine. These reactions are included in mathematical models describing cloud or rain chemistry.[48-50] However, their efficiency in converting chloride ions to chlorine has not been studied explicitly. Graedel and Goldberg[48] included several other aqueous-phase reactions of chloride ions in their model but they considered those reactions to be inefficient in converting chloride to chlorine-containing free radicals.

To summarize, some recent laboratory experiments indicate that chloride ions may indeed be converted in the atmosphere to more reactive forms of chlorine but the efficiencies of these reactions is not understood. In addition, so far, the reactant concentrations used in the experiments were much higher than typical atmospheric concentrations.

Once the kinetic data become available through further experiments. these reactions should be included in tropospheric chemistry models, which then can be exercised to determine the importance of these reactions.

C. Chloride as a Catalyst

Clarke and Radojevic[15,51] studied the influence of chloride ions on SO_2 oxidation rates in sea-salt aerosols and concluded that chloride ions in sea salt catalyze SO_2 oxidation. The oxidation rate (percent per hour) increased with decreasing SO_2 concentration. The maximum observed oxidation rate was 6.4%/h for $(SO_2) = 1$ ppb, RH = 90%, and a salt concentration = 10 $\mu g/m^3$. They proposed the reaction between SO_4^- and chloride (Reaction 7 in Table 6) as one of the possible mechanisms.

V. HEALTH HAZARDS OF A LARGE HCl SOURCE

To explore what the health hazards of HCl emissions from a typical large size utility source might be, we estimated the downwind HCl concentrations from a hypothetical 750-MW coal-fired power plant using realistic stack emission rates (assuming no SO_2 or particulate removal) and stack characteristics which were

Emission rate	30.2 g/s (about 1050 TPY)
Stack height	182 m
Stack diameter	7.6 m
Stack exit velocity	33.5 m/s
Stack exit temperature	433 K

PTPLU, an EPA guideline screening model was, used to estimate the maximum ground level concentration for the worst-case meteorological conditions. The maximum 1-h ground level concentration estimated using this procedure is 7.1 mg/m^3 at a distance of 1600 m from the stack. To put this concentration level in perspective, it is approximately equivalent to the maximum 24-h average HCl concentration of 6.3 mg/m^3 from a 1-year, eight-station data set for the Los Angeles area where there is virtually no coal combustion.

HCl is not considered to be a carcinogen.[52] Although no standard exists for either acute or chronic exposures to HCl, EPA has recommended a reference concentration for protection of public health against long-term (chronic) noncarcinogenic effects.[53] Using the assumption commonly used in screening analysis such as this one, that the annual average concentration is about 1/10 of the maximum 1-h average concentration,[54] we derived the following comparison of stack impacts and EPA reference concentration. The maximum annual average concentration due to the power plant is 0.7 mg/m^3. The EPA recommended reference concentration for long-term effects is 7 mg/m^3.

These results are based upon uncontrolled emission rates and, therefore, the impact when SO_2 and/or particulate matter controls are present, we expect, would be much smaller. Our future research will assess the emission rates and impacts for actual plant operating configurations. Nevertheless, this initial assessment with near-worst-case conditions indicates that HCl emissions from individual coal combustion sources are not likely to pose a hazard to public health.

VI. SUMMARY AND CONCLUSIONS

The goal of our review of sources, atmospheric fate, and ambient concentrations of chlorides in the troposphere was to scope the relative importance and environmental impacts of anthropogenic HCl gas and particulate chloride salt emission in the U.S. The following conclusions emerge from our review and analysis:

1. The emissions of chlorine-containing compounds from fuel combustion sources are in the form of HCl gas or inorganic chloride salts and not chlorine gas which is more reactive and toxic.
2. The reaction of sulfuric and nitric acids with chloride-containing particles (e.g., sea-salt and soil-dust particles) can lead to formation of gaseous HCl. On the other hand, HCl, a volatile gas, can react with atmospheric ammonia to form ammonium chloride particles. The displacement and subsequent condensation of HCl would amount to the transfer of some of the chloride form coarse to fine-mode particles, which would be more effective in scattering light. Because of the ease with which chloride partitioning between particle and gas phases can change in the atmosphere, total chloride (gaseous HCl plus particulate chloride salts) quantities need to be considered in the assessment of emission fluxes, ambient concentrations, and source-receptor relationships.
3. Based upon available information, we estimate the U.S. anthropogenic emissions of HCl are about 690,000 TPY, with coal combustion accounting for about 98%. Total HCl and chloride emissions from human activities, involving fuel combus-

tion and manufacturing, and suspension of soil dust are estimated to be 890,000 TPY: 70% of this flux is emitted by coal-fired electric generating plants. For coal combustion units, our estimates do not account for the presence of SO_2 or particulate control devices, which can remove a substantial fraction of HCl and chloride emissions, according to available data. Therefore, actual emissions from coal combustion are likely to be less than our estimates. Moreover, the implementation of SO_2 controls mandated by the Clean Air Act Amendments of 1990 will further reduce HCl and chloride emissions from coal combustion units.

4. About 8×10^8 to 7×10^9 TPY of chloride are emitted into the atmosphere in the form of sea salt aerosol produced by breaking of waves in the oceans. Approximately 90% of this chloride is deposited back into the oceans and the remaining 10% is carried over the continents, making oceans the largest worldwide source of chloride. In comparison global emissions of HCl and chloride from human activities, we estimate, are 4×10^6 TPY. Therefore, on the global scale, emissions from the oceans appear to be about 20 to 200 times larger than those from human activities.
5. There is experimental evidence that chlorides react with other compounds found in the atmosphere (e.g., ozone, nitrogen dioxide, and dinitrogen penta-oxide) to form chemically more reactive chlorine compounds such as chlorine atoms. Some studies have also proposed that chlorides can catalyze the oxidation of SO_2 in aerosols. However, data on the kinetics of these reactions are not available. Once the kinetic data become available from laboratory experiments, the overall importance of these reactions in the atmosphere can be assessed using air quality simulation models.
6. The reported long-term average continental air concentrations of HCl are between 0.2 and 3.0 mg/m^3. In coastal areas like Los Angeles, HCl can account for a substantial fraction of the gas-phase strong acidity even though there are no major anthropogenic sources of HCl or chloride. Long-term average concentrations of 0.1 to about 4.0 mg/m^3 have been reported for particle phase chloride. The contribution of chlorides to total particle mass is small but appreciable. Long-term average chloride concentrations in precipitation display a large spatial variability (0.05 to 2.5 mg/L). For most inland sites, precipitation chloride concentrations are much smaller than concentrations of sulfate and nitrate. In coastal areas, on the other hand, chloride concentrations at some locations are higher than sulfate and nitrate concentrations.
7. To explore what the health hazards of HCl emissions from a typical large size utility source might be, we estimated the downwind HCl concentrations due to a 600-ft-high stack emitting about 1000 TPY of HCl. Using a screening model, the 1-h maximum concentration under worst-case meteorological conditions was estimated to be about 7 mg/m^3, which, to put it in perspective, is roughly equal to the maximum 24-h HCl concentration in the Los Angeles area. These estimates indicate that the long-term average concentrations due to the modeled emission source (of the order of 0.7 mg/m^3), would be about an order of magnitude smaller than the reference concentration recommended by EPA (7 mg/m^3) for the protection of public health against chronic noncarcinogenic effects. These results are based upon uncontrolled emission rates and, therefore, the impacts when SO_2 and/or particulate matter controls are present are expected

to be much smaller. Our future research will assess the emission rates and impacts for actual plant operating configurations. Nevertheless, our exploratory assessment with near-worst-case assumptions indicates that, in general, HCl emissions from individual coal combustion sources are not likely to pose a hazard to public health.

ACKNOWLEDGMENTS

We are grateful to Dr. Christian Seigneur of ENSR Consulting and Engineering for assistance with modeling and public health criteria.

REFERENCES

1. Lightowlers, P. J. and Cape, J. N., Sources and fate of atmospheric HCl in the UK and Western Europe, *Atmos. Environ.*, 22, 7, 1988.
2. Gibbs, W. H., The nature of chlorine in coal and its behavior during combustion, in *Corrosion Resistant Materials for Coal Combustion Systems*, Meadowcroft, D. B. and Manning, M. I., Eds., Applied Science, New York, 1983.
3. Solomon, P. A., Fall, T., Salmon, L., Lin, P., Vasquez, F., and Cass, G. R., Acquisition of Acid Vapor and Aerosol Concentration Data for Use in Dry Deposition Studies in the South Coast Air Basin, Report No. EQL 25, California Institute of Technology, Pasadena, CA, 1988.
4. Robbins, R. C., Cadle, R. D., and Eckhardt, D. L., The conversion of sodium chloride to hydrochloric acid in the atmosphere, *J. Meteorol.*, 16, 53, 1959.
5. Eriksson, E., The yearly circulation of chloride and sulfur in nature: meteorological, geochemical and pedological implications. II, *Tellus*, 12, 63, 1960.
6. NRC, *Medical and Biologic Effects of Environmental Pollutants — Chlorine and Hydrogen Chloride*, The National Research Council, National Academy of Sciences, Washington, DC, 1976.
7. Hitchcock, D., Spiller, L. L., and Wilson, W. E., Sulfuric acid aerosols and HCl release in coastal atmospheres: evidence of rapid formation of sulfuric acid particulates, *Atmos. Environ.*, 14, 165, 1980.
8. Clegg, S. L. and Brimblecombe, P., Equilibrium partial pressures of strong acids over concentrated saline solutions. II. HCl, *Atmos. Environ.*, 22, 117, 1988.
9. Hildemann, L. M., Russell, A. G., and Cass, G. R., Ammonia and nitric acid concentrations in equilibrium with atmospheric aerosols: experiment vs. theory, *Atmos. Environ.*, 18, 1737, 1984.
10. Charlson, R. J., Lovelock, J. E., Andreae, M. E., and Warren, S. G., Oceanic phytoplankton, atmospheric sulfur, cloud albedo and climate, *Nature*, 326, 655, 1987.
11. Schwartz, S. E., Are global cloud albedo and climate controlled by marine phytoplankton?, *Nature*, 336, 441, 1988.
12. Wigley, T. M. L., Possible climate change due to SO_2-derived cloud condensation nuclei, *Nature*, 339, 365, 1989.
13. Slingo, A., Sensitivity of the earth's radiation budget to changes in low clouds, *Nature*, 343, 49, 1990.

14. Kreidenweis, S., Penner, J. E., Yin, F., and Seinfeld, J. H., The effects of dimethyl sulfide upon marine aerosol concentrations, *Atmos. Environ.*, 25A, 2501, 1991.
15. Clarke, A. G. and Radojevic, M., Chloride ion effects on the aqueous oxidation of SO_2, *Atmos. Environ.*, 17, 617, 1983.
16. Johnson, M. G. and Stevenson, W. H., Overview of proposed air emission standards and guidelines for municipal waste combustors, *J. Air Waste Manage. Assoc.*, 40, 932, 1990.
17. Hall, B., Lindqivst, O., and Ljungstrom, E., Mercury chemistry in simulated flue gases related to waste incineration conditions, *Environ. Sci. Technol.*, 24, 108, 1990.
18. Saeger, M., Langstaff, J., Walters, R., et al., The 1985 NAPAP Emissions Inventory (Version 2): Development of the Annual Data and Modelers' Tapes, Alliance Technologies Corporation, Chapel Hill, NC, prepared for the U.S. Environmental Protection Agency (EPA-600/7-89-012a), Research Triangle Park, NC, 1989.
19. Miller, M., Private communication, Electric Power Research Institute, Palo Alto, CA, 1990.
20. Clegg, S. L. and Brimblecombe, P., Potential degassing of hydrogen chloride from acidified sodium chloride droplets, *Atmos. Environ.*, 19, 465, 1985.
21. Wall, S. M., John, W., and Ondo, J. L., Measurement of aerosol size distributions for nitrate and major ionic species, *Atmos. Environ.*, 22, 1649, 1988.
22. Eldering, A., Soloman, P. A., Salmon, L. G., Fall, T., and Cass, G. R., Hydrochloric acid: a regional perspective on concentrations and formation in the atmosphere of Southern California, *Atmos. Environ.*, 25A, 2091, 1991.
23. Radian Corporation, Air Emissions Species Manual, Volume II — Particulate Matter Species Profiles, Prepared for the U.S. Environmental Protection Agency (EPA-450/2-90-001), Research Triangle Park, NC, 1990.
24. Erickson, D. J., III and Duce, R. A., On the global flux of atmospheric sea salt, *J. Geophys. Res.*, 93, 14079, 1988.
25. Erickson, D. J., III, Private communication, National Center for Atmospheric Research, Boulder, CO, 1991.
26. Blanchard, D. C., Private communication, State University of New York, Albany, NY, 1991.
27. Grosjean, D., Liquid chromatography analysis of chloride and nitrate with negative ultraviolet detection: ambient levels and relative abundance of gas-phase inorganic and organic acids in southern California, *Environ. Sci. Technol.*, 24, 77, 1990.
28. Harrison, R. M. and Allen, A. G., Measurements of atmospheric HNO_3, HCl and associated species on a small network in eastern England, *Atmos. Environ.*, 24A, 369, 1990.
29. Allen, A. G., Harrison, R. M., and Erisman, J.-M., Field measurements of the dissociation of ammonium nitrate and ammonium chloride aerosols, *Atmos. Environ.*, 23, 1591, 1989.
30. Sturges, W. T. and Harrison, R. M., The use of nylon filter to collect HCl: efficiencies, interferences and ambient concentrations, *Atmos. Environ.*, 23, 1987, 1989.
31. Solomon, P. A. and Moyers, J. L., A chemical characterization of wintertime haze in Phoenix, Arizona, *Atmos. Environ.*, 20, 207, 1986.
32. Wagner, G. H. and Steele, K. F., Na^+/Cl^- ratios in rain across the USA, 1982–86, *Tellus*, 41B, 444, 1989.
33. Saxena, P., Arcado, T. D., Marler, B. L., and Altshuler, S. L., Acid Precipitation Chemistry in an Urban Plume, presented at the 80th Annu. Meeting of the Air Pollution Control Association, June 21–26, New York, 1987.

34. Pio, C. A. and Harrison, R. M., The equilibrium of ammonium chloride aerosol with gaseous hydrochloric acid and ammonia under tropospheric conditions, *Atmos. Environ.*, 21, 1243, 1987a.
35. Pio, C. A. and Harrison, R. M., Vapour pressure of ammonium chloride aerosol: effect of temperature and humidity, *Atmos. Environ.*, 21, 2711, 1987b.
36. Pilinis, C. and Seinfeld, J. H., Continued development of a general equilibrium model for inorganic multicomponent atmospheric aerosols, *Atmos. Environ.*, 21, 2453, 1987.
37. Harrison, R. M. and Sturges, W. T., Physico-chemical speciation and transformation reactions of particulate atmospheric nitrogen and sulphur compounds, *Atmos. Environ.*, 18, 1829, 1984.
38. Harrison, R. M. and Pio, C. A., Major ion composition and chemical associations of inorganic atmospheric aerosols, *Environ. Sci. Technol.*, 17, 169, 1983.
39. Yoshizumi, K. and Okita, T., Quantitative estimation of sodium- and ammonium-nitrate, ammonium chloride and ammonium sulphate in ambient particulate matter, *J. Air Pollut. Control Assoc.*, 33, 224, 1983.
40. Hara, H., Kato, T., and Matsushita, H., The mechanism of seasonal variation in the size distributions of atmospheric chloride and nitrate aerosol in Tokyo, *Bull. Chem. Soc. Jpn.*, 62, 2643, 1989.
41. Wexler, A. S. and Seinfeld, J. H., The distribution of ammonium salts among a size and composition dispersed aerosol, *Atmos. Environ.*, 24A, 1231, 1990.
42. Harrison, R. M., Sturges, W. T., Kitto, A.-M. N., and Li, Y., Kinetics of evaporation of ammonium chloride and ammonium nitrate aerosols, *Atmos. Environ.*, 24A, 1883, 1990.
43. Wexler, A. S. and Seinfeld, J. H., Second-generation inorganic aerosol model, *Atmos. Environ.*, 25A, 2731, 1991.
44. Behnke, W. and Zetzsch, C., *Simulation of the Tropospheric Production of Atomic Cl with Consideration of the Impact of Aerosols*, Fraunhofer-Institut fur Toxicologie und Aerosolforschung, Hanover, Germany, 1988.
45. Yeatts, L. R. B., Jr. and Taube, H., The kinetics of the reaction of ozone and chloride ion in acid aqueous solution, *J. Am. Chem. Soc.*, 71, 4100, 1949.
46. Finlayson-Pitts, B. J. and Johnson, S. N., The reaction of NO_2 with NaBr: possible source of BrNO in polluted marine atmospheres, *Atmos. Environ.*, 22, 1107, 1988.
47. Finlayson-Pitts, B. J., Ezell, M. J., and Pitts, J. N., Jr., Formation of chemically active chlorine compounds by reactions of atmospheric NaCl particles with gaseous N_2O_5 and $ClONO_2$, *Nature*, 337, 241, 1989.
48. Graedel, T. E. and Goldberg, K. I., Kinetic studies of raindrop chemistry. I. Inorganic and organic processes, *J. Geophys. Res.* 88, 10865, 1983.
49. Chameides, W. L., The photochemistry of a remote marine stratiform cloud, *J. Geophys. Res.*, 89, 4739, 1984.
50. Seigneur, C. and Saxena, P., A theoretical investigation of sulfate formation in clouds, *Atmos. Environ.*, 22, 101, 1988.
51. Clarke, A. G. and Radojevic, M., Oxidation rates of SO_2 in sea-water and sea-salt aerosols, *Atmos. Environ.*, 18, 2761, 1984.
52. Von Burg, R. et al., *Toxicology Profiles*, Vol. 1, prepared for the Electric Power Research Institute by ENSR Consulting and Engineering, Alameda, CA, 1991.
53. IRIS, *Integrated Risk Information System*, U.S. Environmental Protection Agency, Washington, DC, 1992.

54. BAAQMD, *Permit Modeling Guidance*, Bay Area Air Quality Management District, San Francisco, 1988.
55. Mueller, P. K. and Hidy, G. M., The Sulfate Regional Experiment: Report of Findings, Vol. 2, Report EA-1901, Electric Power Research Institute, Palo Alto, CA, 1983.
56. Watson, J. G., Chow, J. C., et al., *The 1987–88 Metro Denver Brown Cloud Study*, Vol. I, II, and III, Desert Research Institute (8810 1F3), Reno, NV, 1988.
57. Ludwig, J. and Klemm, O., Acidity of size-fractionated aerosol particles, *Water Air Soil Pollut.*, 49, 35, 1990.
58. Cofer, W. R., III, Collins, V. G., and Talbot, R. W., Improved aqueous scrubber for collection of soluble atmospheric trace gases, *Environ. Sci. Technol.*, 19, 557,1985.
59. Farmer, J. C. and Dawson. G. A., *J. Geophys. Res.*, 87, 8931, 1982.
60. Matusca, P., Schwartz, B., and Bachmann, K., Measurements of diurnal concentration variations of gaseous HCl in air in sub-nanogram range, *Atmos. Environ.*, 18, 1667, 1984.
61. Cicerone, R. J., Halogens in the atmosphere, *Rev. Geophys. Space Phys.*, 19, 123, 1981.
62. Topol, L. E. and Cleary, P., UAPSP Data Displays for 1979 through 1987, UAPSP 116, Electric Power Research Institute, Palo Alto, CA, 1989.
63. ARB, 4th Annu. Report to the Governor and the Legislature on the Air Resources Board's Acid Deposition Research and Monitoring Program, California Air Resources Board, Sacramento, CA, 1986.
64. Neta, P. and Huie, R. E., Rate constants for reactions of NO_3 radicals in aqueous solutions, *J. Phys. Chem.*, 90, 4644, 1986.

Ambient Air Toxics Data from California's Toxic Air Contaminant Monitoring Program

William V. Loscutoff
Michael W. Poore

I. INTRODUCTION

In 1984, legislation in California became effective requiring the California Air Resources Board (CARB) to identify and control compounds as Toxic Air Contaminants (TACs).[1] This enabling legislation resulted in the CARB's Toxic Air Contaminant Program. The Program required that sampling and analytical techniques be developed to demonstrate the presence of chemical compounds in ambient air at concentrations sufficient to be of concern. For many potential TACs, high health-risk estimates made it necessary to measure these compounds at sub-part-per-trillion concentrations. In addition, since the data would be used for regulatory action, it was necessary to develop quality control and quality assurance procedures to demonstrate the accuracy and precision of the measurements.

The CARB presently operates a monitoring system of 22 sites which effectively covers the major population centers throughout the state (Figure 1); 24-h composite samples are collected at each site on a once-every-12-day schedule. In addition, the CARB has instituted a "rover" monitoring station program. This "rover" program was implemented to determine ambient air concentrations of TACs in areas without a nearby monitoring site. Data from this program will provide the CARB with guidance as to the need for additional permanent sites in other California locations. In addition to the 22-site TAC network, 2 air quality management districts, the San Francisco Bay Area and the South Coast Air Quality Management Districts, also have Toxics monitoring networks. The most extensive of these, the 11-site Bay Area network, was recently reviewed by Levaggi and Siu.[2] These networks provide California regulatory decision makers with a wealth of information on concentrations, trends, and possible public health risks.

Implementation of a monitoring network as geographically extensive and demanding in terms of the variety of target compounds as is the CARB network has required substantial innovation in sampling system design. In addition to those sampling systems that were already in place to support the criteria pollutant network [total suspended particulate (Hi-Vol) and size selective inlet (PM10) samplers], the CARB staff designed two innovative sampling systems to deal with the special requirements associated with sub-part-per-billion concentration volatile and nonvolatile ambient air sampling.

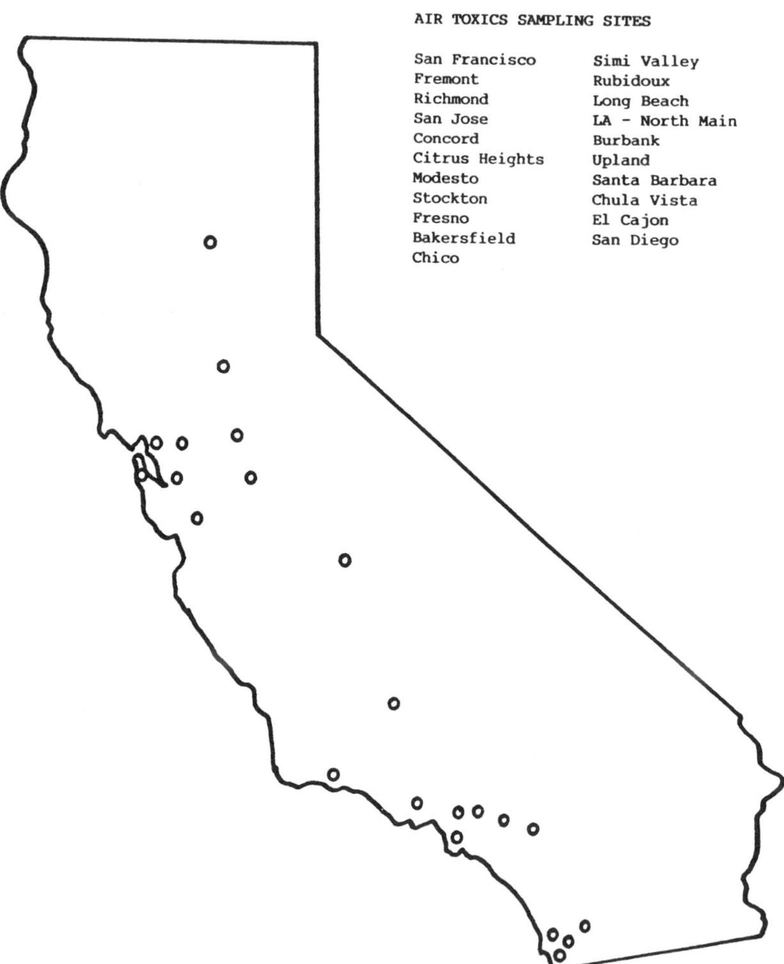

Figure 1. Air Resources Board Toxics Network monitoring sites (1991).

The compounds for which the CARB presently monitors and the limits of detection (LOD) for each compound are listed in Tables 1, 2, and 3. Table 1 is a list of eight halogenated substances, 1,3-butadiene, two aldehydes, and two ketones. Table 2 is a list of the eight toxic metals in the program, and Table 3 is a list of aromatic and polyaromatic hydrocarbons of interest. In most cases, analytical meth-

TABLE I
TOXIC COMPOUNDS LIST (1991)
LIMITS OF DETECTION

TOXIC COMPOUND NAME	LIMIT OF DETECTION (PPB)
DICHLOROMETHANE	1.0
CHLOROFORM	0.02
ETHYLENE DICHLORIDE	0.20
1,1,1-TRICHLOROETHANE	0.01
CARBON TETRACHLORIDE	0.02
TRICHLOROETHYLENE	0.02
ETHYLENE DIBROMIDE	0.01
PERCHLOROETHYLENE	0.01
1,3-BUTADIENE	0.04
ACETALDEHYDE	0.10
FORMALDEHYDE	0.10
METHYL ETHYL KETONE	0.10
METHYL ISOBUTYL KETONE	0.10

TABLE II
TOXIC COMPOUNDS LIST (1991)
LIMITS OF DETECTION

TOXIC COMPOUND NAME	LIMIT OF DETECTION
ARSENIC	0.4 ng/m3
BERYLLIUM	0.02 ng/m3
CADMIUM	0.2 ng/m3
LEAD	3 ng/m3
HEXAVALENT CHROMIUM	0.2 ng/m3
TOTAL CHROMIUM	1 ng/m3
NICKEL	1 ng/m3
MANGANESE	1 ng/m3

odology for these compounds did not exist for ambient air. The methods have been developed as adaptations of NIOSH or other environmental methods, or they have been created from basic principles.

Although the ability to develop imaginative sampling and analysis techniques have been important to the success of the California Toxics program, from the regulatory standpoint the quality control and quality assurance programs have been critical. Quality control is an activity of the sampling and analysis personnel to insure that the data reported are of a known precision and accuracy and reflect the environmental conditions. In 1986, the CARB let a research contract with Battelle, Columbus[3] to critically evaluate the sampling, analysis, and quality control procedures being used at that time. As a result of this study, many of the procedures used were independently validated or, where deficiencies were noted, corrective action was taken. An example of this is the switch from the use of Tedlar bags for whole-air collection to Summa-polished stainless steel canisters. A second result of the study

TABLE III
TOXIC COMPOUNDS LIST (1991)
LIMITS OF DETECTION

TOXIC COMPOUND NAME	LIMIT OF DETECTION (PPB)
BENZENE	0.5
TOLUENE	0.2
ETHYL BENZENE	0.6
1,4-XYLENE	0.5
1,3-XYLENE	0.6
1,2-XYLENE	0.1
CHLOROBENZENE	0.1
STYRENE	0.1
1,3-DICHLOROBENZENE	0.2
1,2-DICHLOROBENZENE	0.1
1,4-DICHLOROBENZENE	0.2
BENZO(A)PYRENE	0.05 ng/m3
BENZO(B)FLUORANTHENE	0.05 ng/m3
BENZO(K)FLUORANTHENE	0.05 ng/m3
DIBENZ[a,h]ANTHRACENE	0.05 ng/m3
BENZO[ghi]PERYLENE	0.05 ng/m3
INDENO[1,2,3-cd]PYRENE	0.05 ng/m3

was the intensive effort the CARB has made to support the development of ambient level toxics standards by the National Institute of Standards and Technology (NIST, formerly NBS). Prior to 1986, no such standards existed. As a result of a 4-year program, the CARB now has NIST certified reference and calibration standards with concentrations at or near average ambient concentrations.

The quality assurance function represents an independent review of all monitoring activity to insure that data quality objectives are being met. Routine quality assurance activities are a part of the CARB TAC network and have been described elsewhere.[4] Additional activities are system audits to determine that quality-control procedures are being followed and performance audits. Performance audits are conducted quarterly to check the precision and accuracy of the analytical process. High-pressure cylinders are submitted to the laboratories in California (including those of the participating districts) for analysis. The contents and concentrations of these cylinders are similar to ambient air and have been certified by NIST. The most innovative quality assurance procedure, to our knowledge practiced only in California, is the Through-the-Probe (TTP) audit.[5] This procedure is performed by installing a high concentration, NIST certified gas cylinder in a dynamic dilution system and directing the resultant gas into the inlet probe of the sampling system. These audits are performed at regular sampling sites on scheduled sampling days. The resultant 24-h composite audit sample is then submitted to the laboratory for analysis. The laboratory staff has no knowledge that the sample represents an audit. The audit represents a characterization of the accuracy of the entire monitoring system.

Although the accuracy of the entire monitoring system can be described by the TTP audits, routine precision checks are available through the collocated sampling systems in the network. The CARB staff operate collocated sampling sites at Concord, Bakersfield, and Rubidoux, CA. A collocated site is one at which two independent samplers are utilized, and the two samples collected are sent to the same laboratory. In 1990, the overall precision for whole-air samples was ±30%. The overall accuracy, as represented by the TTP audits, was ±35%. The larger "accuracy" value represents the total of the overall precision and the precision with which the audit cylinder can be diluted. When considering that measurements are generally being made at concentrations of 0.10 ppb, or less, we feel that the overall precision and accuracy is acceptable.

The following section of this paper will deal with measured data from the system. Although, for many of the compounds, there are data available since 1985, only data from 1989 and 1990 will be presented. Earlier data may be referred to in comments concerning trends.

II. 1989 AND 1990 AMBIENT TOXICS DATA

A. Annual Averages and Trends

Tables 4 and 5 contain the annual average concentrations of TACs in California for the calendar years 1989 and 1990. Although approximately 6 years' worth of data is available on many of the compounds, the results of just the last two complete years are presented here. As shall be demonstrated later, such factors as seasonality and meteorology can confuse the interpretation of trends when extrapolated from only 2 years. The annual averages have been calculated as the arithmetic mean, with values below the detection limit being assigned a value of one half the detection limit. In cases where there is a significant number of observations below the detection limit, the annual average will appear to be substantially below the detection limit for that compound. In addition to annual averages, the tables contain the highest and lowest measured value for each compound, the number of measurements, and the percentage of the measurements above the LOD. This represents almost 19,000 measurements over the 2 years. During the years 1989 and 1990, ethylene dichloride (1,2-dichloroethane) was not measured in ambient air at a concentration greater than the LOD (0.2 ppbv). Ethylene dibromide was measured above the LOD (0.01 ppbv) only 14% of the time in 1990, with a maximum observed concentration of 0.07 ppbv. These two compounds were commonly used as additives in leaded gasoline. The low observed concentrations are a result of the phase out of leaded gasoline in California and restricted agricultural use.

The two commonly used solvents, methyl chloroform and dichloromethane, have been found in relatively high concentrations consistently throughout the state. Methyl chloroform has been measured at levels of 20 to 30 ppbv at various locations and especially in the South Coast area, and exhibits the highest annual average concen-

Table IV
Toxic Data Summary for 1989
All Sites

COMPOUND	CONCENTRATION, PPB AVERAGE	MAXIMUM	MINIMUM	# OBS.	LOD	% ABOVE LOD
Halocarbons						
Dichloromethane	1.07	6.7	< 1	449	1	33
Chloroform	0.037	0.22	< 0.02	446	0.02	82
Carbon Tetrachloride	0.116	0.34	0.06	444	0.02	100
1,1,1 Trichloroethane	1.91	31	0.09	396	0.01	100
Ethylene Dichloride	0.100	0.3	< 0.2	436	0.2	0
Perchloroethylene	0.396	14	0.03	448	0.01	100
Trichloroethylene	0.184	2.7	< 0.01	440	0.01	93
Ethylene Dibromide	0.005	0.02	< 0.01	451	0.01	2
Aromatics						
Benzene	2.58	16	< 0.5	436	0.5	95
Other						
1,3 - Butadiene	0.37	2.4	< 0.04	435	0.04	98
Formaldehyde	3.74	24	0.24	585	0.1	100
Acetaldehyde	2.24	13	0.11	584	0.1	100
Benzo(b)Fluoranthene *	0.617	12	< 0.05	324	0.05	81
Benzo(k)Fluoranthene *	0.196	3.6	< 0.05	324	0.05	48
Benzo(a)Pyrene *	0.614	14	< 0.05	415	0.05	60

NOTE:

1. Concentrations at < LOD are replaced with one half of LOD FOR estimating of AVERAGES.

2. * => Concentrations in ng/m3

tration of any of the halogenated compounds measured. Due to the use of methyl chloroform in large industrial operations, small-scale degreasing and automotive shops, and consumer products, it is not surprising that methyl chloroform measurements exhibit the largest variability of any of the TACs monitored. The concentrations of dichloromethane, used both as an industrial and consumer product solvent, are also high and extremely variable. These compounds have been part of the monitoring program since its inception and show a gradual increase in ambient concentrations due to increased use.

Chloroform, a solvent not widely used in California, is thought to be released into the atmosphere primarily from chlorination processes in water and wastewater treatment. This is confirmed by the very low concentrations measured in the atmosphere. In general, concentrations throughout the state are between 0.02 and 0.05 ppbv, with annual averages of approximately 0.03 ppbv. The highest concentrations measured

Table V
Toxic Data Summary for 1990
All Sites

COMPOUND	CONCENTRATION, PPB AVERAGE	MAXIMUM	MINIMUM	# OBS.	LOD	% ABOVE LOD
Halocarbons						
Dichloromethane	1.07	11	< 1	557	1	34
Chloroform	0.029	0.32	< 0.02	575	0.02	61
Carbon Tetrachloride	0.131	0.30	0.06	572	0.02	100
1,1,1 Trichloroethane	1.69	23.00	0.09	577	0.01	100
Ethylene Dichloride	0.100	< 0.2	< 0.2	579	0.2	0
Perchloroethylene	0.275	5.00	0.03	579	0.01	100
Trichloroethylene	0.118	11	< 0.01	570	0.01	91
Ethylene Dibromide	0.007	0.07	< 0.01	575	0.01	14
Aromatics						
Benzene	2.78	12.0	< 0.5	579	0.5	99
Toluene	4.36	29.0	3.0	345	0.2	100
Ethyl Benzene	0.524	3.9	< 0.6	335	0.6	34
1,2 - Xylene	0.748	6.2	< 0.1	345	0.1	97
1,3 - Xylene	1.46	12.0	< 0.6	345	0.6	81
1,4 - Xylene	0.591	4.8	< 0.5	339	0.5	49
Styrene	0.348	1.9	< 0.1	317	0.1	95
Chlorobenzene	0.058	0.3	< 0.1	345	0.1	8
1,2-Dichlorobenzene	0.086	0.6	< 0.1	340	0.1	30
1,3-Dichlorobenzene	0.101	0.3	< 0.2	344	0.2	< 1
1,4\-Dichlorobenzene	0.142	1.3	< 0.2	333	0.2	20
Other						
1,3 - Butadiene	0.313	2.40	< 0.04	435	0.04	97
Formaldehyde	2.11	11.4	< 0.1	585	0.1	100
Acetaldehyde	1.70	8.3	< 0.1	584	0.1	100
Benzo(b)Fluoranthene *	0.443	22	< 0.05	324	0.05	87
Benzo(k)Fluoranthene *	0.182	9.6	< 0.05	324	0.05	52
Benzo(a)Pyrene *	0.382	23	< 0.05	415	0.05	64

NOTE:

1. Concentrations at < LOD are replaced with one half of LOD FOR estimating of AVERAGES.

2. * => Concentrations in ng/m3

(0.22 ppbv in 1989 and 0.32 ppbv in 1990) were at the North Long Beach site. The second highest concentrations occurred there as well. This would indicate a point source in the Long Beach area.

Carbon tetrachloride concentrations found in California are generally between 0.10 and 0 .14 ppbv. High concentrations, up to 0.4 ppbv, are occasionally measured at the Concord site. However, at all other sites, the measured values are consistently similar to reported global background levels.[6]

Trichloroethylene, a solvent used in metal-cleaning operations, is being gradually phased out in favor of methyl chloroform and chlorofluorocarbon solvents. As expected, the measured concentrations have decreased since the monitoring program started. The annual average concentration of trichloroethylene has decreased by 36% from 1989 to 1990. The annual average for 1990, however, is biased artificially high by an exceptional event: one sample collected at the Los Angeles-North Main site with a concentration of 11 ppbv. If this value were removed from the database, the annual average concentration for 1990 would be 0.099 ppbv, a decrease of 46% over 1989.

Perchloroethylene is used extensively in dry cleaning operations. As such, it could be expected that concentrations in California would be fairly consistent from location to location. This, however, is not the case. All samples collected in 1989 and 1990 had measurable concentrations of perchloroethylene above the LOD (0.01 ppbv), with the annual average concentration in the South Coast Air Basin at a level nearly three times that of other areas. This is due, primarily, to elevated concentrations at the Los Angeles and North Long Beach sites. This, coupled with the comments about high levels of methyl chloroform and trichloroethylene, would indicate that both sites are impacted by localized industrial use of solvents. Elsewhere, however, the trend of the average annual concentrations for perchloroethylene has been downward for the last 6 years, with a decrease of approximately 30% from 1989 to 1990. This trend may be caused by better emission controls for dry cleaning operations and improvements in recovery and disposal techniques.

Benzene concentrations in California have not exhibited an identifiable trend over the last 6 years. The concentrations do not differ significantly from site to site. There are no known significant stationary sources within the state, and it has been proposed that the ambient air concentrations are due solely to vehicular emissions. Levaggi and Siu[2] demonstrated a consistent relationship between carbon monoxide and benzene from data taken from the Bay Area Air Quality District toxics program. Furthermore, they reported an ambient air benzene:toluene ratio of approximately 2:1, similar to that found in auto exhaust. This is confirmed by our statewide data. In addition, Poore and LaPurga[7] demonstrated a significant correlation between benzene and 1,3-butadiene, another compound which has vehicular emissions as its only known source in California.

The California Air Resources Board has been performing monitoring activities for 1,3-butadiene since 1987. It is the only organization known with such a comprehensive 1,3-butadiene database. Since 1,3-butadiene concentrations in ambient air represent a significant health risk, this database has allowed for effective decision making. As mentioned above, Poore and LaPurga developed a monitoring method for 1,3-butadiene in 1987. After 2 years of gathering monitoring data, they reported a benzene/1,3-butadiene correlation (correlation coefficient = 0.90) with a regression equation of 1,3-butadiene = 0.163(benzene) – 0.13. A review of the 1989 and 1990 data indicates that this relationship has not changed significantly.

It has been estimated that the major source of formaldehyde and acetaldehyde in ambient air is through photochemical oxidation of hydrocarbons. In California, directly emitted aldehydes are from the combustion of fossil fuels and refinery processes. The CARB has been performing monitoring for formaldehyde and acetaldehyde since 1988. The data show a downward trend for both compounds (formaldehyde, 43% reduction; acetaldehyde, 24% reduction). With only 2 years of data, it is unclear if this trend is indicative, or related to meteorology. Since approximately 90% of atmospheric formaldehyde is from photooxidation, it is possible that the mild summers experienced in California in 1989 and 1990 may have contributed to the decrease. Since only 40 to 60% of the atmospheric acetaldehyde is from photooxidation, the smaller reduction in the acetaldehyde concentration would tend to support this theory.

In 1988, the CARB instituted a monitoring program for benzo(a)pyrene. Since benzo(a)pyrene (BaP) is, at normal temperatures, a nonvolatile polyaromatic hydrocarbon (PAH) and can be efficiently trapped on filter media, the existing PM10 monitoring system was used to collect the samples. Since there is only 2 years of complete data available, trend analysis would be very difficult. In addition, the task is complicated by the fact that annual averages are completely dominated by concentrations during the winter months (October through March). Concentrations during the spring and summer months are close to or below the detection limit (0.05 ng/m^3).

B. Seasonality

Monthly average concentrations for benzene, acetaldehyde, and BaP for the period January 1989 to December 1990 are displayed in Figure 2. These three compounds were chosen to illustrate three classes of compounds: compounds directly emitted at approximately the same rate throughout the year (benzene), compounds which have a photooxidation component (acetaldehyde), and compounds with large variations in emission rates throughout the year. Note that the concentration units for benzene and acetaldehyde are ppbv, while the concentration units for BaP are nanograms per cubic meter.

The monthly average concentrations for benzene peak in the December and January time period, with minimum concentrations during the summer months. The ratio of peak to valley is approximately 3.5. This is typical of any primarily emitted pollutant and is due to lower atmospheric mixing layers in the winter. As you can see in the BaP plot, this effect is strongly influenced by increased emissions during the winter from such things as fireplaces and agricultural burning. The peak to valley ratio for BaP is approximately a factor of 50. Although annual averages are used in trend and exposure analyses, it should be recognized that, for limited information, meteorology and seasonal emissions can dominate such information.

Acetaldehyde, on the other hand, does not exhibit as strong a seasonality pattern as the other two pollutants. The peak to valley ratio is approximately 2.0. Although the low mixing layer causes some elevation of the winter-time concentrations,

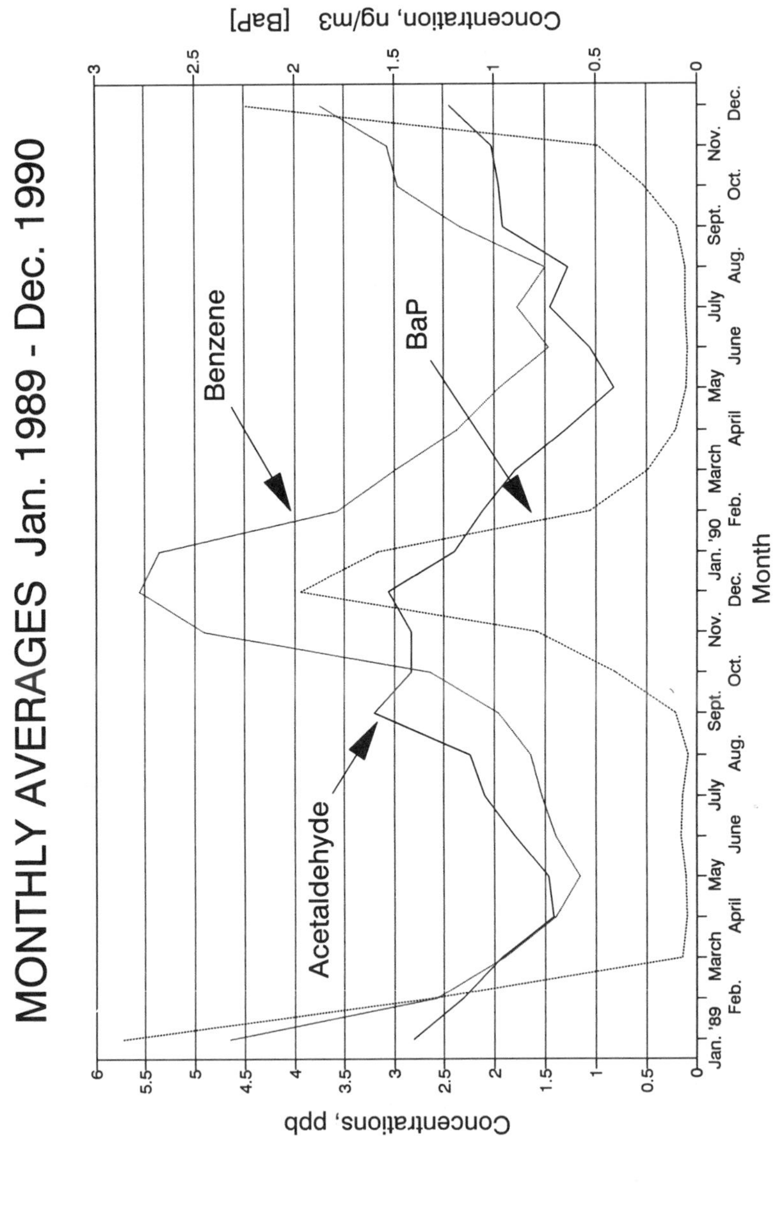

Figure 2. Monthly averages, January 1989 through December 1990.

Table VI

TOXIC'S DATA SUMMARY

Compound	Bay Area Sites	South Coast Sites	Fresno
	Mean	Mean	Mean
Butadiene	0.289	0.444	0.440
Formaldehyde	2.780	4.319	4.067
Acetaldehyde	1.652	2.818	2.433
Dichloromethane	1.059	1.785	0.790
Chloroform	0.030	0.041	0.022
Carbon Tetrachloride	0.125	0.128	0.121
1,1,1-Trichloroethane	1.574	3.560	0.526
1,2-Dichloroethane (EDC)	0.100	0.101	0.100
Tetrachloroethene (Perc)	0.231	0.615	0.152
Trichloroethene (TCE)	0.161	0.235	0.059
Benzene	2.305	3.699	2.999
Toluene	4.105	6.040	4.128
Ethyl benzene	0.510	0.646	0.571
o-Xylene	0.712	0.940	0.783
m-Xylene	1.390	1.845	1.483
p-Xylene	0.567	0.733	0.624
Styrene	0.358	0.445	0.341
Chlorobenzene	0.054	0.055	0.058
o-Dichlorobenzene	0.089	0.087	0.094
m-Dichlorobenzene	0.101	0.100	0.100
p-Dichlorobenzene	0.195	0.147	0.117
Benzo(b)fluoranthene	0.464	0.479	1.636
Benzo(k)fluoranthene	0.148	0.148	0.512
Benzo(a)pyrene	0.468	0.345	1.820
1,2-Dibromoethane (EDB)	0.006	0.006	0.006

photooxidation during the summer causes the levels to be higher than one would expect due simply to emissions.

Carbon tetrachloride, not shown here, is very unusual in that its atmospheric concentration is not affected by season. The measured concentration of carbon tetrachloride is between 0.10 and 0.14 ppbv at almost all locations and seasons. Since meteorology does not effect the measured levels, we believe this confirms that we are measuring world background levels.

C. Interbasin Comparisons

Table 6 is a summary of 1989 and 1990 data for three regions: the San Francisco Bay Area, the South Coast Air Basin, and the Fresno monitoring site. The Fresno site was chosen to represent the California Central Valley. As discussed above, average

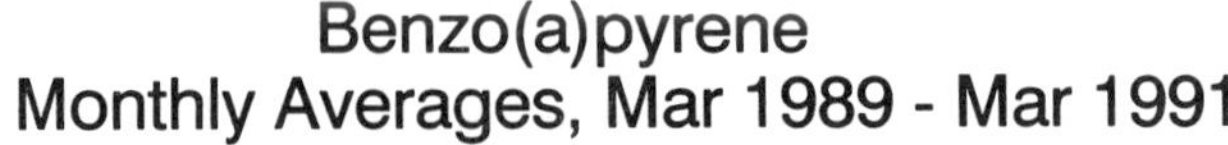

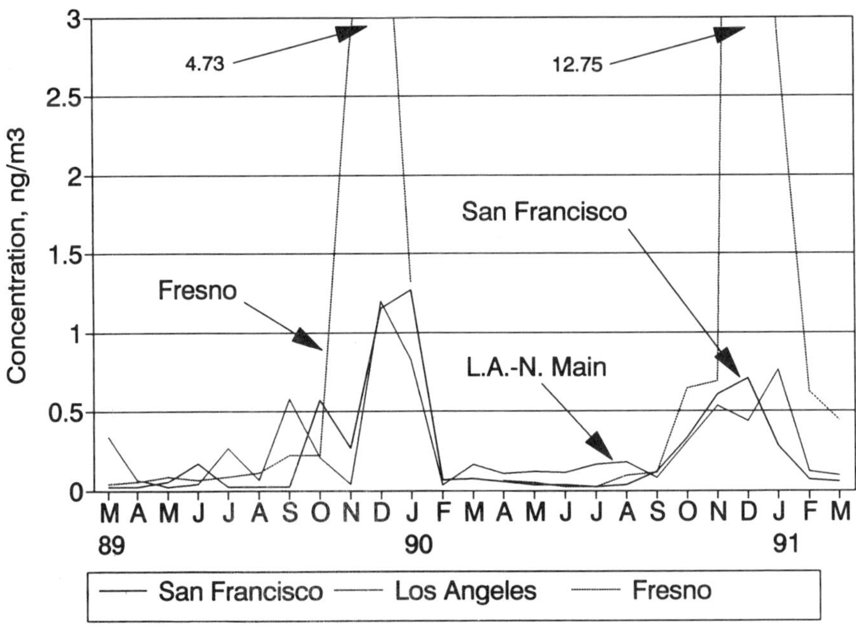

Figure 3. Monthly averages, March 1989 through March 1991.

concentrations for methyl chloroform, trichloroethylene, and perchloroethylene in the South Coast are substantially higher than in the Bay Area or Fresno. This is consistent with emission inventory estimates. With the exception of these three pollutants, it is interesting to note that average concentrations in the South Coast and in Fresno are not, in general, substantially different. The pollutant concentrations in the San Francisco Bay Area, on the other hand, are lower than both areas, especially for those pollutants associated with vehicular emissions.

The one pollutant category for which the California Central Valley dominates the state averages is the polyaromatic hydrocarbons. As can be seen from Table 6, the average concentrations of BaP, benzo(b)fluoranthene, and benzo(k)fluoranthene at the Fresno site are four times those found in the Bay Area and five times those found in the South Coast. As shown in Figure 3, the peak Fresno concentrations during the winter months are more than ten times higher than those found, for example, at Los Angeles or San Francisco.

III. CONCLUSIONS

Since California state law mandating a TAC program went into effect in 1984, the CARB has successfully implemented a 22-site, 38-compound TAC monitoring network for the sampling and analysis of 24-h composite ambient air samples collected every 12 days. Working with a commercial vendor, the CARB staff has developed innovative sampling instrumentation for whole-air, semivolatile, reactive, and nonvolatile compounds. After developing sensitive and sophisticated analytical methods, limits of detection in ambient air have been lowered to levels equivalent to a one-in-a-million population risk for most compounds.

Extensive quality assurance measures have been implemented to define and insure data quality. All of these efforts and innovations have led to California having the most extensive air toxics database in the world.

REFERENCES

1. California Health and Safety Code, Division 26 Air Resources, Part 2, Chapter 3.5, Section 39650 et. seq.
2. Levaggi, D. A. and Siu, W., Gaseous Toxics Monitoring in the San Francisco Bay Area: A Review and Assessment of Four Years of Data, presented at the 84th Annu. Meeting of the Air and Waste Management Association, Paper #91-78.10, June 16–21, 1991.
3. Evaluation and Improvement of Methods for the Sampling and Analysis of Selected Toxic Air Contaminants, Contract #CARB863-R-0235, Battelle, Columbus Division, final report, April 1987.
4. Oslund, W. E., Implementing a Quality Assurance Program and Analysis of Ambient Air Toxics Compounds, Annu. EPA/AWMA Symposium on the Measurement of Toxic Air Pollutants, Raleigh, NC, April 1986.
5. Dunwoody, C. and Effa, R. C., A Review of the Accuracy and Precision of the Toxic Air Contaminant Program of the California Air Resources Board, Annu. EPA/AWMA Symposium on the Measurement of Toxic Air Contaminants, Raleigh, NC, May 1990.
6. Grimsrud, E. P. and Rasmussen, R. A., Survey and Analysis of Halocarbons in the Atmosphere by Chromatography-Mass Spectrometry, *Atmos. Environ.*, 9, 1014, 1975.
7. Poore, M. W. and LaPurga, N., Sampling and Analysis of 1,3-Butadiene in Ambient Air, 82nd Annu. Meeting of the Air and Waste Management Association, 1989.

Comparison of Chemical Composition of Fly Ash Particles Collected in the Plume and Stack of a Coal-Fired Power Plant

George M. Sverdrup
Jane C. Chuang
Laurence Slivon
Andrew R. McFarland
John A. Cooper
Robert W. Garber
Blakeman S. Smith

I. ABSTRACT

Preliminary studies on collection techniques for fly ash particles were conducted. The chemical composition of particulate matter collected in the near-stack plume (within 1 km of the stack) and in the stack of a coal-fired power plant is compared. Samples were composed of particles with diameters less than 10 μm. Plume particulate matter was collected using a helicopter-borne high-volume sampler. Stack samples were collected at actual stack conditions and also under diluted, cooled conditions using a dilution sampler. Organic and inorganic measurements were made on the samples. A benzene/methanol (90:10) solvent was used to extract material from the samples. The extracts of plume samples were partitioned into a nonpolar cyclohexane-soluble fraction and a polar methanol-soluble fraction. The extracts of diluted-stack samples were partitioned into cyclohexane, dichloromethane, and aqueous fractions. Selected sample extracts of plume and diluted-stack samples were fractionated into 12 fractions by high-pressure liquid chromatography (HPLC) for bioassay testing. The results showed that significant amounts of the organic material were present in the polar fractions from both liquid-liquid partitioning and HPLC fractionation. The cyclohexane fractions were analyzed by gas chromatography/mass spectrometry (GC/MS) to determine major components, polycyclic aromatic hydrocarbons (PAHs), and nitro-PAH.

II. INTRODUCTION

The influence of fly ash emissions from coal-fired power plants on public health is unknown. At the time this project was undertaken (1985 to 1989), data on the biological activity of airborne power plant plume particles were scant. Although considerable physical and chemical data had been obtained on fly ash from inside power plants, such data on plume samples were fragmentary, and the uncertainty in the data was high. These data suggested that plume particles may contain more adsorbed compounds than particles from inside the plant (e.g., stack or electrostatic precipitator hoppers). Adsorption of compounds onto plume particles had raised the twofold question of how plume and in-plant samples of particles compare in terms of biological activity, and ultimately what their relevance is for public health. This and other key questions led the Electric Power Research Institute (EPRI) to investigate the biological effects of plume fly ash (BEPFA).

A critical element in addressing these questions is technology to collect representative samples of particulate matter from the plume and stack of power plants that are large enough to permit biological, physical, and chemical testing. Development of this technology was the principal component of preliminary studies on collection, storage, chemistry, and physics of fly ash. Results of storage studies are reported in Reference 1. This paper summarizes the results of studies on collection.

III. HOST POWER PLANT

The Tennessee Valley Authority (TVA) Cumberland Fossil Plant was the host power plant for the project. The Cumberland plant is a base load, coal-fired plant with two identical 1300-MW units. Pulverized Kentucky #9 bituminous coal (3% sulfur, 10% ash) is burned in a water-tube wall furnace with opposed-wall firing. Electrostatic precipitators (ESPs) reduce the normal particle mass concentration from 4.5 g/m^3 at the entrance to the ESP to about 0.02 g/m^3 at its exit.

IV. EXPERIMENTAL DESIGN

To address the project's objectives, concurrent collections of particles from the plume, upwind air, stack, ESP hoppers, and pulverized coal injectors were made in the Cumberland Field Study (CFS). These collections and related analysis are listed in Table 1. Plume collections were made with a high-volume particle sampler fitted to a helicopter. Because the plume samples are comprised of both electric utility emissions and ambient aerosol, simultaneous upwind collections of ambient aerosol were made using a small fixed-wing aircraft to permit estimation of the utility contribution to the collected plume particulate matter. A stack dilution sampler was used to sample a stream of hot stack gas, dilute it with filtered ambient air inside the sampler, and collect the diluted gas stream. This technology was used to investigate whether or not dilution-sampler technology can provide samples that are acceptable surrogates for plume samples. Hot stack samples were collected to provide a reference for investigating the extent of adsorption of compounds onto fly ash particles as the particles cool in the plume or dilution sampler. Samples were collected from an ESP hopper and pulverized feed coal to document operation of the power plant.

Table 1. Measurements Made for the CFS

Measurement	Collection					
	Plume	Upwind	Diluted stack	Stack	ESP	Coal
Bulk Particles[a]						
PAH	X	X	X	X	X	
Nitro-PAH	X		X	X	X	
Sulfur compounds	X					
Elements	X		X	X	X	X
Individual particles[a]						
Size distribution	X	X	X	X	X	
Morphology	X	X	X	X	X	
Elements	X	X	X	X	X	
Gas-phase PAH	X		X	X		

[a] Particle diameter less than 10 μm.

The chemical and physical properties listed in Table 1 were analyzed. Properties of bulk particle samples, individual particles, and the gas-phase component of the semivolatile PAHs were measured. In this context bulk particle samples refer to a wide range in quantity of material — from a few milligrams on a filter to a kilogram of ESP hopper ash. These measurements were selected because of their relevance to possible biological activity or their importance in gaining an improved understanding of PFA characteristics. For relevance to public health, collections and analyses were restricted to those particles that possessed aerodynamic diameters less than about 10 mm. Results of organic and sulfate measurements are reported herein.

V. COLLECTION METHODS

Collection methods and sample handling procedures are summarized in this section. Reference 2 contains details.

A. Plume

Three primary goals for the plume sampler were to (1) maximize the quantity of collected plume particulate matter comprised of particles with diameters less than 10 mm, (2) ensure compatibility of the collected samples with the analytical techniques used for biological, chemical, and physical analyses, and (3) minimize artifact formation in the samples.

A schematic diagram of the plume sampler is shown in Figure 1. An air sample was drawn through the inlet at a rate of 9.1 m^3/min by a propane-powered engine housed in the fuselage of the helicopter. The sample stream was decelerated in the transition tube, drawn through a horizontal elutriator to remove particles with diameters greater than 10 mm from the air stream, and then drawn into the filter housing. A bank of seven quartz fiber filters (each with an effective filtration area of 406 cm^2)

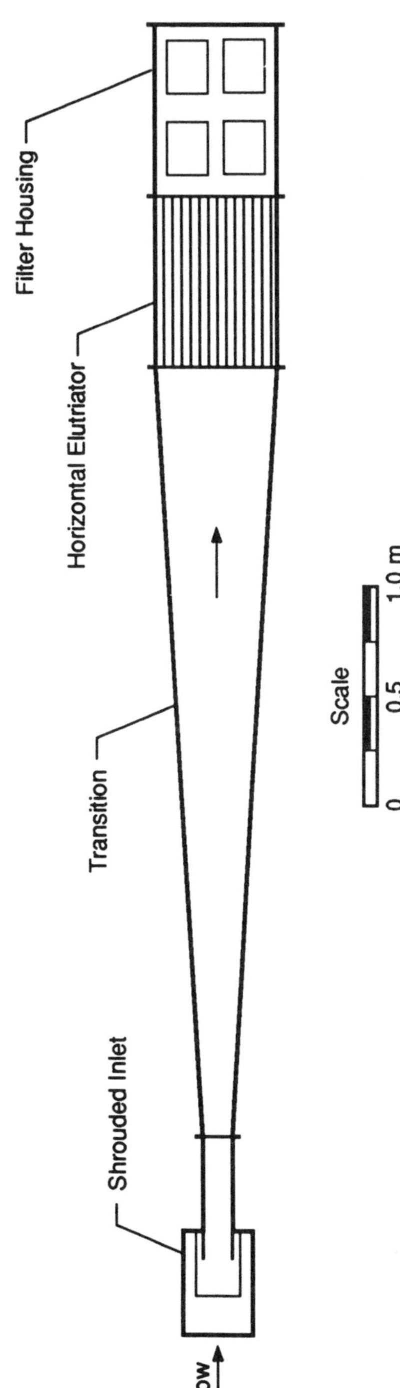

Figure 1. Schematic diagram of sampler.

collected material destined for organic analyses. A single Teflon® filter of the same dimensions collected material for elemental analyses. In addition, 47-mm diameter Teflon® filters collected particles for microscopy[3] and a cascade impactor collected material for determination of the size distribution of sulfur compounds.[4]

To maximize the quantity of collected plume particulate matter in these preliminary studies, the helicopter sampled the plume by flying in the plume along its axis from a point about 1 km downwind of the stack to a distance about 200 m from the stack.

B. Upwind Aerosol

A Cessna 172 aircraft was used to provide samples of ambient aerosol particles from upwind of the stacks. Particles were collected on a quartz fiber filter. Also onboard the aircraft were two 47-mm diameter filter samplers for microscopy.

C. Diluted Stack

The dilution source sampler was modified to create a stainless steel system suitable for organic measurements.[5,6] Two quartz fiber filters collected material for organic analysis, and Teflon® and two-membrane filters were used to collect material for elemental analysis and microscopy. The sampler was operated with a dilution ratio of about 30:1. The collection temperature was in the range 20 to 25°C. The collection period was 5 h.

D. Sample Handling

The quartz fiber filters were cleaned in the laboratory by baking them in a muffle furnace for 16 h at 400°C. Aluminum foil was cleaned by the same procedure. The filters were packaged in aluminum foil until their use. In the field, the filters were removed from their package, weighed, and placed in filter holders just before sampling commenced. Following collection the filters were removed from their holders, weighed at ambient conditions, folded using a clean forceps, placed in a clean glass Petri dish, sealed, and then stored at –79°C until the time of their analysis.

A collection was defined as one period of time (4 to 5 h) in which sampling took place. A total of 21 collections were made during the period September 11 through October 7, 1987. Eight plume collections were made with concurrent upwind and diluted-stack sampling. In order to maximize the quantity of material that could be made available for biological assays, collected material was combined to form pooled samples. For example, collections 12 and 14; 16 and 18; and 13, 20, and 21 were combined to form pooled samples C, D, and E. Collection 11, an upwind collection involving both the helicopter and Cessna, constitutes pool F. The data discussed in this paper are primarily results for each of plume, upwind aerosol, and diluted-stack pooled samples C, D, and E.

VI. ANALYTICAL METHODS

A. Extraction

All quartz fiber filters from a single collection (seven filters for plume samples, two filters for diluted-stack samples, and one filter for upwind samples) were Soxhlet extracted with benzene/methanol (90:10, v:v) (BM) for 16 h. The use of BM as extracting solvent was selected by SRI International based on its previous studies that showed BM extracts of PFA have the highest mutagenicity as compared to other solvents examined. The filters were then extracted with water for an additional 8 h. The BM extracts were concentrated by Kuderna-Danish (K-D) evaporation. The water extracts were concentrated by rotary evaporation for sulfate analysis. Extracts from individual collections were combined to form pooled samples. The BM extracts of plume and diluted-stack samples were used for liquid-liquid partitioning, HPLC fractionation, sulfate analysis, and bioassay testing. The BM extracts of upwind aerosol samples were used for chemical analysis and bioassay testing.

1. Liquid-Liquid Partitioning

The BM extracts of the pooled plume samples were partitioned with cyclohexane ($C\text{-}C_6$) into two layers — $C\text{-}C_6$-soluble portion and methanol (MEOH)-soluble portion. The BM extract of diluted-stack sample was adjusted to pH of 4 to 6 by adding $NaHCO_3$ solution. The extract was then partitioned with $C\text{-}C_6$ into a $C\text{-}C_6$ fraction and an aqueous fraction. The aqueous fraction was further partitioned with dichloromethane (DCM) into a DCM fraction and an aqueous fraction.

2. HPLC Fractionation

The BM extracts of plume and diluted-stack samples were fractionated on a semipreparative (10 × 250 mm) HPLC column packed with Rainin Dynamax® silica. The eluate was monitored for UV absorbance at 254 nm at a flow rate of 3.6 mL/min. The gradient condition started with hexane/DCM (95:5) for 10 min, followed by a gradient to 100% DCM in 10 min and a 10-min hold, then to 100% acetonitrile in 10 min and a 10-min hold, and finally to 100% MEOH in 1.0 min and a 20-min hold. A Gibson 201 fraction collector was programmed to collect 12 5-min fractions; 16 replicates of the plume sample (253 mg) and 36 replicates of the diluted-stack sample (521 mg) were processed.

3. GC/MS

Sample extracts were analyzed by gas chromatography/mass spectrometry (GC/MS) with electron impact (EI) to determine PAH and major components and by GC/MS with negative chemical ionization (NCI) to determine nitro-PAH. A Finnigan

Table 2. Organic Extractable Mass of the Pooled Plume, Upwind Aerosol, and Diluted-Stack Samples

	Mass concentration (μg/m³)[a]				
	Plume			Diluted stack	
Pool	EM	Mass	Upwind EM	EM	Mass
C	240	340	3.6	2800	3100
D	340	440	3.0	2700	3600
E	390	660	5.3	2400	3700
Average	320 ± 80	480 ± 160	4 ± 1	2600 ± 200	3500 ± 300

[a] Mass concentration at 25°C and 1 atm; EM = extractable mass.

TSQ-45 GC/MS/MS system operated at GC/MS mode and an INCOS 2300 data system were employed. Methane with a 150-eV electron beam was the reagent gas for NCI. The MS was operated at EI mode in the full scan (FS) mode to determine major components present in each type of sample. The MS was operated in the selected ion monitoring (SIM) mode to determine concentrations of PAH and nitro-PAH. The GC column was an ultra No. 2, fused silica, capillary column (50 m × 0.322 mm; 0.51-mm film thickness, Hewlett-Packard Company). Identification of PAH and nitro-PAH was based on their GC retention times relative to that of the corresponding internal standards (9-phenylanthracene for PAH, and 1-nitropyrene-D_9 for nitro-PAH). Quantification of target compounds was based on comparisons of the respective integrated ion current responses of the target ions to that of the corresponding internal standards, with calibration response curves of each target compound generated from standard analyses.

VII. RESULTS

A. Extractable Mass

The BM extractable mass from pooled plume, upwind aerosol, and diluted-stack filter samples is summarized in Table 2. Data are presented as extractable organic mass and total mass per volume of air to indicate the physical conditions in which sampling took place. The average plume dilution was a factor of 60 compared to the stack conditions. The diluted-stack data reflect an operating dilution ratio for the sampler of about 30. Note that total mass data were not obtained for the upwind samples that were collected with the small aircraft. On the one mission that the helicopter flew upwind (pool F), the total mass concentration was 40 mg/m³, and extractable organic mass was 10% of the total.

The plume samples consisted of about 68% extractable mass (BM solvent), whereas the diluted-stack samples were comprised of about 77% extractable mass. In contrast, the ESP hopper ash contained only 0.42% extractable mass, and the two

Table 3. Mass Distribution of Liquid-Liquid Partitioning of Pooled Plume and Diluted-Stack BM Sample Extracts

Sample	Fraction mass (mg)		
	Pool C	Pool D	Pool E
Plume			
Unfractionated BM extract	360	400	800
C-C_6 fraction	6.7	4.1	8.2
MEOH fraction	300	380	740
Diluted stack			
Unfractionated BM extract	600	600	600
C-C_6 fraction	0.44	1.4	1.6
DCM fraction	0.60	0.60	0.56
Aqueous fraction	NA[a]	NA	NA

[a] NA denotes that the residue weight measurements were not determined.

pooled hot stack samples (pools A and B collected earlier in the field study) had extractable mass components of 7.5 and 25%, respectively.

The average sulfate content of the plume and diluted-stack filter samples was 8.2 ± 2.4% and 21 ± 2%. Most of the sulfate was removed in the BM extraction, and the subsequent aqueous extraction yielded less than either 10% (plume) or 5% (diluted stack) of the total sulfate content. These results show, as expected, an increased sulfate fraction in particulate matter compared to the feed coal. Both extractable organic material and sulfate are consistent with accumulation of compounds on the surface layer of fly ash as it cools upon exit from the stacks.

Results from both the liquid-liquid partitioning (Table 3) and HPLC fractionation show that the BM extracts of plume and diluted-stack samples were comprised predominantly of polar components. The methanol portion of the BM solvent effectively removed much of the polar material along with the relatively nonpolar material extracted by the benzene. The BM extracts of plume (pH of 4 to 6) and diluted stack (pH of 2 to 4) were acidic. Due to the polar and acidic nature of the BM extracts, the BM extracts were fractionated into polar and nonpolar fractions by using liquid-liquid partitioning prior to PAH and nitro-PAH analysis.

As shown in Table 3, the nonpolar components only represent a small portion of the BM extracts for both plume and diluted-stack sample extracts. The mass distribution of HPLC fractions showed that 98% of the mass existed in the four most polar fractions. The PAH should be eluted in the second fraction based on the HPLC retention times of selected PAH. This fraction contained less than 1% of total mass in both plume and diluted-stack samples.

B. PAH Concentration

The PAH data from pooled plume, upwind aerosol, and diluted-stack filter samples are summarized in Table 4. Comparison of plume and upwind PAH data reveals no discernible difference in PAH concentrations. Only the particle-bound

Table 4. Particulate PAH Concentrations in the Pooled Plume, Upwind Aerosol, and Diluted-Stack Samples

	Plume concentration (ng/m)[a]			Upwind aerosol (ng/m^3)		Diluted stack (ng/m^3)		
Compound	**Pool C**	**Pool D**	**Pool E**	**Pools C and D**	**Pool E**	**Pool C**	**Pool D**	**Pool E**
Phenanthrene	0.26	0.36	0.24	0.22	0.20	0.98	0.61	0.37
Anthracene	0.022	0.037	0.023	0.019	0.021	0.075	0.049	0.13
Fluoranthene	0.080	0.11	0.044	0.10	0.087	0.18	0.14	0.10
Pyrene	0.017	0.025	0.023	0.063	0.056	0.16	0.091	0.073
Cyclopenta[*c,d*]pyrene	ND[b]	ND	ND	ND	ND	ND	ND	ND
Benz[*a*]anthracene	0.015	0.013	0.0092	BQL[c]	0.018	BQL[d]	BQL[e]	BQL[f]
Chrysene	0.091	0.051	0.044	0.042	0.056	BQL	0.074	BQL
Benzofluoranthenes	0.046	0.036	0.018	0.032	0.041	BQL	0.064	0.10
Benzo[*e*]pyrene	BQL[g]	0.015	0.0048	0.018	0.021	BQL	BQL	BQL
Benzo[*a*]pyrene	BQL[g]	0.0097	BQL[h]	BQL[c]	BQL[i]	BQL	BQL	BQL
Indeno[1,2,3-*c,d*]pyrene	BQL[g]	0.016	0.0056	BQL[c]	BQL[i]	BQL	BQL	BQL
Benzo[*g,h,i*]perylene	BQL[g]	0.016	0.0056	0.018	0.016	BQL	BQL	BQL

[a] Determined from the cyclohexane fraction of pooled samples C, D, and E.
[b] ND = Not detected.
[c] BQL = Below quantification limit of 0.013 ng/m^3.
[d] BQL = Below quantification limit of 0.047 ng/m^3.
[e] BQL = Below quantification limit of 0.044 ng/m^3.
[f] BQL = Below quantification limit of 0.041 ng/m^3.
[g] BQL = Below quantification limit of 0.0068 ng/m^3.
[h] BQL = Below quantification limit of 0.0048 ng/m^3.
[i] BQL = Below quantification limit of 0.016 ng/m^3.

Table 5. Particulate Nitro-PAH Concentrations in Pooled Plume Samples

Compound	Concentration (ng/m^3)		
	Pool C	Pool D	Pool E
1-Nitronaphthalene	0.035	0.0030	0.0018
2-Nitronaphthalene	0.0026	0.0020	0.0017
Nitroanthracene isomer[a]	0.55	0.22	0.21
9-Nitroanthracene	0.0051	0.0055	0.0048
9-Nitrophenanthrene	0.014	0.019	0.0068
2,3-Nitrofluoranthene[b]	0.010	0.034	0.0024
1-Nitropyrene	0.0083	0.023	0.018

[a] Instrument response for 9-nitroanthracene used to estimate the concentration of the unidentified isomer.
[b] The two compounds, 2-nitrofluoranthene and 3-nitrofluoranthene are coeluted.

PAH concentrations were measured on the filter samples. Based on a previous study,[7] approximately 30 to 90% of three-ring PAHs and 30 to 60% of four-ring PAHs are present in the vapor form and are not retained by filters during sampling. Therefore, the total three- to four-ring PAH concentrations present in the plume, diluted-stack, and upwind samples were expected to be higher than reported here. The results also show that there was no significant difference in PAH concentrations among the plume and diluted-stack samples. Note that, in the results described before (Table 2), much higher amounts of extractable mass were found in diluted-stack as compared to plume samples. These results suggest that the presence of PAH did not contribute to the difference of the extractable mass between the plume and diluted-stack sample. This difference is mainly due to the unknown polar components.

C. Nitro-PAH Concentration

The nitro-PAH data from pooled plume filter samples are summarized in Table 5. Financial constraints limited determination of nitro-PAH to analysis of the plume samples. Strong direct-acting mutagens such as 2,3-nitrofluoranthene and 1-nitropyrene were found in the plume samples. These compounds have been identified in both ambient and indoor air samples in other studies.[8,9] In general, the levels of nitro-PAH concentrations in the plume samples were similar to those found in suburban ambient air.[8,9]

D. Identification of Nontarget Organic Compounds

The C-C_6 fractions of the plume and diluted-stack sample extracts were also analyzed by GC/MS to identify the organic components in addition to PAH and nitro-PAH. Results are shown in Table 6. Organic components found in the plume samples are alkyl benzene, alkanes, organic acids, and organic acid esters that are very similar to the components present in diluted-stack samples. Note that a homologous series of unknown components with molecular weights of 180 and 194 was detected in the

Table 6. **Compounds Identified in the Nonpolar Fraction of Particulate Samples**

Compound	Approximate concentration (ng/m^3)[a]		
	Plume	Upwind	Diluted stack
Tetrachloroethane	2	2	—
C_2, alkyl benzenes[b]	1	2	1+
C_3, alkyl benzenes	3	—	2
C_3, methoxyl benzenes	3	—	—
C_4, alkyl benzenes	3	—	1
Aliphatic hydrocarbons	1	2	1
Fatty acids[c]	1	3	1+
Fatty acid esters	2	3	1+
Sulfuric acid[b]	2	—	1
Sulfuric acid dimethyl ester	2	—	1
Benzoic acid[c]	3	—	1
Benzoic acid methyl ester	3	—	1
Benzene acetic acid[c]	4	—	—
Benzene acetic acid methyl ester	4	—	—
C_1, alkyl benzoic acid[c]	4	—	—
C_1, alkyl benzoic acid methyl ester	4	—	—
Methoxy benzene acetic acid[c]	3	—	2
Methoxy benzene acetic acid methyl ester	3	—	2
Benzene sulfonic acid[c]	3	—	1+
Benzene sulfonic acid methyl ester	3	—	1+
Phenyl benzene acetaldehyde	4	—	1
Benzene tricarboxylic acid[c]	3	—	1+
Benzene tricarboxylic acid trimethyl ester	3	—	1+
Phthalates	1	3	1
Nitrobenzene	2	—	2
Biphenyl	3	4	2
C_1, alkyl biphenyls	3	—	1
Unknown homologs of MW 180 and MW 194	3	—	—
Benzothiazole	—	—	1
Sulfonyl bisbenzene	—	—	1
Dichloro benzoic acid[c]	—	—	1
Dichloro benzoic acid methyl ester	—	—	1
Benzene dicarboxylic acid[c]	—	—	1
Benzene dicarboxylic acid dimethyl ester	—	—	1
Phenyl benzene acetic acid[c]	—	—	1
Phenyl benzene acetic acid methyl ester	—	—	1
Unknown nitrogen-containing compound of MW 221	—	—	1+

[a]

Concentration code	Estimated concentration range (ng/m^3)
1+	> 100,000
1	10,000–99,999
2	1,000–9,000
3	100–999
4	10–99

[b] The terms C_n refer to n carbon atoms present in groups on the parent compound. The data do not indicate, for example, if an alkyl group contains all methyl groups or combinations of alkyl side chains of varying lengths.

[c] Free acids were identified as acid esters in the extracts because some fatty acids may have been converted to their fatty acid esters during Soxhlet extraction with the benzene/methanol solvent.

plume samples but not in the diluted-stack samples. A group of nitrogen-containing compounds with molecular weights of 221 was only found in the diluted-stack samples. The results reported here only represent the nonpolar partition (cyclohexane fraction) of the plume and diluted-stack samples. None of the methanol fractions of these samples was analyzed by GC/MS because most of the polar components do not elute from the GC column.

VIII. DISCUSSION

Discussion of the results of these preliminary sampling studies centers on three questions.

1. Can a particle collector that operates on the principle of high-volume air filtration be used successfully with a helicopter to collect representative samples of plume particulate matter that are of sufficient quantity to permit biological, chemical, and physical testing?

The plume sampler collected an average of 410 ± 210 mg of particulate matter from the near-stack plume on each sampling mission. When pooled together these samples yielded a sufficient quantity of material for a battery of tests which included gravimetric, GC/MS, instrumental neutron activation analysis, microscopy, and Kado modification of the Ames assay.

2. Are samples collected with a stack dilution sampler an adequate surrogate for plume fly ash samples?

Comparing the mass fractions of plume and diluted-stack samples that are soluble in solvents of various polarities, it appears that the diluted-stack samples are at least qualitatively similar to near-stack plume samples. The diluted-stack samples possess higher sulfate concentrations which may be the result of condensing more sulfuric acid within the sampler compared to the open atmosphere.

Considering the organic data, several compounds were identified in the nonpolar fraction of both the plume and diluted-stack samples. Present in both samples were alkylbenzenes, alkanes, organic acids, and organic acid esters. Only qualitative comparisons can be made regarding comparability of PAH concentrations in the two types of samples. The measured concentrations were very low, especially for compounds heavier than chrysene. The PAH results from the two types of samples are qualitatively similar.

3. What are the chemical and physical characteristics of particulate matter in a power plant plume within 1 km of the stack?

This summary of the CFS focuses on the organic composition of plume samples; inorganic and physical characteristics were measured but are not reported herein. The plume samples were largely comprised of polar material: the nonpolar (cyclohexane-soluble) fraction of the BM fraction of the plume samples was only 1.4%. Sulfate comprised 8.2% of the sample mass, further indicating that the bulk of the polar fraction of the BM extract likely consisted of a combination of organic compounds and water.

The PAH and nitro-PAH results leave open the possibility that artifact generation occurred in the samples. The combination of very low values of PAH and relatively high ratios of nitro-PAH to PAH raise the concern. However, the data from the plume, diluted-stack, and upwind samples are internally consistent for PAH. If the Cumberland plant was generating very low levels of PAH, little PAH would have been available to accumulate on particle surfaces. Studies on the effects of combining and concentrating large groups of samples that contain complex acidic chemical mixtures were not undertaken in this project. The samples were processed in a batch mode according to preestablished protocols.

IX. RECOMMENDATIONS

The chemistry of plume particulate matter is not yet sufficiently understood to permit assessment of electric utility contribution to air toxics in the vicinity of coal-fired power plants. This preliminary plume study demonstrated that sufficient quantities of plume particulate matter can be collected for a battery of analyses. A series of further plume studies is recommended in which samples are collected in the near-stack plume as well as further downwind. Such a study is required to evaluate the possible influence of emitted sulfur and nitrogen oxides on particle samples during collection.

Further work is needed on protocols to extract several groups of samples and then combine and concentrate their extracts into a single pool of material. Characterization of various fractions of polarity in the sample extracts, especially the polar components, is needed throughout the sample preparation process.

Additional testing of source dilution sampler technology is warranted. The lack of observed PAH enhancement in the CFS diluted-stack samples compared to earlier projections needs confirmation by additional testing before this technology is routinely used.

ACKNOWLEDGMENT

This work was carried out under RP2482-5 for the Electric Power Research Institute.

REFERENCES

1. Sverdrup, G. M., Buxton, B. E., Chuang, J. C., and Casuccio, G. S., Determination of optimal storage conditions for particles, *Environ. Sci. Technol.,* 24, 1186, 1990.
2. Biologic effects of plume fly ash: Preliminary studies — collection, storage, chemistry, physics, Report from Battelle to Electric Power Research Institute, RP2482-5, 1991.
3. Sverdrup, G. M., Casuccio, G. S., and Henderson, B. C., Characterization of the trace element concentration and morphology of stack and near-field plume fly ash particles, Reprint 89-71.2, 82nd Annual Meeting of the Air and Waste Management Association, Anaheim, CA, June 25–30, 1989.
4. Sverdrup, G. M., Measurement of acid aerosols in the near-field plume of a coal-fired power plant, presented before the Division of Environmental Chemistry, American Chemical Society, Miami, September 10–15, 1989.
5. Sousa, J. A., Houck, J. E., Cooper, J. A., and Daisey, J. M., The mutagenic activity of particulate organic matter collected with a dilution sampler at coal-fired power plants, *JAPCA,* 37, 1439, 1987.
6. Houck, J. E., Cooper, J. A., and Larsen, E. R., Dilution sampling for chemical receptor source fingerprinting, Paper 82-61M.2, presented at the 75th APCA Annual Meeting, New Orleans, June 1982.
7. Coutant, R. W., Brown, L., Chuang, J. C., and Riggin, R. G., Phase distribution and artifact formation in ambient air sampling for polynuclear aromatic hydrocarbons, *Atmos. Environ.,* 22, 409, 1988.
8. Chuang, J. C., Mack, G. A., Petersen, B. A., and Wilson, N. K., Identification and quantification of nitro-polynuclear aromatic hydrocarbons in ambient and indoor air particulate samples, in *9th Int. Symp. on Chemical Analyses and Biological Fate, Polynuclear Aromatic Hydrocarbons,* 1986, 155.
9. Chuang, J. C., Mack, G. A., Kuhlman, M. R., and Wilson, N. K., Polycyclic aroamtic hydrocarbons and their derivatives in indoor and outdoor air in an eight-home study, *Atmos. Environ.,* 258, 369, 1991.

Chapter 4
Health and Environmental Information and Models

Chairs: Donald B. Porcella, EPRI
Ronald E. Wyzga, EPRI

An A Priori Screening Methodology to Identify Air Toxics of Potential Ecological Concern: A Methodology for Quantitative Risk Assessment

George E. Taylor, Jr.

I. INTRODUCTION

An airborne pollutant is defined as any chemical of anthropogenic origin in the atmosphere that has the potential to influence human health and/or welfare.[1] This definition encompasses many of the estimated 65,000 chemicals that are released into the environment from an array of point sources (e.g., power plants, waste incinerators, chemical plants, oil refineries), area sources (e.g., agriculture, forestry), and mobile sources (e.g., motor vehicles, aircraft).[2] Although pollutants are chemically diverse in their physical and chemical attributes, they can be categorized as industrial organics (e.g., polycyclic aromatic hydrocarbons, polychlorinated biphenyls), pesticides (e.g., toxaphene, atrazine), inorganic gases (e.g., sulfur oxides, ammonia), and trace metals (e.g., mercury, vanadium, lead). The U.S. Environmental Protection Agency has recently compiled a list of toxic chemicals, with a primary focus on their potential to affect human health.[3] The new amendments to the Clean Air Act identify 189 airborne chemicals that have the potential to adversely affect human health and welfare. For many of these chemicals, experimental data regarding their ecological toxicity do not exist.

Once released to the atmosphere, pollutants undergo a variety of physical, chemical, and photochemical processes that determine their transport and fate (Figure 1).[1] As a consequence, the atmosphere serves a dual function as transporter of chemicals and large stirred reactor. The former results in chemicals being dispersed at distances far removed from the source, whereas the latter chemically alters the pollutants. These atmospheric reactions can either degrade chemicals to rudimentary inorganic components (e.g., ozone to oxygen), thus serving as an effective sink, or transform the chemicals into secondary polluants, sometimes of greater toxicity (e.g., oxides of nitrogen to peroxyacetyl nitrate).

While the physical and chemical dynamics of the atmosphere have received attention with respect to the fate of airborne toxics,[4] an analogous role of the biosphere as a dynamic reservoir influencing the transport, transformation, and fate of chemicals is less well documented. The processes that govern the transport of an airborne chemical from the atmosphere to the biosphere are also of critical

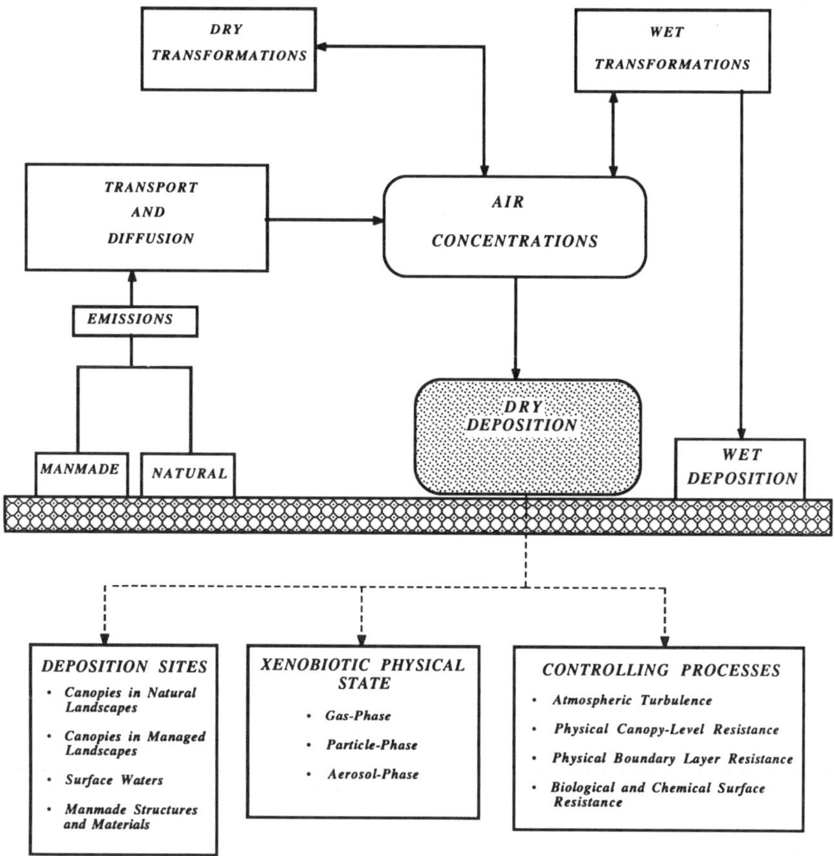

Figure 1. Schematic diagram of the gas-phase (atmosphere and biosphere) processes governing the transport, decomposition, and fate of airborne toxic chemicals within the atmosphere and biosphere. (Adapted from Schroeder, W. H. and Lane, D. A., *Environ. Sci. Technol.,* 22, 240, 1988.) The diagram highlights the role of dry deposition processes in governing the capacity of terrestrial vegetation to serve as a sink for xenobiotics, with attention to processes operating at the level of whole canopies, individual leaves, and the leaf interior. The lower third of the figure partitions the relevant mechanics of dry deposition for xenobiotics into the (1) sites of deposition in the biosphere, (2) physical state of the xenobiotic, and (3) biological, chemical, and physical processes controlling deposition.

importance to its fate, and an understanding of those processes is essential to the development of strategies for quantitative risk assessment and risk management of air toxics.[5]

One of the more important first principles of pollutant-biosphere interactions is that, while there are a number of factors contributing to the ecological toxicity of an airborne chemical, one of the most crucial is the efficacy with which the chemical is partitioned from the atmosphere to the biosphere. The flux of an air toxic compound across this interface is a prerequisite to the compound (1) entering environmental pathways in terrestrial landscapes (biogeochemistry) and (2) eliciting a toxicological response in biota. It is therefore proposed that a mechanistically based model of the processes governing the partitioning of airborne toxicants between the atmosphere and vegetation can be used as a screening methodology to identify those compounds with the highest probability of ecological concern.

This paper presents a model that could serve as a basis for an a priori screening methodology and discusses the model's applicability using two case studies. The paper addresses (1) gas-phase air toxics, (2) atmosphere-to-plant canopy pathway in general and atmosphere-to-individual leaf pathway specifically, and (3) transport and fate of air toxics rather than biological effects on plants and animals. Because the focus is vegetated landscapes (natural and intensively managed), the conclusions are not applicable to landscapes in which the interface between the atmosphere and earth's surface is dominated by surface water or soils.

The objective of this paper is to identify a set of physiochemical parameters from which predictions can be offered with respect to the probability that a given compound will be transported from the atmosphere to the biosphere. The elucidation of this structure-activity relationship would permit a first-principles approach to screening for airborne pollutants that have the potential for human health effects (through introduction into the food chain) or ecological effects and for which experimental data are not available. This analysis is an initial part of any quantitative ecological risk assessment activity for gas-phase pollutants. As an example, two case studies are presented for specific air toxic chemicals. The first is toluene, a regionally and globally distributed pollutant in the atmosphere originating from mobile and industrial sources. The second is methyl isothiocyanate, a volatile pesticide that is locally elevated during soil application and accidental spills and thus has the potential to affect nontarget species.

II. MODEL DEVELOPMENT

A. First Principles

The transport of gaseous chemicals from the atmosphere to plant canopies is governed by the interplay of turbulent and molecular process, with the former dictating movement within the free troposphere and plant canopy, and the latter operating at the level of the leaf surface and leaf interior (Figure 1).[6,7] Whereas the path length of a gas-phase air toxic compound being transported to the leaf surface is on the order of 10^0 m, this path length is several orders of magnitude greater than the subsequent segment between the leaf surface and leaf interior ($\leq 10^{-4}$ m). For most airborne chemicals, path length is inversely proportional to the degree of control that segment exerts on flux, so that processes operating on the leaf surface or within the leaf interior control the ease with which many gas-phase air toxics deposit to plant canopies.[8]

Because the transport of gases at this interface is via diffusion, first principles dictate that molecular processes alone determine flux. Accordingly, the effectiveness of a given canopy to function as a sink for an array of air toxics in the atmosphere reflects the molecular attributes of the chemicals, which are gas specific. If these physical and chemical attributes can be identified, a basis for predicting the flux of any gas-phase chemical in the atmosphere can be developed.

At the level of the individual leaf, a gas-phase air toxic compound may be deposited to one of two surfaces. The first is the external surface of the leaf, which is cuticular in nature and chemically hydrophobic. However, in most ambient conditions the cuticular surface is likely to be altered by a surface layer of water deposited as rain or cloud water or adsorption of water (H_2O) vapor molecules from the atmosphere. Accordingly, while this surface is chemically lipophilic in nature, the ambient environment results in a composite surface, exposing deposition sites that are both lipophilic and hydrophilic.

The second site of deposition is tissues of the leaf interior.[6,7] Unlike the planar nature of the leaf surface, the leaf interior consists of a stacked array of tissue layers that present an area for deposition that is 15 to 40 times larger than that of the leaf surface. Access to the leaf interior is controlled by the porosity of the stomate, which is a consequence of guard-cell physiology and strongly responsive to environmental stimuli such as light, temperature, vapor pressure deficit, soil water availability, and the concentration of atmospheric pollutants, including carbon dioxide, ozone, and sulfur dioxide (SO_2). The deposition sites within the leaf interior are hydrated, extracellular surfaces either in the substomatal cavity or at progressively greater depths in the leaf interior. It is important to recognize that the gas-phase pathway from the free troposphere is continuous with the gas phase of the intercellular spaces of leaf interior. Consequently, unlike deposition onto the leaf surface, deposition of a gas-phase air toxic compound at sites within the leaf interior is controlled by two processes, one operating at the guard cell and effectively regulating access to the leaf interior and the other operating at the interface between the intercellular space and hydrated cell surface of the mesophyll cells.

B. The Analogue Resistance Model

The flux of gaseous pollutants to individual leaves is driven by the chemical potential gradient between the atmosphere and deposition sites either on exterior surfaces or cells of the leaf interior.[1,6] Flux (J) can be represented as the ratio of the chemical potential gradient (ΔC) to the sum of physical, chemical, and biological resistances (ΣR_L) to diffusion along the pathway. The ΔC is calculated as the difference between the gas-phase concentration in the free troposphere (C_a) and that at the deposition site (C_i). The ΣR_L represents the aerodynamic (R_a), leaf boundary layer (R_b), and leaf surface (R_s) resistances operating in series, with the last (R_s) comprised of parallel pathways to the leaf surface ($J_{SURFACE}$) and leaf interior ($J_{INTERIOR}$). The R_s component represents processes in both the gas (stomate) and liquid (cell surfaces within the leaf interior) phases. Mathematically, the relationship between J, ΔC, and ΣR_L is expressed in a form analogous to Ohm's law (for resistances in series):

$$J = [\, C_a - C_i \,] \cdot [\, R_a + R_b + R_s \,]^{-1} \tag{1}$$

The units of diffusive resistance at the individual leaf level (R_L) are s cm^{-1}, calculated as the ratio of concentration to deposition. At the whole canopy level, deposition of a gas-phase air toxic to a landscape is quantified in terms of deposition velocity (V_d), expressed in units of cm s^{-1}. Whereas the reciprocal of resistance (leaf conductance or g_l) has the same units as V_d (cm s^{-1}), g_l and V_d characterize different levels of biological organization and are not quantitatively relatable without recognition of the canopy's leaf area index, nonfoliar sites of deposition, and the role of aerodynamic boundary-layer resistance in governing deposition.[8]

For determining the flux of an air toxic chemical in the gas phase to individual leaves using the analogue resistance model, ΔC is commonly set equal to C_a, assuming C_i approaches zero.[9] The ΣR_L is calculated by analogy to H_2O vapor and assuming that the liquid phase resistance approaches zero. This approach quantifies the influence of R_b and R_s to pollutant deposition, and any residual resistance is attributed to physiological and/or biochemical processes affecting the ΔC of the gas or the diffusive resistance of the pathway. This residual is usually referred to as the mesophyll resistance, although it may not be physically located on or within the mesophyll cells of the leaf interior.

For many airborne chemicals in the gas phase for which experimental data are not available, the application of the analogue resistance model has two major sources of uncertainty. The first is the assumption of C_i approaching zero, so that ΔC is equivalent to C_a, the concentration in the free troposphere. Because for many air toxics C_i is likely to be greater than zero, that assumption will result in an overestimate of flux. Second, the analogue resistance model assumes that the pathway to the exterior leaf surface (i.e., pathway parallel to that leading to the leaf interior) is short circuited. This assumption results in an underestimate, affecting specifically those air toxic compounds for which the leaf surface is the principal site of deposition (e.g., nitric acid vapor, high molecular weight organics).

The a priori model takes into account both of these factors, thus eliminating sources of significant uncertainty in predicting the potential transport of air toxics between the atmosphere and plant canopies.

C. Conceptual Basis for the A Priori Model

The transport of gas-phase air toxics from the stomate on the leaf surface through the intercellular space to cell surfaces in the leaf interior can be represented by the two-layer, stagnant film model of Danckwerts.[10] That model has been applied in atmosphere-surface exchange processes by Liss and Slater[11] and has also been utilized in atmosphere-leaf exchange of trace gases.[12] In the Danckwerts model, the flux of any species is governed by molecular processes in the gas and liquid phase, and diffusion and chemical partitioning of molecules across the phase interface is taken into account (e.g., Henry's law coefficient). In many applications in the environmental sciences, the concept of a stagnant layer is unrealistic because turbulence in both phases eliminates the stagnant layer. However, for atmosphere-leaf exchange, consecutive stagnant layers exist because the stomate creates an environment in the leaf interior that is separated from the turbulence exterior to the leaf boundary layer, and the apoplasm is uncoupled from the symplast's bulk transport processes.

D. Source of the Data for the A Priori Model

The a priori model is based on flux data for a suite of sulfur-containing gases that vary substantially in their physical and chemical properties (Figure 2).[13] The gases are sulfur dioxide (SO_2), hydrogen sulfide (H_2S), carbonyl sulfide (COS), methyl mercaptan (CH_3SH), and carbon disulfide (CS_2). The most obvious differences among the gases lie in their H_2O solubility (30-fold), molecular size (2.5-fold), and diffusivity in air (1.5-fold). Accordingly, while all the gases are sulfur containing, they offer a range of attributes with which to evaluate the role of physiochemical properties in governing the flux of gas-phase chemicals across the atmosphere-leaf interface.

The flux data were generated in a controlled environment in which three representative plant species (*Glycine max* L., or soybean; *Phaseolus vulgaris* L., or bush bean; and *Lycopersicon esculentum,* or tomato) were exposed to each gas individually at a concentration of 5 μmol m^{-3} (0.12 μL L^{-1}) under near-ambient daylight conditions. Because the exposure system operated as a continuously stirred tank reactor, mass balance techniques were used to calculate the reactivity of each gas normalized for foliar surface area (projected leaf area in units of m^2) and time (h) as total flux ($J_{TOTAL} = J_{SURFACE} + J_{INTERIOR}$) in units of μmol gas m^{-2} h^{-1}. The concentration of each gas was maintained well below the threshold for physiological effects possible for this length of exposure ($\leq$3 h) in order to avoid a negative feedback.

While the plant species differed significantly in their reactivity with the gases, the pattern of response was the same (Table 1, Figure 3). On average, the fluxes were greatest for *G. max,* intermediate for *P. vulgaris,* and least for *L. esculentum.* Mean

SELECT PHYSICAL AND CHEMICAL PROPERTIES OF SULFUR GASES

GAS	STRUCTURE	MOLECULAR WEIGHT (g)	MOLECULAR SIZE (10^{-10} m)	DIFFUSIVITY RATIO (Dgas/Dwater)	WATER SOLUBILITY (µmoles/cm^3)
SO_2	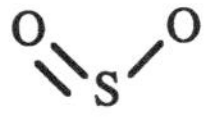 O=S–O	64	8.9	0.53	1218
H_2S	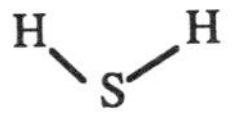 H–S–H	34	12.5	0.73	210
COS	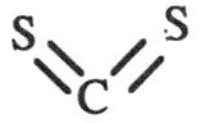 S=C=S	60	11.0	0.49	37
CH_3SH	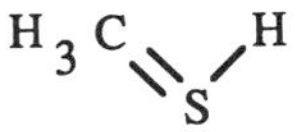 H_3C=S–H	48	22.9	0.61	257
CS_2	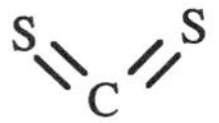 S=C=S	76	10.8	0.48	36

Figure 2. Comparison of the chemical structure and physiochemical attributes of the five sulfur-containing gases.

Table 1. Total Flux ($J_{TOTAL} = J_{SURFACE} + J_{INTERIOR}$) for Each Sulfur-Containing Gas in Each of the Three Species and the Arithmetic Mean

	Total flux (μmol m^{-2} h^{-1})			
Sulfur gas	***P. vulgaris***	***G. max***	***L. esculentum***	**Mean**
SO_2	58.0 (1.3)	75.2 (1.2)	48.2 (1.2)	60.5 (1.2)
H_2S	35.2 (1.5)	64.1 (1.2)	30.6 (1.5)	43.3 (1.4)
COS	24.5 (1.1)	55.7 (1.3)	6.6 (1.8)	29.0 (1.4)
CH_3SH	11.5 (1.8)	40.9 (1.2)	11.1 (1.2)	21.2 (1.4)
CS_2	10.2 (1.8)	21.5 (1.4)	3.7 (1.2)	11.8 (1.5)

Note: The parenthetical data are geometric means, from which the upper and lower 95% confidence intervals can be calculated by multiplication and division, respectively.

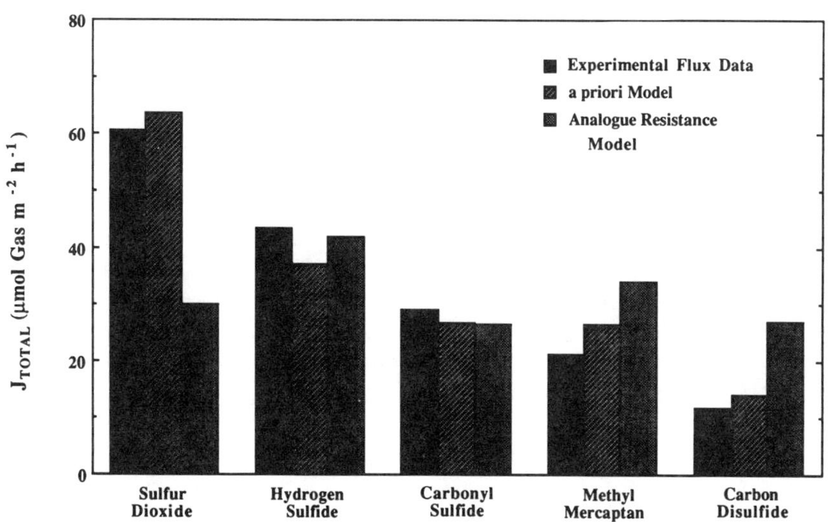

Figure 3. Mean total flux of each sulfur-containing gas as derived from controlled environment studies and the corresponding estimates of flux calculated using the a priori model and the analogue resistance model.

total flux of the sulfur gas was recorded in the following hierarchy (parenthetical data are mean fluxes in units of μmol m^{-2} h^{-1}):

SO_2	>	H_2S	>	COS	>	CH_3SH	>	CS_2
(60.5)		(43.3)		(29.0)		(21.2)		(11.8)

Mean fluxes across the three species ranged by a factor of 5 from a minimum of 11.8 μmol m^{-2} h^{-1} for CS_2 to a maximum of 60.5 μmol m^{-2} h^{-1} for SO_2. It is concluded that, when the atmosphere contains gas-phase chemicals at comparable volumetric concentrations, leaves of representative plant canopies will function as sinks for each gas but that the effectiveness of the foliage will differ significantly among the gases in a consistent pattern. Given the underpinnings of the two-layer, stagnant film model, this consistent pattern must reflect the differences in physiochemical properties among the five sulfur-containing gases.

E. Model Formulation and Parameters

To evaluate the role that the physical and chemical attributes (i.e., molecular diameter, solubility in H_2O, diffusivity in air) play in governing flux and to establish the parameters for the a priori model, stepwise regression analysis was performed using the criterion of a statistically significant reduction of the residual sum of squares for addition of an independent parameter to the model. Initially, the model was evaluated using the independent variables in an untransformed state. This model indicated that only H_2O solubility met the requirements for inclusion and that solubility alone could account for 61% of the variation.

The model was subsequently run using transformed data for H_2O solubility (ln H_2O solubility) and molecular diameter (diameter$^{1/2}$) based on previous studies in biochemistry that suggested these transformations were appropriate.[13,14] Use of the transformed data resulted in a more statistically robust regression. Molecular diameter alone accounted for 48% of the variation in flux and the inclusion of H_2O solubility accounted for an additional 46%. In the same regression, the addition of diffusivity in air did not reduce significantly the residual sum of squares, accounting for less than 3% of the variation in flux. The resulting equation for estimating J_{TOTAL} in units of μmol m^{-2} h^{-1} is as follows:

$$J_{TOTAL} = 61.0 + 8.8\ [\ln H_2O \text{ solubility}] - 20.0\ [\text{molecular diameter}]^{1/2} \quad (2)$$

The standard error of the coefficients in the equation were 2.3 for [ln H_2O solubility] and 4.9 for [molecular diameter]$^{1/2}$. The effect of increasing the H_2O solubility of the gas was to enhance its flux. Conversely, molecular diameter was inversely correlated with flux, so that large molecular structure impeded uptake. It is concluded that 94% of the variation among the sulfur-containing gases in the total flux to foliar surfaces can be accounted for solely by differences in their H_2O solubility and molecular diameters; the degree of fit of the model to the data is shown in Figure 3.

In comparison with the analogue model approach, the a priori model provides a more scientifically sound and statistically robust fit to the data (r^2 of 0.97 vs. a r_2 of 0.07). This is most evident at the extremes of H_2O solubility among the sulfur-containing gases. The analogue resistance model tends to oversestimate J_{TOTAL} for gases of low H_2O solubility (e.g., CS_2) while underestimating it for gases of high H_2O solubility (e.g., SO_2).

F. Mechanistic Foundation of the Model

The fact that molecular size and H_2O solubility govern the flux of the sulfur-containing gases has a strong mechanistic basis. After diffusing into the intercellular space of the leaf interior, gas molecules partition across the gas-to-liquid interface on the cell surfaces at a rate determined by their H_2O solubility. Hill demonstrated that 50% of the variation in trace gas uptake by plant canopies could be explained by H_2O solubility in general and Henry's law coefficient specifically.[15]

Molecular size also plays a role along the length of the diffusion path, although the mechanisms are not as well documented as that for H_2O solubility.[12] Nobel proposed that molecular diameter plays a role in both the gas and liquid phase.[16] In the case of the gas phase, flux at the leaf surface is bidirectional, with influxing gases (e.g., gas-phase chemicals) colliding with effluxing gases that are volatilizing from cell surfaces within the leaf interior (e.g., H_2O, volatile organics). Collisions among bidirectionally diffusing chemical species will be far more common for large than for small molecules, and it is estimated from principles of physical chemistry that the largest molecule among the sulfur-containing gases (methyl mercaptan) experienced a collision frequency nearly 2.5 times greater than any of the other gases and that this impeded its flux into the leaf interior. The significance of binary molecular collisions is particularly relevant in atmosphere-leaf exchange processes because of the role that guard cells play in physically constraining the channel through which gases must diffuse.

III. APPLICATION OF THE MODEL: CASE STUDIES

The a priori model has potential for a number of applications in the study of the transport and fate of gas-phase air toxics from the atmosphere to the biosphere; case studies of two such chemicals are presented here. The first is the pesticide methyl isothiocyanate, which exhibits a high vapor pressure and is readily volatilized. Even though the compound is organic, many of its physiochemical properties are similar to those of the more soluble sulfur-containing gases, notably its H_2O solubility and molecular weight. In contrast, the second case study is toluene, a more regionally and globally elevated air toxic whose physiochemical properties differ significantly from those of the sulfur-containing gases, principally in its H_2O solubility. The gas originates from a variety of industrial processes and is more elevated in urban than in remote air sheds. It is important to recognize that experimental data do not exist for either chemical regarding the efficacy with which they react with plant canopies in either managed or natural ecosystems.

A. Methyl Isothiocyanate at the Atmosphere-Biosphere Interface

As a pesticide, methyl isothiocyanate is used to control soil fungi, soil insects, and nematodes and is thus found in agricultural landscapes.[17] The chemical is typically applied as a soil drench, but because of its volatility it functions as a soil fumigant. This same attribute results in methyl isothiocyanate being volatilized to the free troposphere, after which it is transported locally and regionally. The chemical attributes of methyl isothiocyanate relative to estimating its flux to vegetation are as follows: chemical structure of CH_3-N=C=S, molecular weight of 73.12 g, H_2O solubility of 1040 μmol cm^{-3}, molecular diameter of 24.82×10^{-10} m, and diffusivity relative to that of H_2O of 0.50. In the environment the chemical exhibits an abbreviated half-life of $\leq$3 days, decomposing to an array of inorganic and organic gases.

Using the model (Equation 2), J_{TOTAL} for methyl isothiocyanate is 22.2 μmol m^{-2} h^{-1} at an ambient concentration of 5 μmol m^{-3} (0.12 μL L^{-1}). This value places it at an intermediate level relative to the sulfur-containing gases, being $\leq$50% of values for SO_2 and H_2S and roughly equivalent to that of CH_3SH (Table 1). Based on these data, tissue-level concentrations of the chemical can be estimated by scaling in space (leaf surface to leaf weight) and time (hour to day). For example, assuming a 10-h exposure to the chemical at a mean concentration of 5 μmol m^{-3}, the chemical's concentration in foliar tissue on a dry weight basis is 22 μg g^{-1}. This initial value for methyl isothiocyanate in foliage would decay to below detectable levels (e.g., 10 ng g^{-1}) in $\leq$30 days.

The alternative approach to estimating flux to vegetation, based on the analogue resistance model (Equation 1), assumes a representative R_l to H_2O vapor of 2.5 cm s^{-1}, no deposition to the leaf surface, gas-phase diffusivity ratio to H_2O of 0.50, and $C_i = 0$. The resulting estimate of methyl isothiocyanate flux is 32.4 μmol m^{-2} h^{-1}, 46% higher than that calculated using the a priori technique. The corresponding dry-weight concentration approaches 36 μg g^{-1}. The difference in the estimates derived from the two techniques is attributed to molecular processes controlling the diffusion of the gas at the leaf surface and within the leaf interior. It is proposed that the single most important attribute is the chemical's molecular diameter, which impedes the diffusion of methyl isothiocyanate in both the gas (i.e., through the stomate and intercellular spaces of the leaf interior) and the liquid phase (at the mesophyll surface).

B. Toluene at the Atmosphere-Biosphere Interface

Toluene is a common organic emission from many different industrial sources; it is listed as one of the air toxic compounds identified by the U.S. Environmental Protection Agency in the new amendments to the Clean Air Act. The principal industrial processes in which toluene is emitted include coal combustion and manufacturing of solvents for paints, resins, and gasoline additives. As such the chemical originates from a variety of point and mobile sources, but the major sources lie in industrial regions of the world. The compound has a molecular weight of 92.13 g, H_2O solubility of 7.3 μmol cm^{-3}, molecular diameter of 9.48×10^{-10} m, and a diffusivity relative to that of H_2O of 0.44. Compared to the sulfur-containing gases, toluene is less soluble (e.g., two orders of magnitude less than SO_2) and has a higher molecular weight but comparable molecular size.

Using the analogue resistance model as presented above (Equation 1) and assuming a gas-phase toluene concentration of 40 nmol m^{-3} (~1 nL L^{-1}), the flux of toluene is 250 nmol m^{-2} h^{-1}. After a growing season of 210 days, this flux results in a tissue-level toluene concentration of 80 μg g^{-1} in the absence of any biodegradation (a simplification). Assuming a leaf area index of 5, the flux of toluene is scaled to the landscape level, resulting in a growing season input of toluene at 30 mol ha^{-1} $year^{-1}$ due to dry deposition alone.

The alternative approach, using the a priori model (Equation 2), results in a toluene flux (J_{TOTAL}) of 15 nmol m^{-2} h^{-1} at an ambient concentration of 40 nmol m^{-3}. This estimate is more than an order of magnitude less than that provided by the analogue resistance approach. Over the course of a full growing season, the flux would approach 5 μg g^{-1} in the tissue on a dry-weight basis. By a method similar to that above, J_{TOTAL} is scaled to that of the landscape level, resulting in a growing-season toluene deposition of 2 mol ha^{-1} $year^{-1}$.

C. Model Limitation, Assumptions, and Future Development

The a priori model has some limitations. The foremost of these reflects the limited suite of gases from which the model was based; while the sulfur-containing gases were highly dissimilar in their physiochemical attributes, they represent a narrow range of molecular size and H_2O solubilities. It is suspected that further experimental study with a broader suite of inorganic and organic gases would not alter the role of molecular diameter and H_2O solubility, but it would change the coefficients.

The most significant modification and improvement in the model would arise from focusing on gases that are less H_2O soluble. For example, in its present form the model cannot accommodate mercury vapor (Hg^0) as a gas-phase air toxic ($J_{TOTAL} = 0$) whose solubility is four orders of magnitude less than that of CS_2. However, the alternative approach, using the analogue resistance technique, estimates a Hg^0 dry deposition that is two orders of magnitude higher than that reported under field conditions.[18] The tendency for the analogue resistance approach to overestimate flux for relatively insoluble gases is illustrated in Figure 3 by the disproportionately higher J_{TOTAL} for CH_3SH and CS_2.

While dry deposition is one of the most important processes governing the flux of gas-phase air toxics from the atmosphere to the biosphere, there are other mechanisms that also result in toxics being transferred from the atmosphere to the biosphere. These include dry deposition via sedimentation and impaction of aerosols and particles as well as wet deposition of cloud water (via interception) and rainfall. The relative importance of the roles that these processes play will vary as a function not only of the particular chemical but also of the environment. It is proposed that, for many of the gas-phase air toxics whose H_2O solubility is low, the role of dry deposition exceeds that of wet deposition.

The a priori model assumes that the leaf is coupled directly to the free troposphere. In reality, both atmospheric and canopy-level processes affect the transport pathway, and for some gas-phase species these processes may control flux. However, in general, the role that atmosphere and canopy-level processes play is inversely proportional to the chemical's H_2O solubility and reactivity in the liquid phase, so that these canopy-level processes are likely to be less important in controlling the flux of gas-phase air toxics.

One of the model's principal assets is the ability to scale observations in space and time to address a number of concerns in environmental science. With appropriate attention to scaling factors,[19] the model could predict J_{TOTAL} to regional and global scales for a variety of chemicals for which data under field or laboratory conditions do not exist. Similarly, the model provides a basis for comparing the roles of the atmosphere and the biosphere in controlling the half-life of air toxics in the atmosphere. For example, it is well established that the hydroxyl radical dictates the half-life for many organic pollutants in urban air sheds. It is proposed that the biosphere functions in the same capacity and may be as significant a scavenging process as atmospheric reactions.

IV. SUMMARY

The physiochemical and physiological factors governing the atmosphere-to-individual-leaf flux of gas-phase air toxics have been reasonably well characterized from a conceptual and first principles perspective. While physiological processes (e.g., stomatal porosity) influence transport across this interface, for a given suite of air toxic compounds two properties that govern flux can be used in the a priori model to predict transport: the chemical's molecular diameter and H_2O solubility. In the case of molecular size, the pattern is one of declining transport with increasing size of the molecule. For H_2O solubility (as estimated using Henry's law coefficient), the ease of transport from the atmosphere to vegetation increases as solubility rises. The a priori model provides a technique for estimating deposition of a gas-phase air toxic compound to plant canopies independent of field observations, and the values can be scaled in space (e.g., hectare) and time (e.g., growing season). With further development of the model, the role of plant canopies could be evaluated for use in (1) estimating regional and global budgets for air toxics and (2) calculating more realistic half-lives for chemicals in the atmosphere, using both photochemical and biological scavenging processes.

ACKNOWLEDGMENTS

This manuscript benefited from reviews by Dr. Jerry Barker and Ms. Julie Pierson. The author acknowledges with appreciation support during manuscript preparation from the College of Agriculture, University of Nevada System.

REFERENCES

1. Schroeder, W. H. and Lane, D. A., The fate of toxic airborne pollutants, *Environ. Sci. Technol.*, 22, 240, 1988.
2. Moser, T. J., Barker, J. R., and Tingey, D T., Long-range atmospheric transport and deposition of anthropogenic contaminants and their potential effects on terrestrial ecosystems, in *Biogeochemistry and Environmental Change: Global Impacts of Human Activity*, Dunnette, D. and O'Brien, R. Eds., American Chemical Society, Washington, DC, 1992, 134.
3. U.S. Environmental Protection Agency, *Toxics in the Community: National and Local Perspective*, EPA560/4-90-017, U.S. Government Printing Office, Washington, DC, 1990.
4. Grosjean, D., Atmospheric chemistry of toxic contaminants. I. Reaction rates and atmospheric persistence, *J. Air Waste Manage. Assoc.*, 40, 1397, 1990.
5. Travis, C. C. and Hester, S. T., Global chemical pollution, *Environ. Sci. Technol.*, 25, 814, 1991.
6. Taylor, G. E., Jr., Hanson, P. J., and Baldocchi, D. D., Pollutant deposition to individual leaves and plant canopies: sites of regulation and relationship to injury, in *Assessment of Crop Loss from Air Pollutants*, Heck, W. W., Taylor, O. C., and Tingey, D. T., Eds., Elsevier Applied Science, New York, 1988, 227.
7. Hosker, R. P., Jr. and Lindberg, S. L., Review: atmospheric deposition and plant assimilation of gases and particles, *Atmos. Environ.*, 25, 1615, 1982.
8. Taylor, G. E., Jr. and Hanson, P. J., Forest trees and tropospheric ozone: role of canopy deposition and leaf uptake in developing exposure-response relationships, *Agric. Ecosys. Environ.*, 42, 255, 1992.
9. Fowler, D., Cape, J. N., and Unsworth, M. H., Deposition of atmospheric pollutants on forests, *Phil. Trans. R. Soc. London B*, 324, 247, 1989.
10. Danckwerts, P. V., *Gas-Liquid Reactions*, McGraw-Hill, New York, 1970.
11. Liss, P. S. and Slater, P. G., Flux of gases across the air-sea interface, *Nature*, 247, 181, 1974.
12. Taylor, G. E., Jr., Application of the two-layer stagnant film model to atmosphere-leaf exchange of trace gases, in *Precipitation Scavenging and Atmosphere-Surface Exchange Processes*, Schwartz, S. E. and Slinn, S. G. N., Eds., Hemisphere, New York, 1992, 1069.

13. Taylor, G. E., Jr., McLaughlin, S. B., Shriner, D. S., and Selvidge, W. J., The flux of sulfur-containing gases to vegetation, *Atmos. Environ.*, 17, 789, 1983.
14. Bull, H. B., *An Introduction to Physical Biochemistry*, F.A. Davis, Philadelphia, 1964.
15. Hill, A. C., Vegetation: a sink for atmospheric pollutants, *J. Air Pollut. Control Assoc.*, 21, 341, 1971.
16. Nobel, P. S., *Physiochemical and Environmental Plant Physiology*, W.H. Freeman, San Francisco, 1991.
17. Anon., Summary of toxicity data on methyl isothiocyanate (MITC), *J. Pesticide Sci.*, 15, 297, 1990.
18. Lindberg, S. E., Meyers, T. P., Taylor, G. E., Jr., Turner, R. R., and Schroeder, W. H., Atmosphere/surface exchange of mercury in a forest: results of modeling and gradient approaches, *J. Geophys. Res.*, 97, 2519, 1992.
19. Hanson, P. J. and Lindberg, S. L., Dry deposition of reactive nitrogen compounds: a review of leaf, canopy and non-foliar measurements, *Atmos. Environ.*, 25, 1615, 1991.

Complex Ecological Assessment Influence of Pulp Mills and Power Plant Emissions on the Environment

Albert M. Beim
Elena I. Grosheva
Boris K. Pavlov

Air pollution by potentially toxic emissions of pulp mills together with power plants became an object of attention of regulatory agencies and of ecology specialists. A necessity arose of elaborating scientifically based ecological norms for various chemical compounds (and their combinations) in steam and gas emissions (dust, sulfur dioxide, nitrogen oxides, chlorine, hydrogen sulfide, organic sulfides, terpenes, and others). A long-term program of complex ecological investigations has been compiled, including (1) studies of transport, accumulation, and circulation of potential contaminants in the environment and (2) study of influence of pollutants on ecosystem structure. Investigations have been carried out within the area of an operating kraft pulp mill and power plant equipped with orderly discharge sources.

In the course of implementing the first task, the results of the land pollution by the dust particles of the mill emissions, data were obtained on dust-particle flow intensity to the adjoining mountainous-forest landscapes. Heavy metals contents were measured in the atmospheric precipitation samples, the main source of which is burned coal ash. Precipitations and soil pH, and the contents in them (chlorides, sulfates, nitrites, sodium, ammonium, and other ions) were determined. Changes of soil geochemical characteristics were registered within the zone of the maximum dust-particle outfalls and its "deoxidation" was observed: pH increased by 30 to 36%.

When addressing the second problem, long-term (over 15 years) observations were carried out regarding the condition of flora and fauna surrounding the plant. The sanitary state of the forests in the emissions zone and in the control areas (background) was studied by comparing the data on dendrochronology and morphometry of coniferous species. The accumulation of certain chemical elements in the plant organs was studied. Samples of aerosols, soils, and plants were analyzed for the contents of Hg, Cd, As, Pb, Cu, Zn, Fe, Co, Cr, and other microelements.

The cryptoindication method was used to detect the zones of influence of pollutants on plant societies. On the whole, no radical changes of the forest state in the polluted territories were noted, although facts were registered indicating appearance of new tendencies in the development of coniferous species under the influence of emissions.

When studying the ecology of small mammals and birds, most attention was given to the dynamics of populational, morphophysiological, and biochemical characteristics from the point of view of their possible use as bioindicators of emissions. A similar problem was solved by the entomological investigations. Changes in correlation between species and vital forms, and also dynamics of abundance, and other faunistic indices were employed as the basic criteria.

Results of the investigations carried out may serve as a basis for elaborating (1) a rational system of bioindication of pollutants at organism and ecosystem levels and (2) scientifically based criteria for ecological reglementation of chemical-substance contents in emissions of pulp mills and power plants.

Neurotoxic Risks of Methylmercury

Bernard Weiss

The dominant risk assessment models rely upon cases. For example, the linearized multistage model is deigned to predict the number of cancers expected for each increment of exposure. In the realm of neurotoxicity, another perspective prevails; the most crucial questions arise from exposures at levels too low to cause flagrant neurological signs such as mental retardation. Instead, the most common criteria come from performance assays such as intelligence tests. Because cases cannot be defined, risks at low environmental levels need to be posed in terms of degree of functional deficit rather than incidence.

Methylmercury is an example of such an agent. In fact, it is such a cogent example that it became a prototype in neurotoxicology. A massive study aimed at the reproducibility of behavioral endpoints in six different laboratories, the Collaborative Behavioral Teratology Study,[1] selected methylmercury as one of the two compounds for such a test.

Methylmercury is a potent poison, known for over 125 years to act primarily on the central nervous system. Until the 1950s, methylmercury poisoning was viewed primarily as an industrial hazard because it was used mainly as a fungicide, and workers employed in seed-dressing operations served as its main victim.[2] In Sweden, it also was identified as an ecological hazard; large numbers of seed-eating species, such as pheasant, were found poisoned during or after periods of sowing with treated grain.[3] Swedish scientists also documented sharp rises in the tissues of fish and fish-eating birds after the introduction of mercurials for agriculture, but other sources of mercury also provided significant contributions.

The first indications that methylmercury might represent an even broader ecological hazard for humans came from Japan. Minamata is a fishing village on the southernmost island of the Japanese chain, Kyushu. An outbreak of a mysterious neurological disease in the 1950s finally led investigators to a plant that manufactured acetaldehyde. The process discharged methylmercury, adventitiously formed, into Minamata Bay, where it contaminated fish and shellfish consumed by fishermen and their families.[4]

The major signs of methylmercury poisoning appear in Table 1. Paresthesia, or numbness and tingling, tends to be the earliest indication of excessive exposure, and served as the critical effect in Swedish calculations of the Acceptable Daily Intake. These calculations were based on adults. The issue of developmental neurotoxicity erupted later. Even as late as 1970, the risk of intrauterine toxic effects took the form of speculation, although, of the 359 children born in the Minamata area between 1955

Table 1. Indices of Methylmercury Toxicity

Sensory
Paresthesia
Pain in limbs
Visual disturbances (constriction)
Hearing disturbances
Astereognosis
Motor
Disturbances of gait
Weakness, unsteadiness of legs, falling
Thick, slurred speech (dysarthria)
Tremor
Other
Headaches
Rashes
"Mental disturbance"

and 1959, 23 showed signs of cerebral palsy, an incidence 10 times greater than expected and which could only be attributed to fetal exposure.[4] Two children who died exhibited pathological changes in the brain consistent with methylmercury poisoning. One reason for the tentative response to prenatal risk was another episode in Japan, in the mid 1960s, in Niigata. There, no instances of fetal neurotoxicity appeared.

One nagging question about Minamata remained. Mothers who displayed little evidence of adverse effects delivered severely handicapped children. Because the exposure measures were so crude and tardy, and the latencies to visible damage so prolonged, maternal and fetal dose-response relationships remained elusive. It was not until the winter of 1971/1972 that they emerged with astonishing clarity.

Drought struck Iraq in the summer of 1971, virtually eradicating the wheat crop. To restore stocks for the following year, the Iraqi government ordered about 90,000 tons of seed grain, largely of the robust and drought-resistant Mexipak variety. It also specified, probably through a transmission error, that the seeds be treated with a methylmercury fungicide. The grain arrived after the planting season, however. Although Iraqi farmers had been warned that the grain was meant only for planting, the labels were printed in Spanish and English and hungry families, desperate for food, baked the tainted grain into bread. Ingestion of the bread, in the winter of 1971/1972, triggered the largest mass chemical disaster in history.[5] Although only 450 deaths occurred in hospitals, a later analysis suggested a more likely figure of 5000, with 50,000 significantly affected.[6]

Rochester scientists, led by Thomas W. Clarkson, established a laboratory in Baghdad at the request of the Iraqi government, which faced a growing panic. Clarkson and his colleagues began a massive effort to collect exposure data. These efforts culminated in the most complete dose-response information even gathered on

Table 2. Comparison of EC_{50}s in mgHg/kg for Nonpregnant Adults and Prenatal Exposures Based on Maternal Body Burden

	Adult exposures			Prenatal exposures	
	Hockey stick	Logit		Hockey stick	Logit
Paresthesia	2.1	2.1	**Delayed walking**	0.52	0.56
Ataxia	2.7	3.0	**CNS signs**	1.23	1.01

From Clarkson, T. W. et al., in *Measurement of Risk*, Berg, G. G. and Maillie, H. D., Eds., Plenum Press, New York, 1981, 111.

a toxic material in humans and greatly expanded our ability to quantify the risks associated with methylmercury exposure.

As they surveyed the countryside, the investigators were struck by mounting evidence, still obscure in Minamata, of enhanced fetal vulnerability. Hospital admissions of infants exposed prenatally indicated marked neurological damage similar to the manifestations observed in Minamata. As in Japan, mothers with mild symptoms such as transient paresthesias delivered severely handicapped children. For confirmation, Rochester neurologists and their Iraqi colleagues undertook a study of children in the affected areas. They queried mothers about developmental milestones; they examined children at home, outdoors, or in local dispensaries; and, they secured samples of hair, cutting small bundles from the scalp. The indispensable contribution to the project came from the discovery that methylmercury blood levels, which reflect consumption, are directly correlated with mercury levels in scalp hair. The methylmercury is incorporated into the hair follicle as the shaft forms. Because scalp hair grows at the nearly constant rate of 1.1 cm/month, with a hair:blood mercury ratio of about 250:1, segmental analysis recapitulates the history of methylmercury exposure. A hair 24 cm long can be used to reconstruct a 2-year history. With this invaluable resource, maternal consumption of methylmercury in the contaminated bread could be correlated with the results of neurological examinations in children exposed during gestation.[7] Assuming a hair:blood ratio of 250, and that 5% of the body store is contained in the blood compartment, the total body burden is estimated from the formula

$$r = \text{(hair concentration)/(whole body concentration)} \qquad (1)$$

given r = 175, which is based on radioactive tracer data in human volunteers.

Clarkson and colleagues compared adult and prenatal susceptibility to methylmercury by comparing EC_{50}s for the signs and symptoms listed in Table 2. Motor retardation was defined as failure to walk by 18 months of age, and central nervous system (CNS) signs refer to speech retardation, abnormal reflexes, seizures, and other neurological abnormalities. Table 2 suggests that the fetal brain is three to four times more sensitive to methylmercury than the adult brain. The gap would be much

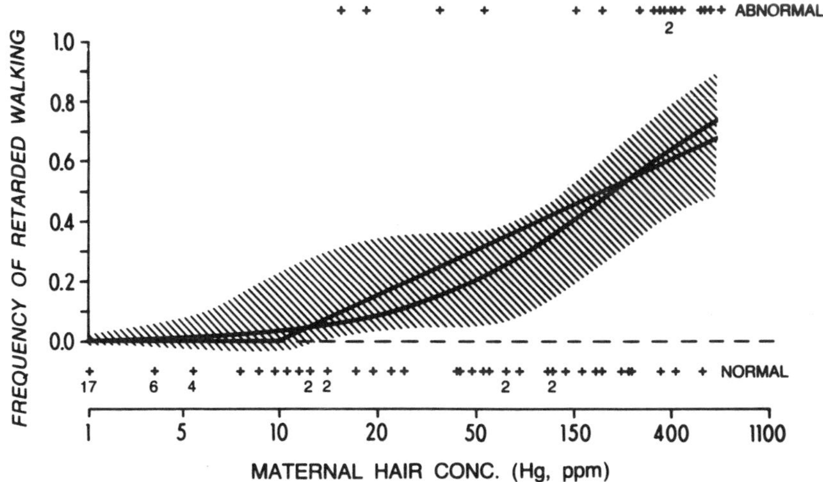

Figure 1. Logit and hockey-stick functions showing dose-response relationships between maternal mercury concentrations and retarded walking in offspring. Upper markers show abnormal cases (walking after 18 months). Lower markers show normal cases. Hatched areas include 95% confidence limits. (From Cox, C., Clarkson, T. W., Marsh, D. O., and Amin-Zaki, L., *Environ. Res.*, 48, 318, 1989. With permission.)

wider, however, were severity to be factored into the comparison. Transient paresthesias in mothers are hardly comparable to a cerebral palsy syndrome in their children. By the standard of severity, the EC_{50} values in Table 2 are gross underestimates of the differences in vulnerability between the fetal and the mature brain.

More complete documentation and additional subjects, for a total of 81 mother-infant pairs, permitted a more extensive statistical analysis.[8] Figure 1 shows logit and hockey-stick fits to the data for delayed walking. Both models indicate a rise in incidence beginning at about 10 ppm in hair. Such a level may be attained in populations that depend upon fish as a major source of protein.

Substantial methylmercury levels in certain fish species or in particular areas began to provoke questions some time ago about the potential risks of fish consumption, and led the U.S. Food and Drug Administration to endorse a level of 1.0 ppm as a limit or action level. Investigators in other parts of the world where fish consumption is greater, or where fish bore higher flesh levels, also addressed the issue of risk. New Zealand became a focus because certain population groups such as the Maori and other native Pacific Islanders tend to consume fish with concentrations up to several ppm, a practice that yields hair levels of 6 ppm and above in a substantial number of individuals.[9] Such a level, specified by the World Health Organization as equivalent to a provisional tolerable weekly intake from fish, was attained in about 20% of adults eating three or more such meals per week. Most of

Table 3. Psychological Test Scores at Age 6 Years and Maternal Hg Hair Levels

Test	Mercury level (ppm)			
	>6	3–6	<3 (high)	<3 (low)
TOLD-S[a]	72.8	77.3	74.5	79.6
WISC-P[b]	95.9	98.2	96.1	100.0
WISC-F[c]	90.6	94.2	91.0	96.0
McC-PF[d]	53.8	56.5	56.9	58.3
McC-MO[e]	59.1	62.1	63.0	61.1

[a] Test of language development, spoken.
[b] Wechsler intelligence scale, performance.
[c] Wechsler intelligence scale; full IQ.
[d] McCarthy scales of children's abilities, perceptual.
[e] McCarthy scales, motoric.

From Kjellstrom, T., Kennedy, P., Wallis, S., and Stewart, A., Physical and Mental Development of Children with Prenatal Exposure to Mercury from Fish. Stage 2. Interviews and Psychological Tests at Age 6, National Swedish Environmental Protection Board, Solna, 1989.

the fish came from fish-and-chips shops and consisted of species, such as shark, carrying high methylmercury levels.

Eventually, out of a large cohort, 31 children whose mother's hair level reached or exceeded 6 ppm, and 31 reference children matched for ethnicity and other important variables, were selected for study at the age of 4 years. The Denver Developmental Screening Test (DSST), used as a routine inventory for identifying children with deviant developmental patterns, provided the criteria of developmental neurotoxicity. The DSST is scored as *Abnormal, Questionable,* or *Normal* on the basis of clusters of items in different functional groups such as maturity of motor performance. About 50% of the high mercury children and 17% of the reference children gave scores defined as Abnormal or Questionable on the DSST; a statistically significant dose-response relationship, based on hair levels, was evident as well.

Because the DSST is relatively crude, yielding only the categories noted above, the investigators undertook a second-stage examination at 6 years of age.[10] This time, in addition to 61 high exposure children, they also included 3 other reference groups: one with high maternal fish consumption and hair levels of 3 to 6 ppm; one with high maternal consumption but hair levels of 0 to 3 ppm; and one with low maternal fish consumption and hair levels of 0 to 3 ppm. The groups were matched for ethnicity, place, and duration of residence in New Zealand, age, and smoking habits. The DSST was replaced by a battery of standardized tests consisting of the Test of Oral Language Development (TOLD), the Wechsler Intelligence Scales for Children-Revised (WISC-R), and the McCarthy Scales of Children's Abilities.

On the basis of a multiple regression analysis, the authors concluded that a consistent association existed between prenatal methylmercury exposure and test

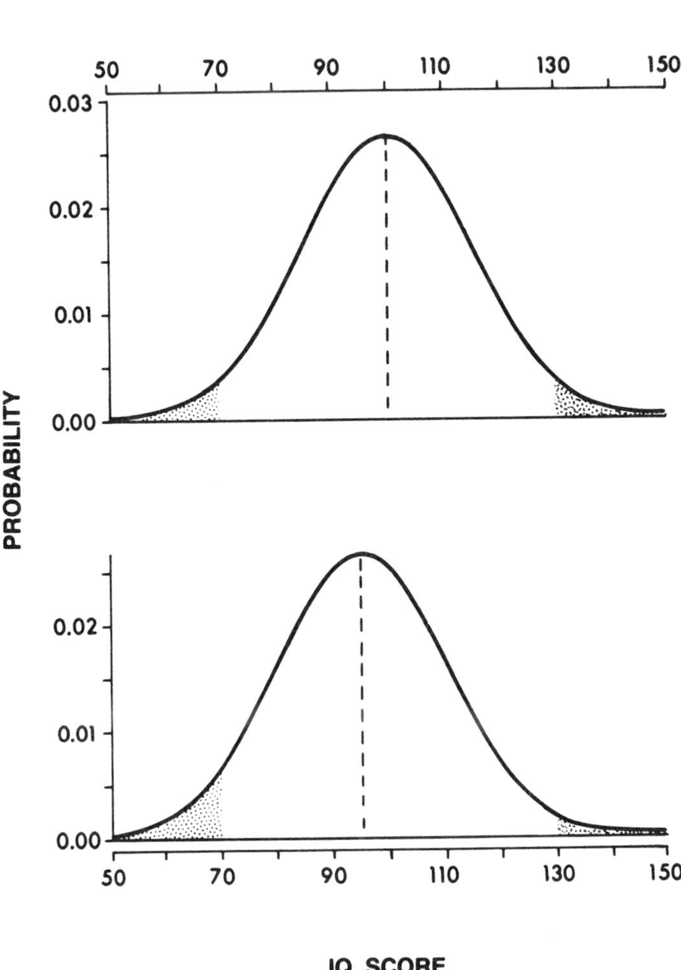

Figure 2. Consequences of shift in IQ. The upper curve depicts a distribution with a mean of 100 and standard deviation of 15 (as on the Stanford-Binet and Wechsler Intelligence Scale for Children). In a population of 100 million, 2.3 million will score above 130 and an equivalent number will score below 70, as shown by the hatched areas. If the mean is shifted by 5%, only 990,000 will score above 130. (From Weiss, B., *Trends Pharmacol. Sci.*, 9, 59, 1988. With permission.)

Table 4. Methylmercury Risk Calculations: Fish Intake Correspondence

6 ppm	21 oz (600 g) tuna per week (at 0.3 ppm)
	180 μg MeHg per week
FDA action level	1 ppm (e.g., swordfish)
6 ppm in hair	6.4 oz/week (one meal, at action level)
12 ppm in hair	6.4 oz/week of 2 ppm fish (shark, pike)

performance but that the effect is relatively weak, accounting for only a small proportion of the total variance. However, there are other aspects to consider. Table 3 lists the group scores on the various tests and test components. Note the consistently lower scores by the group whose maternal hair levels exceeded 6 ppm. In addition, a close examination of the WISC-R distributions indicated skewing toward low scores in the group with hair levels above 6 ppm.

Even a small shift in the IQ distribution holds important implications, as I have noted for lead.[11] A seminal study performed in Boston traced three groups of children categorized on the basis of cord blood lead concentrations.[12] All came from intact, upper middle-class homes and none showed any evidence of developmental retardation. The high-lead group had a mean a 14.5 mg/dL, which, not long ago, had been considered moderate or even low and without notable risk. Tests at 6-month intervals, based on the Bayley Scales of Infant Development, demonstrated an 8% difference, at age 24 months, in scores between the low and medium groups and the high lead group, although all the groups scored above the standardized mean.

To view this in perspective, note that the standardized mean for intelligence tests such as the WISC and the Stanford-Binet is 100, with a standard deviation of 15. In a population of 100 million, therefore, 2.3 million will yield IQ scores above 130, as indicated by the hatched sections of Figure 2. If the population mean is reduced by 5%, to 95, only 990 thousand will attain an IQ of 130, and the number scoring below 70 is proportionally inflated. Such a shift constitutes a grave societal more than an individual risk, perhaps, but still needs to be incorporated into risk assessment doctrine. Besides limiting the number of high-performance individuals, it also shifts more individuals into an area, IQ scores below 70, that mandates expensive remedial programs in many school districts.

What level of fish intake corresponds those hair concentrations, implied by the data from Iraq and New Zealand, at which the risk of developmental deviations begins to rise above background? Table 4 suggests that markedly aberrant consumption is not necessary. One tuna sandwich daily will bring hair levels into that range; so will ingestion of one meal weekly, on average, of species such as swordfish and snapper, which generally exceed 1 ppm. Large pike from certain Adirondack lakes and from certain areas of the Great Lakes, and shark as well, may reach 2 ppm, with the consequences depicted in Table 4.

The resulting concentrations in brain, of course, are our ultimate concern. Data from monkeys and rodents indicate that fetal brain levels exceed maternal levels. A fetal:maternal ratio of 2:1 has been noted in monkeys,[13] and we have seen fetal:maternal

Table 5. Methylmercury Risk Calculations: Maternal Hair and Fetal Brain Levels

6 ppm in maternal hair	24 ppb in maternal blood 48 ppb in fetal blood 0.24 ppm in fetal brain
1–2 ppm in fetal brain	Mitotic arrest
20 ppm in maternal hair	0.8 ppm in fetal brain

blood levels in mice of about the same magnitude. In primates, brain:blood ratios of up to 5:1 and obtained.[14] Table 5 demonstrates the outcome of these relationships. Assumed brain levels for humans are based on a brain:blood ratio of 5:1, close to the 6:1 value calculated by Magos.[15] With evidence (1984) that mitotic arrest occurs in the immature mouse brain at methylmercury concentrations of 1 to 2 ppm,[16] a possible coupling between the New Zealand and Iraqi findings and their mechanistic bases is not difficult to discern.

Epidemiological data typically pose problems of interpretation. Assessments of developmental neurotoxicity are enveloped in a nexus of confounding variables such as parental IQs and educational attainments, social environment, and others. Despite the ability of statistical techniques to compensate for such variables, as in multiple regressions models, critics of the lead literature continue to maintain that the complications are still excessively intrusive. For this reason, confirmation from experimental studies with laboratory animals, which also indicate behavioral deficits at low exposure levels of lead, has supplied further confidence in the human data.

Rather few experimental studies have explored exposure levels of methylmercury low enough to bring them into the range provoking questions about human risk. Those that indicate adverse effects at low levels have relied primarily on complex learned behavior as the criterion. One study in rats reported deficits in operant performance as an aftermath of prenatal exposure.[17] The rats had been trained to press levers on a schedule of food reinforcement that required high rates of responding. The experimenters administered four doses of either 0.05 or 0.01 mg/kg methylmercury, or a total of 0.02 or 0.04 mg/kg. Methylmercury is almost completely absorbed, so that the equivalent body burden for a 60-kg female would be 12.0 or 2.4 mg. According to formula (1), the resultant human hair concentrations would be (175)(0.2) = 35 ppm or (175)(0.04) = 7 ppm. Although marked differences between human and rat pharmacokinetics have been noted, this is still an intriguing finding. Another series of studies in rats[18] surveyed a variety of behavioral endpoints in animals exposed during gestation via drinking water supplied to the dams, which resulted in daily intakes of 0, 0.3, 0.9, or 2.7 mg/kg. The most sensitive index of exposure, by far, turned out to be an operant task requiring the rat to emit alternate presses on right and left levers in the experimental chamber.

Studies in non-human primates should offer the most cogent basis for extrapolation to humans. Exposure during gestation or infancy has been documented to produced abnormalities in visual function, and deficits on tasks, such as visual

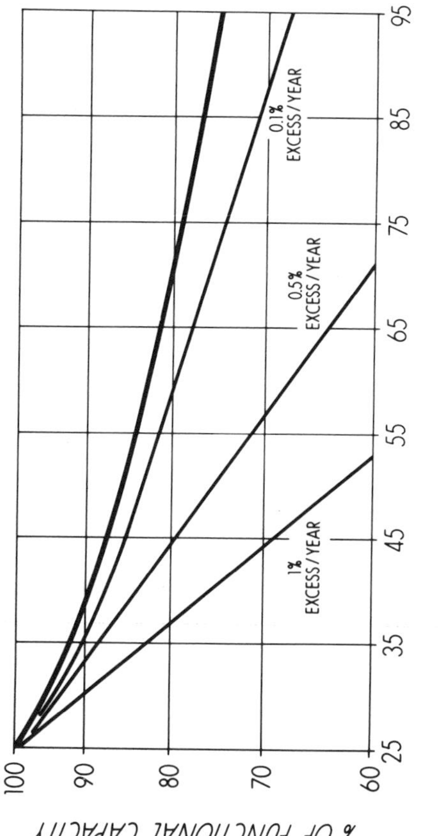

Figure 3. The heavy line traces the course of brain aging as reflected by neuronal cell density, oxygen consumption, and glucose consumption; age 25 years is taken as baseline. The rate of brain aging with superimposed accelerations of 0.1, 1.0, and 5.0% annually is given by the additional lines. (From Weiss, B. and Simon, W., *Behavioral Toxicology*, Weiss, B. and Laties, V. G., Eds., Plenum Press, New York, 1975, 429. With permission.)

recognition memory, adapted from techniques used to test human infants.[19] However, these findings are based on calculated or assumed brain levels during development of 3 to 11 ppm, considerably above those held to present a significant risk to the human fetus.[20]

Unlike severe neurological abnormalities, subtle deficits are difficult to discern in the infant. Mild motor and intellectual retardation do not become apparent until much later in life. The full impact of fetal exposure to methylmercury may be obscure at first but emerge more clearly with time.[21] In fact, some evidence suggests that the most profound expression of neurotoxicity may emerge during advanced age, when the compensatory capacity of the brain is overwhelmed by earlier damage. A lifetime experiment with mice demonstrated the onset of neurological and behavioral deficits, in mice exposed prenatally, beginning with middle age and accelerating as the mice aged further.[22]

The implications of such a process are depicted in Figure 3. It shows the projected decline of brain function, based on neuronal cell density, glucose uptake, and oxygen consumption, after the age of 25. Imposing a rate increment of as little as 0.1%, as might result from methylmercury ingestion over a lifetime, brings an arbitrary decline, such as 20%, several years earlier.[23] Damage incurred during brain development would produce another effect: beginning the decline with a lower baseline value. The ultimate effect would be similar.

The predominant neurotoxic risks of methylmercury, as we learned from lead, lie in their subtleties, not in the prospect of overt poisoning. Effects such as reduced scores on intelligence tests and accelerated aging are easy to overlook from a risk perspective governed by cases. However, their societal imprint may be even more devastating. Determining their scope will require a greater investment in low-dose, longitudinal research than we so far have been willing to expend. We should also be prepared for surprises. The following exchange took place after the paper by Westöö[3] at the 1968 Rochester Conference on Environmental Toxicity. Robert Risebrough, University of California, Berkeley, asked "Why should mercury be a problem in the northern countries rather than here in the more temperate areas?" Fredrik Berglund, National Institute of Public Health, Stockholm, replied "I think one reason that this problem does not exist in the United States with mercury is that the levels are not known...I feel, personally, that the problem also exists here as it does in other parts of the world but it is not recognized."

ACKNOWLEDGMENTS

Preparation supported in part by grants ES-01247 and ES-05433 from the National Institute of Environmental Health Sciences.

REFERENCES

1. Buelke-Sam, J., Kimmel, C. A., Adams, J., Nelson, C. J., Vorhees, C. V., Wright, D. C., St. Omer, V., Korol, B. A., Butcher, R. E., Geyer, M. A., Holson, J. F., Kutscher, C. L., and Wayner, M. J., Collaborative behavioral teratology study: results, *Neurobehav. Toxicol. Teratol.,* 7, 591, 1985.
2. Hunter, D. S., *The Diseases of Occupations,* 5th ed., British Universities Press, London, 1975.
3. Westöö, G., Methylmercury compounds in animal foods, in *Chemical Fallout,* Miller, M. W. and Berg, G. G., Eds., Charles C Thomas, Springfield, IL, 1969, 75.
4. Tsubaki, F. and Irukayama, K., Eds., *Minamata Disease,* Elsevier, New York, 1977.
5. Bakir, F., Damluji, S. F., Amin-Zaki, L., Murtadha, M., Khalidi, A., Al-Rawi, N. J., Tikriti, S., Dhahir, H. I., Clarkson, T. W., Smith, J. C., and Doherty, R. A., Methylmercury poisoning in Iraq, *Science,* 181, 230, 1973.
6. Greenwood, M. R., Methylmercury poisoning in Iraq. An epidemiological study of the 1971–72 outbreak, *J. Appl. Toxicol.,* 5(3), 148, 1985.
7. Clarkson, T. W., Cox, C., Marsh, D. O., Myers, G. J., Al-Tikriti, S. K., Amin-Zaki, L., and Dabbagh, A. R., Dose-response relationships for adult and prenatal exposure to methylmercury, in *Measurement of Risk,* Berg, G. G. and Maillie, H. D., Eds., Plenum Press, New York, 1981, 111.
8. Cox, C., Clarkson, T. W., Marsh, D. O., and Amin-Zaki, L., Dose-response analysis of infants prenatally exposed to methylmercury: an application of a single compartment model to single-strand hair analysis, *Environ. Res.,* 48, 318, 1989.
9. Kjellstrom, T., Kennedy, P., Wallis, S., and Mantell, C., Physical and Mental Development of Children with Prenatal Exposure to Mercury from Fish. Stage 1. Preliminary Tests at Age 4, (report no. 3080), National Swedish Environmental Research Board, Solna, 1986.
10. Kjellstrom, T., Kennedy, P., Wallis, S., Stewart, A., Friberg, L., Lind, B., Wutherspoon, T., and Mantell, C., Physical and Mental Development of Children with Prenatal Exposure to Mercury from Fish. Stage 2. Interviews and Psychological Tests at Age 6, Rep. No. 3642, National Swedish Environmental Protection Board, Solna, 1989.
11. Weiss, B., Neurobehavioral toxicity as a basis for risk assessment, *Trends Pharmacol. Sci.,* 9, 59, 1988.
12. Bellinger, D., Leviton, A., Waternaux, C., Needleman, H., and Rabinowitz, M., Longitudinal analyses of prenatal and postnatal lead exposure and early cognitive development, *N. Engl. J. Med.,* 316, 1037, 1087.
13. Mottet, N. K., Shaw, C. M., and Burbacher, T. M., Health risks from increases in methylmercury exposure, *Environ. Health Perspect.,* 63, 133, 1985.
14. Evans, H. L., Garman, R. H., and Weiss, B., Methylmercury: exposure duration and regional distribution as determinants of neurotoxicity in nonhuman primates, *Toxicol. Appl. Pharmacol.,* 41, 15, 1977.
15. Magos, L., The absorption, distribution and excretion of methylmercury, in *The Toxicity of Methylmercury,* Eccles, C. U. and Annau, Z., Eds., Johns Hopkins, Baltimore, 1987, 24.
16. Rodier, P. M., Ashner, M., and Sager, P. R., Mitotic arrest in the developing CNS after prenatal exposure to methylmercury, *Neurobehav. Toxicol. Teratol.,* 6, 379, 1984.
17. Bornhausen, M., Müsch, H. R., and Greim, H., Operant behavior performance changes in rats after prenatal methylmercury exposure, *Toxicol. Appl. Pharmacol.,* 56, 305, 1980.

18. Elsner, J., Suter, K. E., Ulbrich, B., and Schreiner, G., Testing strategies in behavioral teratology. IV. Review and general conclusions, *Neurobehav. Toxicol. Teratol.,* 8, 585, 1986.
19. Gunderson, V. M., Grant-Webster, K. S., Burbacher, T. M., and Mottet, N. K., Visual recognition memory deficits in methylmercury-exposed *Macaca fascicularis* infants, *Neurotoxicol. Teratol.,* 10, 373, 1988.
20. Burbacher, T. M., Rodier, P. M., and Weiss, B., Methylmercury developmental neurotoxicity: a comparison of effects in humans and animals, *Neurotoxicol. Teratol.,* 12, 191, 1990.
21. Marsh, D. O., Dose-response relationship in humans: methylmercury epidemics in Japan and Iraq, in *The Toxicity of Methylmercury*, Eccles, C. U. and Annau, Z., Eds., Johns Hopkins, Baltimore, 1987, 45.
22. Spyker, J. M., Behavioral teratology and toxicology, in *Behavioral Toxicology,* Weiss, B. and Laties, V. G., Eds., Plenum Press, New York, 1975, 311.
23. Weiss, B. and Simon, W., Quantitative perspectives on the long-term toxicity of methylmercury and similar positions, in *Behavioral Toxicology,* Weiss, B. and Laties, V. G., Eds., Plenum Press, New York, 1975, 429.

Evidence for the Carcinogenicity of Nickel

L. S. Goldstein

I. INTRODUCTION

Nickel is released into the environment from the lithosphere, industrial processes involving the refining of nickel-containing ore, and the combustion of fossil fuels including coal and oil. Nickel is most likely an essential element but is toxic at high doses. Acute nickel toxicity is manifest mostly as an allergic contact dermatitis. Prolonged exposure to nickel-containing dusts is associated with increased incidence of cancer. The regulatory standard for nickel is based on cancer induction.

Although nickel is found in both coal and oil fly ash, there are no data indicating that prolonged exposure to ambient levels results in an increased risk of cancer. Nevertheless, because their emission levels of nickel compounds exceed the limits set forth in the Clean Air Act Amendments of 1990, electric power plants are subject to regulation. The Act does not specify the chemical form of nickel but lists "nickel compounds".

In this paper I outline some of the data upon which the determination of nickel carcinogenicity was made. I will then relate these data to the potential risks posed by fly ash from stack emissions and I will outline recommendations for additional studies to advance our understanding of potential carcinogenicity due to exposure to fly ash.

A. Human Studies

Epidemiologic data come mainly from four studies on the effects of exposure to dusts in nickel refineries, and one from an electroplating plant.

In 1982 Enterline and Marsh[1] reported on a cohort of 1855 workers employed by the International Nickel Company's manufacturing plant in Huntington, WV; 8 lung cancer deaths and 2 nasal cancer deaths were reported in the 133 deaths of workers in the manufacturing process and who were exposed to air levels of nickel ranging from 0.01 to 5.0 mg/m^3.

In another study,[2] 495 workers involved in a sinter plant in an Ontario nickel refinery were followed. These workers were exposed to air levels in excess of 50 mg/m^3; 37 lung cancer deaths were found among the 85 cohort members who died before the initial followup in 1981. A second followup in 1990[3] revealed 63 lung cancer deaths and 6 nasal cancer deaths among those workers with 15 years or more since first exposure.

In 1977 Doll et al.[4] reported on 967 workers employed at a nickel refinery in Clydach, Wales. There were 137 lung cancer deaths out of a total of 612 deaths. In a followup study, Peto et al.[5] found a higher lung cancer mortality among workers classified into high-exposure job titles than among workers in low-exposure job titles.

A Norwegian study[6,7] of nickel refinery workers found 82 lung cancer cases among 2247 workers. No data on the level of exposure were available.

An analysis of these and other[8] data led the International Agency for Research on Cancer (IARC) to conclude that there was *sufficient evidence* in humans for the carcinogenicity of nickel sulfite and the combinations of nickel sulfides and oxides found in the nickel refining industry.[8] The U.S. EPA concluded that there is *sufficient evidence* that nickel refinery dust and nickel subsulfide are carcinogenic in humans.[9] Nickel compounds are identified as hazardous pollutants in the Clean Air Act Amendments of 1990.

B. Animal Studies

Laboratory studies using inhalation[10] or intratracheal[11,12] administration of nickel subsulfide found increased rates of cancer induction. Nickel subsulfide accounts for over 50% of the compounds found in flue dusts at nickel refineries. Ottolenghi exposed Fischer 344 rats to 6 h/day for 70 to 80 weeks and found 29% had neoplastic changes in the lungs including adenomas, adenocarcinomas, squamous cell carcinomas, and fibrosarcomas. These studies used a small number of animals in each study group and had a high level of mortality during the second year. It is not clear to what degree the lung tumors contributed to mortality.

Two studies using an intratracheally administered nickel subsulfide found evidence for tumor induction in rats,[12] but not mice.[11] Induced lung cancers were found for inhaled nickel carbonyl[13] but not nickel oxide.[14] Studies to determine the rate of cancer induction in animals inhaling nickel subsulfide, nickel oxide, or nickel sulfate are now under way using the protocol recommendations of the National Toxicology Program.

Many other studies investigating carcinogenicity due to specific nickel compounds have been reported but few have found an effect. Tumors are found primarily at the site where the compounds are administered, e.g., at the injection site. The relevance of these studies to risk assessment is not clear.

II. NICKEL FROM UTILITY EMISSIONS

A. Introduction

Unlike the nickel refining process that takes place at temperatures less than 400°F, combustion in a power plant involves temperatures in excess of 1500°F. It is reasonable to assume that the predominant nickel species will be different for the two processes, and this supposition has been borne out.[15,16] Analysis of fly ash from oil-fired boiler revealed that 60 to 100% of the nickel was water soluble and predominantly in the form of nickel sulfate. The insoluble fraction, when found, was predominantly nickel oxide. These are the predominant forms in coal fly ash although the ratio of soluble-to-insoluble components differs from that found in oil-fired ash. There is no evidence that nickel subsulfide is produced, and in fact the formation of this compound is not favored at high reaction temperatures. Thus, the nickel found in electric utility-generated ash is very different from that found from certain processes of nickel refining where nickel subsulfide is a major component.[17]

Since the predominant compounds in the ash from coal- and oil-fired power plants are nickel sulfate and nickel oxide, the question to be addressed in whether these forms of nickel in coal and oil ash are hazardous to human health. This will be addressed in terms of the components of a risk assessment: hazard identification, dose response, and exposure assessment.

B. Hazard Identification

The designation of nickel sulfate as a probable carcinogen is based on studies of workers in a nickel plating facility.[8,18] These workers were exposed to a variety of hazardous pollutants including chromium, and it is uncertain whether the agent of interest is really the nickel sulfate. Attempts to demonstrate carcinogenicity due to nickel sulfate in animals have been successful only by intraperitoneal injection of a high dose;[8] oral intake of nickel sulfate at doses high enough to cause weight-gain decrements showed no evidence of cancer induction in rats or dogs.[19]

Several studies have looked at nickel oxide. Wehner et al.[20] could not find evidence for tumorigenicity for nickel oxide in hamsters exposed via inhalation. Similar results were reported for fly ash artificially enriched with nickel.[21] Studies in rats were inconclusive.[14]

Nickel in fly ash is associated with combustion particles and it is not evident that nickel would be bioavailable to the organism. Wehner et al.[20] exposed hamsters to fly ash sufficient to cause pneumoconiosis but could detect no increased rate of tumor formation. Data on the mean residence time of the various nickel compounds should come from the 2-year carcinogenicity study mentioned earlier.

Overall, these data strongly suggest that, while nickel is a pollutant from utility operations, the forms of nickel emitted may represent a lesser hazard than that for nickel dusts and nickel subsulfide.

C. Dose Response

The epidemiologic studies cited as evidence of the carcinogenicity of nickel subsulfide and nickel dusts were for worker exposures at concentrations up to 400 mg/m^3. In contrast, the ambient level of Ni in the Southern California Air Basin ranges from 0.0028 to 0.0231 mg/m^3,[22] a difference of over seven orders of magnitude. It is not clear what the bioavailability is at low doses and whether a linear dose-response relationship is appropriate for such a pronounced range.

One can calculate the effective dose in humans from animal data. In the nickel subsulfide study of Ottolenghi et al., the daily dose of 970 μg/m^3 represents the equivalent of a human lifetime exposure of 51 μg/m^3. Thus, animal experiments predict cancer induction at doses (2 to 20) $\times$ 10^3 higher than that found in the ambient environment.

The estimated dietary intake of nickel is 156 mg/day.[23]

The date suggest a substantial reliance on linearity that is based on only a few very high dose points. The shape of the dose-response curve has not been established.

D. Exposure

While inhalation is the major route of nickel exposure in occupational settings, dietary ingestion appears to be the major route for ambient exposure. Furthermore, body burdens would appear to be overwhelmed by the dietary route when one considers ambient nickel levels (see above).

It is not at all clear that when the nickel is found as part of a complex fly ash particle that it is bioavailable to the organism. The studies of Wehner et al.[21] suggest that a substantial amount may not be available since studies using nickel-enriched fly ash (6%), a level substantially in excess of that normally found in fly ash, found no excess cancers even though there were significant effects of the dust in the lungs.

III. RECOMMENDATIONS FOR FUTURE WORK

Animal and epidemiologic studies should address certain very basic questions about nickel carcinogenicity:

- Are all nickel compounds equally potent? Data from the ongoing inhalation studies should address this as far as cancer induction is concerned. However, these data should be buttressed with results from bioavailability studies and cellular/molecular mechanistic studies.
- What is the response of cells and tissues to low doses of nickel? Various cellular markers associated with relatively high exposures to certain nickel compounds have been used to address mechanistic questions. It should be possible to modify these assays to make them sensitive detectors of low doses of nickel, and thereby form the basis of a reproductible bioassay for nickel exposures at low doses.
- Is the nickel found in fly ash bioavailable? The connection between exposure to a hazardous pollutant and an adverse health effect is bridged by demonstrating that toxic substances are released in the body. To date there is no evidence that the fly-ash-bound nickel is bioavailable.

- How does nickel subsulfide or other nickel compounds act to induce cancers? Despite intensive work in this area very few answers are available. Does nickel act genotoxically or epigenetically? Is Ni^{2+} the ultimate carcinogen? Does Ni act indirectly to enhance other carcinogenic processes?

IV. SUMMARY

There are considerable gaps in our knowledge of nickel carcinogenesis in general, and the potential hazard to the population posed by nickel-containing fly ash in particular. Research directed to reduce these uncertainties will lead to a reasonable regulatory finding that fulfills the mandate of the U.S. EPA to protect the public health.

REFERENCES

1. Enterline, P. E. and Marsh, G. M., Mortality among workers in a nickel refinery and alloy manufacturing plant in West Virginia, *J. Natl. Cancer Inst.,* 68, 925, 1982.
2. Chovil, A., Sutherland, R. B., and Halliday, M., Respiratory cancer in a cohort of nickel sinter plant workers, *Br. J. Ind. Med.,* 38, 327, 1981.
3. Roberts, R. S., Julian, J. A., Sweezey, D., Muir, D., Shannon, H. S., and Mastromatteo, E., A study of mortality in workers engaged in the mining, smelting, and refining of nickel, *Toxicol. Ind. Health,* 5, 957, 1989.
4. Doll, R., Matthews, J. D., and Morgan, L. G., Cancers of the lung and nasal sinuses in nickel workers: a reassessment of the period of risk, *Br. J. Ind. Med.,* 34, 102, 1977.
5. Peto, J., Cuckle, H., Doll, R., Hermon, C., and Morgan, L. G., Respiratory cancer mortality of Welsh nickel refinery workers, in *Nickel in the Human Environment: Proceedings of a Joint Symposium held at IARC,* Lyon, France, Sunderman, F. W., Ed., International Agency for Research on Cancer, 53, 37, March 8–11, 1983.
6. Magnus, K., Andersen, A., and Hogetveit, A. C., Cancer of respiratory organs among workers at a nickel refinery in Norway, *Int. J. Cancer,* 30, 681, 1982.
7. Pedersen, E., Hogetveit, A. C., and Andersen, A., Cancer of respiratory organs among workers at a nickel refinery in Norway, *Int. J. Cancer,* 12, 32, 1973.
8. IARC Monographs on the Evaluation of Carcinogenic Risks to Humans: Chromium, Nickel and Welding, IARC, Lyon, 49, 257, 1990.
9. EPA, Health Assessment Document for Nickel — Final Draft, U.S. Environmental Protection Agency, Research Triangle Park, NC, EPA 600, 8, 012F, 1985.
10. Ottolenghi, A. D., Haseman, J. K., Payne, W. W., Falk, H. L., and MacFarland, H. N., Inhalation studies of nickel subsulfide in pulmonary carcinogenesis of rats, *J. Natl. Cancer Inst.,* 54, 1165, 1974.
11. Fisher, G. L., Chrisp, C. E., and McNeill, D. A., Lifetime effects of intratracheally instilled nickel subsulfide on B6C3F1 mice, *Environ. Res.,* 40, 313, 1986.
12. Pott, F., Ziem, U., Reifter, F. J., Ernst, H., and Mohr, U., Carcinogenicity studies on fibres, metal compounds, and some other dusts in rats, *Exp. Pathol.,* 32, 129, 1987.

13. Sunderman, F. W., Donnelly, A. J., West, B., and Kincaid, J. F., Nickel poisoning. IX. Carcinogenesis in rats exposed to nickel carbonyl, *AMA Arch. Ind. Health,* 20, 36, 1959.
14. Takenaka, S., Hochrainer, D., and Oldiges, H., Alveolar proteinosis induced in rats by long-term inhalation of nickel oxide, in *Progress in Nickel Toxicology — Proceedings of the Third International Conference on Nickel Metabolism and Toxicology,* Brown, S. S. and Sunderman, F. W., Eds., Blackwell Scientific, Boston, 1984, 89.
15. Gendreau, R. M., Jakobsen, R. J., and Henry, W. M., Fourier transform infrared spectroscopy for inorganic compound speciation, *Environ. Sci. Technol.,* 18, 990, 1980.
16. Henry, W. M. and Knapp, K. T., Compound forms of fossil fuel fly ash emissions, *Environ. Sci. Technol.,* 14, 450, 1980.
17. Filman, J. W. P., Metal carcinogenesis. II. A study on the carcinogenic activity of cobalt, copper, iron, and nickel compounds, *Cancer Res.,* 22, 158, 1962.
18. International Committee on Nickel Carcinogenesis in Man, ICNCM, *Scand. J. Work Environ. Health,* 16, 1, 1990.
19. Ambrose, A. M., Larson, P. S., Borzelleca, J. F., and Hennigar, G. R., Long term toxicologic assessment of nickel in rats and dogs, *J. Food Sci. Technol.,* 13, 181, 1976.
20. Wehner, A. P., Busch, R. H., Olson, R. J., and Craig, D. K., Chronic inhalation of nickel oxide and cigarette smoke by hamsters, *Am. Ind. Hyg. Assoc. J.,* 36, 801, 1975.
21. Wehner, A. P., Moss, O. R., Milliman, E. M., Dagle, G. E., and Schirmer, R. E., Acute and subchronic inhalation exposures of hamsters to nickel-enriched fly ash, *Environ. Res.,* 19, 355, 1979.
22. State of California, Technical support document: Proposed Identifications of Nickel as a Toxic Air Contaminant, Part A Report — Public Exposure to, Sources of, and Atmospheric Fate of Nickel in California, Air Resources Board, Stationary Source Division, Sacramento, June 1990a.
23. Bennett, B. G., Exposure assessment for metals involved in carcinogenesis, in *Environmental Carcinogens: Selected Methods of Analysis,* Vol. 8, O'Neill, I. K., Schuller, P., and Fishbein, L., Eds., International Agency for Research on Cancer, Lyon, France, 71, 1986.

Carcinogenic Risks of Arsenic in Fly Ash

Ronald E. Wyzga
Janice W. Yager

I. INTRODUCTION

A. Emissions

Arsenic is found in coal and oil. When these are combusted, minute quantities of arsenic are emitted into the atmosphere. The quantities emitted depend upon the arsenic content in fuel and upon the characteristics of the power plant. Concentrations of arsenic in coal and in oil vary considerably with arsenic concentrations as low as 0.02 ppm and as high as 357 ppm reported in bituminous coals;[1] the range and maximum concentration for oil are considerably less, with reported concentrations ranging from 0.0024 ppm to 1.11 ppm. Concentrations in the ambient environment around a power plant reflect the fuel content, the configuration of the power plant, including stack height and the extent of particulate controls, and local meteorology. For this reason, no typical concentrations can be given, but they are expected to be small. Figure 1 gives a plot of the annual arsenic concentration around a small coal-fired power plant; the maximum annual arsenic concentration is 7.82×10^{-6} μg/m^3.[2]

A study commissioned by the U.S. EPA[1] investigated the health risks associated with power plant emissions of several substances. Of the trace metals studied, the highest risks were associated with arsenic. Current emissions levels for arsenic yielded estimates of 0.7 cancers per year in the entire U.S. from coal-fired power plants and 0.3 cancers per year in the U.S. from oil-fired power plants. Maximum individual lifetime risks from arsenic were estimated to be 1×10^{-4} in the case of coal-fired power plants. These estimates are derived from the U.S. EPA's inhalation unit risk estimate (lifetime risk of cancer for per microgram per cubic meter continuous lifetime exposure) for arsenic; the latter estimate is based upon exposures and observed cancers in metal smelting operations.[3]

B. Characteristics of Arsenic in Fly Ash

Knowledge of the nature of arsenic exposure from power plants is limited, but the exposure characteristics are likely to be quite different from those of smelting operations. Limited analysis of fly ash from U.S. coal-fired power plants suggests that the arsenic is in valence state V and is located both on the surface of fly ash

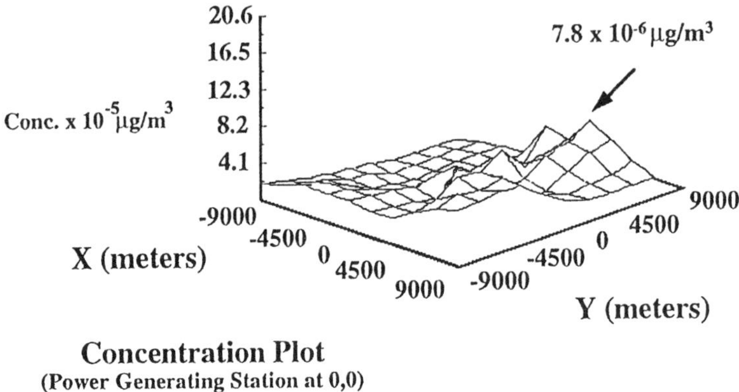

Figure 1. Example of incremental arsenic concentrations from power plant source.

particles as well as in a silica matrix throughout the particle.[4] In smelting operations, arsenic is largely in valence state III and is often present in a fumelike or oxide particle. Exposure levels are much higher than those observed in or around power plants. Some contrary evidence for the characteristics of arsenic in fly ash exists, however. One consideration of the physicochemical processes has suggested that the prevalent valence state of arsenic in fly ash may be III.[5] Others have suggested that the valence state of arsenic in oil fly ash may be different from that in coal fly ash.[6] Clearly, additional analyses are needed to characterize arsenic in fly ash.

II. EVIDENCE FOR HUMAN CARCINOGENICITY

A. Epidemiologic Studies

The health effect of greatest regulatory concern for arsenic is cancer. The evidence for the carcinogenicity of arsenic is based upon several epidemiology studies. Significantly increased mortality from respiratory cancer has been observed in several studies of copper smelter workers.[7-10] In these work settings, exposure to high concentrations of arsenic trioxide (As III) occurred; in the majority of the studies, airborne time-weighted average concentrations of arsenic were estimated to range from 0.4 to 62 mg/m^3. Concomitant exposure to a complex mixture of other compounds also occurred in these workplaces. However, consideration of confounders, exposure to complex mixtures in the working environment, and smoking could not account for the excess lung cancer mortality. The data from these copper smelter studies have been used by the U.S. EPA to derive a unit risk estimate for arsenic exposures by inhalation.[3] Studies of workers manufacturing and using ar senic pesticide mixtures (principally lead arsenate, calcium arsenate, sodium arsenate, and arsenic trioxide) have also shown an excess mortality from respiratory cancer.[11-13]

Studies of populations exposed to high arsenic concentrations in drinking water (ranging from 0.4 to 1.8 mg/L) in a small geographic area of Taiwan have shown an increase in skin cancers;[14,15] more recently, increases in bladder, kidney, lung, and liver cancer have also been reported among the exposed population.[6] Excess skin disorders and lesions have also been reported among populations who drink water with high arsenic levels in Mexico and Chile;[17,18] however, such findings have not been replicated in studies of U.S. populations exposed to high levels of arsenic in drinking water.[19] The U.S. studies were limited by the small size of the populations exposed. The principal form of arsenic in drinking water is in valence state V. The role of confounders, such as other contaminants in water, diet, and nutritional status, in the drinking water studies in unclear. The U.S. EPA has used the Taiwanese skin cancer data to derive a unit risk estimate (lifetime risk of cancer per μg/L lifetime exposure) for arsenic exposure by ingestion through drinking water.[3] More recent analyses, incorporating the cancer incidence data for internal organ sites, have suggested that the unit risk estimates should be increased.[20]

B. Other Evidence

By and large, animal studies have not been successful in demonstrating the carcinogenicity of arsenic compounds. One study of exposure to a mixture of arsenic trioxide, sulfuric acid, and fine particulates administered by intratracheal instillation to hamsters did show an increase in respiratory carcinomas; however, only when adenomas were included was the increase in tumors statistically significant.[21] This method of exposure results in a substantial delivered dose, and the extent to which one can extrapolate from this mixture is unknown.

Arsenic is inactive or extremely weak in its ability to *directly* induce gene mutations (in contrast to most initiating chemical carcinogens) in a number of test systems.[22-25] However, both arsenite (As III) and arsenate (As V) have been shown to induce chromosomal aberrations and sister chromatid exchanges in cultured human and animal cells *in vitro*. Arsenite is about tenfold more effective than arsenate.[26]

It has been suggested that arsenic may act as a cocarcinogen rather than an initiator or promoter.[27] Possible mechanism(s) for this role need to be further investigated; among the possibilities are that arsenite (As III) may act by inhibiting DNA repair enzymes, inducing gene amplification or hypomethylation of DNA.[28] Production of oxygen radicals that then damage DNA is another possible mechanism of action for arsenic. It would be particularly instructive to learn if arsenic from fly ash is equally implicated in these mechanisms as arsenic in other mixtures, matrices, or valence states. Another factor that may play a considerable role is the presence of other compounds in the exposure. These could inhibit or amplify alternative types of mechanism. Competition for binding among the various trace elements in fly ash needs to be investigated in this context.

Considerable information is known about the metabolism of arsenic by the liver methyl-transferase enzymes. As III is the substrate for methylation, and As V must be reduced to As III before it can get into the cell and be methylated; however, some As V is eliminated directly.[29,30] Methylation results in a relative detoxification of inorganic arsenic and increases the rate of arsenic excretion. Conversely, inhibition of the methylation capacity results in an increasing amount of arsenic being retained in the body.[31] The concentration and availability of thiol groups and other substances (e.g., choline) is also important for optimum methylation.[32,33] Concentration of these substances in the body is influenced by diet; a diet low in protein- or choline-containing compounds can lead to a higher probability that methylation will be inhibited. There is some evidence that at high doses of arsenic (500 to 1000 μg/day in humans) the arsenic methylation pathway may become saturated, presumably leading to greater toxicity.[31,34,35] This in turn would imply that there is a nonlinear dose response for arsenic.

C. Uncertainties

Clearly there is evidence that under certain exposure conditions arsenic is a human carcinogen. At issue is *whether* exposure to arsenic in fly ash is carcinogenic and *what* is the quantitative risk associated with these exposures. Current evidence is based upon much higher arsenic exposures than anticipated from fly ash. There is some evidence suggesting that at high exposure levels the detoxification metabolism of arsenic in the human body becomes saturated; at lower levels this mechanism is quite efficient.[35] The inhalation evidence is based upon exposures to arsenic in copper smelters. This arsenic is likely of a different valence state from that in fly ash; this in turn may indicate a somewhat different metabolism and acute toxicity activity. It is also contained in a particle matrix that possesses different size and solubility characteristics than fly ash. The drinking water studies are presumably based upon arsenic exposures to a similar valence state as that which may exist in coal fly ash, but the routes of exposure, arsenic solubilities, and exposed populations are very different. There is evidence that diets deficient in protein- and choline-containing foodstuffs, as apparently existed in Taiwan, can lead to depletion of thiol-containing compounds in the body that are necessary for the detoxification of arsenic.[32,33] This could eventually result in the retention of larger amounts of arsenic in the body.

We need to understand the ramifications of the differences between the arsenic exposures and metabolism for fly ash and for the circumstances in which we have firm carcinogenic evidence. Better understanding will allow better extrapolation of risks to exposures from arsenic in fly ash than is currently possible.

D. Research Issues and Needs

Several areas of needed research are evident. We still lack a comprehensive characterization of arsenic in fly ash with respect to (1) particle size distribution; (2) the relative distribution and concentration of total arsenic, arsenic III, and arsenic V on both the surface and interior matrix of fly ash particles of varying sizes; and (3) determination of the elemental composition, particularly for other trace elements. The latter may be of particular concern because the presence of other trace elements may interfere with the binding of arsenic to cells within human organs. For example, it has been hypothesized that the ratio of arsenic to selenium may be an important predictor of arsenic toxicity.[27]

Studies of metabolism are key to understanding the relative difference in the toxicological behavior of arsenic in fly ash and from other sources, such as copper smelter particulates and drinking water. In the past, animal studies have proven to be highly successful in the study of the metabolism and excretion of arsenic. These studies can be used to study the bioavailability and pharmacokinetics (absorption, lung deposition and clearance, distribution, metabolism, and excretion) of arsenic as it exists in the complex coal fly ash matrix, in copper smelter particulates, or even in drinking water.

The animal studies can be buttressed by epidemiology studies which obtain biomarkers of arsenic exposure and metabolism from various populations exposed to arsenic, from fly ash and other sources. These studies are likely to elucidate a number of important scientific issues with regard to human response. For example, the effect of a relatively high intake of arsenic on methylating capacity (with implications for linear or nonlinear kinetics), effect of exposure on genetic biomarkers, and effects of diet and other factors (sex, age, other exposures) on methylating capacity can be examined.

The animal and human studies would allow the development of pharmacokinetic models, partially validated by human data, that can be used to predict changes in the bioavailability or kinetics that may occur with alternative exposure scenarios, e.g., changes in route of exposure, valence state, exposure matrix, or population characteristics. Estimates (or relative estimates) of the relevant biological or internal doses can then be derived for the populations exposed to arsenic in fly ash and from other sources for which we can estimate risk (e.g., smelter workers, exposed Taiwanese subpopulation). For example, assume the pharmacokinetic model were to indicate that X% of the arsenic associated with fly ash exposure were detoxified and not bound to relevant organ tissue, whereas this were only true for Y% of the arsenic exposure among smelter workers. Then, for a comparable ambient concentration, the risks of arsenic exposure from fly ash are the fraction X/Y of those for smelter workers. Similarly, adjustments could be made for nonlinearities found in the dose-response relationship.

Support for a nonlinear dose response could also come from studies of the possible mechanism(s) of action for arsenic. These could suggest relevant mechanism(s) for arsenic in fly ash or the true nature of the dose response.

III. CONCLUSIONS

Evidence from epidemiological studies clearly indicts arsenic as a human carcinogen. Arsenic is also present in fossil fuels and the resulting by-products of their combustion. Current estimates of the human health risks from this arsenic exposure are based upon very different exposure circumstances from those associated with arsenic in fly ash. These involve differences in valence state, matrix, solubility, concentration, mixture, and/or exposure and population characteristics. Information can be obtained to understand the implications of these differences and to factor them into subsequent quantitative risk assessments. This will result in more accurate estimates of the risk from exposure to fly ash arsenic.

REFERENCES

1. U.S. Environmental Protection Agency, Estimating Air Toxics Emissions from Coal and Oil Combustion Sources, EPA-450/2-89-001, Washington, DC, 1989.
2. Gratt, L. B. and Dusetzina, M., A Numerical Comparison of the Human Exposure Model and the Air Emission Risk Assessment Model, presented at the 80th Annual Meeting of the Air Pollution Control Association, 87-98.3, New York, June 21-26, 1987.
3. U.S. Environmental Protection Agency, Health Assessment Document for Inorganic Arsenic, Final Report, EPA-600/8-83-021F, Washington, DC, 1984.
4. Electric Power Research Institute, Fly Ash Exposure in Coal-Fired Power Plants, Draft Final Summary Report 256-043-07-00 submitted by Radian, September 1991.
5. Bezacinsky, M., Pilatova, B., Jirele, V., and Bencko, B., To the problem of trace elements and hydrocarbons emissions from combustion of coal, *J. Hyg. Epidem. Microbiol. Immunol.*, 28, 129, 1984.
6. Eatough, D. J., Lee, M. L., Later, D. W., Richter, B. E., Eatough, N. L., and Hansen, L. D., Dimethyl sulfate in particulate matter from coal- and oil-fired power plants, *Environ. Sci. Technol.*, 15, 1502, 1981.
7. Lee, A. M. and Fraumeni, J. F., Arsenic and respiratory cancer in man: an occupational study, *J. Natl. Cancer Inst.*, 42, 1045, 1969.
8. Axelson, O., Dahlgren, E., Jansson, C. D., et al., Arsenic exposure and mortality: a case referent study from a Swedish copper smelter, *Br. J. Ind. Med.*, 35, 8, 1978.
9. Pinto, S. S., Henderson, V., and Enterline, P. E., Mortality experience of arsenic-exposed workers, *Arch. Environ. Health*, 33, 325, 1978.

10. Enterline, P. E. and Marsh, G. M., Cancer among workers exposed to arsenic and other substances in a copper smelter, *Am. J. Epidemiol.,* 116, 895, 1982.
11. Roth, F., The sequelae of chronic arsenic poisoning in Moselle vintners, *Ger. Med. Mon.,* 2, 172, 1957.
12. Ott, M. G., Holder, B. B., and Gordon, H. I., Respiratory cancer and occupational exposure to arsenicals, *Arch. Environ. Health,* 29, 250, 1974.
13. Matanoski, G., Landau, E., Tonascia, J., Lazar, C., Elliot, E., McEnroe, W., and King, K., Cancer mortality in an industrial area of Baltimore, *Environ. Res.,* 25, 8, 1981.
14. Tseng, W. P., Chu, H. M., How, S. W., Fong, J. M., Lin, C. S., and Yeh, S., Prevalence of skin cancer in an endemic area of chronic arsenicism in Taiwan, *J. Natl. Cancer Inst.,* 40(3), 453, 1968.
15. Tseng, W. P., Effects and dose-response relationships of skin cancer and blackfoot disease with arsenic, *Environ. Health Perspect.,* 19, 109, 1977.
16. Chen, C. J., Chuang, Y. C., You, S. L., et al., A retrospective study on malignant neoplasms of bladder, lung, and liver in blackfoot disease endemic area in Taiwan, *Br. J. Cancer,* 53, 399, 1986.
17. Cebrian, M. E., Albores, A., Aguilar, M., and Blakely, E., Chronic arsenic poisoning in the north of Mexico, *Hum. Toxicol.,* 2, 121, 1983.
18. Zaldivar, R., Arsenic contamination of drinking water and foodstuffs causing endemic chronic poisoning, *Beitr. Path. Bd.,* 151, 384, 1974.
19. Andelman, J. B. and Barnett, M., Feasibility Study to Resolve Questions on the Relationship of Arsenic in Drinking Water to Skin Cancer, University of Pittsburgh, Center for Environmental Epidemiology, Technical Report 84-8, Pittsburgh, PA, 1984.
20. Gibb, H. and Chen, C., Is inhaled arsenic carcinogenic for sites other than the lung?, in *Assessment of Inhalation Hazards,* Bates, D., Dungworth, D., Lee, P., McClellan, R., and Roe, F., Eds., Springer-Verlag, Berlin, 1989, 171.
21. Pershagen, G., Nordberg, G., and Bjorklund, N. E., Experimental evidence on the pulmonary carcinogenicity of arsenic trioxide, *Arch. Toxicol.,* Suppl. 7, 403, 1984.
22. Rossman, T. G., Stone, D., Molina, M., et al., Absence of arsenite mutagenicity in *E. coli* and Chinese hamster cells, *Environ. Mutagen.,* 2, 371, 1980.
23. Nordenson, I., Sweins, A., and Beckman, L., Chromosome aberrations in cultured human lymphocytes exposed to trivalent and pentavalent arsenic, *Scand. J. Work Environ. Health,* 7, 277, 1981.
24. Nakamuro, M. and Sayato, Y., Comparative studies of chromosomal aberrations induced by trivalent and pentavalent arsenic, *Mutat. Res.,* 88, 73, 1981.
25. Singh, I., Induction of reverse mutation and mitotic gene conversion by some metal compounds in *Saccharomyces cerevisiae, Mutat. Res.,* 117, 149, 1983.
26. Larramendy, M. L., Popescu, N. C., and DiPaolo, J., Induction of inorganic metal salts of sister chromatid exchanges and chromosome aberrations in human and Syrian hamster strains, *Environ. Mutagen.,* 3, 597, 1981.

Toxic Constituents of Coal Fly Ash

Jeffrey B. Hicks

I. INTRODUCTION

Coal fly ash is known to contain potentially toxic chemicals, such as metals, crystalline minerals, products of incomplete combustion, and radio isotopes.[1] This paper presents the methods, results, and observations of a fly ash exposure characterization study conducted at six coal-fired power plants located in different regions of the U.S. The objective of the study was to characterize worker exposures to fly ash and its potentially toxic constituents during the operation and maintenance of these power plants. Some significant findings were revealed by this study that have important implications for power plant worker health and for potential air toxic health concerns associated with fly ash emissions from power plants.

As part of this worker exposure study, a large number of bulk material and airborne fly ash samples were collected during a variety of activities conducted at these power plants and from different work areas. The air samples and bulk material were subjected to several different chemical, mineral, and radio chemical analyses to characterize the presence of elements, minerals, organic compounds, and radioactive elements.

The primary goal of the study was to characterize worker exposures at representative power plants under common work activities throughout the U.S. Particular emphasis was placed on selecting plants using different coal types, as this was anticipated to be a major determinant of the composition of the fly ash workers would encounter. Secondary emphasis was placed on the type of samples collected and their subsequent analysis, since information gathered during a literature review and meetings with industry representatives indicated that particular concern existed over the exposure to crystalline silica, arsenic, and other trace elements, during normal plant operations and maintenance activities.

The results of the study present a good characterization of the types and constituents of fly ash exposures encountered by the diverse cross section of workers at power plants; the information also indicates important implications concerning toxic compounds in fly ash. These include the toxicity associated with arsenic present in fly ash, its valence form, and its bioavailability within the fly ash matrix. The presence of detectable quantities of crystalline silica is also an important health issue, since this mineral may present a pulmonary fibrosis and a cancer hazard. This paper provides information concerning the methods and results from this exposure evaluation, as well as detailed information concerning the assessment and exposure to arsenic and crystalline silica from coal fly ash.

II. STUDY METHODS

A. Site Selection

Since a primary goal of the study was to characterize workplace exposures in representative power plant environments, selecting the power plants to be included in the study was an important task. Information was gathered from the available literature and from discussions held with electric utility companies to determine what characteristics would effect exposures to fly ash and its toxic constituents. This revealed that highly variable exposures to fly ash were commonplace within the industry. Routine (daily) exposures at relatively low concentrations were encountered by individuals engaged in operation, inspection, and preventative maintenance of fly ash collection, conveyance, and disposal systems. More intermittent, but significantly higher exposures were encountered by individuals performing maintenance tasks such as work inside the boiler or fly ash cleanup. Approximately equal numbers of employees are engaged in plant operations as compared to maintenance, and maintenance workers for the most part are exposed intermittently, over discreet periods of time during maintenance outages. Based on the information gathered, six different power plants were selected for inclusion in this study. The type of coal used at each plant and other characteristics are summarized on Table 1.

B. Exposure Evaluation Methods

National Institute for Occupational Safety and Health (NIOSH) methods were selected for air sampling and analysis whenever possible. Bulk fly ash materials were collected in selected work areas, and subjected to similar analytical techniques. Table 2 presents a list of the various analytes evaluated during this study, and Table 3 presents the sampling and analytical methods employed. Modified or alternative methods were occasionally employed where NIOSH methods were not available. These included particle characterization, particle size distribution, and radio chemical analysis. Modified EPA, or ASTM, recognized methods were used in these situations.

Air and bulk samples were collected from Sites 1 through 4 during two separate site visits, individually focused on normal plant operations and maintenance activities. Single plant visits were conducted at Sites 5 and 6, and additional visits were made to Sites 2 and 4 for further clarification studies.

The results from the field studies were compared to recognized occupational health criteria, such as permissible exposure limits (PELs) established by the Occupational Health and Safety Administration (OSHA). These comparisons were made to determine regulatory compliance and the potential health impacts associated with the measured exposures. Throughout all sampling and analytical procedures performed for this study, rigorous quality assurance/quality control procedures were

Table 1. Characteristics of Selected Plant Sites

Plant site	Coal type	BTUs per pound of coal	Percent ash by weight following combustion	Boiler type	Number of units	Total MW	Collection type (no. of units)	Fly ash: Conveying system	Fly ash: Storage system	Disposal method
1	Western sub-bituminous	11,500	11–12	Pulverized coal-balanced draft	3	1,320	ESP (2), Baghouse(1)	Pressurized dry	Ash silo	Landfill or off-site sale
2	Western sub-bituminous	9,500	12	Pulverized coal-balanced draft	4	2,000	ESP (4)	Pressurized dry	Ash silo	Landfill or off-site sale
3	Interior bituminous	13.000	8	Pulverized coal-balanced draft	8	870	ESP (8)	Pressurized dry	Ash silo	Landfill or off-site sale
4	Eastern bituminous	11,500	12	Pulverized coal-balanced draft	3	1,700	ESP (3)	Pressurized dry	Ash silo	Landfill or off-site sale
5	Southern Gulf coast lignite	6,800	11	Pulverized coal-balanced draft	1	720	ESP	Pressurized dry	Ash silo	Mixed with scrubber; sludge; transported to landfill
6	Northern central lignite	6,753	14	Pulverized coal-balanced draft	2	910	Baghouse	Pressurized dry	Ash silo	Mixed with dry scrubber particulate prior to baghouse; transported from silo to landfill

Note: MW = megawatt, ESP = electrostatic precipitator, and BTU = British thermal units.

Table 2. Substances and Parameters Selected for Monitoring during Phase II Survey

Substance/parameter	Reasons for monitoring
Respirable particulates	General index of fly ash exposure; respirable size fraction of interest because of lung deposition characteristics
Total particulate	General index of fly ash exposure; total particulate airborne concentration of interest for use in estimating individual airborne trace element concentrations based upon compositional analysis
Respirable quartz/total quartz	Quartz content of fly ash is significant; exposure criteria for quartz-containing dusts may be exceeded when airborne fly ash concentrations are elevated
Arsenic	Due to its relative toxicity, arsenic has a very low exposure criterion; overexposure to arsenic may occur, depending upon its concentration in the fly ash
Particle size analysis	Particle size analysis is desirable to determine the extent to which airborne fly ash particles are respirable
Particle description	Microscopic analysis is desirable in order to characterize the collected particulate and to determine the extent to which non-fly ash particles (e.g., coal dust and dirt) may be present as contaminants
Other trace elements	Other trace elements may present an exposure hazard, depending upon their concentration in the fly ash; some metals, such as lead, have low exposure criteria; 20 elements were included in the analysis
Polynuclear aromatic hydrocarbons (PAHs)	Analyses have shown that PAHs may be present on fly ash particles; 17 PAH species are included
Radionuclides	Analyses have shown that radionuclides may be present in fly ash particles; monitoring is desirable to determine the level of activity in airborne fly ash particulates; analysis considered gross a, b, and radium 226

Table 3. Sampling and Analytical Methods

Parameter	Sample type	Sampling device (for air samples)	Collection media	Flow rate	Sample analysis	Comment
Respirable particulate	Personal and area	Personal air-sampling pump	Preweighed polyvinyl chloride (PVC) filter, 5.0-μm pore size, 10-mm nylon cyclone size	1.7 LPM	Gravimetric (NIOSH method 06000)[a]	Nylon cyclone effectively screens out particles >10 μm; captured particles represent potential deep lung deposition
Total particulate	Personal and area	Personal air-sampling pump	Matched-weight mixed cellulose ester (MCE) filters, 0.8-μm pore size	2.0 LPM	Gravimetric (NIOSH method 0500)[a]	Useful for comparison with regulatory limit
Respirable quartz	Personal and area	Personal air-sampling pump	PVC filter, 5.0-μm pore size, 10-mm nylon cyclone size selector	1.7 LPM	X-ray diffraction (NIOSH method 7500)[a]	Nylon cyclone effectively screens out particulate ≥10 μm; captured particulates are weighed and then analyzed for quartz by X-ray diffraction
Total quartz	Personal, area, and bulk	Personal air-sampling pump	PVC filter for air samples, 5.0-μm pore size, 40-mL sample jar for bulk samples	2.0 LPM	X-ray diffraction (NIOSH method 7500)[a]	Captured particles are weighed and then analyzed for quartz by X-ray diffraction
Arsenic	Personal, area and bulk	Personal air-sampling pump	Matched-weight MCE filters, 0.8-μm pore size, 40-mL sample jar for bulk samples	2.0 LPM	Atomic absorption (modified NIOSH method 7901)[a]	Known to concentrate during coal combustion; low exposure limit; phase III investigated appropriate analytical technique;
Arsine	Personal and area	Personal air-sampling pump	Activated-charcoal solid sorbent tube	0.2 LPM	Atomic absorption (NIOSH method 6001)[a]	Collected during phase III only

Table 3. Sampling and Analytical Methods (continued)

Arsenic trioxide	Area	Personal air-sampling pump	NA_2CO_3-impregnated MCE filter, 0.8-µm pore size and backup pad	2.0 LPM	Atomic absorption (NIOSH method 7901)[a]	Inorganic forms of arsenic; limited data on their presence; collected during phase III only
Selenium	Personal and area	Personal air-sampling pump	Matched-weight MCE filters, 0.8-µm pore size	2.0 LPM	Atmoc absorption (modified NIOSH method 7901)[a]	Relatively toxic, potentially present in lignite fly ash, phase III only
Trace elements	Area and bulk	Personal air-sampling pump	Preweighed PVC filter for air samples, 5.0-µm pore size, 40-mL sample jar for bulk samples	4.0 LPM	ICAP (NIOSH method 7300)[a]	To measure concentrations of other elements in the fly ash matrices
Radionuclides	Area and bulk	High volume air-sampling pump	Glass fiber filter for air samples, 8 × 10 inch, 40-mL sample jar for bulk samples	20 SCFM	Radiochemical analysis (gross α, gross β, radium 226)	Literature indicates potential presence in fly ash
Particle size analysis	Area	High volume air-sampling pump	Preweighed glass fiber filter, 5-stage Anderson® Model 235 cascade impactor	20 SCFM	Gravimetric	Selects airborne particles according to aerodynamic diameter; effective cuts at 10.2, 4.2, 2.1, 1.4, and 0.73 mm
Particle description	Area and bulk	Personal air-sampling pump	Open-faced cassette, mixed cellulose ester filter for air samples, 0.8-µm pore size, 40-mL sample jar for bulk samples	2.0 LPM	Microscopic evaluation (polarized light)	Useful in quantifying the content of fly ash in the total particulate and the presence of other particles (coal dust, mineral dust, fibers, metal particles, etc.)

Table 3. Sampling and Analytical Methods (continued)

Polynuclear aromatic hydrocarbons	Area and bulk	Personal air-sampling pump	Zeflour® filter followed by XAD-2 resin sorbent tube for air samples, 40-mL sample jar for bulk samples	2.0 LPM	Gas chromatography with flame ionization detector (GC/FID) NIOSH method 5515[a]	Literature indicates potential presence in fly ash from incomplete combustion of fuels

Note: LPM = liter per minute, SCFM = standard cubic feet per minute, AA-GF= atomic absorption-graphite furnace, and ICAP = inductively coupled plasma are atomic emission spectroscopy.

[a] NIOSH Manual of Analytical Methods, 3rd ed., U.S. Department of Health and Human Services (DHS), Publication No. 84-100, 1984.

followed as samples were collected and analyzed. The detailed QA/QC program and other facets of the overall study methodology are available to the reader through the Phase I,[1] II,[2] and III[3] reports available through EPRI.

III. EXPOSURE PATTERNS AND ANALYTICAL RESULTS

A. General Observations and Results

Over 1500 air and bulk samples were collected as part of this study, with up to 43 different analyses or parameters of interest. Due to the extensive number of data points, and the primary relevance to worker exposure characterization, a detailed presentation of these results is not included in this paper. Summarized information is presented, with emphasis on the analytes presenting the most significance related to air toxic emissions and public health.

- Polynuclear aromatic hydrocarbons (PAHs): Particulate and vapor phase samples collected for these compounds were analyzed for 17 different species. In all areas sampled, the concentrations were very low. Fluorene was the most frequently detected compound (59% of samples), with a concentration range of 1.7 to 4.0 mg/m^3. Naphthalene was detected in 50% of the air samples ranging from 0.08 to 40 mg/m^3. The remaining species were detected much less frequently, present in less than 10% of the remaining samples and at concentrations less than 2 mg/m^3. Bulk samples did not reveal the presence of any PAH species.
- Radionuclides: The results for gross a, gross b, and radium-226 analysis revealed detectable, but relatively low, radionuclide concentrations in most samples. Of all samples, 6% revealed concentrations above maximum permissible concentrations established by the Nuclear Regulatory Commission at plants using subbituminous and bituminous coals. Of the samples from lignite-fired plants, 40% exceeded this criteria value, suggesting that lignite coal deposits present a potential radionuclide exposure hazard to workers as compared to coal fired facilities.
- Particle size and characteristics: In the majority of the areas at each plant studied, fly ash represented the majority of the airborne particles. Airborne size distributions varied considerably from plant site to plant site, and from area to area at each plant. The mass medium diameter generally ranged from 0.1 to 50 mm, with the majority of sample sites between 1.0 and 10 mm in diameter.
- Trace elements: Samples were collected and subjected to broad spectrum analysis for 15 different elements. When comparing the airborne concentrations to the OSHA PELs (permissible exposure limit), only two samples (less than 1% of all these types of air samples) revealed aluminum concentrations that exceeded the OSHA PEL. Other elements analyzed by this technique were uniformly below the OSHA limits.
- Arsenic and crystalline silica: The analytes yielding the exposures most frequently above the PELs are arsenic and crystalline silica. Specific information is presented below.

IV. ARSENIC

Arsenic exposures associated with coal fly ash is of special interest. Arsenic presents potential significant health concerns including potential acute toxicity as well as carcinogenic risk. OSHA has a very restrictive exposure limit established for arsenic.

The arsenic content in the fly ash was observed to be significantly different at the six plants studied. Very limited amounts of arsenic were detected in the air samples collected from the lignite coal plants, indicating that it is not a element of significant concern from the combustion of this coal type. Noticeable differences were present in the western subbituminous, interior, and eastern bituminous coal-burning facilities, with the highest concentrations noted in the site burning eastern bituminous coal (Site 4). Table 4 presents the arsenic content detected in collected ash air samples at various locations at plant sites 1 through 4. Figure 1 presents average arsenic concentrations by site, showing that Sites 2 and 4 exceeded that of Sites 1 and 3; Site 4 had the highest levels. This suggests that ash from eastern bituminous coal may present a more significant arsenic concern as compared to other coal sources.

Some unusual observations concerning the concentrations of arsenic detected and the analytical results were noted:

- The highest airborne concentrations of arsenic were several hundred micrograms per cubic meter of air, with the highest being 380 $\mu g/m^3$. These were noted in personal samples from individuals working in high fly ash dust concentrations, primarily at Site 4. These concentrations are well above the OSHA PEL (10 mg/m^3) and, due to the acute toxicity characteristics of arsenic, this level of exposure on a daily basis could cause acute adverse symptoms in workers. No reports of any acute arsenic poisonings were reported by workers or other individuals in the electric utility industry. It is not clear why this level of exposure did not produce acute responses.
- Co-located samples analyzed by two different techniques — inductively coupled plasma arc emission spectroscopy (ICAP) and atomic absorption-graphite furnace (AA-GF) — revealed significant discrepancies. These side-by-side sample results are presented on Table 5. This observation stimulated further investigation into the appropriate analytical procedure that should be used for arsenic determinations in coal fly ash. Additional details follow.

A specimen of National Bureau of Standard Reference Material 1633A "Trace Elements in Coal Fly Ash" was obtained and subjected to a variety of analyses. These included the standard methods employed during the study as well as additional digestion procedures and analytical methods. The analytical methods employed included atomic absorption-hydride generation (AA-HG), AA-GF, and ICAP. The various digestion procedures performed included NIOSH Method 7300 (standard for ICAP), a modification of this method that involved digesting more material in the same volume of digestion solution, NIOSH Method 7901 (standard for AA-GF), enhanced digestion using acid solutions and microwave heating, and "acid bomb

Table 4. Arsenic Content of Fly Ash Samples

Site number	Location	No. of samples	Arsenic content in fly ash: Range (µg/g)	Mean	SD
1	Boiler	10	2.1–240	52	82
	ESP	9	0.71–15	10	4.4
	ESP hopper	3	4.9–55	22	29
2	Ash silo	6	28–77	48	16
	Boiler	5	9.0–65	35	24
	ESP	8	46–148	72	33
	ESP hopper	1	39	NC	NC
	Preheater	3	9.0–33	25	13
3	Ash silo	5	94–350	238	110
	Boiler	19	0.0–480	197	160
	Economizer	2	67–240	154	120
	Preheater	10	50–460	143	120
4	Ash silo	1	310	NC	NC
	Boiler	21	4.0–6800	940	1500
	Economizer	9	23–400	170	114
	ESP	10	40–630	270	165
	ESP hopper	5	140–200	180	25

Note: µg/mg = micrograms of arsenic per gram of collected fly ash (equivalent to parts per million by weight [ppmw]), SD = standard deviation, and NC = not calculated.

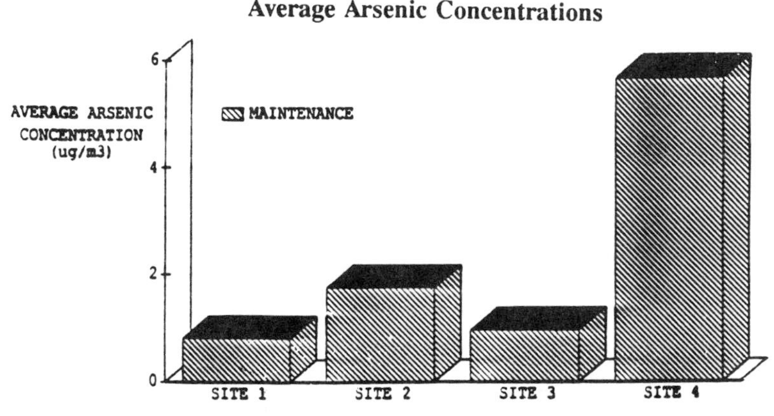

Figure 1. Average arsenic concentrations.

Table 5. Comparison of Co-Located Arsenic Results from Two Analytical Methods (ICAP and AA-Graphite Furnace)

Site number	Location	Arsenic content (μg/g) ICAP	AA-GF	Comments
1	Unit 1 boiler	140	<3	Slag from base of coal burners
	Unit 1 superheat curtains	370	23	Hard slag removed from tubes
	Unit 1 superheat curtains	<5	<3	Soft slag removed from tubes
	Unit 1 ESP inlet	<5	<1	
	Unit 1 ESP hopper	<5	4	
	Coal sample	<5	<3	
2	Unit 3 boiler penthouse	<5	9	
	Unit 3 air preheater	96	9	
	Unit 3 ESP	<5	60	
	Unit 3 ash silo	110	28	
	Coal sample	<5	<3	
3	Unit 7 boiler penthouse	120	55	
	Unit 7 boiler	<5	33	Slag from top of coal burners
	Unit 7 boiler	130	42	Slag from base of coal burners
	Unit 7 superheat curtains	220	120	
	Unit 7 economizer	1400	67	
	Unit 7 air preheater	120	110	
	South ash silo, penthouse	150	350	
	South ash silo	230	94	
	Coal sample	<5	3	
4	Unit 3 boiler	110	14	Slag from base of coal burners
	Unit 3 boiler	<5	4	Slag from top of coal burners
	Unit 3 superheat curtains	<5	52	Slag removed from tubes
	Unit 3 reheat tubes	1700	1300	Slag removed from tubes
	Unit 3 economizer	590	400	
	Unit 3 economizer hopper	140	59	Undisturbed fly ash from hopper
	Unit 3 economizer hopper	<5	23	Reddish chunks of fly ash
	Coal sample	<5	17	
	National Bureau of Standards 1633A	81	120	Target concentration 145 μg/g

Arsenic content of dust sample is less than the analytical detection limit. Actual arsenic concentration is less than the reported concentration.

digestion" that included enhancing the digestion procedure following acid treatment with heating in a sealed container.

The results of these various matrix analyses indicated that NIOSH Method 7300 using ICAP does not produce accurate or repeatable arsenic results with the fly ash matrix. It is possible that the matrix may produce interfering signals that affect the analytical outcome. The AA-GF and AA-HG methods performed acceptably. The microwave digestion procedure did not produce accurate results; arsenic recoveries

Table 6. Arsenic Valence State

Sample ID	As (III)	As (V)
NBS SRM 1633[a]	<0.2[b]	0.2
NBS SRM 1633[c]	ND[d]	100
QC sample[c]	40	60
QC sample expected value	50	50

[a] As milligrams per gram of arsenic on the particle surface.
[b] <0.2 % = (by weight) limit of detection for arsenic by ESCA.
[c] As percentage of arsenic atoms in the sample.
[d] ND = not detected.

were low using this method. The acid bomb digestion followed by AA-HG yielded results with high accuracy and precision.

Due to the interest in determining arsenic oxidation or valence state in coal fly ash, samples of the NBS fly ash specimen were subjected to electron spectroscopy for chemical analysis (ESCA). This instrumental analysis permits determination of the arsenic valence form and the relative location of the detected arsenic on the ash particle (i.e., surface vs. matrix). These results are presented on Table 6. The results revealed that all of the detected arsenic is present in the pentavalent [As(V)] form, and all of the delectable arsenic is on the surface of the particle. The carcinogenic form of arsenic is believed to be the As(III) form, not the As(V) form,[4] although due to the absence of a good animal model this has not been confirmed.

V. CRYSTALLINE SILICA

The presence of crystalline silica in coal fly ash is of interest due to its potential toxicity. Long-term exposures to crystalline silica present in natural materials is known to be associated with a degenerative lung disorder, known as silicosis. This is the basis for the exposure limits established by OSHA. More recently, crystalline silica has generated additional concerns due to evidence suggesting it causes cancer.[5] In California, crystalline silica is considered an air toxic material.[6]

During this study, exposures to and analysis of "respirable" fly ash air samples for crystalline silica were performed at each power plant studied. Analysis was performed by the standard X-ray diffraction technique, which is not especially sensitive. The analytical detection limit by this method is 1% by weight. Crystalline silica in the form of a-quartz was detected at four of the six plants studied. The two lignite plants did not reveal detectable concentrations of crystalline silica. Exposure to this analyte represented the most frequent values above the OSHA PELs, as compared to any of the other analytes studied.

VI. DISCUSSION

This study revealed that power plant workers may be exposed to excessive concentrations of coal fly ash during some work activities, with specific reference to arsenic and quartz. Questions remain concerning the toxicity of these fly ash constituents, their bioavailability in human tissues, and the methods available to accurately measure their presence. The degree of excessive exposures observed, and the potential health effects (i.e., chronic effects — lung scarring and cancer), will promote continued interest in the health effects to individuals exposed to fly ash. Animal exposure assays using coal fly ash have not produced the expected pulmonary lesions observed with quartz.[7] It is believed that the smooth vitrified surface of fly ash particles may modify the expected quartz toxicity. Discussions with electric utility medial officers yielded no reports of silicosis or X-ray evidence suggesting even mild forms of this lung scarring disorder.[8]

Exposure to other potentially hazardous constituents of fly ash, including PAHs, trace elements other than arsenic, and radionuclides, generally did not suggest these presented health hazards to the workforces studied. The methods employed in this study for the measurement of elements (other than arsenic) are subject to some doubt due to the discrepancies observed in the analysis of arsenic by ICAP, which was the method used for the other elements examined.

This study was limited in some respects, thus making broad conclusions is problematic. Several trends are apparent from the data set suggesting important implications, such as the variations of arsenic in coal fly ash from different coal types, the similarity of crystalline silica content and exposures in the plants studied, and the lack of arsenic and crystalline silica exposures at plants fired with lignite. To clearly define the apparent differences detected during this project, additional information and studies are needed to broaden the understanding of exposures workers encounter. When considering the implications the findings in this study represent to ambient "air toxic" issues, the toxicity of the potentially hazardous constituents are of key importance.

The research questions stimulated by this study include (1) what are the human health hazards presented by exposure to coal fly ash and its potentially toxic constituents? (2) what is the bioavailability of components of fly ash such as arsenic and crystalline silica? (3) Is the form of arsenic found in coal fly ash similar in toxicity as compared to other, well-studied forms, such as arsenic trioxide? (4) does the vitrified form of fly ash affect the toxicity of its constituents? (5) what are accurate sampling and analytical methods to determine the presence of metal elements in coal fly ash that may present health concerns, such as chromium, cadmium, nickel, etc.? Responding to these questions will help determine the significance and health consequences of exposure to coal fly ash.

REFERENCES

1. Electric Power Research Institute, Airborne Fly Ash Particulate in Utility Work Areas — Phase I, EPRI RP2222-2, Palo Alto, CA, 1988.
2. Electric Power Research Institute, Characterization and measurement of Airborne Fly Ash Encountered in Selected Coal Fired Power Plants — Phase II, EPRI RP2222-2, Palo Alto, CA, 1988.
3. Electric Power Research Institute, Characterization and measurement of Airborne Fly Ash Encountered in Selected Coal Fired Power Plants — Phase III, EPRI RP2222-2, Palo Alto, CA, 1989.
4. U.S. Environmental Protection Agency, Health Assessment Document for Inorganic Arsenic, Final Report, Research Triangle Park, NC, 1984.
5. International Agency for Research on Cancer, IARC Monographs, V. 42, Lyon, 1987.
6. California Health and Safety Code 44360 et seq. Air Toxic Hot Spots Information and Assessment Act of 1987.
7. Raask, E. and Schilling, C. J., Research findings on the toxicity of quartz relevant to pulverized fuel ash, *Ann. Occup. Hyg.*, 23, 147, 1980.
8. Weyzen, W., personal communication, 1990.

Chapter 5

Models for Risk Assessment, Management, and Design

Chair: Hugh E. Evans, PowerGen
Co-Chair: Leonard Levin, EPRI

Partitioning of Organic Pollutants in the Environment

Yoram Cohen

I. ABSTRACT

The distribution of chemical pollutants throughout the various environmental compartments (e.g., air, water, soil, and biota) is the result of complex physical, chemical, and biological processes. Human health risks related to the presence of various pollutants in the environment depend on the degree of multimedia exposure of humans to these chemicals. Thus, an integrated approach to estimating exposure to organic pollutants is required. In order to meet this demand, a comprehensive dynamic screening-level multimedia transport model and a dynamic multimedia total dose (MTD) model were developed.. The dynamic MTD analysis enables the prediction of the time-dependent dose and dose rate for the human receptor resulting from multipathway exposures. Results are presented from case studies on the multimedia distribution of toxic air pollutants and multipathway exposures in the Los Angeles area and in the vicinity of an oil refinery located on the east coast.

II. INTRODUCTION

One of the ultimate purposes of environmental pollution control is to reduce potential or existing risks. This calls for the determination of health risks, ecological risks, and welfare risks which are directly related to the exposure of human and various other ecological receptors to environmental chemicals. Exposure, in turn, is a function of contaminant concentrations in various environmental media that affect both the receptor directly, and various components of the food chain. Pollutant concentrations can in principle be obtained either by pollutant fate and transport modeling or by field measurements. The above tasks require an understanding of the complex physical, chemical, and biological processes that govern the movement of pollutants among the different environmental media, exposure pathways, and the health effects and other environmental effects associated with exposure to chemical contaminants.

The assessments of risks due to the exposure of a receptor (usually a biological receptor) to pollutants is generally determined from appropriate dose-response relations. In order to utilize dose-response relations to predict the expected response of a target receptor, one must first determine the dose. The dose in turn is directly related to the exposure of the receptor to the given agent. The exposure is a function, among

Table 1. Summary of Major Intermedia Transport Processes

Transport from atmosphere to soil and water
Dry deposition of gaseous and particulate pollutants
Adsorption onto particle matter and subsequent dry and wet deposition
Rain scavenging of gases and particles
Infiltration
Runoff
Transport from water to atmosphere, sediment, suspended solids, and biota
Evaporation
Aerosol formation at the air/water interface
Sorption by sediment and suspended solids
Sedimentation and resuspension of solids
Uptake and release by biota
Transport from soil to atmosphere, water, sediment, and biota
Volatilization from soil and vegetation
Dissociation in rain water which is associated with infiltration and runoff
Leaching to groundwater
Adsorption on soil particles and transport by runoff or wind erosion
Resuspension of contaminated soil particles by wind
Uptake by biomass such as microorganisms, plants, and animals

other factors, of the pollutant concentration in various environmental media that affect the receptor directly or indirectly.[1-7]

Extensive reviews of various approaches to determining multimedia distribution of pollutants have been given in the literature and there is a growing list of studies that have focused on multimedia analysis of the distributions of pollutants in the environment.[8-15]

II. MULTIMEDIA PARTITIONING OF ENVIRONMENTAL POLLUTANTS

A. General Considerations

Pollutants move across environmental phase boundaries as the result of complex intermedia transport processes. A list of some of the major intermedia transport processes among the various nonbiological environmental compartments (e.g., air, water, soil) is given in Table 1. Detailed discussions of intermedia transfer processes can be found in a number of references on the subject area.[9,10,12-16]

The distribution of chemical pollutants in the environment can be estimated using appropriate mass transfer models which are based on the principle of mass conservation. The global scheme for the multimedia modeling approach is shown schematically in Figure 1. The input variables include media properties, physicochemical and thermodynamic properties of the pollutant under consideration, climatic conditions, source emissions, and initial and background concentrations. The major information items required for multimedia partitioning analysis can be classified as given in Table 2. Information is needed to describe the geographical area where the exposure assessment is to be conducted. Such information includes topographic information

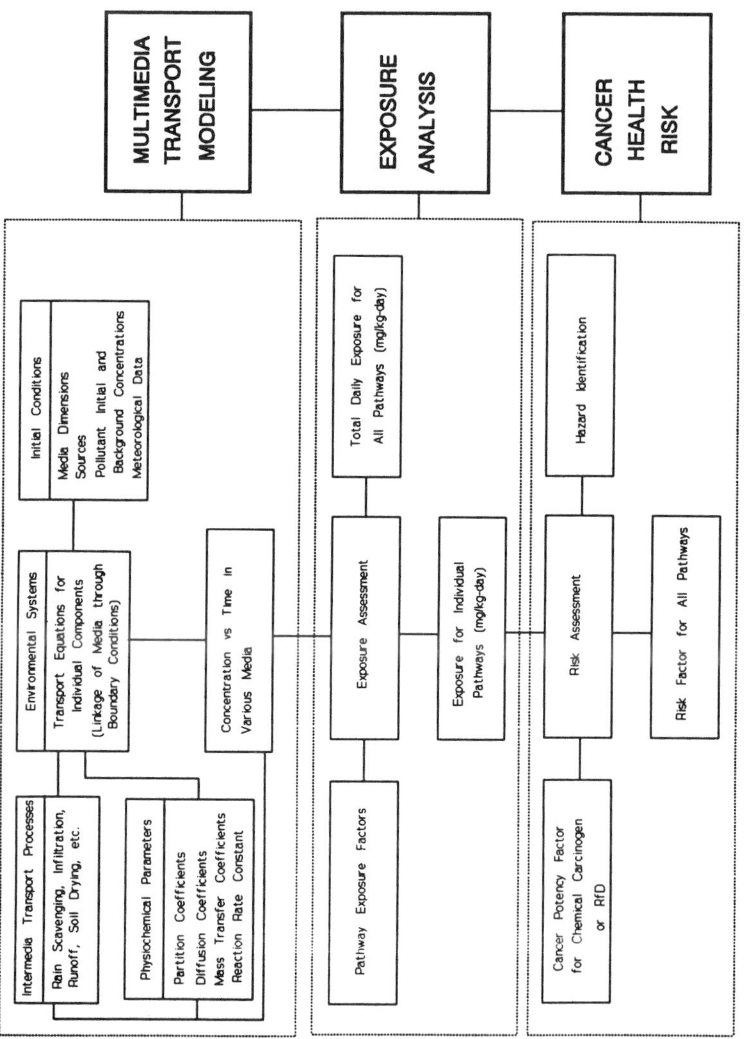

Figure 1. Global scheme of the multimedia analysis approach.

Table 2. Chemical Partitioning and Exposure Assessment: Major Information Needs and Analysis Steps

Definition and characterization of environmental media
- Topographical information for the region of interest
- Physicochemical properties of the environmental media
- Meteorological parameters

Source characterization

Determination of pollutant concentrations in the multimedia environment
- Multimedia transport and fate modeling
- Determination of pollutant partitioning

Receptors
- Characterization of exposed population (e.g., location, demographic characteristics)
- Population time-activity patterns
- Dietary habits
- Physiological properties of the receptor

Total dose assessment
- Determination of contaminant intake by receptor via all pathways

Table 3. Major Environmental Media

Nonbiological compartment	Biological compartment
Atmosphere	Aquatic animals
Surface water	Terrestrial animals
Sediment	Vegetation
Surface water	Food (solid and liquid)
Soil	
Groundwater	

for the region of interest, physicochemical properties of the environmental media (e.g., properties of the soil and sediment, depth of the water body), and meteorological parameters (wind conditions, mixing heights, etc.). The environmental system is described by a collection of media that can also be categorized into two distinct groups: nonbiological and biological environmental compartments, and the human receptor as a separate environmental compartment as indicated in Table 3. The above major compartments can be further subdivided into subclasses as deemed necessary. Although the properties of the above environmental media are not pollutant specific, their physicochemical properties can affect the pollutant interaction with these compartments. For example, the fraction of organic carbon in the soil has a direct effect on the sorption of organic pollutants by the soil matrix. Similarly, the pH of the soil can affect the partitioning of inorganics into the soil matrix. It is also known, for example, that the fraction of fat in beef and milk has a direct correlation with the amount of the chemical that can be retained in these environmental media. It is important to note that the required model parameters are often model specific and excellent references that focus on the evaluation of chemical and media parameters for determining pollutant distribution in the environments have been described in a number of available references.[3-17] Finally, in addition to the above, intermedia transport as well as transformation processes must be specified (see Table 1).

IV. MULTIMEDIA TRANSPORT AND FATE MODELING

A. General Considerations

It is important to realize that the main driving force in any multimedia model is the source (i.e., environmental emissions) of the chemical under consideration. Meaningful predictions from any multimedia model require the identification and quantification of all sources of the chemical. The pertinent information for each source should include (1) total emission rate of the chemical to the particular environmental medium, (2) the physical state of the chemical (dissolved, particulate, or gaseous state), and (3) the specific molecular form or speciation.

Given the above information, transport equations for individual environmental media are formulated and coupled through the appropriate boundary conditions. The standard output from the model is in the form of transient spatial concentration profiles for each of the environmental compartments under consideration. The resulting concentration fields, for example, can be coupled with exposure and risk-assessment models (Figure 1).

Environmental multimedia fate and transport models may be either generic or site specific. The level of complexity of the model and its input and output data vary depending on the spatial and temporal scales of interest, the release scenario, the lifetime of the chemical, and the environmental transport rates. In each case the model has to be adapted to mimic the characteristics of the problem. There are three major classes of models:

1. Multimedia-compartmental (MCM) models
2. Spatial-multimedia (SM) models
3. Spatial-multimedia-compartmental (SMCM) models

MCM models consist of a collection of uniform (i.e., well-mixed) compartments, each representing a different sector of the environment. In contrast, SM models involved detail modeling of temporal and spatial variations. In the alternative, SMCM models represent a combination of both uniform and nonuniform compartments. When a site-specific predictive capability is needed spatial models are necessary. Such models require the solution of complex four-dimensional space-time partial differential equations by appropriate numerical techniques. Additionally, spatial models require a considerable meteorological and hydrological database. In contrast, the solution of compartmental models, with only well-mixed compartments, requires only the solution of time-dependent ordinary differential equations and a modest amount of input data.

B. Multimedia Compartmental Models

The need for simple multimedia screening level models led to the development of the MCM modeling approach.[7-9] In this approach, all the environmental media under consideration are assumed to be well mixed. In these models, estimated concentrations in the air and water compartments can be reasonably approximated for cases involving area or dispersed atmospheric emissions.[9,17-20] The soil environment, however, is highly nonuniform, so estimates from compartmental models are likely to be inaccurate. Compartmental models have been introduced at different levels of complexity depending of whether convection (advection) streams among compartments, rain scavenging, transformations, and time-dependent sources are included in the model formulation.[9]

Compartmental models generally assume that all compartments are well mixed. The compartmental mass balance equations are generally of the following form:

$$\frac{d(V_i C_i)}{dt} = \sum_{j=1}^{N} A_{ij} K_{oij} (C_{ij} * - C_i + \sum_{j=1}^{N} \Omega_{ij} + \sum_{j=1}^{N} (Q_{ji} C_j - Q_{ij} C_i) + R_i + S_i \qquad (1)$$

where i = 1,..., N-J $i \neq j$, with the initial conditions

$$C_i = C_i(0) \text{ at } t = 0 \qquad (2)$$

where V_i is the volume of compartment i (m^3), C_i is the concentration of the species of interest in compartment i ($gmol/m^3$), N is the total number of compartments, J is the total number of nonuniform compartments, A_{ij} is the interfacial area between compartments i and j (m^2), K_{oij} is the ith side overall mass transfer coefficient between compartments i and j (m/h), $C_{ij}*$ is the pollutant concentration in compartment i which would be in equilibrium with compartment j and it can be expressed as $C_{ij}* = C_j H_{ij}$, where H_{ij} is the compartment i to j partition coefficient, R_i denotes the production or degradation rate of chemical (gmol/h), S_i is the source strength (gmol/h), and Q_{ij} and Q_{ji} are the volumetric flow rate (m^3/h) from compartment i to compartment j, and compartment j to compartment i, respectively. The term on the left-hand side represents the rate of accumulation of chemical in compartment i. The first term on the right-hand side of Equation 1 represents the rate of the chemical mass entering or exiting compartment i by interfacial mass transfer. The second term accounts for additional intermedia transport processes between compartments i and j such as dry deposition of particle-bound organics and rain scavenging. For example, the loss of pollutant mass in the air compartment resulting from rain scavenging can be incorporated via the second term of Equation 1, e.g.,

$$\Omega_{aw} + \Omega_{as} = -R * (A_{aw} + A_{as}) \lambda_g * C_a H_{wa} \qquad (3)$$

where the subscripts a, w, and s represent the air, water, and soil compartments, respectively; R* is the precipitation rate (e.g., mm/h); A_{as} and A_{aw} are the air/soil and air/water interfacial areas, respectively; C_a is the pollutant concentration in the air compartment; and H_{wa} is the dimensionless chemical water/air partition coefficient (i.e., $H_{wa} = C_w/C_a$, at equilibrium). λ_g* is a dimensionless gaseous rain scavenging ratio; it describes the efficiency by which rain drops can remove contaminants from the air, and varies between 0 and 1 for chemicals which are nonreactive in the aqueous phase.[13] Contaminants entering to the water compartment due to rainfall are also incorporated by using the second term of Equation 1, e.g.,

$$\Omega_{aw} = R * A_{aw}\lambda_g * C_a H_{wa} + f_{ro}(R_o)L_{ro}H_{sh}\lambda_g * C_a H_{wa} \tag{4}$$

The terms on the right-hand side of Equation 4 account for the amount of pollutant added to the water compartment by wet deposition and surface runoff, respectively. R_o is the runoff rate (m/h), f_{ro} is the fraction of runoff which will reach surface waters, L_{ro} is the length perpendicular (m) to the direction of runoff flow, and H_{sh} is the height of the sheet of runoff water (m).

The third term in Equation 1 accounts for the net convective flow of the chemical into compartment i. The fourth term represents the production/degradation by chemical reaction. Finally, the last term is the net input into the compartment from source emissions.

Compartmental models that employ only uniform (box-type) compartments can be expressed by the following compact form:

$$\frac{dC}{dt} = \mathbf{K}C + Q + S \tag{5}$$

Initial condition:

$$C(0) = C_o \tag{6}$$

in which C is the concentration vector and Q and S are the advection flows and source vectors and **K** is the parameter matrix. For the case of time invariant, **K**, Q, and S Equation 5 can be solved analytically. When **K** is time variant, a numerical solution is more appropriate in most cases. Although a model based on the assumption of uniform compartments may be inherently inappropriate for some compartments (e.g., soil and sediment), the simplicity of the approach makes it useful for obtaining approximate assessment of pollutant distribution as well as for the study of parameter sensitivities.

C. Spatial Multimedia Models

Spatial models are desired when a site-specific predictive capability is required. Such models are designed to provide one-, two-, and three-dimensional spatial resolution of chemical concentration profiles. There are two categories of spatial multimedia models: (1) partially coupled, integrated multimedia models and (2) composite multimedia models. The partially coupled models consist of single-medium models that are solved sequentially. Output from different submodules (e.g., air, water) is shared and managed usually by a central executive program. Since the modules are fixed by the model developer the general applicability of the integrated model is limited. The partially coupled modeling approach as implemented by the remedial Action Priority System (RAPS) methodology[8,21] is useful for screening purposes and ranking of chemical waste sites. The UTM-TOX model[22] and the Air, Land, Water Analysis System (ALWAS) model[23] are examples of comprehensive multimedia models which consist of integrated single-medium models.

Composite multimedia models consist of individual pathway models selected by the user. The individual components are not connected. The user may, however, run the models sequentially and use the output file from a given pathway as part of the output file for the subsequent pathway. The above approach does not allow for specific feedback among the different environmental media. The user may, however, analyze the individual modules and allow for manual feedback, at least as a first level of an approximation. The composite multimedia models require extensive input data, user expertise, and computer resources. They can however be adapted to handle site-specific problems. Examples of composite models include land disposal restriction (LDR) methodology,[24] the chemical migration risk assessment (CMRA) methodology,[25,26] and the multimedia contaminant environmental exposure assessment (MCEA) methodology.[27,28] More recently, the U.S. EPA has provided a compilation of existing single and multimedia models through the Geographical Exposure Modeling System (GEMS). Reviews of the above multimedia models have been published by Cohen[9,13] and Onishi et al.[8]

Spatial multimedia models can be represented by the following general formulation:

$$\frac{\partial^\alpha C_i}{\partial t} + \nabla \cdot vC^\alpha{}_i = -\nabla \cdot J^\alpha{}_i + R^\alpha{}_i + N^\alpha{}_i + S^\alpha{}_i \tag{7}$$

where i = 1,...N

Initial condition:

$$C^\alpha{}_i(0, x) = C^\alpha{}_{io}(x) \tag{8}$$

Boundary conditions

$$\text{Type A:} \quad C^\alpha{}_i(t, x^*) = C^\alpha{}_i \tag{9}$$

$$\text{Type B:} \quad -K_D \nabla C^\alpha{}_i \,|_{i/j} = K_{ij}(C^\alpha{}_{j\infty} - C^\alpha{}_j) + J^\alpha_R \tag{10}$$

The subscripts i and j refer to compartments i and a bordering compartment j, i/j denotes the interface between compartments i and j, and α designates the particular chemical species under consideration. The velocity vector is designated by v, J_i^α is the diffusive flux (molecular or turbulent) of component α in compartment i, R_i^α is the rate of chemical or biotransformations per unit volume, S_i^α is the source strength (e.g., g/cm^3 s), and N_i^α is the rate of physical transformations (e.g., adsorption in the soil matrix) per unit volume of a given compartment, and K_{ij} is the phase i-side mass transfer coefficient. Constitutive equations are required for N_i^α, R_i^α, and J_i^α. Finally, $C_{j\infty}^\alpha$ is the concentration of the chemical in the bulk region of compartment j. Multimedia models as represented by Equation 10 require the numerical solution of complex partial differential equations of the convective-diffusion type for each of the nonuniform compartments under consideration — coupled by the appropriate boundary conditions. As the degree of desired spatial resolution increases the required input of meteorological and hydrological data also increases.

Through the use of boundary conditions (Equations 9 and 10), it is recognized that the uniform and nonuniform compartments are not necessarily in dynamic equilibrium at the boundaries, and thus more appropriate flux boundary conditions must be applied at the top and bottom boundaries of the soil compartment. For example, the additional flux term J_R^α, which for the case of the soil environment includes the deposition of airborne pollutants, the infiltration of rain water, and the interfacial mass transfer constrained by equilibrium have to be incorporated to the top flux boundary condition for the soil matrix (see Equation 10).

D. Spatial-Multimedia-Compartmental (SMCM) Models

More recently, Cohen et al.[16] introduced the spatial-multimedia-compartmental (SMCM) approach in which the soil and sediment compartments are treated as nonuniform compartments. Thus, this approach can accommodate processes such as runoff, precipitation scavenging, and soil drying. In the SMCM approach partial and ordinary differential equations that describe the different compartments are solved simultaneously with the appropriate boundary conditions. The SMCM model has an improved resolution compared to the simple compartmental models as well as a capability of describing a variety of intermedia processes. The SMCM approach can be extended to include spatial variability of specific compartments while retaining a box-type description for some compartments, thereby allowing for improved accuracy and reduced model complexity.

V. MULTIMEDIA EXAMPLES

The multimedia partitioning of a number of chemical pollutants has been recently evaluated using the NCITR spatial-compartmental (SMCM) multimedia fate and transport model. The structure of the SMCM model is illustrated in Figure 2 and the main features of the model are listed in Table 4. The details of the SMCM model are

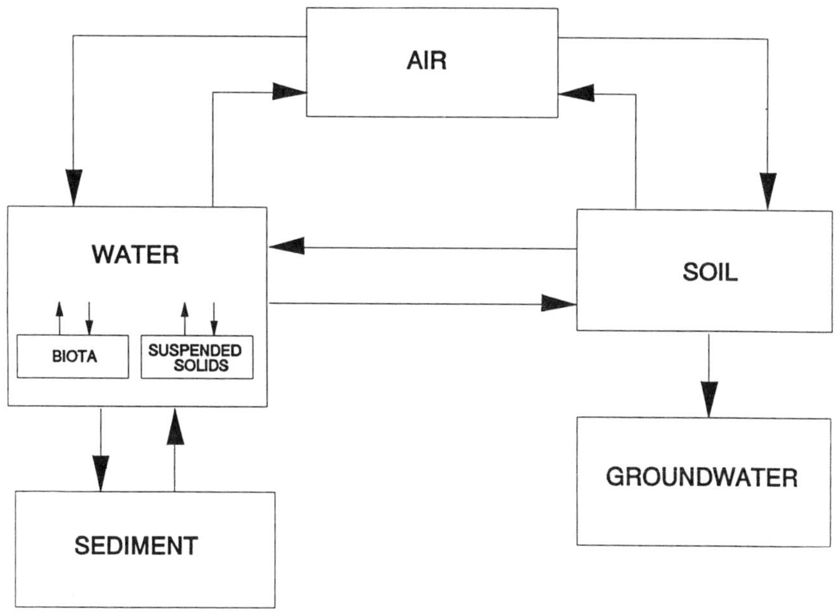

Figure 2. Schematic illustration of the SMCM model.

described elsewhere.[17] A few examples that illustrate the multimedia partitioning of air toxics are described below.

In the first and second examples, the multimedia distribution of benzene and trichloroethylene (TCE) was evaluated for the Santa Clara Valley (California) given estimates of the total emissions in the Santa Clara region. The results as indicated in Table 5 are in reasonable agreement, for a screening-level analysis, with the reported average concentrations for the air compartment. As expected, much of the mass of these volatile chemicals is in the air compartment. The results also suggest that monitoring the concentrations of benzene and TCE in the soil and water compartments may be difficult given the low expected concentrations.

In the third example, the multimedia distributions for benzene and methyl-tert-butyl ether (MTBE) emitted from the Amoco Yorktown oil refinery, located near the Chesapeake Bay, were evaluated by considering the Toxics Release Inventory (TRI) reported emissions from the facility and mobile sources.[29] The atmospheric concentrations for benzene and MTBE were estimated to be 2.2 and 0.14 $\mu g/m^3$, respectively. The contribution of the Amoco Yorktown Refinery to the environmental concentrations are directly related to the level of emissions from the refinery relative to the total emissions. For example, it was estimated that the Yorktown Refinery contributed about 36 and 100% to the annual average air concentration of benzene and MTBE, respectively, due to total sources in the modeled region. This behavior is expected given that the contribution of the Yorktown Refinery to the total emissions of benzene and MTBE in the modeled region was also about 36 and 100%,

Table 4. Main Features of the SMCM Model

The SMCM is a user-friendly software package that
- Can be used to answer "what if" type questions
- Allows for rapid scenario changes
- Minimizes data input
- Provides a graphical output display for quick scenario analysis
- Provides specific online help for input data fields
- Provides a menu system for user selection of data input, simulation execution, plotting, and printing a summary report of the calculated results
- Allows the software to be run on IBM-PC/XT/AT-compatible computers
- Allows an inexperienced user to run the SMCM software with virtually no background in transport phenomena

The SMCM model applies a new modeling approach that
- Makes use of both uniform (air, water, biota, suspended solid) and nonuniform compartments (soil and sediment)
- Allows for mass exchange of pollutant between the air compartment and its surrounding atmospheric environment; the water compartment is also treated in a similar way
- Treats nonuniform compartments as an unsteady state, one-dimensional diffusion- type equation with convection and chemical reaction
- Incorporates the simulation of a chemical buried in the soil compartment
- Considers a variety of source types and allows the user to select and input source data through the data input screens
- Applies flux boundary conditions for nonuniform compartments; although groundwater is not treated as a compartment in the SMCM model, flux condition at the bottom boundary of the soil compartment can be incorporated to account for the chemical transport to groundwater

The SMCM model accounts for the effects of rainfall and temperature on the environmental transport of pollutants
- The SMCM has a rain generation module which can generate rainfall in the form of a single event of specified intensity and duration, or randomly distribute rainfall within specified levels of rainfall intensity, duration, and total rainfall
- The transport processes associated with rainfall such as rain scavenging, infiltration, runoff, and soil drying are simulated by a water balance method which uses theoretically based correlations
- User-supplied average monthly temperatures are used to construct average daily temperatures

Provides accurate and reliable parameter estimation methods
- Physicochemical parameters such as mass transfer coefficients, diffusion coefficient, and partition coefficient are estimated using theoretical methods and empirical correlations; the user can input partition coefficients and diffusion coefficients if known; these will override any model-estimated values
- Temperature variations of diffusivities, partition coefficients, mass transfer coefficients, and reaction rate constants are included by either internal predictions or via user-input data
- Production or degradation rates are treated as first-order reactions

respectively. For example, for benzene about 97% of the total amount of benzene that accumulated in the modeled environment is found to be present in the atmospheric phase (Figure 3B). About 1.8% of the total amount of benzene that accumulated in the modeled environment is present in the water compartment with the remainder distributed to the soil, sediment, suspended solids, and biota (Figure 3B). In contrast, MTBE, which is significantly more water soluble than benzene, partitions preferentially into the water phase with about 73% of the MTBE that accumulated in the modeled environment present in the water compartment, while only about 25%

Table 5. Comparison of Steady-State Results from the SMCM Model and Results of the IEMP Air Monitoring Program in the Santa Clara Valley

Compartment	Predicted concentration (gmol/m³, × 10⁸)	Predicted concentration (μg/m³)	Mass (kg)	Monitored concentration (μg/m³)
	Benzene[a]			
Air	4.6	3.6	9.5×10^3	7.7[b]
	12.7	9.9	2.6×10^4	
Water	13.3	10.4	1.8	—
	36.6	28.6	4.9	
Soil	21.5	16.8	117.2	—
	59.5	46.5	339.6	—
Sediment	7.3	5.7	8.6×10^{-3}	—
	20.5	16.0	2.5×10^{-2}	
Biota	61.6	48.1	4.2×10^{-6}	—
	169.0	132.0	1.1×10^{-5}	
Suspended soilds	33.1	25.9	2.2×10^{-5}	—
	91.0	71.1	6.1×10^{-5}	
	Trichloroethylene			
Air	1.7	2.2	5.9×10^3	1.8[b]
Water	3.4	4.5	0.8	—
Soil	8.8	11.5	67.6	—
Sediment	3.6	3.5	6.1×10^{-3}	—
Biota	39.0	51.2	4.4×10^{-6}	—
Suspended solids	20.7	27.2	2.4×10^{-5}	—

[a] The first and second rows of predicted benzene concentrations in each compartment are based on benzene background concentrations of 1 and 6.04 ppb, respectively.
[b] Average concentration during the Santa Clara Screening Program (Versar Inc., 1985).

accumulated in the air compartment (Figure 4B). It is also interesting to note that rain scavenging leads to fluctuations in the soil concentration of benzene and MTBE as shown in Figures 3A and 4A. Also, starting with a pristine environment, simulation results shown in Figures 3A and 4A reveal that the concentration in all media rises quickly to their "steady-state" value. Also shown in Figure 3A and 4A is the decline in the environmental concentration upon the elimination of source emissions. It is apparent from the above results that the recovery of the environment from benzene contamination is faster than for MTBE.

IV. CONCLUSIONS

The future development of multimedia models for the assessment of pollutant transport and fate will only be feasible if modeling efforts will be coupled with multimedia monitoring studies. Presently, the user may select from among the various classes of models described earlier in this paper. As more efficient computational methods are developed and with the increase in computational hardware it

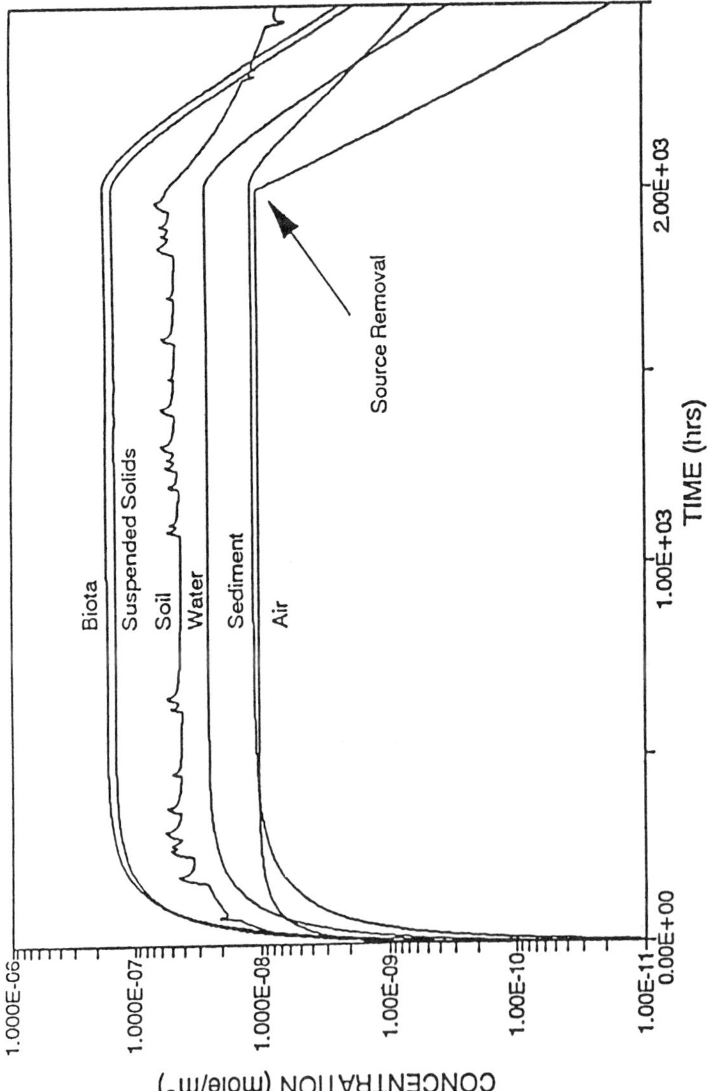

CONCENTRATION (mole/m3)
1.000E-06
1.000E-07
1.000E-08
1.00E-09
1.00E-10
1.00E-11
0.00E+00
1.00E+03
2.00E+03
TIME (hrs)
Biota
Suspended Solids
Soil
Water
Sediment
Air
Source Removal

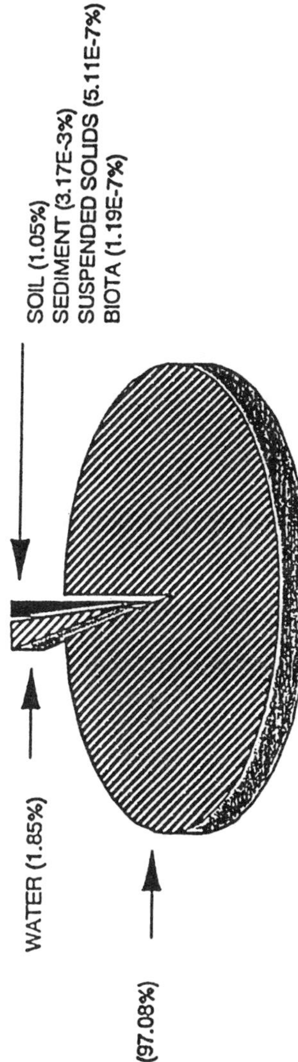

Figure 3A and B. Distribution of benzene in the environment surrounding the Yorktown refinery.

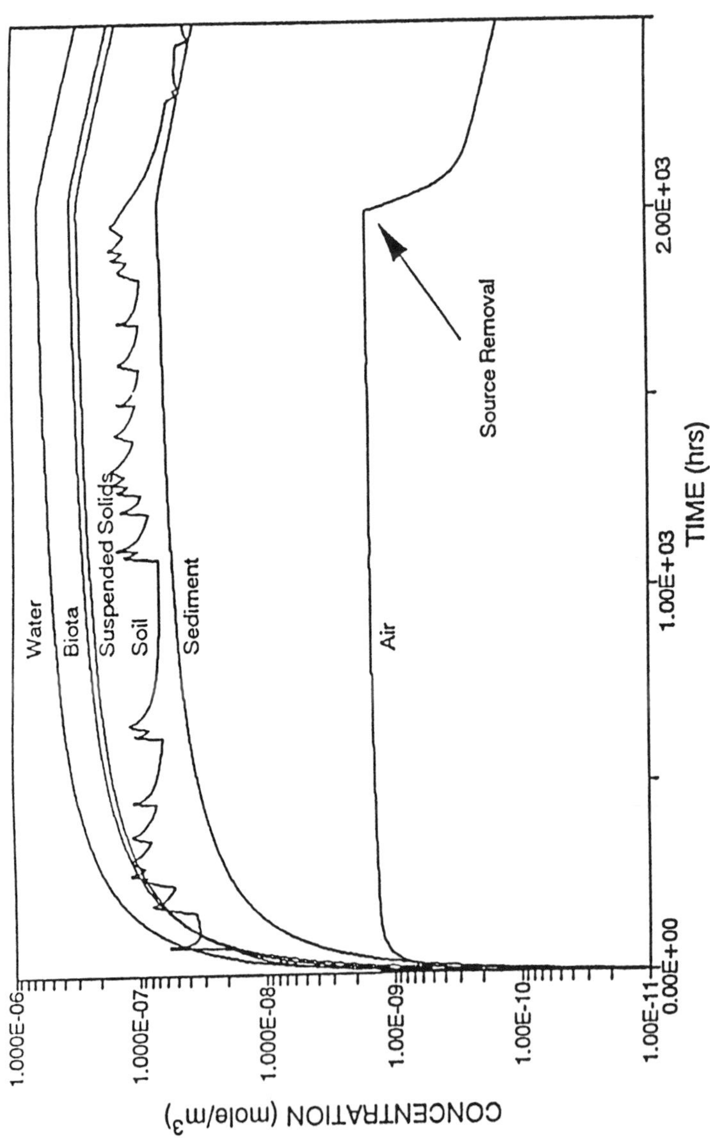

CONCENTRATION (mole/m^3)
1.000E-06
1.000E-07
1.000E-08
1.00E-09
1.00E-10
1.00E-11
0.00E+00
1.00E+03
2.00E+03
TIME (hrs)
Water
Biota
Suspended Solids
Soil
Sediment
Air
Source Removal

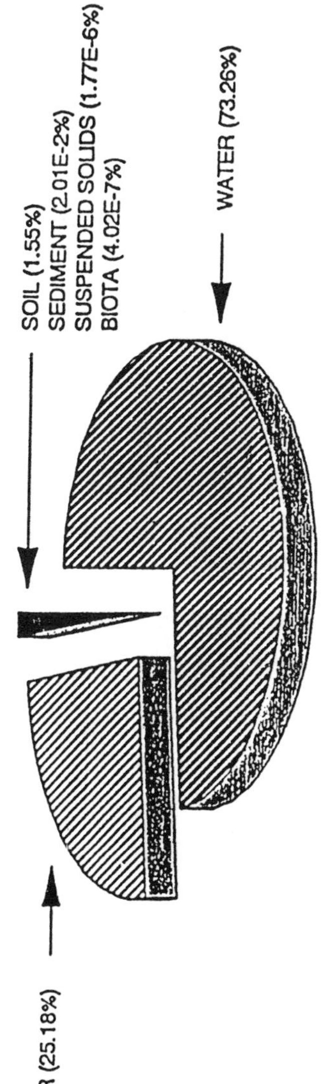

Figure 4A and B. Distribution of MTBE in the environment surrounding the Yorktown refinery.

may become feasible to develop multimedia models of increased spatial resolution for site-specific applications. Also, it must be emphasized that the accuracy of any fate and transport model relies heavily on the accuracy of the description of intermedia transport processes and chemical and biological transformations. There are still numerous knowledge gaps in the area of intermedia transport and as future advancements in this area are made the quality of multimedia models will also improve.

ACKNOWLEDGMENTS

The completion of this paper was funded in part by the U.S. EPA Grant CR-812771 and the University of California Toxic Substances Research and Teaching Program.

REFERENCES

1. EPA/A&WMA, *Total Exposure Assessment Methodology,* Proc. EPA/A&WMA Specialty Conference, November 1989, Las Vegas, Air & Waste Management Association Association, Pittsburgh, PA, 1989.
2. National Research Council, *Human Exposure Assessment for Airborne Pollutant — Advances and Opportunities,* National Academy Press, Washington, DC, 1991.
3. South Coast Air Quality Management District, Multi-Pathway Health Risk Assessment: Input Parameters Guidance Document, prepared for the South Coast Air Quality Management District under Contract #8798 by Clement Associates, Inc., June 1988.
4. Air Resources Board and Department of Health Services (State of California), Health Risk Assessment Guidelines for Nonhazardous Waste Incinerators, August 1990.
5. EPA, *Methods for Assessing Exposure to Chemical Substances, Vol. 8,* EPA/560/5-85-008, Washington, D.C., September 1986.
6. EPA, Guidelines for estimating exposures, *Fed. Reg.,* 51, 34092, 1988.
7. McKone, E. M. and Layton, D. W., Screening the potential risks of toxic substances using a multimedia compartment model: estimation of human exposure, *Regulatory Toxicol. Pharmacol.,* 6, 359, 1986.
8. Onishi, Y., Shuyler, L., and Cohen, Y., Multimedia modeling of toxic substances, in *Proc. Int. Symp. on Water Quality Modeling of Agricultural Non-Point Sources,* Utah State University, Logan, UT, 479–502, 1990.
9. Cohen, Y., Organic pollutant transport, *Environ. Sci. Technol.,* 20, 538, 1986.
10. Cohen, Y., Modeling of pollutant transport and accumulation in a multimedia environment, in *Geochemical and Hydrologic Processes and their Protection: The Agenda for Long Term Research and Development,* Draggan, S., Cohrssen, J. J., and Morrison, R. E., Eds., Praeger, New York, 1987.
11. Allen, D., Kaplan, I., and Cohen, Y., Eds., *Intermediate Pollutant Transport: Modeling and Field Measurements,* Plenum Press, New York, 1989.
12. Lyman, W. J., Reehl, W. F., and Rosenblatt, D. H., *Handbook of Chemical Property Estimation Methods: Environmental Behavior of Organic Compounds,* McGraw-Hill, New York, 1982.

13. Cohen, Y., Intermedia transport modeling in multimedia systems, in *Pollutants in a Multimedia Environment,* Cohen, Y., Ed., Plenum Press, New York, 1986, 7.
14. Cohn, Y. and Ryan, P. A., Multimedia modeling of environmental transport: trichloroethylene test case, *Environ. Sci. Technol.,* 9, 412, 1985.
15. Ryan, P. and Cohen, Y., Multimedia transport of particle bound organics: benzo(a)pyrene test case, *Chemosphere,* 15, 21, 1986.
16. Cohen, Y., Tsai, W., Chetty, S. L., and Mayer, G. J., Dynamic partitioning of organic chemicals in regional environments: a multimedia screening-level modeling approach, *Environ. Sci. Technol.,* 24, 1549, 1990.
17. Thibodeaux, L. J., *Chemodynamics,* John Wiley & Sons, New York, 1979.
18. Mackay, D., Paterson, S., and Joy, M., Applications of fugacity models to the estimation of chemical distribution and persistence in the environment, in *Fate of Chemicals in the Environment,* (ACS Symp. Ser. No. 225), Swann, R. L. and Eschenroeder, A., Eds., American Chemical Society, Washington, DC, 1982.
19. Mackay, D. and Paterson, S., Fugacity revisited, *Environ. Sci. Technol.,* 16, 654a, 1982.
20. Mackay, D. and Paterson, S., Calculating fugacity, *Environ. Sci. Technol.,* 15, 106, 1981.
21. Whelan, G., Strenge, D. L., Droppo, J. G., Jr., Steelman, B. L., and Buck, J. W., *The Remedial Action Priority System (RAPS): Mathematical Formulation,* DOE/RL/87-09/ PNL-6200, Pacific Northwest Laboratory, Richland, WA, 1987.
22. Patterson, M. R., Sworski, T. J., Sjoreen, A. L., Browman, M. G., Coutant, C. C., Hetrick, D. M., Murphy, E. D., and Raridon, R. J., *A User's Manual for UTM-TOX: A Unified Transport Model,* ORNL-6064, Oak Ridge National Laboratory, Oak Ridge, TN, 1984.
23. Tucker, W. A., Eschenroeder, A. G., and Magil, G. C., Air, Land, Water Analysis System (ALWAS): A Multimedia Model for Assessing the Effect of Airborne Toxic Substances on Surface Quality, first draft report, prepared by Arthur D. Little for Environmental Research Laboratory, EPA, Athens, GA, 1982.
24. EPA, Hazardous Waste Management System Land Disposal Restrictions: Proposed Rule, Part III, U.S. EPA CFR (260), *Fed. Reg.,* 1601, 1986.
25. Onishi, Y., Brown, S. M., Olsen, A. R., et al., Assessment Methodology for Overland and Instream Migration and Risk Assessment of Pesticides, Battelle, Pacific Northwest Laboratories, Richland, WA, 1979.
26. Onishi, Y., Olsen, A. R., Parkhurst, M. A., and Whelan, G., Computer-based environmental exposure and risk assessment methodology for hazardous materials, *J. Hazard. Mater.,* 10, 389, 1985.
27. Onishi, Y., Whelan, G., and Skaggs, R. L., Development of a Multimedia Radionuclide Exposure Model for Low-Level Management, PNL-3370, Pacific Northwest Laboratory, Richland, WA, 1982.
28. Onishi, Y., Yabusaki, S. B., Cole, C. R., Davis, W. E., and Whelan, G., Multimedia Contaminant Environmental Exposure Assessment (MCEA) Methodology for Coal-Fired Power Plants, Vol. 1 and 2, Battelle, Pacific Northwest Laboratory, Richland, WA, 1982.
29. Cohen, Y., Allen, D. T., Clay, R. E., Rosselot, K., Tsai, W., Klee, H., Jr., and Blewitt, D. N., Multimedia Assessment of Emissions (MAE) from the Amoco Corporation Yorktown Refinery (Amoco/EPA Pollution Prevention Project), presented at the Air and Waste Management Association 84th Annual Meeting and Exhibition, Vancouver, British Columbia, 1991.

Effects of Deposition Velocities on Predicted Health Risks

Robert G. Vranka
Gayle E. Watkin

I. INTRODUCTION

A comprehensive health risk assessment should incorporate all potential pathways of exposure to toxic chemicals including inhalation and noninhalation routes (e.g., ingestion of soil, water, impacted plant or animal products, or dermal contact with chemicals). A key factor for calculating potential health risks from exposure via the noninhalation route is the consideration of deposition of toxic substances onto surfaces, and subsequent exposure to the chemicals by direct or indirect contact. Deposition velocities that are used in these risk assessments can significantly affect the calculated exposure levels at specific receptors. In fact, deposition velocities can often drive the exposure and risk assessment.

A factor that is related to deposition is the concept of plume depletion. In plume depletion, pollutants are deposited at the ground surface as the plume is transported downwind, thus reducing ground-level air concentrations relative to the values that would be observed if deposition were not occurring. Typically, plume depletion is ignored in dispersion models that are used in risk assessments.

This paper examines the effects of deposition velocities on predicted exposure concentration levels and the results of multipathway risk assessments.

II. PRINCIPLE ASSUMPTIONS

A. Deposition Velocity

Deposition velocity is a mass-transfer boundary condition at the air-surface interface in atmospheric diffusion and transport models. It accounts for the total mass transfer from the air to the ground surface within the lowest few meters of the atmosphere, and is the ratio of the deposition flux divided by the airborne pollutant concentration per unit volume at some height above that surface layer. The deposition velocity is usually reported in units of meters per second (m/s) and can be expressed as

$$V_d = \frac{F}{v} \tag{1}$$

where Vd = deposition velocity (m/s), F = flux to the surface in grams per square meter per second (g/m^2 s), and v = ambient concentration above the surface in grams per cubic meter (g/m^3).

Several variables affect deposition rates for particles. These variables include properties related to the particles, properties related to micrometeorology, and properties related to the surface. These variables are described below.

1. Particle Properties

For most particles, the deposition velocity is dependent on gravitational settling which in turn is a function of the particle diameter and particle density. However, for extremely small particles (i.e., particles with diameters <0.1 μm), deposition is less affected by particle density and more affected by the amount of Brownian motion exhibited. Brownian motion is the irregular random motion exhibited by small particles due to bombardment by surrounding molecules in the air. Brownian diffusion coefficients increase considerably as particle sizes decrease. Therefore, Brownian diffusion is more important in affecting the rate of final deposition for small particles, although particle inertia and particle eddy diffusion are important phenomena in transporting particles to the ground surface.

Thus, when calculating deposition rates in multipathway risk assessments, it is important that particle size and density be included in the analysis.

2. Meteorological and Surface Properties

Certain properties of the surface and meteorological parameters can affect the rate of particle deposition. These include the aerodynamic surface roughness height, Z_0. An increase in surface roughness will increase turbulence near the surface and will enhance deposition. Therefore, surface roughness height can be considered a micrometeorological variable that is driven by the surface properties downwind of a plume. It has been observed that deposition velocities can vary significantly with changing surface properties, and site-specific surface conditions should be incorporated into the calculation of deposition velocities when conducting a multipathway risk assessment.

B. Plume Depletion

As indicated previously, pollutants are deposited at the ground as the plume is transported downwind. Thus, airborne concentrations at ground level would be reduced when compared with values that are observed if deposition does not occur.

Dispersion models that are used in exposure and risk assessments should take this into account when ambient air concentrations are calculated above a deposition surface. Plume depletion can be especially important for large particles because of greater settling and deposition velocities. Plume depletion can also be important if the

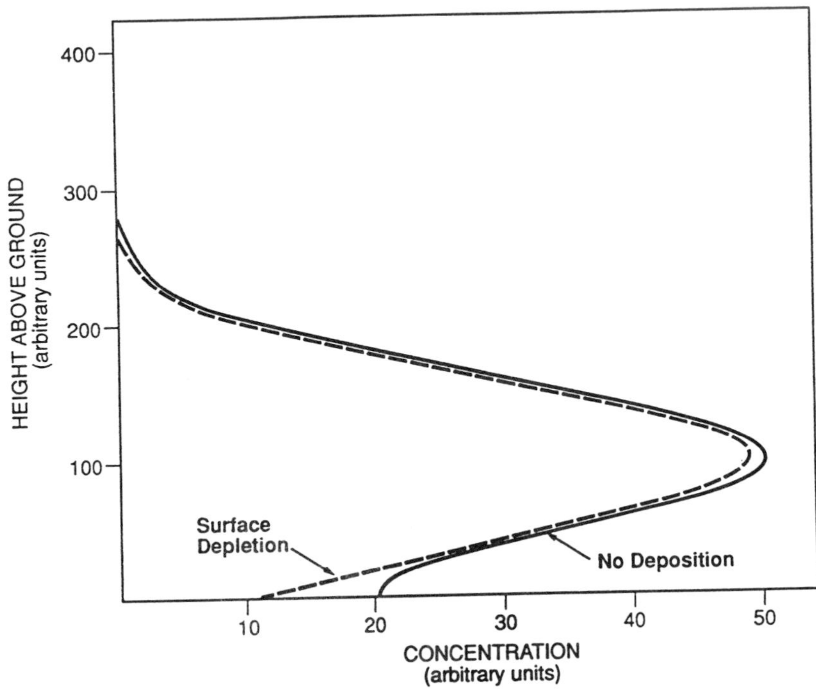

Figure 1. Airborne concentrations as a function of height for sources with and without plume depletion from dry deposition. (Adapted from Ermak, D. L., *Atmos. Environ.*, 2, 21, 1977. With permission.)

plume travels large distances. A simplified source-depletion approach is used to correct for deposition in the Industrial Source Complex Short-Term (ISCST) model in which the source strength is adjusted. This approach reduces the ambient concentration proportionally throughout the vertical cross-section of the plume rather than at the ground.

A more realistic surface depletion approach has been adopted by Ermak[1] in which ambient air concentrations are reduced near the surface to account for deposition at the ground. Figure 1 shows the effect of surface depletion for a typical source that emits particles. It shows that ambient air concentrations near the ground surface would be reduced when compared to no deposition, while these concentrations would not change at greater elevations above the surface. Surface depletion, as defined by Ermak, should be incorporated into the calculation of ambient air concentrations when estimating multipathway exposures in a risk assessment.

III. SCREENING APPROACHES TO DEPOSITION

Typical exposure and risk-assessment studies will incorporate models that use site-specific meteorological data in order to provide more accurate simulations. However, these studies rarely account for site-specific deposition and plume-depletion factors when estimating multipathway risks. Instead, general screening deposition velocities are used, and plume depletion is not considered. When calculating multipathway risks with screening deposition velocities, several simplifying assumptions are often adopted to estimate deposition and subsequent noninhalation exposure.

First, a single uniform deposition velocity is assumed for all particles regardless of particle size and/or ground surface properties. In order to account for all possible conditions, conservative deposition velocities of 0.02 and 0.05 m/s are typically recommended in screening studies for controlled stack emissions and for fugitive emissions, respectively.[2] Based on information from Sehmel and Hodgson,[3] typical particle diameters associated with these deposition rates are 20 and 30 μm, respectively. These mean particle diameters are much larger than emissions from most controlled industrial processes and generally would greatly overestimate deposition rates at receptors. Therefore, exposure by noninhalation pathways would also be overestimated.

A second factor that is assumed when screening deposition velocities are used is that surface depletion is neglected when calculating ambient air concentrations. Thus, exposure levels are counted twice by including the same particulate matter both in inhalation and noninhalation pathways.

The calculated risks for such facilities would therefore be unrealistically high because of the conservative deposition calculations and because of the lack of consideration of plume depletion.

IV. REFINED APPROACHES

Refined methods for calculating deposition in multipathway exposure assessments generally use the analytical solution derived by Ermak.[1] This method allows one to account for both gravitational settling and surface depletion. One model that incorporates Ermak's method is the Fugitive Dust Model (FDM).[4] This model incorporates detailed deposition algorithms into standard dispersion equations to calculate deposition fluxes at the ground surface and ground-level ambient air concentrations.

Gravitational settling velocities and deposition velocities are calculated in the model based on the work of Sehmel and Hodgson,[3] in which relationships were established between deposition velocities and particle diameter, friction velocity, aerodynamic surface roughness height, and particle density. The deposition relationships were based on wind-tunnel observations of surface mass-transfer resistances for depositing particles.

Based on these observations, an empirical relationship of deposition velocity vs. particle diameter was derived by Sehmel. Figure 2 shows a typical plot of this relationship. Deposition velocities can be estimated for each size and density class

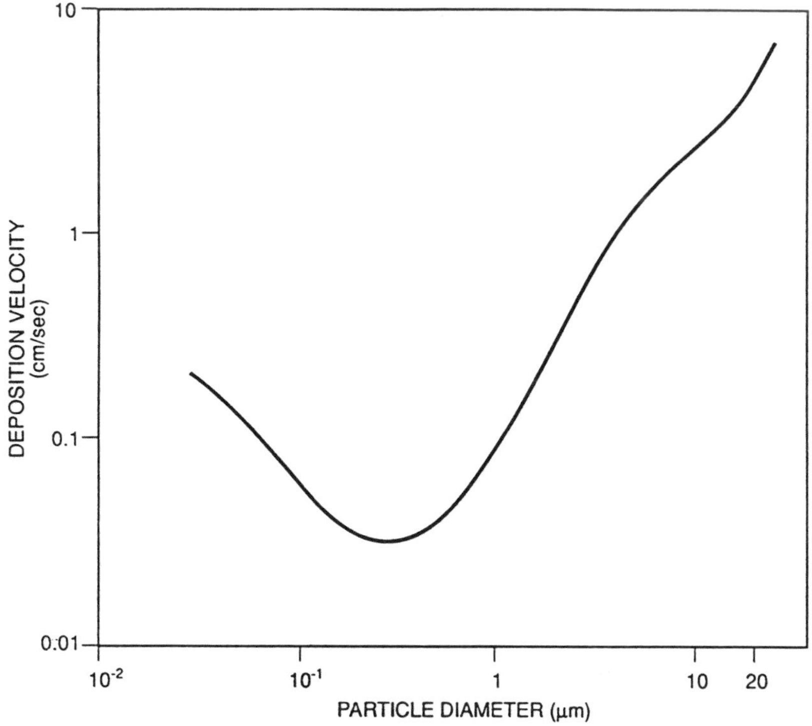

Figure 2. Predicted deposition velocities as a function of particle size for a particle density of 2.5 g/cm^3. (Adapted from Sehmel, G. A. and Hodgson, W. H., Model for Predicting Dry Deposition of Particles and Gases to Environmental Surfaces, PNL-SA-6721, Battelle Pacific Northwest Laboratories, Richland, WA, 1978.)

of particles for each hour of meteorological data based on these observations. This is in contrast to the screening process that uses only one deposition velocity for all particles under all meteorological conditions.

Deposition velocities were calculated in the FDM model for particles emitted from a controlled combustion stack under neutral atmospheric conditions. In this case, the particle density was assumed to be 2.5 g/cm^3 and particle sizes averaged from 2.5 to 7.5 µm, which is conservatively large for a source with particulate controls.

The mean deposition velocity as calculated by FDM was 0.2 cm/s or about one order of magnitude lower than the screening deposition velocity of 2 cm/s generally assumed for risk assessments. The effect of this change in deposition velocity on multipathway risks for a typical emission source was examined and is described in detail below.

V. EFFECTS ON MULTIPATHWAY RISK ASSESSMENTS

For sources that are dominated by emissions in the form of particulate matter, the predicted multipathway risks would be greatly influenced by the deposition velocities used in the dispersion model as well as by plume depletion.

To test this point, a comparison was made between two risk assessments in which one used screening deposition velocities as described above and no plume depletion, and the other used deposition velocities as calculated by Sehmel and Hodgson's[3] method with plume depletion at the surface. All other parameters such as emissions and operating rates remained the same for both cases. Thus, the differences in calculated risk were due to changes in the deposition velocities.

For both cases, multipathway risks were calculated by using the risk assessment model ACE 2588.[5] This model, which was designed for calculating risks in compliance with California's Air Toxics "Hot Spots" Assessment Act (AB 2588), uses output concentrations from the dispersion/deposition model to calculate multipathway risks at a receptor. It adopts methods consistent with guidelines established by the California Department of Toxic Substances Control (DTSC), formerly the Department of Health Services.

The comparative risks were evaluated for an industrial source that emits particulate matter such as polycyclic aromatic hydrocarbons (PAH), tetrachlorodibenzo-*p*-dioxin (TCDD), arsenic, beryllium, and chromium. As indicated previously, sources with emissions dominated by toxic particulate matter emissions would be expected to be affected most by deposition in the risk assessment.

The ACE 2588 model evaluates potential health risks from inhalation of the chemicals and it uses the deposition velocities to predict potential risks associated with subsequent dermal contact and incidental ingestion of contaminated soil, plants, or animals.

The multipathway risks for the case using screening deposition velocities are shown in Table 1. The results indicate that about 96% of the total risk is contributed by the noninhalation pathways.

The calculated risks for the case in which refined deposition velocities were used in the exposure calculations are reported in Table 2. In this case, the noninhalation pathways contribute only about 51% of the total risk to a given receptor. Also, the total predicted risks with the refined deposition velocities are only 13% of the total predicted with the screening deposition velocities.

As indicated above, for the case in which refined deposition velocities were used in the calculations, surface depletion was also considered when calculating airborne concentrations at the surface. However, the airborne concentration did not change significantly (approximately 1%) from the "no depletion" screening case because of the lower deposition velocities. If the particle sizes were larger, the deposition rates would increase and airborne concentrations would be reduced at the surface because of greater surface depletion.

Table 1. Predicted Cancer Risk by Pathway Assuming Screening Deposition Velocities

Pollutant	Inhalation	Dermal contact	Soil ingestion	Plant ingestion	Animal ingestion	Sum
PAH	2.1×10^{-7}	1.3×10^{-7}	2.7×10^{-7}	1.5×10^{-6}	0.0	2.1×10^{-6}
TCDD	1.5×10^{-7}	2.2×10^{-7}	3.0×10^{-7}	5.2×10^{-7}	4.1×10^{-5}	4.2×10^{-5}
As	8.3×10^{-7}	2.3×10^{-8}	1.5×10^{-6}	5.5×10^{-7}	2.9×10^{-7}	3.2×10^{-6}
Be	6.0×10^{-7}	5.9×10^{-8}	3.8×10^{-6}	1.2×10^{-6}	3.5×10^{-7}	6×10^{-6}
Cr	4.9×10^{-7}	0.0	0.0	1.8×10^{-7}	9.9×10^{-6}	1.1×10^{-5}
Sum	2.5×10^{-6}	4.3×10^{-7}	5.9×10^{-6}	3.9×10^{-6}	5.1×10^{-5}	6.4×10^{-5}

Table 2. Predicted Cancer Risk by Pathway Assuming Refined Deposition Velocities

Pollutant	Inhalation	Dermal contact	Soil ingestion	Plant ingestion	Animal ingestion	Sum
PAH	2.1×10^{-7}	1.3×10^{-8}	2.7×10^{-8}	1.5×10^{-7}	0.0	4×10^{-7}
TCDD	1.5×10^{-7}	2.2×10^{-8}	3.0×10^{-8}	5.2×10^{-8}	4.1×10^{-6}	4.3×10^{-6}
As	8.3×10^{-7}	2.3×10^{-9}	1.5×10^{-7}	5.5×10^{-8}	2.9×10^{-8}	1.07×10^{-6}
Be	6.0×10^{-7}	5.9×10^{-9}	3.8×10^{-7}	1.2×10^{-7}	3.5×10^{-7}	1.02×10^{-6}
Cr	4.9×10^{-7}	0.0	0.0	1.8×10^{-8}	9.9×10^{-7}	1.5×10^{-6}
Sum	2.5×10^{-6}	4.3×10^{-8}	5.9×10^{-7}	3.9×10^{-7}	5.1×10^{-6}	8.6×10^{-6}

This example illustrates that significantly lower multipathway risks can be calculated if refined deposition velocities are used to conduct multipathway risk assessments. Using the case presented above, one can see that deposition velocities serve a key role in the outcome of air toxic risk assessments.

Because these assessments are becoming more common, and are often serving as regulatory decision-making tools, it is important that more realistic, refined approaches be used to predict potential health risks from facility emissions so that sound, technical decisions can be made regarding a facility. As discussed in detail in this paper, the use of refined deposition velocities and consideration of plume depletion for larger particles are important in conducting more refined, facility-specific air toxics risk assessments.

ACKNOWLEDGMENT

Additional modeling support was provided by Craig Nicholls of Harding Lawson Associates.

REFERENCES

1. Ermak, D. L., An analytical model for air pollutant transport and deposition from a point source, *Atmos. Environ.*, 2, 231, 1977.
2. California Air Pollution Control Officers Association (CAPCOA), Air Toxics "Hot Spots" Program, Risk Assessment Guidelines, Prepared by the AB 2588 Risk Assessment Committee of CAPCOA, 1991.
3. Sehmel, G. A. and Hodgson, W. H., Model for Predicting Dry Deposition of Particles and Gases to Environmental Surfaces, PNL-SA-6721, Battelle Pacific Northwest Laboratories, Richland WA, 1978.
4. Winges, K. D., User's Guide for the Fugitive Dust Model (FDM), EPA-910/9-88-202R, U.S. Environmental Protection Agency, Washington, D.C., 1990.
5. California Air Pollution Control Officers Association (CAPCOA), User's Guide to Assessment of Chemical Exposure for AB 2588 (ACE2588) Computer Model, 1991.

Improvements to the EPA's Human Exposure Model

Warren Peters

The original Human Exposure Model (HEM or HEM-I), developed in 1979, continues to be an effective tool for screening point sources of hazardous air pollutants and ranking individual sources and source categories in terms of their relative carcinogenic risks. The model was designed to efficiently screen a large number of sources inexpensively and quickly. HEM, on occasion, has been used in rule-making decisions and supporting studies. Many opportunities for public comment on HEM have been provided through publication of proposed rules in the *Federal Register* and summary reports of studies. Many of the comments and criticisms of HEM focus on a simplification of assumptions inherent in a model designed to be used as a screening tool.

The most important of these comments are

1. HEM-I is not user friendly. It is not easy for people with limited HEM-I and computer experience to use. This is an important consideration for state and local air pollution control agencies and for many private parties that would like to use HEM-I.
2. The dispersion algorithm is no longer state of the science.
3. The use of predicted ambient concentrations at the fenceline or the residence, as the basis for exposure estimation, is also uncertain. People are mobile and do not remain at "their residences" 100% of the time. Additionally, the exposed population does not breathe ambient (outdoor) air for the entire duration of exposure.
4. Uncertainty is not explicitly treated. This is needed because uncertainties in risk assessment and risk characterization may be large. Quantification of uncertainty will also be important for estimating the conservatism that is often associated with carcinogenic risk assessments.

The HEM is being revised and released to the public in stages defined as distinct program versions. The present version of the improved HEM (HEM-II version 1.5) addresses the first three comments listed above. It also enables users to present selected outputs graphically with bar charts or two- or three-dimensional graphs. The HEM-II is user friendly. The user is prompted for all input data by a well-designed, logical series of screens.

The HEM-II contains an EPA-approved, validated model, the Industrial Source Complex Long-Term Model (ISCLT), that has been accepted for many industrial source modeling applications. For those situations where ISCLT is not appropriate (e.g., if the facility is sited in a complex terrain), HEM-II provides the user with the means to transfer to HEM-II a concentration file that has been created by the more appropriate dispersion model. The transferred file is then used by HEM-II to estimate population exposure and risk, etc.

The HEM-II provides the user with the option of moving the exposed population "off of their front porches". This is achieved by the user defining up to ten places, or areas of unique concentrations, called microenvironments. Microenvironments may include indoors at home, indoors at work, in transit, mobility (specifically migration out of the study area), etc. The user defines an indoor-to-outdoor concentration ratio typical of the microenvironment in question, the percentage of the exposed population to be assigned to the microenvironment, and the amount of time, on an annual basis, estimated to be spent in each microenvironment.

The next version of HEM-II (2.0) will address uncertainty via a Monte Carlo analysis of the input parameters that have the potential to contribute the most uncertainty to the exposure and risk results. The parameters, under consideration at this time, include emission rate, cancer potency estimate, microenvironment concentrations, amount of time spent in microenvironments, and the amount of time people reside at the primary residences. Version 2.0 is now available.

Review of Version 1.5 and a preliminary Version 2.0 (if not the completed 2.0), is expected to be conducted by the National Academy of Science as part of their review of risk assessment methodologies required by the Clean Air Act Amendments of 1990.

Summary of Case Studies Using EPRI's Air Toxics Risk Analysis Framework: Results and Research Needs

Leonard Levin
Katherine Connor
David Room
Lawrence Gratt

I. BACKGROUND

Title III of the 1990 Clean Air Act Amendments (CAAA) calls for the U.S. EPA to undertake a study of the hazards to public health posed by emissions of 190 pollutants from electric utility steam-generating units.[1] Under the law as written, EPA must complete its study and report to Congress on its findings by November 1993. Based on the findings of the study, EPA will determine whether regulation of power plant emissions of the substances is required.

In response to these provisions, the Electric Power Research Institute (EPRI) initiated a research effort aimed at synthesizing current and ongoing air toxics research, the Air Toxics Comprehensive Risk Evaluation (or CORE) project. The overall objective of this effort is to perform a comprehensive analysis of the potential health risks likely to be posed by electric utility emissions of the listed air toxics after compliance with other sections of the Clean Air Act (e.g., provisions for control of NO_x, SO_2, and particulate) has been achieved. The final comprehensive risk evaluation will explicitly examine key uncertainties in the risk analysis methodology, including those due to alternative future compliance scenarios and those related to specific assumptions used in the risk analysis.

II. CASE STUDY SYNTHESIS

This paper presents the methodology and initial results of EPRI's CORE project. This task involved examining previous utility case studies that utilized either or both of EPRI's air toxics risk models, AERAM (Air Emissions Risk Assessment Model) and AirTox (Air Toxics Risk Management Model). The case studies include air toxics risk assessments that have been conducted for 21 generating facilities over the past 7 years. The case studies represent *modeling* exercises incorporating the best available information at the time the study was conducted. In general, the case studies relied on information available from the utility or data published in the literature. No new field measurements were conducted by EPRI as part of these studies.

The case study synthesis provides a retrospective analysis of what has been done to date, as well as provides insights into potential future research needs. Key activities in the synthesis included defining a common scope and set of assumptions to help put the case studies on a consistent basis for comparisons, assessing the principal drivers of risk, and identifying data and research gaps. The goal of this report is to summarize what has been learned from the previous case studies, so that we can begin to answer the following questions:

- What are the major sources of risk from power plant toxic emissions?
- What conclusions can be drawn regarding the risks posed by different categories of power plants?
- How comprehensive has the case study coverage been to date with respect to plant configuration, size, fuel burned, and urban vs. rural setting?
- Which uncertainties have the largest impact on risk estimates?
- Where are the data gaps?
- What new case studies or research might help address specific issues?

The insights gained from the case study synthesis will provide input to ongoing air toxics research and help define subsequent tasks of the CORE project.

III. THE AERAM AND AIRTOX MODELS

The case studies were performed using two EPRI models: AERAM and AirTox. AERAM provides a methodology for assessing inhalation cancer risk due to emissions of hazardous air pollutants from coal- and oil-fired plants.[2] It includes detailed modeling of the emission and transport of air toxics, using the Industrial Source Complex Long Term (ISCLT) model, and provides a quantitative assessment of the cancer risk resulting from human exposure to air toxics. The AirTox model provides a useful framework for structuring a comprehensive air toxics analysis and examining the impacts of a range of regulatory and/or control alternatives.[3] AirTox permits the dynamic representation of emissions and risks and facilitates performing extensive scenario and "what if" analyses. Uncertainty in key parameters can be explicitly modeled in AirTox, allowing for the computation of *distributions* as well as point estimates for risk.

EPRI's comprehensive Air Toxics Risk Analysis Framework uses the outputs of the AERAM model as inputs to the AirTox model (Figure 1). Multiple runs of the AERAM model are used to develop emissions profiles and ambient air concentrations used in AirTox. AERAM results are combined with data from other sources to define uncertainty and changes over time for the operating and control scenarios that are modeled in the AirTox framework. For example, plant emissions may be uncertain due to variability in fuel characteristics, uncertainty in modeling parameters, or

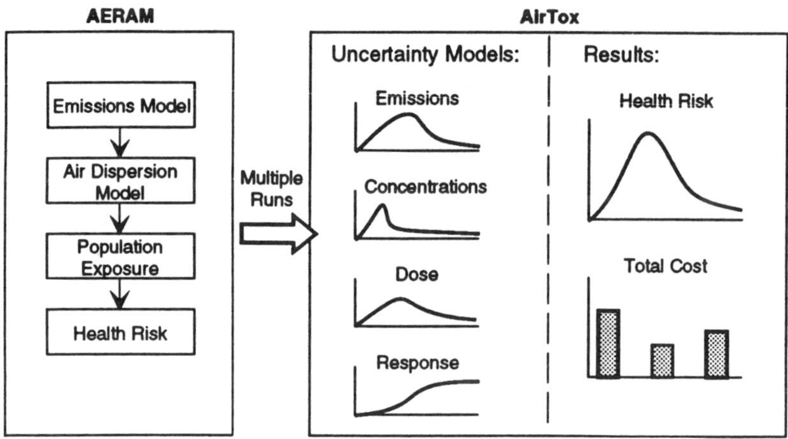

Figure 1. Using AERAM and AirTox.

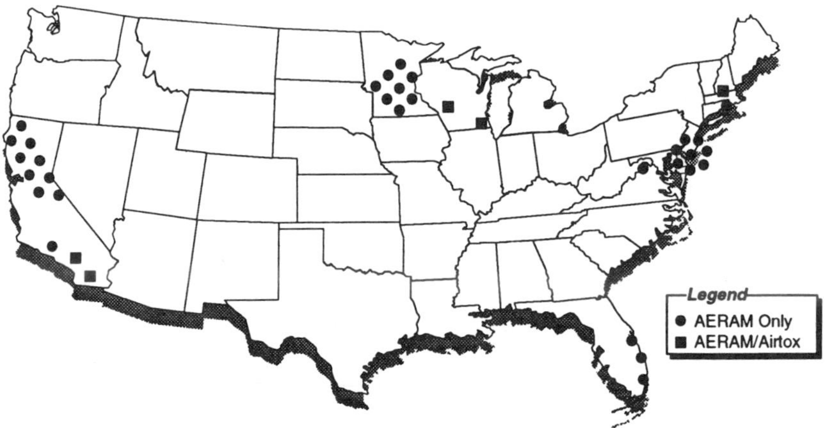

Figure 2. Locations of case studies.

variability in emissions measurements. In addition, emission patterns may change over time due to changes in plant loading and performance, fuel changes, or equipment modifications. This information is input into AirTox which then calculates the total health risk and cost associated with each operating or control scenario.

IV. PREVIOUS CASE STUDIES

The geographic distribution of the case studies is shown in Figure 2. The case studies have examined many plant configuration, fuel, operating, and control scenarios. For coal-fired plants, the majority of the case studies have examined plants burning bituminous coal with electrostatic precipitators for particulate control, and with no controls for SO_2 or NO_x emissions. This particular plant configuration currently represents close to half (48.4%) of national coal-fired capacity. For oil-fired plants, most of the case studies have addressed plants burning residual oil with no particulate, SO_2, or NO_x controls. These plants make up about 32% of total oil-fired capacity. The case studies also have examined four nonfossil-fired or alternative fuel plants.

While the case studies conducted to date are relatively representative of current fossil-fired capacity by plant configuration, most of the cases that examined enhanced particulate control using fabric filters, or the impacts of SO_2 and NO_x controls, represent hypothetical scenarios. As the distribution of national capacity in the future is likely to include more plants with advanced particulate, SO_2, and NO_x controls, additional work is needed to characterize the toxic emissions and risks from these configurations.

V. BASELINE SCOPE

Over the period that AERAM and AirTox studies have been performed, risk analysis practices have evolved. As a result, many assumptions made in earlier studies may have been treated somewhat differently in later studies. In addition, EPA default values for performing risk analyses, in particular dose-response values, have changed substantially in the last decade. In order to compare the case studies in a common framework, it was necessary to put them on a consistent basis. Consistency across case studies allowed us to make direct comparisons of the resulting health risks. To do this, a baseline scope and set of assumptions were defined for all of the case studies. The baseline analysis includes the following:

- Power plant emissions of arsenic, beryllium, cadmium, chromium, formaldehyde, and polycyclic organic matter (POM) are included.
- Of total chromium emissions, 1% is assumed to be the carcinogenic form, hexavalent chromium (Cr VI).
- Risks are computed for the most exposed individual, or MEI. This scenario assumes that a person spends 70 years in the location with the highest air concentrations due to emissions from the plant.
- Incremental lifetime cancer risk is based on exposure through the inhalation pathway only.
- Unit risk values published by EPA or the California Department of Health Services are used to calculate lifetime inhalation cancer risks.[4,5] Using unit risks provides a plausible upper-bound estimate for the incremental probability of contracting cancer as a result of a lifetime exposure to 1 $\mu g/m^3$ of a specific chemical. Because unit risks are generally accepted and used by the regulatory community, we have used them as the basis for our cancer risk estimates.[6]

VI. SENSITIVITY ANALYSES

In addition to the baseline case, sensitivity analyses were conducted to determine the impact on the predicted risks of changes in key assumptions. The sensitivity analyses represent key issues in the national debate that may have a significant impact on power plant health risk estimates.

A. Nickel

We considered a scenario in which all power plant nickel emissions were assumed to be carcinogenic. While most of the research indicates that only certain nickel compounds, including nickel subsulfide and nickel refinery dust, are carcinogenic, the California Air Resources Board (CARB) has proposed identifying metallic nickel and all inorganic nickel compounds as human carcinogens and toxic air contaminants.[7] For the sensitivity analysis, the unit risk value used for all nickel emissions was the proposed California value of 2.6×10^{-4} $(\mu g/m^3)^{-1}$.

B. Chromium

Two scenarios were defined to examine the sensitivity of the risks to assumptions regarding the percentage of hexavalent chromium (Cr VI) in power plant emissions. A scenario was defined in which 5% of the chromium emitted from utility fossil fuel-fired boilers was Cr VI. Available data for fossil-fired combustion sources indicate that Cr VI is <1% of total chromium emissions.[8] However, the difficulty of obtaining accurate measurements, combined with recent changes in measurement techniques, has raised concerns that the percentage of Cr VI in power plant emissions may be as much as 5% or higher. Therefore, a scenario in which all power plant chromium emissions are assumed to be carcinogenic (i.e., 100% Cr VI) also was examined. This scenario was included based on a recent decision by EPA's Office of Solid Waste recommending that all chromium species be considered as toxic as Cr VI.

C. Arsenic

Currently, the toxicity of arsenic by ingestion is being reviewed within EPA for the purposes of revising the drinking water standard. Since changes in the drinking water standard may lead to subsequent review and revision of the inhalation unit risk factor for arsenic, a scenario was examined in which the arsenic unit risk factor was multiplied by 10.

VII. RISK RESULTS AND COMPARISONS

This section describes the results of the AERAM and AirTox risk analyses for the 21 generating stations studied. It should be emphasized that the case studies do not

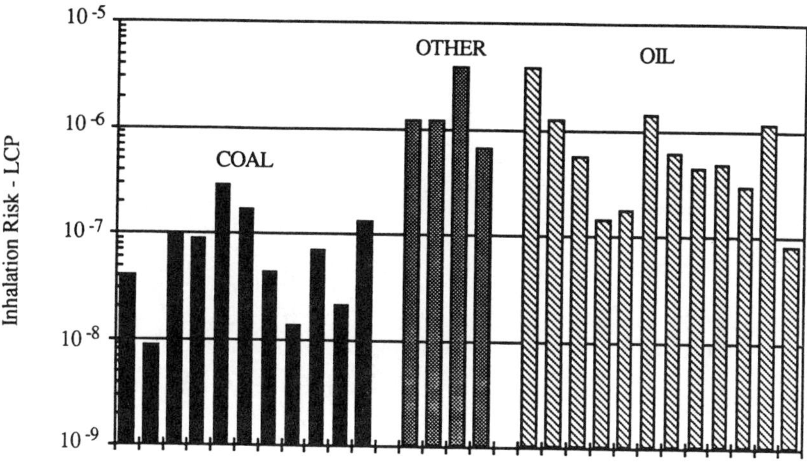

Figure 3. Baseline MEI inhalation cancer risks.

comprise a representative cross-section of the industry. Thus, while this synthesis is useful for beginning to identify potential trends and key uncertainties, it does not provide a solid basis for industry-wide extrapolations.

The case study results are presented in terms of the excess individual lifetime cancer probability (LCP), that is, the incremental chance over a 70-year lifetime that an individual will contract cancer due to inhalation exposure to the substances studied. The risks presented are for the person exposed to the highest predicted concentration levels (or the maximum exposed individual), and represent the sum of the risks for all of the carcinogens studied.

In existing EPA and state regulations, various risk levels have been defined (both explicitly and implicitly) as "acceptable", that is, not requiring any risk management action. For sources currently subject to control under the CAAA, measures beyond Maximum Achievable Control Technology (MACT) may be required if carcinogenic risks exceed a 1×10^{-6} incremental lifetime risk threshold. Although this standard may not apply to steam electric stations, we have used the 1×10^{-6} risk level throughout this report as a common threshold against which to compare various risk results.

A. Baseline Risks

The lifetime inhalation cancer risks for all of the plants studied are shown in Figure 3. The risks include the risks from the selected carcinogens (arsenic, beryllium, cadmium, chromium, formaldehyde, and POMs) under the baseline assumptions. The risks depend on the annual load for the particular plant. None of the coal plants exceeds a risk level of 1×10^{-6} excess individual LCP. Most of the "other" plants and one-third of the oil plants exceed this level.

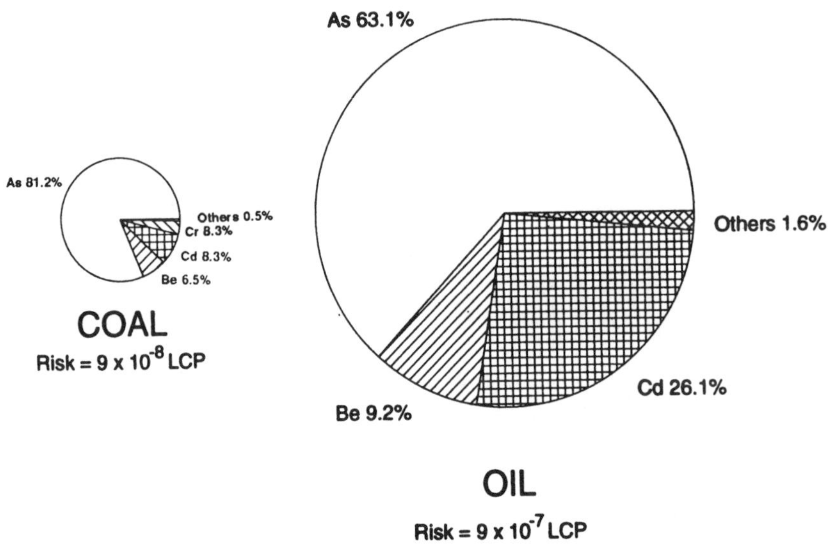

Figure 4. Contribution to risk by chemical.

B. Contribution to Risk by Chemical

The contribution to risk by each chemical depends on both the relative emissions and the toxicity of the chemical. A series of analyses was performed examining the average contributions to risk for the fossil-fired plants under baseline assumptions. This analysis provides an initial indication of which chemicals are likely to be the drivers of utility risk estimates.

As shown in Figure 4, arsenic is the major contributor to risk from both coal- and oil-fired plants, followed by cadmium and beryllium. For the plants studied, the inhalation risks from the oil plants are about an order of magnitude greater than for the coal plants. Considering the size distribution of the particulates emitted from oil-fired plants would potentially decrease the inhalation risk by a factor of two (based on studies showing that only about 50% of the mass in the plume is in the respirable fraction).[9,10]

C. Sensitivity Analysis

This analysis examined the sensitivity of the risk estimates to potential changes in key assumptions for the average coal- and oil-fired plant. Hence, the "average plant" represents the average of plants studied rather than the typical plant in the electric utility industry.

Figure 5 presents the impacts of changes in key assumptions on the estimated risks for the average coal-fired power plant. In all cases, the risk from the average coal-fired plant did not exceed a risk of 1×10^{-6}. The risks are not sensitive to

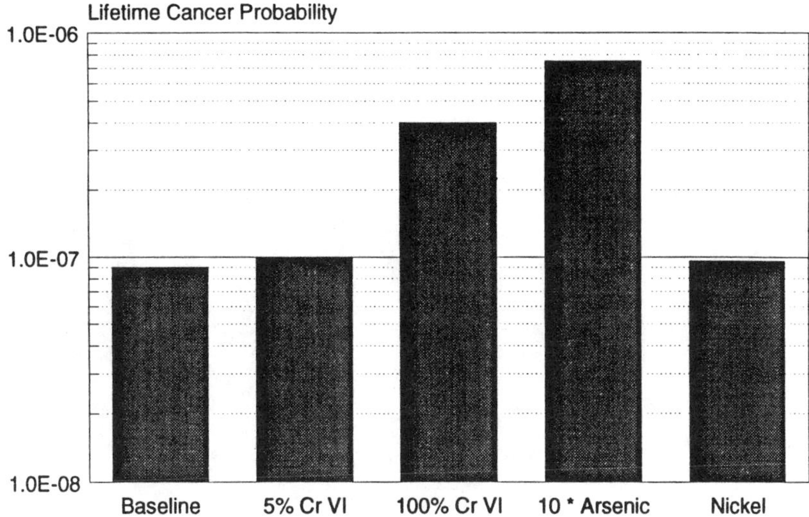

Figure 5. Sensitivity analysis for "average" coal-fired plant.

assumptions about nickel carcinogenicity or the assumption that up to 5% of chromium emissions are hexavalent. If, however, all of the chromium emissions are assumed to be hexavalent, the risks increase by a factor of 5 with respect to the base case, and chromium is then responsible for about 75% of the risk. If the toxicity of arsenic is increased by a factor of 10, total risk increases by a factor of about 8 and approximately 98% of the risk is due to arsenic, vs. 81% in the base case.

Figure 6 presents the impacts of changes in key assumptions on the estimated risks from the average oil-fired power plant. For the plants studied, the use of any alternative assumption — that all chromium is hexavalent, that nickel is a carcinogen, or that the toxicity is 10 times as great — results in the estimated risks reaching or exceeding a risk of 1×10^{-6}. Risks for the coal plants were not sensitive to assumptions regarding nickel while assuming nickel is a carcinogen increases total oil-fired risks by a factor of 3, with about 70% of the risk then due to nickel.

D. Uncertainty Analysis

The decision–tree structure of AirTox allows the explicit representation of uncertainty in the key variables of a risk assessment. For four case studies using both the AERAM and AirTox models, the combined impact of uncertainty in the utility emission and control effectiveness variables was examined. The ranges of uncertainty in utility emission and control effectiveness variables may be due to uncertainties in fuel characteristics, trace element enrichment in fly ash, the combustion process, or operating uncertainties. In general, the combined uncertainty range for trace metal content and enrichment factor spanned less than one order of magnitude.

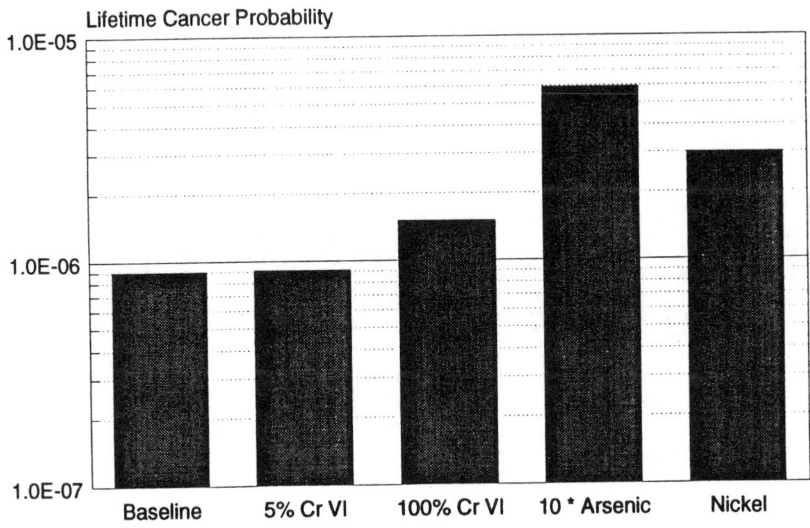

Figure 6. Sensitivity analysis for "average" oil-fired plant.

The published emission factors data for formaldehyde and POM were relatively sparse for utility boilers; thus, the uncertainty range for these compounds spanned several orders of magnitude.

The control effectiveness data were based on estimates provided by the host utility or data reported in the literature. Control effectiveness was assumed to be the same for all trace metals, with no removal provided for POMs or formaldehyde. The largest uncertainty was in the baseline assumptions of control effectiveness across plants. For example, one coal-fired utility reported that their current electrostatic precipitators (ESPs) provide 99.4% removal, as compared to 95% removal reported by the other coal-fired plant. Similarly, the two coal plants assumed that hypothetical fabric filter controls would be five times more effective than did the oil-fired plants. While there are relatively large differences in the baseline effectiveness assumptions, the assumed uncertainty ranges around the baseline for a particular control scenario and plant were no greater than about 10% of the nominal (or best-guess) value.

Figure 7 shows the MEI risk ranges for the two coal-fired plants analyzed using AirTox under current and alternative future control scenarios. The high-risk estimate represents the upper bound on risks given the nominal dispersion, exposure, and dose-response assumptions. The low-risk estimate represents the lower bound with these same assumptions. The dividing line on the bar represents the nominal risk estimate for the plant. The risks have been normalized to a 100-MW annual load to facilitate cross-plant comparisons.

Even considering uncertainty in plant emissions and control effectiveness, none of the MEI risks for the coal-fired plants studied exceeded the 1×10^{-6} risk level. Figure 7 also highlights the impact of the baseline assumptions for control effectiveness. The estimated risk with ESPs is almost an order of magnitude higher for Coal

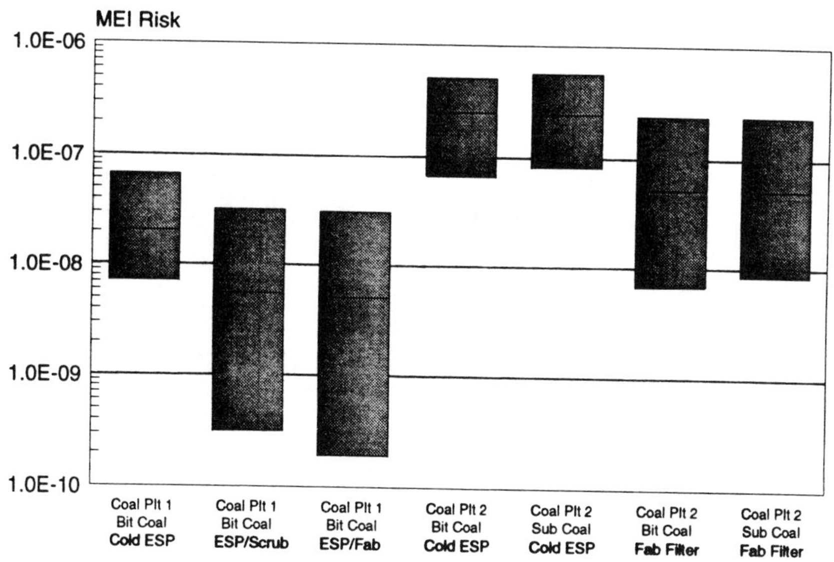

Figure 7. MEI risk ranges — coal-fired power plants. Reflects range in MEI risk estimate for a normalized 100-MW plant. Ranges include uncertainty in fuel characteristics and control effectiveness.

Plant 2 than for Coal Plant 1. This is primarily due to the differences in baseline ESP removal efficiency, described above.

Figure 8 shows the MEI risk ranges for the two oil-fired plants studied with AirTox, both under current operations and with alternative fuels and controls (again normalized to a 100-MWe annual load). If the emission factors and fuel trace element contents assume their high-end values, the estimated risks from the oil-fired plants studied may approach or exceed the 1×10^{-6} level. The worst-case MEI risk for Oil Plant 2 without controls exceeds the 1×10^{-6} risk level. This plant represents an older residual oil-fired plant located in an urban area. This figure shows relatively similar nominal risks for distillate and residual oils without controls. This may be due, in part, to the high POM emission factor used for distillate oil. The current EPA emission factor for distillate oils is based on measurements for relatively old, nonutility boilers. With a better estimate for POM emissions, it is likely that the risks for distillate oil (in general a higher-quality fuel) would be less than those for residual oil.

VIII. INSIGHTS AND FUTURE NEEDS

The Case Study Synthesis represents a critical first step in EPRI's CORE project. In this task, the results of a wide variety of previous utility case studies were reviewed for consistency and analyzed to begin to identify both potential trends and important uncertainties in utility risk assessments. While the case studies include air toxics risk assessments performed for over 21 generating facilities in the past 7 years, they do

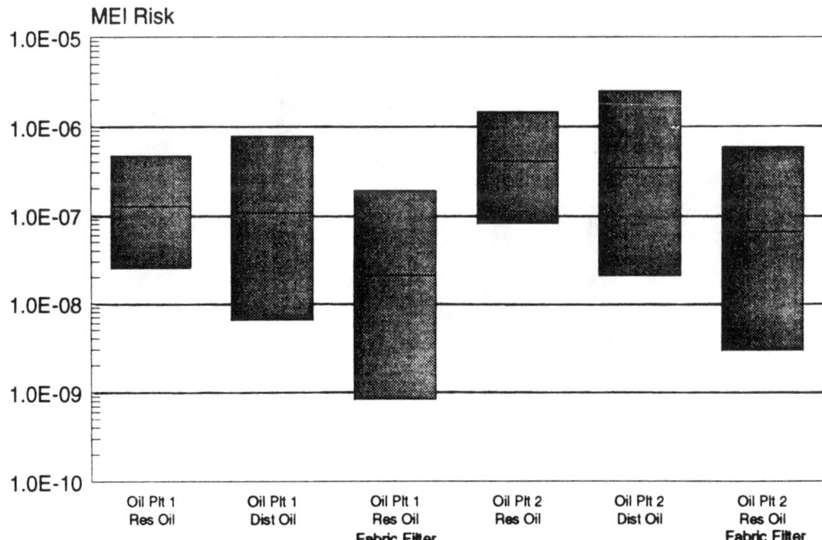

Figure 8. MEI risk ranges— oil-fired power plants. Reflects range in MEI risk estimate for a normalized 100-MW plant. Ranges include uncertainty in fuel characteristics and control effectiveness.

not present a comprehensive or representative picture of the potential risks associated with toxic emissions from electric utilities. The distribution of plants (by configuration, fuel, and setting) and the scope of this analysis are limited by the scope of the previous work and the data available at the time the studies were conducted. As new data and research results become available, these initial results will be revisited and new case studies incorporated in the project scope.

The review of this subset of plants provides preliminary insights into potentially important assumptions and key drivers of risk for fossil-fired power plants, as summarized below.

- For most plants studied, total MEI risk was less than 1×10^{-6} LCP. The plants with the highest risks were residual oil-fired plants in urban locations.
- Under baseline dose-response assumptions the inhalation risks from controlled coal-fired plants (with ESPs) are slightly less than the risks from uncontrolled oil-fired plants.
- The major contributor to risk from both the coal- and oil-fired plants studied is arsenic (30 to 80%), followed by cadmium and beryllium.
- If nickel is considered a carcinogen, the MEI risk from uncontrolled oil-fired plants approaches 1×10^{-6} LCP, and nickel is responsible for 60 to 70% of the risk.
- The risks are not sensitive to the percentage of Cr VI in total chromium emissions within the range of 1 to 5%. If all chromium is assumed to be Cr VI, chromium

becomes the dominant contributor to risk for coal-fired plants and a major contributor for oil-fired plants.

- Modeled fabric filters reduce the risks from both oil- and coal-fired power plants by approximately 80%.

A. FUTURE RESEARCH NEEDS

This initial synthesis provides useful insights into several areas where additional information and analysis could potentially improve the accuracy and representativeness of the preliminary risk assessment results. Specifically, tasks are under way to gather information on and examine the implications of

- Additional chemicals — including radionuclides, mercury, and possibly other carcinogens and noncarcinogens listed in the CAAA.
- Additional plant types — including larger plants and plants with additional SO_2, NO_x, or other control technologies.
- Power plant emissions — including better data on the range of trace metal contents in various fuels, the quantity and species present in POM emissions from oil-fired plants, the species or compounds of As, Cr, and Ni in power plant emissions, and their enrichment in fly ash.
- Control effectiveness — including better estimates for the efficiency of current and potential future particulate control devices and information on the impacts of SO_2 and NO_x controls on toxic emissions.
- Alternative exposure scenarios — such as the effects of including respirable particle sizes only; adding exposure through other media (water, food, soil) and other pathways (ingestion, dermal absorption); and examining the implications of indoor/outdoor activity patterns.
- Dose-response data — including new and evolving information for As, Be, Hg, Ni, and POMs and examining the impacts of dose-response functions other than the unit risk.
- Other sources — including fugitive emissions sources such as tank farms, and nonboiler point sources such as cooling towers.

Subsequent tasks in the CORE project and the results from several ongoing EPRI research efforts will begin to address many of the needs identified above in the near future.

REFERENCES

1. Public Law 101-549, 101st Congress, November 15, 1990.
2. IWG Corp., Air Emissions Risk Assessments Model (AERAM) Manager, User's Guide, San Diego, CA, 1987.
3. Decision Focus Incorporated, User's Guide to the Air Toxics Risk Management Model (AirTox) Version 1.0, Draft Report, Electric Power Research Institute, Palo Alto, CA, 1988.

4. U.S. EPA, Integrated Risk Information System, Washington, D.C., September 1990.
5. California Air Pollution Control Officer's Association, Air Toxics Assessment Manual, Sacramento, CA, October 1987.
6. U.S. EPA, Guidelines for carcinogen risk assessment, *Fed. Reg.*, 51(185), 1986.
7. State of California Air Resources Board, Proposed Identification of Nickel as a Toxic Air Contaminant, Scientific Review Panel Version, Sacramento, CA, April 1991.
8. Hinman, K. et al., Santa Clara Valley Integrated Environmental Management Project, Revised Stage One Report, Environmental Protection Agency, Washington, DC, May 1986.
9. Perera, F. P. and Ahmed, A. K., Respirable Particles, Ballinger, Cambridge, MA, 1979.
10. Piper, B. P., Hersh, S., and Mormile, D. J., Particulate Emissions Characteristics of Oil-Fired Utility Boilers, EPRI CS-1955, Project 1131-1, Final Report, Palo Alto, CA, August 1981.

Characterizing Uncertainty in the Risk Assessment Process

Michael T. Alberts
Fred O. Weyman

I. INTRODUCTION AND BACKGROUND

Reliance on health risk assessment results and techniques to aid in environmental decision making has steadily increased over recent years. Defining clean-up goals for a hazardous waste site, developing emission standards for a toxic air contaminant, or permitting of new air emission sources are but a few examples. In each case, risk assessment results decide or influence a course of action that may result in expending millions of dollars.

Health risk assessment is composed of four basic steps: hazard identification, exposure assessment, dose-response assessment, and risk characterization. In each of the identified steps, the risk assessor must make health protective assumptions and judgments where data are highly variable or inadequate to support a definitive choice. Bias and error introduced as a result of those assumptions and judgments are transferred from one step to the next step, compounding and propagating their effect in the end. Therefore, it is critical that the risk assessor provide an analysis of uncertainty. Unfortunately, uncertainty is seldom adequately addressed. In some instances, uncertainty analysis is neglected altogether, relying only on the blanket statement that the risk is unlikely to be higher and could be as low as zero. By failing to provide an uncertainty analysis in the risk assessment, the risk manager is forced to make a decision based on limited data.

Key sources of uncertainty include the period of exposure, various ingestion rates, microenvironment effects, and dose-response data. Cancer slope factors, a product of dose-response assessments, likely contain the greatest degree of uncertainty. For example, current U.S. Environmental Protection Agency (EPA) policy requires an assumption that any dose level, no matter how small, has an associated risk of contracting cancer, that is, there is no threshold to a cancer response. Guidelines further assume that risk is linear with dose, by using high-dose animal studies or relatively high dose occupational studies as the basis for extrapolation to low doses in humans. Many scientists today (most vocal being Dr. Bruce Ames of the University of California at Berkeley) believe that for many carcinogens there is a threshold to cancer, and that the dose-response curve is not linear with dose. When following current guidelines by assuming no threshold and a linear response, bias and error are introduced; this is further complicated by applying the upper 95% confidence limit of the slope factors derived from animal studies.

Methods for evaluating the uncertainty in risk assessment assumptions are numerous. The simplest procedure is to assume the worst; for each variable for which uncertainty or variability exists, a worst-case value is selected. Until recently, this was the most common method for regulatory applications. The advantage to this approach is an ability to state with confidence that exposure or risk has not been understated. The risk manager can be confident under this approach that the risk will not be any greater. The major drawbacks are an unrealistic result that is difficult for most individuals, and especially the public, to interpret and a mixing of risk assessment and risk management. Often this worst-case approach is accompanied by a brief discussion or laundry list of assumptions and parameters that contain uncertainty. While one or two worst-case assumptions may be consistent with the lifestyles of many individuals, it is highly unlikely that all worst-case assumptions will be appropriate. When the risk manager and reviewers of this information fully understand the compounding effects of a worst-case scenario, it becomes an acceptable approach.

EPA recently revised their risk assessment guidelines for Superfund; the worst-case approach has been eliminated in favor of a concept referred to as the Reasonable Maximum Exposure (RME) scenario.[1] The RME scenario recognizes the compounding effects associated with the worst-case approach, and attempts to define an upper 95th percentile estimate of risk. Major changes in the RME scenario are exposure period (30 years) and median values for certain exposure parameters. Albeit a more reasonable methodology, and EPA is to be commended for taking a positive step forward, it still relies on single high-dose studies (both animal and human occupational) and linear extrapolation for cancer slope factors.

Presenting both a worst-case and a "lower-bound" estimate of risk provides some insight into the problem of uncertainty, but ignores other important concepts. For example, it identifies two extremes of potential health risk without giving any indication of the realism of each, or the distribution of risks between those extremes. As in the worst-case approach, the lower-bound estimate forces the risk assessor to make an assumption or judgment about human behavior, which in itself is highly variable. For many risk assessment assumptions and variables, there is no single right answer. Rather, there are many correct answers that apply to different people.

Stochastic analysis represents another alternative to characterizing uncertainty. Under a stochastic approach, variables and assumptions are assigned probability distributions as opposed to point estimates. These distributions are then carried through the process of calculating risk, yielding a distribution of potential risk estimates, each with a corresponding confidence limit. As an example, Superfund risk assessment assumes an exposure period of 30 years while other regulations rely on a 70-year lifetime. In a stochastic analysis, a distribution would be used for the exposure period rather than selecting a single number. The same may be done for other uncertain variables in the risk assessment including ingestion rates, emissions, and environmental fate data among others.

There are two primary advantages of the stochastic approach. First, it eliminates the compounding effects of using worst-case assumption upon worst-case assumption. Second, it fully recognizes there is no single risk estimate or assumption that applies to everyone and everything. One person may live next to an emission source for 7 years and someone else may live there 25 years. Drawbacks to this approach include securing public acceptability and understanding, lack of data for developing reasonable distributions for some parameters, and difficulty in verifying the results. Fortunately, studies carried out or sponsored by EPA and other federal and state agencies have resulted in a tremendous growth of available statistical data on key risk assessment variables.

II. CASE STUDY OVERVIEW

In 1987 the California Legislature adopted the Air Toxics "Hot Spots" Information and Assessment Act of 1987 (AB 2588). This legislation required certain California facilities to prepare comprehensive air toxics inventories (over 400 compounds were included) and, depending on emission rates, prepare a comprehensive multiple pathway risk assessment. The goals of AB 2588 were to improve the California air toxics inventory, identify any risk "hot spots", and guide future regulatory programs targeting the reduction of toxic emissions. Guidelines were prepared governing the risk assessment methods and data to ensure that all facilities were on an even playing field. These guidelines provided nearly all the necessary data and parameter values to complete the risk assessment calculations. To be certain that risks were not underestimated, the guidelines followed a typical worst-case approach.

Standardizing the risk assessment process under AB 2588 was a necessity in order for the law to achieve its primary goals. However, significant concern was expressed on the part of industry that a worst-case approach would be difficult for the public to comprehend and place in perspective without the proper educational background. Furthermore, releasing the worst-case numbers without the proper background information on the uncertainties and bias inherent in such an approach would unduly alarm the public. As a means of addressing the industry concerns, the cognizant agencies allowed alternative analyses of risk to be presented.

Several of the large industrial facilities elected to prepare stochastic risk assessments to examine the uncertainty in the worst-case approach and provide an alternative estimate of health risk for the hypothetical maximally exposed individual (MEI). The balance of this paper discusses the methods used in the stochastic analysis and compares those results with the worst-case approach required by the AB 2588 guidelines. In addition, the worst-case values were revised in accordance with the EPA RME methodology developed for Superfund as another comparison. The RME values are reasonable approximations only; not all calculations were revised.

III. STOCHASTIC METHODS

A Monte Carlo sampling scheme was used as the basis for preparing the stochastic analysis and evaluating the uncertainty in the worst-case risks. In brief, a Monte Carlo analysis uses a random number generator and a series of mathematical equations to sample values from specified distributions; distributions are described by a mean and standard deviation with an indication of distribution type. For each variable included in the stochastic analysis, a value is selected from the distribution. Exposure and risk is calculated for the first sample set and the results are stored. The process is then repeated a thousand or more times until a large distribution of risk outcomes is developed. Statistical calculations are then performed on the risk distribution to identify confidence limits and risk percentiles.

IV. VARIABLE DISTRIBUTIONS

It would be impractical to include all risk assessment variables in a stochastic analysis. Rather, the focus should be on those variables and parameters that contain significant uncertainty and for which adequate data are available in the literature. Exposure period, slope factors, and various ingestion rates were among the key stochastic variables in the AB 2588 alternative risk assessments. Where correlations existed between one or more variables, one or both variables were held constant (e.g., inhalation volume and body surface area). Distributions for key variables (expressed as a mean value and plus one and minus one standard deviation) are shown in Table 1 along with the worst-case value (guideline) prescribed in the AB 2588 guidelines. Following is a brief discussion of the key variables.

A. Exposure Period

Exposure period refers to the length of time an individual is exposed to emissions from the facility. AB 2588 guidelines, consistent with past and present practice, assume a value of 70 years as a worst-case exposure period. This assumes an individual spends his or her entire life at a single location, never leaving the residence. EPA assumes a 30-year exposure period, representing a 90th percentile for the RME scenario, and 9 years (50th percentile) for a most-likely estimate.[1]

Uncertainty in the exposure period was addressed by developing a distribution for the number of years an individual lives in one location. This approach neglects the time spent away from the home, such as at work or on leisure travel. Data from the U.S. Census Bureau on the frequency and distance of moving events, supplemented with a second study by Long et al., were used to derive the distribution.[2,3] According to Long et al., as much as 29% of moves are within 3 miles, where risk is reduced but not eliminated. The rate at which risk decreased with distance from the facility was used to adjust the moving frequency data to account for local moves.

Table 1. Stochastic Analysis Input Variables

Variable	Mean (x)[a]	x - σ[b]	x + σ[c]	Guideline
Exposure period (years)	9	5	14	70
Soil ingestion (mg/day)	50.5	1.65	154.4	150
Plant ingestion (kg/day)	23	15	31	Unspecified[d]
Root vegetable ingestion (kg/day)	3.8	1.5	6.1	Unspecified[d]
Plant yield (kg/m^2)	0.98	0.53	1.9	2
Soil dust on skin (mg/cm^2)	0.25	0.1	0.6	0.5
Potencies (mg/kg/day)[1]				
Arsenic (inhalation)	7.2	1.5	31.5	11.6
Arsenic (oral)	No distribution available			1.7
Benzene	9.95×10^{-3}	2.9×10^{-3}	3.4×10^{-2}	0.102
Butadiene, 1,3-	1.9×10^{-2}	1.6×10^{-2}	2.3×10^{-2}	0.98
B(a)P (inhalation)	6.11×10^{-3}	9.17×10^{-5}	4.07×10^{-1}	6.11
B(a)P (ingestion)	3.11	1/4	6.9	11.5
Cadmium	6.67	2.25	1.98×10^{1}	14.7
Chromium (VI)	1.2×10^{2}	5.9×10^{1}	2.4×10^{2}	490
Formaldehyde	4.55×10^{-5}	6.83×10^{-7}	3.03×10^{-3}	0.046
Nickel	1.8×10^{-1}	4.9×10^{-2}	7.0×10^{-1}	0.84

[a] Mean value.
[b] Mean minus one standard deviation.
[c] Mean plus one standard deviation.
[d] Consumption rates only were specified. Guidelines allowed application of a locally grown percentage factor, but did not specify any values.

B. Slope Factors

Cancer slope factors contain considerable uncertainty and bias, and hence are prime targets for inclusion in a stochastic analysis. AB 2588 guidelines stipulate the use of "official" slope factors for calculating risk, as do those guidelines of EPA and other agencies. In California, slope factors developed by the state take precedent over the EPA values and are generally slightly higher. When animal data are used as the basis for a slope factor, the extrapolation models provide a most-likely estimate (MLE) and a 95% upper confidence limit (UCL). It is the 95% UCL that is specified as the "official" slope factor. When human epidemiological data form the basis, EPA relies on the MLE, and California DHS often specifies a 95% UCL.

Using slope factor estimates other than "official" values, or using distributions rather than point estimates, constitutes a controversial departure from normal practice. To minimize the controversy, the slope factor distributions were limited to the same animal study or human epidemiology data, as well as the same extrapolation model used by EPA in developing the "official" estimates that appear in the Integrated Risk Information System (IRIS) database. For example, the 95% UCL and the MLE were used in defining the distribution from animal studies. The only exceptions to this rule were for benzene, 1,3-butadiene, and hexavalent chromium. A composite distribution was prepared for benzene using the EPA MLE (which is based on the mean of four MLE values developed from the one-hit model) and the work of Brett et al.[4] The 1,3-butadiene distribution was developed from a risk assessment prepared by the Occupational Safety and Health Administration (OSHA).[5] A composite was prepared for hexavalent chromium by combining the California DHS slope factor with that of EPA. Had alternative bioassays and model results been incorporated in the distributions, mean values and upper 95th percentile estimates would have been significantly lower.

C. Ingestion Rates

A soil ingestion rate distribution was developed from a review of the literature using a geometric mean of several tracer studies and eliminating a safety factor equivalent to one standard deviation applied by the authors.[6] These results agree well with other tracer studies found in the literature.[7,8] A distribution for ingestion of backyard garden produce was developed using data from EPA, the U.S. Department of Agriculture, and other researchers.[9-13] Because of a weak correlation with body weight, the garden ingestion distribution considered only the fraction actually obtained from the garden. A distribution for the amount of dust on the skin, used in assessing exposure via dermal contact, was derived from literature data reported by Clement.[14] Skin surface area, which is highly variable depending on climate and season, was held constant at an average value of 3100 cm^2 to avoid correlation problems with inhalation volume and body weight.

Other variables and assumptions held constant, but which contain significant uncertainties and bias, include root uptake factors, the deposition rates of particulate substances, and methods for calculating soil concentrations resulting from deposition. Sufficient data were not available to develop reasonable distributions for these parameters. Therefore, the conservative worst-case values were used in the stochastic assessment.

V. DISCUSSION OF RESULTS

Stochastic risk results can be displayed in a variety of ways, including cumulative frequency distribution curves or by percentile. Figure 1 compares the EPA RME scenario, and the 50th and 95th percentiles from the stochastic analysis, with the state guideline results. The industrial facilities used for the comparison include four oil refineries and a large cement plant.

As clearly seen in Figure 1, the RME risks (approximated) are consistently less than 30% of the worst-case values. The differences can be attributed largely to the 30-year assumption for exposure in the RME scenario and the differences in California DHS and EPA slope factors for benzene and chromium. EPA's estimates of the slope factor for these substances are 3.5- and 12-fold lower, respectively. Had hexavalent chromium played a larger role in the worst-case risk from these facilities, the differences depicted in Figure 1 would have been greater. Also of interest is the

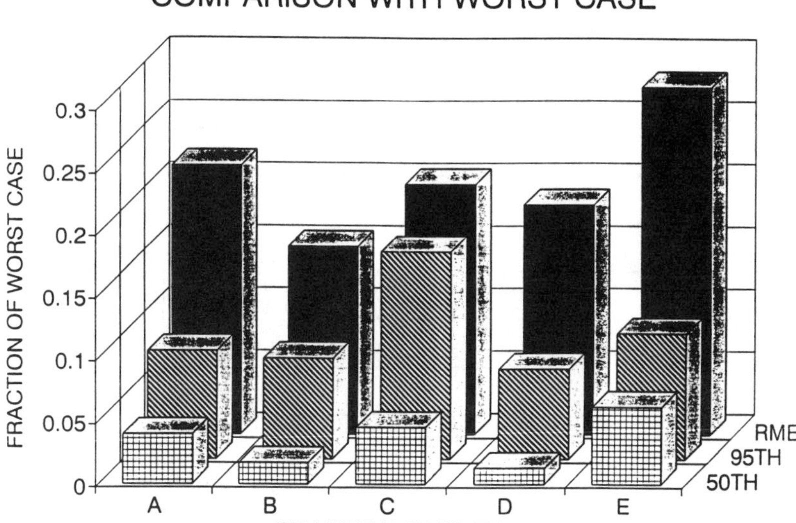

Figure 1. Stochastic and RME scenario risks as a fraction of guideline results.

relative differences between the RME and 95th percentile risks from the stochastic analysis; these values agree reasonably well, indicating the value of the RME approach.

Stochastic results from the Monte Carlo analysis, shown as the 50th and 95th percentiles in Figure 1, were significantly below the worst-case estimates. At the 95th percentile, which could be regarded as an upperbound estimate, risks were 5 to slightly less than 15% of the worst-case value, depending on which substances dominated the risks. At the 50th percentile, representing a median estimate, risks were between 1 and 6% of the worst case. When benzene dominated the worst-case value, the 50th percentiles tended to represent a smaller fraction than did cases where hexavelent chromium was a major contributor.

The results shown in Figure 1, particularly for the 95th percentile, provide a good indication of how error and bias can be compounded under a worst-case approach to risk assessment. If other substances played a more dominant role in the risk, the relative differences between stochastic and worst-case results could be more or less significant. As a good example, the EPA inhalation slope factor distribution for benzo(a)pyrene spans more than three orders of magnitude between the mean and 95% UCL. If other extrapolation models are considered the distribution can span up to nine orders of magnitude.

A cumulative frequency distribution curve for facility D is shown in Figure 2, along with the worst-case estimate. The distribution of cancer risk covers 3 orders of magnitude with a 99th percentile estimate that is approximately 3-fold lower than the worst case; since only 1000 simulations were used in the analysis, the 99th percentile

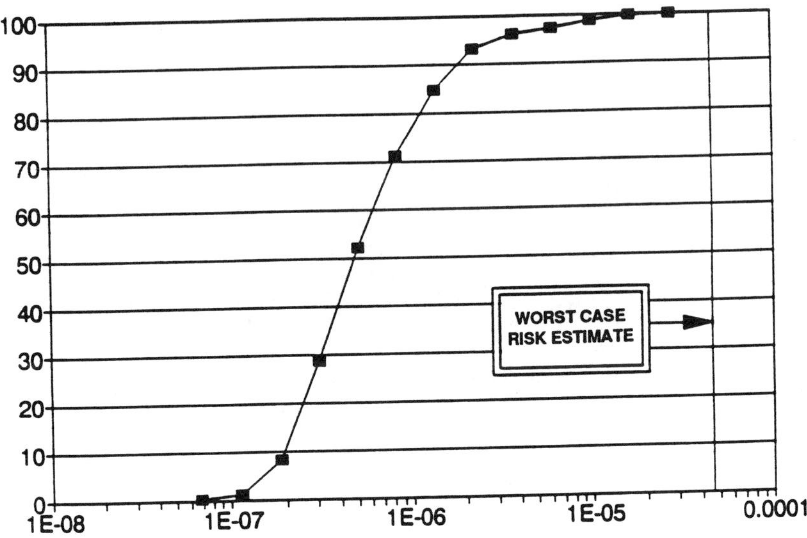

Figure 2. Cumulative frequency distribution for Facility D.

may not be stable or statistically significant. The lowest risk estimate calculated in the Monte Carlo simulation was approximately 8 in 100 million. Because of the limitations in developing the slope factor distributions and following a no-threshold theory, the lower estimate of risk would still be regarded as zero.

Worst-case estimates of risk were generally dominated by benzene and the inhalation pathway. Other important contributors to risk for the refineries were hexavalent chromium and polycyclic aromatic hydrocarbons (PAHs) from combustion sources and 1,3-butadiene. Generally, the relative contributions to risk by pollutant varied only slightly between the 50th and 95th percentiles, and inhalation was the major route of exposure in each case. PAH compounds, hexavalent chromium, and 1,3-butadiene were the only substances for which any notable differences in risk contribution occurred between the 50th and 95th percentiles and the worst case. These differences are due largely to the narrow hexavalent chromium slope factor distribution, the relative importance of PAH ingestion risks at the 95th percentile, and the 1,3-butadiene slope factor distribution developed by OSHA. Figure 3 shows the relative contributions to risk by substance for the worst case, 50th, and 95th percentiles for facility D.

VI. CONCLUSIONS

A worst-case approach to risk assessment provides the advantage of stating with confidence that the risk estimate is not likely to be any higher than this value. However, it provides very little information to the risk manager about how realistic or possible the risk estimate might be, or what the uncertainty is surrounding the end

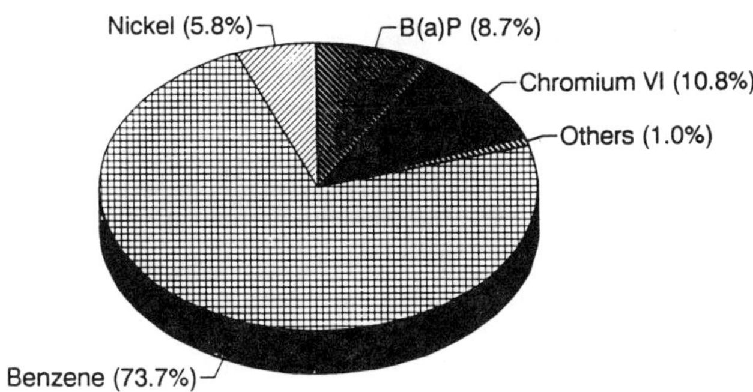

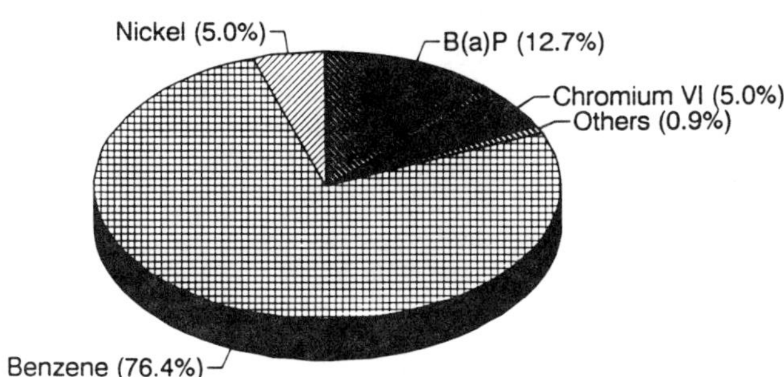

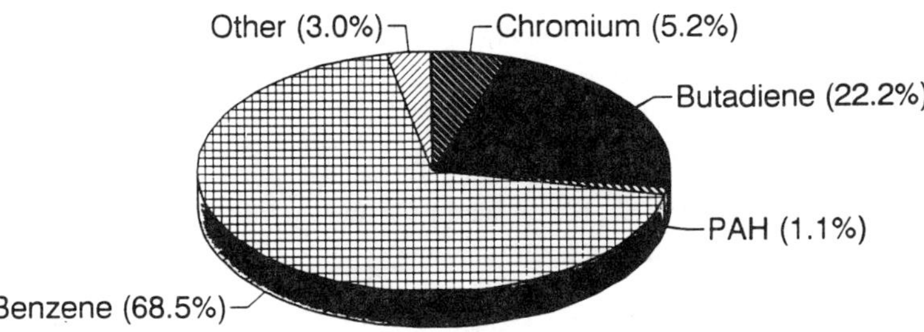

Figure 3. Risk composition for guideline and stochastic results for Facility D.

result. Formulating decisions based strictly on worst-case values is an acceptable approach, provided the limitations of the analysis and the magnitude of uncertainties in the results are both known and understood.

The Monte Carlo results shown in Figure 1 demonstrate the degree of bias and error introduced by a worst-case scenario. Seldom is this type of information provided to the risk manager to assist in making an informed risk management decision. Typically, text discussions are provided highlighting the uncertainties present in the risk estimates, but quantitative estimates provide a more definitive picture.

Finally, the Monte Carlo analyses prepared for the five industrial facilities examined only a small fraction of the bias, error, and uncertainties inherent in risk assessment. Other areas that warrant attention include the no threshold to cancer assumption, biological and physiological modeling, synergistic and antagonistic effects, and environmental fate and transport modeling. As research in these areas continues, and scientific understanding improves, stochastic risk assessment techniques can be used to address these uncertainties as well.

REFERENCES

1. U.S. EPA, Risk Assessment Guidance for Superfund, Volume I, Human Health Evaluation Manual, EPA/540/1-89/002, U.S. Environmental Protection Agency, 1989.
2. U.S. Bureau of the Census, Current Population Reports, Series P-20, No. 430, Geographical Mobility: March 1986 to March 1987, U.S. Government Printing Office, Washington, DC, 1989.
3. Long, L., Tucker, C. J., and Urton, W. L., Measuring migrating distances: self-reporting and indirect methods, *Am. Stat. J.*, 83, 674, 1988.
4. Brett, S. M., Rodericks, J. V., and Chinchilli, V., Review and update of leukemia risk potentially associated with occupational exposure to benzene, *Environ. Health Perspect.*, 82, 267, 1989.
5. Federal Register, Part II, August 10, 1990.
6. Sedman, R. M., The development of applied levels for soil contact: a scenario for the exposure of humans to soil in a residential setting, *Environ. Health Perspect.*, 79, 291, 1989.
7. Clausing, P., Brunekreef, B., and Van Wijnen, J. H., A method for estimating soil ingestion by children, *Int. Arch. Occup. Environ. Health*, 59(1), 73, 1987.
8. Calabrese, E. J., Barnes, R., Stanek, E. J., III, Pastides, H., Gilbert, C. E., Veneman, P., Wang, X., Lasztity, A., and Kostecki, P. T., How much soil do young children ingest: an epidemiological study, *Regul. Toxicol. Pharmacol.*, 10, 123, 1989.
9. U.S. EPA, Exposure Factors Handbook, PB90-106774, Versar, Springfield, VA, 1989.
10. U.S. Department of Agriculture, Food Consumption of Households in the U.S., Household Food Consumption Survey, Report No. 12, 1966.
11. National Gardening Association, Gardens for All, National Garden Survey, The National Gardening Association, Inc., Burlington, VT, 1987.
12. U.S. EPA, *Methodology for Assessing Health Risks Associated with Indirect Exposure to Combustor Emissions*, Office of Health and Environmental Assessment, Washington, DC, 1990.
13. Knott, J. E., *Handbook for Vegetable Growers*, John Wiley & Sons, New York, 1957.
14. Clement Associates, Inc. Multipathway Health Risk Assessment Input Parameters Guidance Document, 1988.

A Probabilistic Emissions Model for Managing Hazardous Air Pollutants

Edward S. Rubin
Michael B. Berkenpas

I. ABSTRACT

Any assessment of control strategies or impacts of potentially hazardous airborne pollutants must begin with the characterization of source emissions. This paper describes a new computer-based model developed for the Electric Power Research Institute (EPRI) to quantify the emissions of chemical species in all gaseous, liquid, and solid streams entering and leaving a plant. User-specified parameters allow the model to be tailored to a variety of power plant configurations and site-specific conditions. A unique feature of the model is that all parameters and input data may be characterized probabilistically so that uncertainties can be analyzed rigorously.

II. INTRODUCTION

The Clean Air Act Amendments (CAAA) of 1990 gave new importance to the control of hazardous air pollutants. Under previous Clean Air Act provisions (Section 112) hazardous pollutants were identified and regulated based on a determination of harm by the U.S. Environmental Protection Agency (EPA). To date, fewer than ten species have been regulated in this manner. Now, 189 chemical species have been named in the new CAAA provisions for air toxics (Title III). Control is required across a broad spectrum of industrial and other sources emitting 10 tons per year (tpy) or more of any one of the 189 listed substances, or 25 tpy or more of any combination of substances. The basis for regulation is the use of "maximum available control technology" (MACT). Additional controls could be required if EPA finds an unacceptable level of remaining risk to public health after MACT is applied.[1]

Electric utilities are not initially subject to these new air toxics requirements. Still, the CAAA requires EPA to perform a study of the hazards to public health reasonably anticipated to occur from emissions of hazardous air pollutants from electric steam-generating units after the imposition of the other requirements of the 1990 Amendments. In the study report, due to Congress by November 15, 1993, EPA also must develop and describe alternate control strategies for hazardous emissions that may warrant regulation. EPA must regulate electric utilities under Title III if "appropriate and necessary" after considering the results of the study.

Two other studies required by the 1990 Amendments also may impact electric utilities. Sections 112(n)(1)(B) and (C) call for studies of mercury emissions from electric utilities and other sources. One of these studies will define threshold mercury exposure for adverse human health effects. The other study addresses the health and environmental effects of deposition of hazardous air pollutants in the Great Lakes, the Chesapeake Bay, Lake Champlain, and U.S. coastal waters. If EPA concludes that further regulation is required because of either of these studies, electric utilities could be included in the air toxics regulations.

The EPRI already has undertaken a program to study utility air toxics and their control. The EPRI program — known as PISCES (Power Plant Integrated Systems: Chemical Emissions Study) — has several major products and activities including (1) a database of published information on trace species for conventional fossil fuel power plants; (2) a probabilistic computer model to estimate power plants emissions; (3) a field monitoring program to collect new data; (4) development of emission control technology selection guidelines; and (5) a sampling and analytical methods reference guide for trace chemical measurements. Descriptions of these and related EPRI activities appear elsewhere.[2-4]

III. THE POWER PLANT ASSESSMENT MODEL

This paper focuses on the development and applications of the power plant chemical assessment model. The purpose of the model is to allow utilities to evaluate the performance of a given plant configuration with respect to multimedia emissions of chemical substances. The model provides estimates of the mass flow rates of all solid, liquid, and gaseous streams emanating from the plant, including quantitative estimates of all trace species emissions.

A unique feature of the model is its ability to characterize uncertainties probabilistically. Any or all model input parameters can be assigned a probability distribution rather than a single value. The combined effect of all input uncertainties then is reflected in an uncertainty distribution for output parameters of interest obtained using Monte Carlo methods. Such distributions give the likelihood of a particular value, in contrast to conventional single-valued estimates. The model can be run using either deterministic values or uncertainty distributions yielding probabilistic results.

Descriptions of the initial model development and applications have been reported previously.[5,6] Here we briefly summarize the model structure and design. We then elaborate on recent developments involving the user interface and linkage with the PISCES database.

A. Model Structure and Data Requirements

The model allows any fossil-fueled power plant to be configured for analysis. Version 1.0 is limited to conventional coal-, oil-, and gas-fired plants employing any

Table 1. Current Technology Options in the Power Plant Model and Interface

Fuel characteristics	Solid waste disposal
Coal	Landfill
Oil	Ponding
Gas	Co-disposal
Boiler systems	Cooling water system
Tangential	Once-through
Wall	Cooling tower
Cyclone	Pond or lake
	Fresh or saline
Particulate controls	Water treatment systems
Precipitator	Plant makeup water
Fabric filter	Plant service water
	Boiler makeup
SO_2 controls	Condensate polisher
Wet Lime FGD	Cooling tower makeup
Wet limestone FGD	Tower slip stream
Lime spray dryer	Ash pond discharge

of the technologies listed in Table 1. Future versions will include additional environmental control technologies and a number of advanced power generation systems (e.g., fluidized beds, gasification combined cycle, etc.).

The model employs fundamental mass and energy balances to compute all system flow rates. Empirical data also are employed where necessary (e.g., in calculating nitrogen oxide emissions).

The underlying model is written in the Demos (Decision modeling system) computer language. This environment provides significant flexibility in configuring power plant designs while providing the capability to conduct probabilistic analyses. Documentation of the underlying model and the Demos computing environment can be found in other reference documents.[7-9] The model is designed to run on the Macintosh operating system.

Utilization of the model requires two types of data. One involves parameters specifying the power plant configuration and performance. The other involves data needed to evaluate the emissions of trace chemical substances. The latter includes information on the concentration of trace species in all plant input streams (including fuel, reagents, water, and air), plus performance data characterizing how each chemical constituent in a given stream is "partitioned" or removed in various plant components or environmental control systems. If a probabilistic analysis is to be performed, plant characteristics, chemical species input quantities, and environmental control system performance characteristics also must be specified probabilistically. The PISCES database, discussed later, provides most of the information needed for trace species analysis.

B. The Graphical Interface

To simplify the use of the model, a new interactive graphical interface has been developed that eliminates the need to master the underlying computer commands

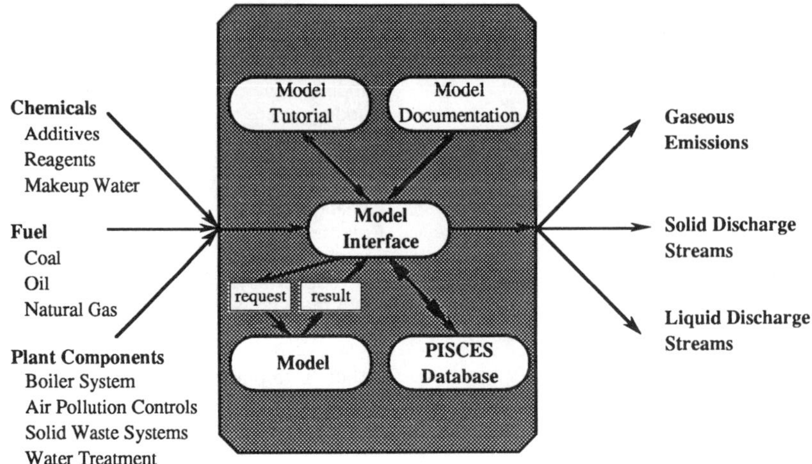

Figure 1. Interaction and communications between the model interface, PISCES database, and model.

normally required for model operation. The graphical interface is a separate program that transmits appropriate commands to the power plant model, and receives executed results for display. The interface program is written in Hypercard, which is a standard software accessory for Macintosh computers.

The complete software package thus involves interactions between three major components: model, interface, and database. Figure 1 shows a schematic of the interactions and intercommunications that are possible with the system inputs and outputs. This passing of messages, requests, and results all happens simultaneously via the graphic interface to the model. Figure 2 shows the interface screen used to begin model operation.

The model interface is divided into three sections: (1) configuring the power plant, (2) setting parameter values, and (3) getting results. These three operations appear near the top of the screen in Figure 2. Each operation is executed with a "button" that can be activated (clicked on) with the computer mouse to move quickly to any of the three major sections of the model.

The first section of the model interface (Figure 2) displays the eight steps involved in configuring the power plant to be analyzed. Each of the eight boxes also is a button that calls up a more detailed menu of options to be selected. For example, clicking on "Flue Gas Cleanup" brings up the screen shown in Figure 3, where SO_2 and particulate control devices are chosen. The interface also is smart enough to prohibit the user from selecting options that are not permitted for technical reasons. Once the plant is configured the interface displays a schematic of the power plant and water system configurations to verify the intended designs. Finally, for trace species analysis, the user selects the species of interest from a graphical menu accessed through Button 8 in Figure 2.

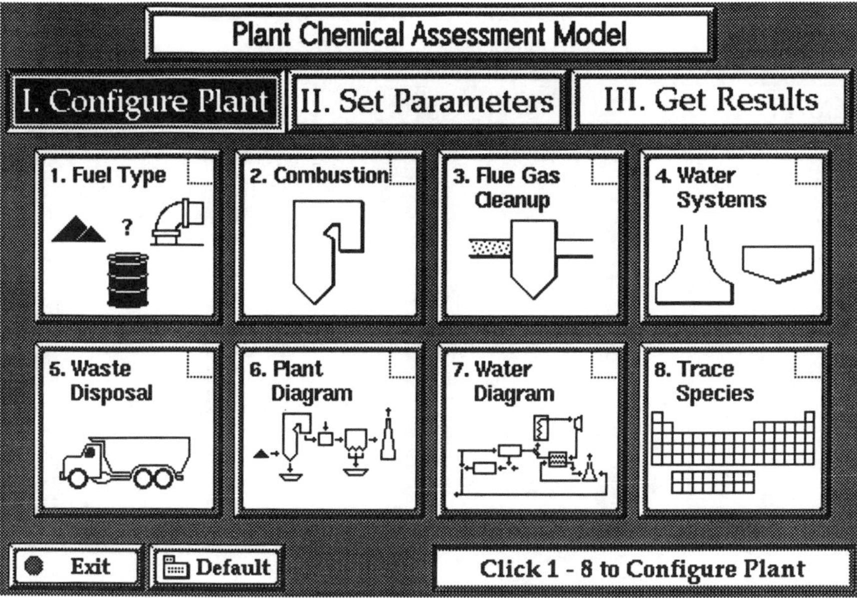

Figure 2. The "Configure Plant" checklist is the first of three major sections of the model. Eight plant areas can be reached from here.

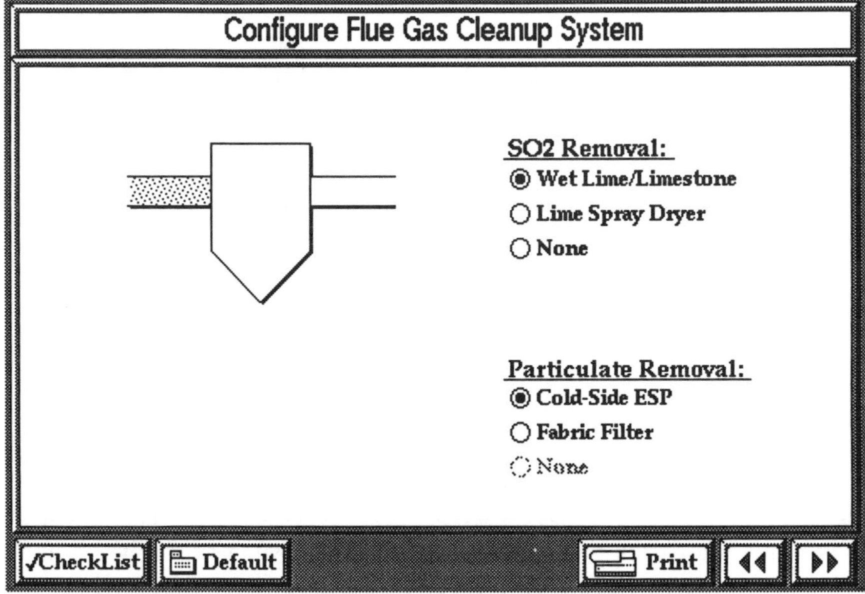

Figure 3. Sulfur and particulate removal devices are configured on this screen.

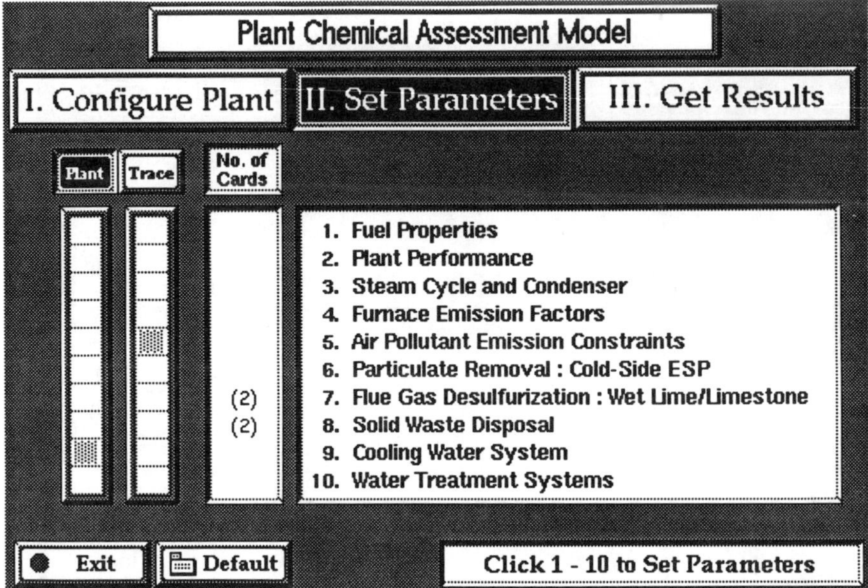

Figure 4. The second checklist has ten major sections and two types of parameters. These 20 elements contain all model input parameter lists.

The second section of the model interface requires setting parameter values for the ten major areas listed in Figure 4. The figure shows two columns requiring model parameter values. One column, labeled "Plant", contains parameters related to the overall plant performance and design. These parameters determine the major flow rates of materials through the power plant. For example, Figure 5 shows the screen for "Plant Performance" parameters, which include plant size and heat rate. The second column in Figure 4, labeled "Trace", refers to additional model parameters that specify the behavior of trace species in a particular power plant section. The "Number of Cards" column shows those areas with more than one screen.

The final section of the interface is used to "Get Results". Figures 6A and B show examples of the "Plant" and "Water" diagrams from which results are obtained. Results may be displayed in graphical, tabular, or diagrammatic form for individual power plant components, or the "Plant Summary" button in the center of Figure 6A can be activated to get overall results for plant inputs and outputs. Results for a particular device appear when clicking on the picture of that device. For example, the plant configuration in Figure 6A shows a coal pile, boiler, air preheater, electrostatic precipitator (ESP), wet flue gas desulfurization (FGD) system, stack, bottom ash pond, and landfill disposal system. These are separate "buttons" that can be activated to get results for that particular component of the plant.

Plant Performance Input Parameters

Parameter Description		Units	Value	Calc	Minimum	Maximum	Default
Gross Electrical Output		(MWg)	540		100	1000	540
Steam Cycle Heat Rate		(Btu/kWh)	7880		6000	10000	7880
Capacity Factor		(%)	75		15	100	75
Excess Air for Furnace		(%)	20		5	30	20
Leakage Air at Preheater		(%)	19		0	40	19
Air Temperature		(°F)	80		60	110	80
Air Pressure		(psia)	14.7		12	15	14.7
Specific Humidity		(lb H2O/lb dry air)	0.018		0	0.03	0.018
Boiler Efficiency		(%)		☒	80	95	Calc
Unaccounted Losses in Boiler		(%)	0.5		0	1	0.5
Temperature Exiting:	Economizer	(°F)	700		600	800	700
	Preheater	(°F)	300		250	400	300
Energy	Pulverizers	(%MWg)	0.6		0	1	0.6
Penalties:	Steam Pumps	(%MWg)	0.65		0.5	0.7	0.65
(Excluding	FD Fans	(%MWg)	1.5		0.7	4	1.5
Gas Cleanup	Cooling System	(%MWg)	0.4		0.4	2	0.4
Systems)	Misc.	(%MWg)	1.3		0	2	1.3

√CheckList | Default | ◉ Value ○ Uncertainty | Print | ◀◀ | ▶▶

Figure 5. Parameters on this screen determine the primary flow rates.

C. Probabilistic Analysis Capability

As noted earlier, a unique feature of the power plant chemical assessment model is its ability to analyze uncertainties probabilistically. Probability distributions are assigned to any plant or trace input parameter using the "Uncertainty" button at the bottom of each parameter input screen. For example, Figure 7 shows the result of choosing "Uncertainty" for the "Plant Performance" card shown earlier in Figure 5. At this point, any parameter on the card can be specified probabilistically by activating the "Edit" button. This brings up the screen illustrated in Figure 8. A pop-up menu allows various types of uncertainty distributions to be entered.

Once all parameter uncertainties are assigned, the "Get Results" screen allows the sampling procedure (Monte Carlo or Latin Hypercube sampling) to be selected, with the desired number of iterations. The combined effect of all uncertain parameters is then evaluated using stochastic simulation methods. Full details of the model operation are described in a comprehensive user's manual.[10]

IV. THE PISCES CHEMICAL DATABASE

Another major element of the PISCES study is the development and compilation of published and other data regarding chemical substances found in streams from

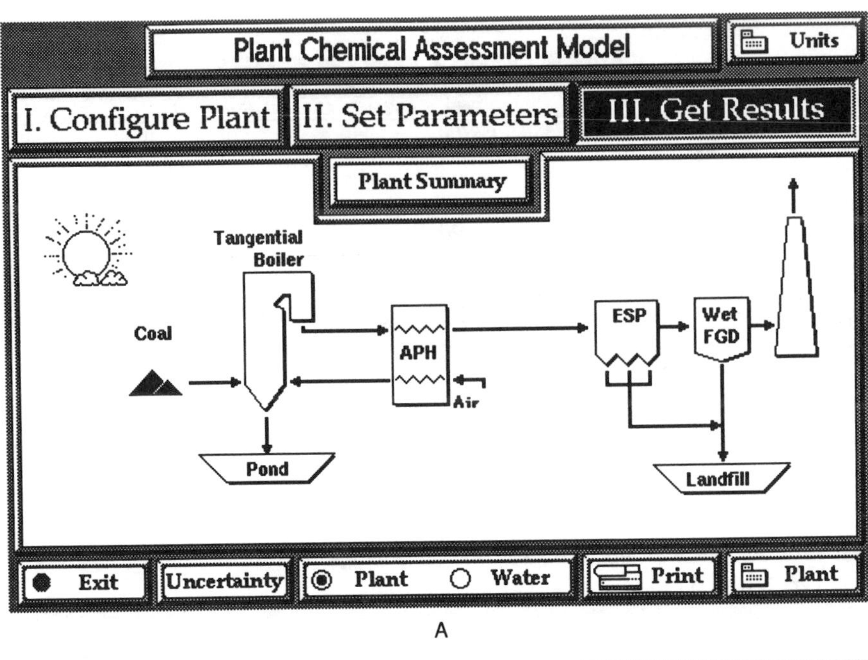

A

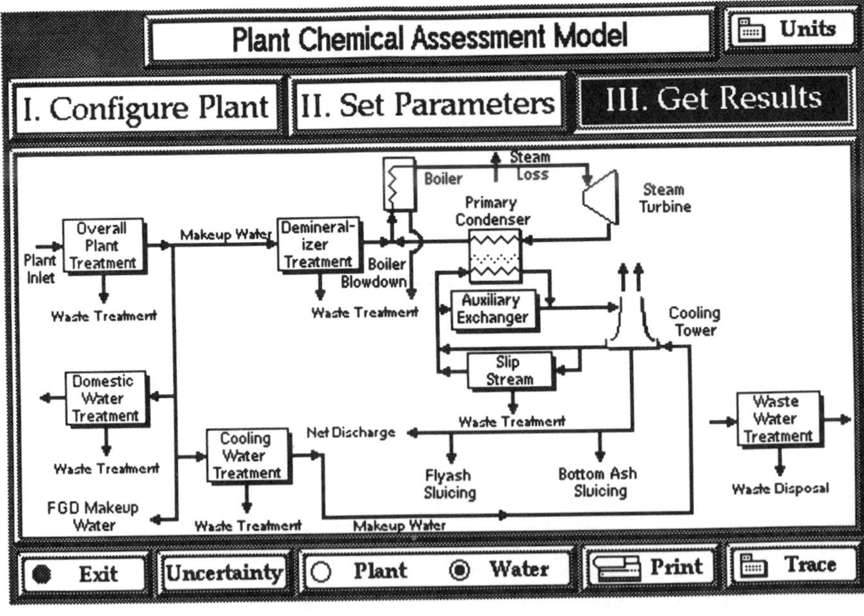

B

Figure 6. These screens are used to obtain results for (A) solid and gaseous streams and (B) plant water streams.

Plant Performance Input Parameters

Parameter Description		Units +/*	Uncertainty	Edit
Gross Electrical Output		(MWg)		○
Steam Cycle Heat Rate		(Btu/kWh)		○
Capacity Factor		(%)		○
Excess Air for Furnace		(%)		○
Leakage Air at Preheater		(%)		○
Air Temperature		(°F)		○
Air Pressure		(psia)		○
Specific Humidity		(lb H2O/lb dry air)		○
Boiler Efficiency		(%)		○
Unaccounted Losses in Boiler		(%)		○
Temperature Exiting:	Economizer	(°F)		○
	Preheater	(°F)		○
Energy	Pulverizers	(%MWg)		○
Penalties:	Steam Pumps	(%MWg)		○
(Excluding	FD Fans	(%MWg)		○
Gas Cleanup	Cooling System	(%MWg)		○
Systems)	Misc.	(%MWg)		○

√CheckList | Default | ○ Value ◉ Uncertainty | Print | ◀◀ | ▶▶

Figure 7. The "Uncertainty" button is used to enter frequency distribution data for any model parameter.

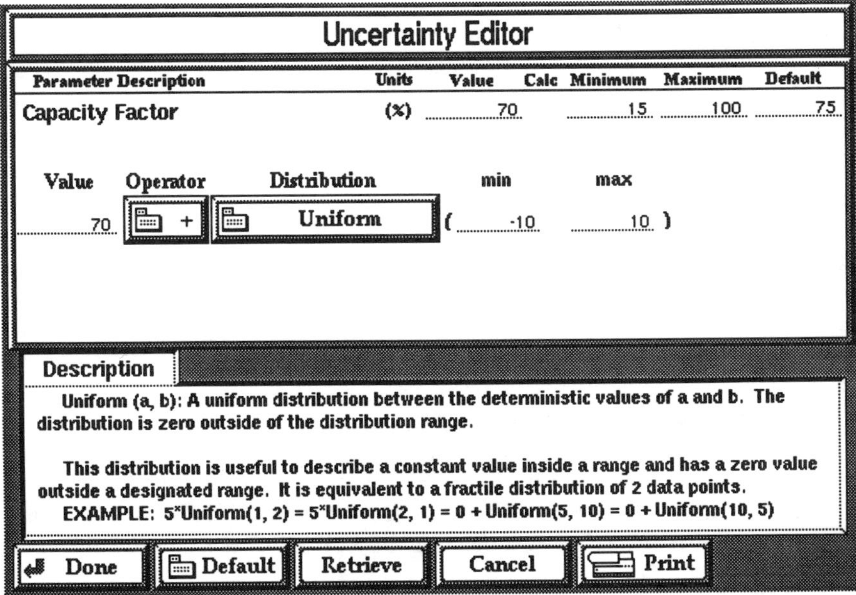

Figure 8. The uncertainty editor builds a distribution for a selected parameter. A pop-up menu contains various options.

Table 2. Selected Trace Species for the Model Database

Ammonia	Chrysene	Mercury
Arsenic	Cobalt	Molybdenum
Barium	Copper	Naphthalene
Benzene	Cyanide	Nickel
Benzo–pyrene	Fluoride	Phosphate
Beryllium	Fluorine	Phosphorus
Cadmium	Formaldehyde	Pyrene
Chloride	HCl	Radium 226
Chlorine	HF	Selenium
Chromium	Lead	Toluene
Chromium–6	Manganese	Vanadium

conventional fossil fuel power systems.[11,12] This database has been developed by Radian Corporation and implemented in a database management system operating on a Sun3 computer workstation. For the chemical assessment model, information from the PISCES database has been downloaded, parsed and restructured for compatibility with the computer model interface. Data files for 33 selected trace species, listed in Table 2, have been created. These species represent key chemicals of potential concern to utilities. The computer model interface can track up to 25 of these species in a single model run.

A. Species Data Files

Frequency distributions for the 33 species have been developed for each trace species input parameter required by the model. There are three types of trace species input parameters: (1) the concentrations of trace species in power plant input streams (i.e., fuel, reagents, water, and air); (2) the concentrations of new trace species created within the power plant (e.g., trace organics); and (3) the partitioning of trace species across various power plant components. In all, there are 33 model input parameters for each trace species, as shown in Table 3.* This table sorts the 33 parameters into the 10 major categories listed on the "Set Parameters" check list screen shown earlier in Figure 4. Note that only a subset of these parameters is used for any given model run. The overall database files, however, contain over 1000 distributions (33 parameters times 33 species). At present, many of these are place-holders awaiting additional data.

B. Interface Access to the Database

Data distributions for each trace species and each model input parameter can be quickly and easily imported to the model directly from data files on the computer. Database retrieval buttons on each trace parameter input screen, shown in Figure 9, are used to perform this task.

* It is merely coincidence that there are both 33 species and 33 parameters.

Table 3. Model Input Parameters for Trace Species

Trace species parameter	Description of parameter	Units
Fuel properties		
Coal	Concentration of trace species in coal	ppmw
Oil	Concentration of trace species in oil	ppmw
Natural gas	Concentration of trace species in natural gas	ppmw
Coal pile runoff	Concentration of trace species in coal pile runoff water	ppmw
Pyrite rejects	Concentration of trace species in pyrite removed from coal pulverizer	ppmw
Plant performance		
Ambient air	Concentration of trace species in ambient air	ppmw
Furnace partition	Ratio of species concentrations: (in coal)/([in bottom ash]×[ash in coal])	Fraction
Steam cycle and condenser		
Boiler cleaning wastes	Concentration of trace species in boiler chemical cleaning waste	ppmw
Furnace emission factors		
Furnace emissions	Concentration of molecular species created inside furnace	ppmw
Fireside cleaning wastes	Concentration of trace species in fireside cleaning waste	ppmw
Air preheater cleaning wastes	Concentration of trace species in air preheater cleaning waste	ppmw
Particulate removal		
ESP partition	Ratio of species concentration: (removed from ESP)/([into ESP]×[TSP efficiency])	Fraction
FF partition	Ratio of species concentration: (into FF-out of FF)/([into FF]×TSP efficiency])	Fraction
Flue gas desulfurization		
Limestone	Concentration of trace species in limestone	ppmw
Lime	Concentration of trace species in lime	ppmw
Wet lime/limestone partition	Ratio of species concentration: (into FGD-out of FGD)/(into FGD)	Fraction
Solid waste disposal		
Bottom ash disposal leachate partition	Ratio of species concentration: (water overflowing pond)/(solids entering pond)	Fraction
Quench evaporation partition	Ratio of species concentration: (evaporated from sluice water)/(in sluice water)	Fraction
Water overflow partition	Ratio of species concentration: (water overflow disposal)/(solids into disposal)	Fraction
Fly ash disposal leachate partition	Ratio of species concentration: (leachate exiting disposal)/(solids into disposal)	Fraction
Cooling water system		
Fresh water	Concentration of trace species in fresh water	ppmw
Saline water	Concentration of trace species in saline water	ppmw
Once-through emissions	Concentration of molecular trace species created inside cooling water system	ppmw
Tower evaporation partition	Ratio of species concentration: (evaporated)/(in recirculating water)	Fraction
Basin sludge	Concentration of trace species in cooling tower basin sludge	ppmw

Table 3. Model Input Parameters for Trace Species (continued)

Trace species parameter	Description of parameter	Units
Water treatment systems		
Plant makeup water partition	Ratio of species concentration: (removed from plant)/(into plant)	Fraction
Zero discharge partition	Ratio of species concentration: (removed from plant)/into plant)	Fraction
Domestic water partition	Ratio of species concentration: (removed from plant)/(into plant)	Fraction
Tower makeup partition	Ratio of species concentration: (removed from plant)/(into plant)	Fraction
Tower slip partition	Ratio of species concentration: (removed from plant)/(into plant)	Fraction
Demineralizer partition	Ratio of species concentration: (removed from demin.)/(entering demin.)	Fraction
Wastewater partition	Ratio of species concentration: (removed from plant)/(into plant)	Fraction
Floor and yard drains	Concentration of trace species in floor and yard drain waste	ppmw

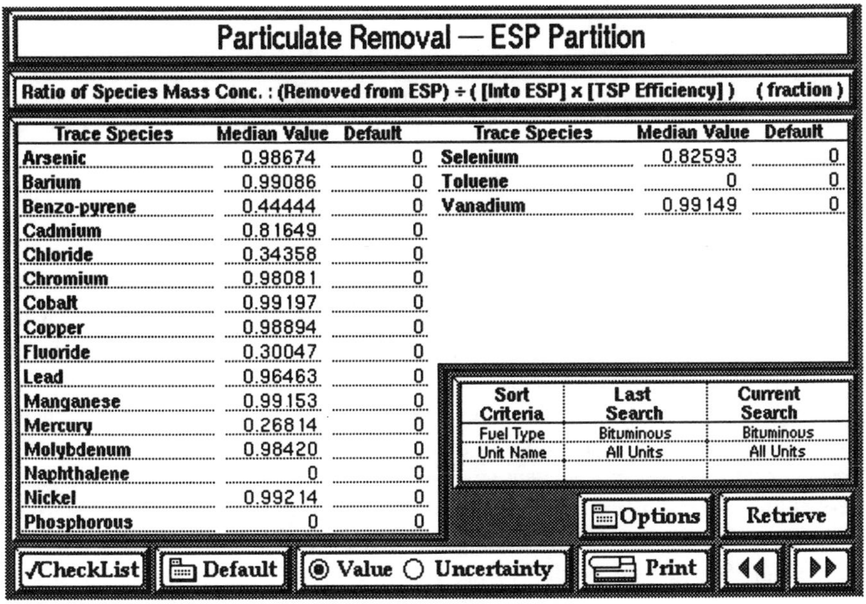

Particulate Removal — ESP Partition

Ratio of Species Mass Conc. : (Removed from ESP) ÷ ([Into ESP] x [TSP Efficiency]) (fraction)

Trace Species	Median Value	Default	Trace Species	Median Value	Default
Arsenic	0.98674	0	Selenium	0.82593	0
Barium	0.99086	0	Toluene	0	0
Benzo-pyrene	0.44444	0	Vanadium	0.99149	0
Cadmium	0.81649	0			
Chloride	0.34358	0			
Chromium	0.98081	0			
Cobalt	0.99197	0			
Copper	0.98894	0			
Fluoride	0.30047	0			
Lead	0.96463	0			
Manganese	0.99153	0			
Mercury	0.26814	0			
Molybdenum	0.98420	0			
Naphthalene	0	0			
Nickel	0.99214	0			
Phosphorous	0	0			

Sort Criteria	Last Search	Current Search
Fuel Type	Bituminous	Bituminous
Unit Name	All Units	All Units

Options Retrieve

√CheckList Default Value Uncertainty Print

Figure 9. Each trace input screen can access the PISCES database files using the area at the bottom-right corner of the screen.

Frequency distributions for each model parameter were derived by sorting the data from the full PISCES database on five criteria: (1) fuel type, (2) boiler type, (3) FGD reagent type, (4) device type, and (5) unit name. Not all five criteria were used for each parameter; the particular sort criteria depended upon what was technically appropriate and how much data were available in the PISCES database. The "Options" button seen in Figure 9 is used to display the sort criteria for which data sets are available. The data for that set of criteria can then be imported using the "Retrieve" button.

The database files for each parameter contain frequency distributions for all 33 species. When data are retrieved, however, only the frequency distributions for those species chosen for an analysis are imported. The "sort criteria" selection further narrows the subset of available data imported to the model. Once entered, these data can be easily edited or modified using the computer model interface. User-specified data also can be entered in lieu of the PISCES data files. Indeed, in instances where the PISCES database does not yet contain parameter information for a particular species, user-specified data are required.

A reference data book summarizes all the PISCES data distributions used for the model interface data files.[13] It also tabulates the results of data sorts for each model parameter and each trace substance.

V. MODEL APPLICATIONS

Any assessment of control strategies or environmental risks for hazardous air pollutants must begin with a characterization of source emissions. Species emission estimates may provide a basis for determining regulatory compliance (e.g., whether emissions exceed some specified value) or may provide input to more comprehensive assessments of effects and risks (such as called for in special studies under the 1990 CAAA). The ability of control technology, fuel choice, and other plant design parameters to affect the emission rates of hazardous pollutants (not only to the atmosphere, but also to water and land) can then be examined where problems are found to exist. The capability to quantify uncertainties is important for judging the robustness of results and the degree of confidence in proposed solutions.

Applications to date of the power plant chemical assessment model have included (1) deterministic studies of ten power plant configurations to benchmark key performance results against independent studies by Bechtel Corporation;[14] (2) illustrative probabilistic studies using early versions of the PISCES database;[5,6] (3) one study of trace species emissions from a European power plant;[15] and (4) six utility-specific case studies carried out as part of the beta-testing of the model. Additional case studies are planned for several power plants visited in EPRI's data acquisition program under PISCES.[3]

Here we present an illustrative example of how the model may be used to quantify hazardous air emissions from a coal-fired power plant equipped with a cold-side ESP and a wet limestone FGD system (the configuration shown earlier in Figure 6A). For convenience, we use the model default parameters for each device, which are based on studies by Bechtel.[14] A set of 19 trace species is selected for illustrative purposes.

The key plant performance parameters are those shown earlier in Figure 5. Figures 10 and 11 show two additional input screens for the fuel type and FGD system characteristics, respectively. Figure 12 shows the median values of trace species in coal using an example frequency distribution for bituminous coal, based on the PISCES database.

A summary table of median input and output flows of the selected trace species is shown in Figure 13. For any plant component of interest more detailed results can be obtained as described earlier. For example, Figures 14 and 15, respectively, show the median performance characteristics and trace species flows for the FGD system. Figures 16 and 17 show the probabilistic results for one species (arsenic) exiting the FGD unit. Both graphical and tabular results are displayed. These two figures show that while the median hourly flow of arsenic is 8.1×10^{-4} lb/h there is a 10% chance (90% cumulative probability) it could be as high as 3.2×10^{-3} lb/h. Flow rates that are judged to be unacceptably high become candidates for control strategies in subsequent analyses.

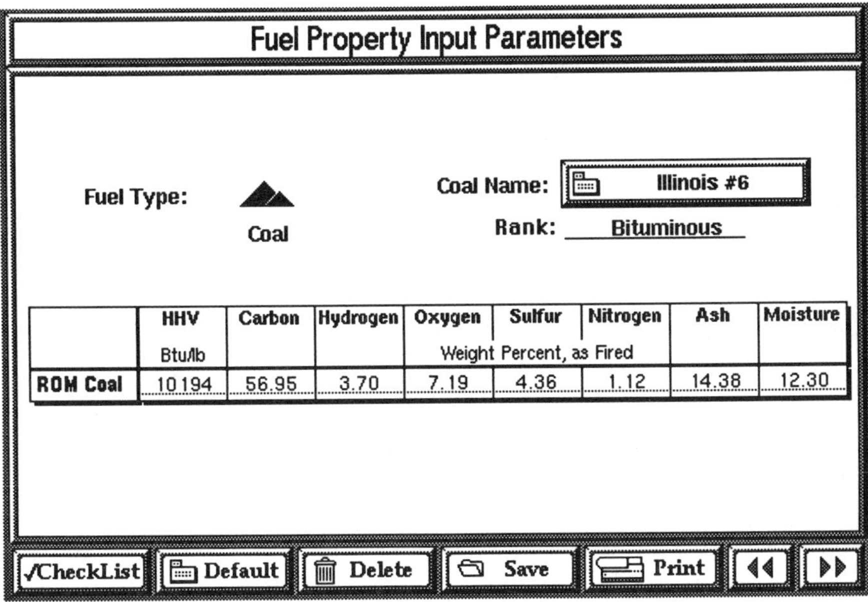

Figure 10. This screen provides an ultimate analysis of the selected fuel.

Wet FGD Input Parameters

Parameter Description	Units	Value	Calc	Minimum	Maximum	Default
Maximum SO2 Removal Efficiency	(%)	95		90	98	95
Actual SO2 Removal Efficiency	(%)		☒	30	98	Calc
Particulate Removal Efficiency	(%)	50		0	100	50
Reagent Stoichiometry	(moles Ca/moles S)	1.15		1	3	1.15
Maximum Absorber Size	(1000 acfm)	425		100	1000	425
Number of Operating Absorbers			☒	1	10	Calc
Number of Spare Absorbers			☒	0	3	Calc
Temperature Rise Across ID Fan	(°F)	14		0	25	14
Gas Temp. Exiting Scrubber	(°F)		☒	125	185	Calc
Gas Temp. After Reheater	(°F)	178		110	225	178
Oxidation of CaSO3 to CaSO4	(%)	90		0	100	90
Entrained Water Past Demister	(% Evap.)	0.79		0.25	1	0.79
Gas Pressure Exiting FGD	(Inches W.G.)	4		4	10	4
Reagent Purity	(wt %)	92.4		80	100	92.4
Water Content of Reagent	(wt %)	0		0	5	0
Reagent Slurry Water	(lb H2O/lb reag.)	3.67		0	10	3.67
Demister Water Usage	(lb H2O/lb reag.)	0		0	10	0
Energy Penalty	(% MWg)		☒	0	5	Calc

√CheckList | Default | ◉ Value ○ Uncertainty | Print

Figure 11. Input parameters for the lime/limestone FGD system are listed on this screen.

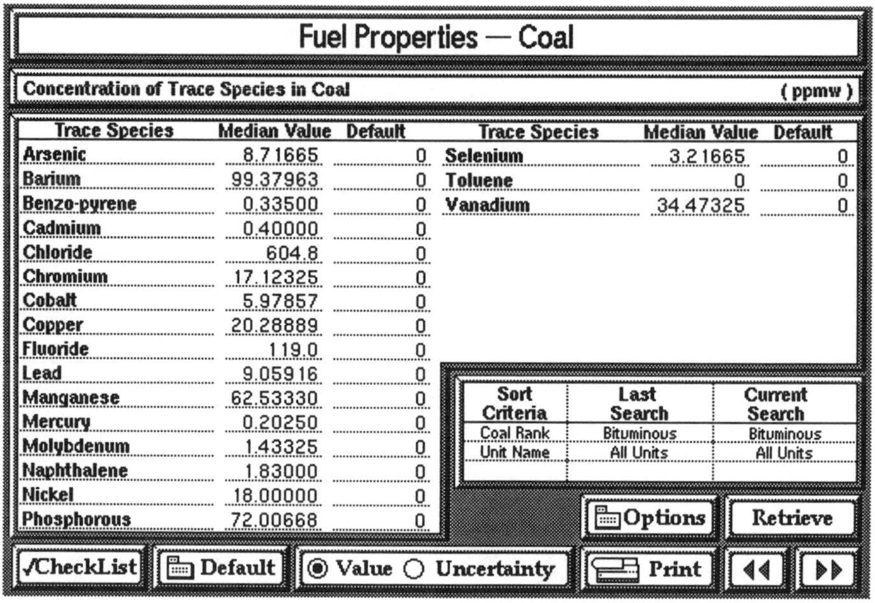

Fuel Properties — Coal

Concentration of Trace Species in Coal (ppmw)

Trace Species	Median Value	Default	Trace Species	Median Value	Default
Arsenic	8.71665	0	Selenium	3.21665	0
Barium	99.37963	0	Toluene	0	0
Benzo-pyrene	0.33500	0	Vanadium	34.47325	0
Cadmium	0.40000	0			
Chloride	604.8	0			
Chromium	17.12325	0			
Cobalt	5.97857	0			
Copper	20.28889	0			
Fluoride	119.0	0			
Lead	9.05916	0			
Manganese	62.53330	0			
Mercury	0.20250	0			
Molybdenum	1.43325	0			
Naphthalene	1.83000	0			
Nickel	18.00000	0			
Phosphorous	72.00668	0			

Sort Criteria	Last Search	Current Search
Coal Rank	Bituminous	Bituminous
Unit Name	All Units	All Units

Options | Retrieve

√CheckList | Default | ◉ Value ○ Uncertainty | Print | ◀◀ | ▶▶

Figure 12. Trace species in coal may be entered automatically from the PISCES database or manually using the graphical interface.

VI. CONCLUSION

The power plant chemical assessment model developed as part of the EPRI PISCES program offers a unique capability for evaluating and managing hazardous air pollutant emissions from electric power plants. The model's user-friendly graphical interface and ability to rigorously evaluate uncertainties make it a powerful analytical tool for general and site-specific studies. The PISCES database is a key complement to the model, offering a rich source of data on trace species quantities and behavior.

As of this writing, the current version of the model is undergoing beta-testing by six utilities in North America and Europe. The results of that testing will be incorporated in an updated release of the model. Subsequent enhancements are scheduled on approximately an annual basis.

The PISCES database also is being revised and updated under the management of Radian Corporation. A revised version of the database compatible with the plant computer model is expected when this revision is completed. Subsequently, the database will be expanded to incorporate new data from the EPRI field sampling program.

Overall Trace Entering | **Overall Trace Flow Rate** lbs/hr | Units

Species	Air Into Furnace	Fuel Into Furnace	Furnace Factors	Intermit. Cleaning Streams	Lime/ Limestone Reagent	Fresh Makeup Water	Saline Cooling Water	Cooling Water Factors	Rainfall Runoff	Total Into Plant
Arsenic	0	4.123	0	0	0.07	132.9	0	0	384u	137.1
Barium	0	46.99	0	0	2.102	1197	0	0	1.69m	1246
Benzo-pyrene	256n	0.158	0	0	0	0.292	0	0	0	0.451
Cadmium	0	0.222	0	0	0.067	0.665	0	0	29.5u	0.954
Chloride	0	285.9	0	0	3.504	5384	0	0	1.012	5675
Chromium	0	8.097	0	0	0.701	7.312	0	0	84.3u	16.11
Cobalt	0	2.827	0	0	0	66.47	0	0	422u	69.3
Copper	0	9.593	0	0	0.224	4.155	0	0	4.17m	13.98
Fluoride	0	56.26	0	0	8.199	58.5	0	0	4.26m	123
Lead	0	4.286	0	0	0.911	15.16	0	0	0	20.35
Manganese	0	29.58	0	0	3.083	11.87	0	0	0.089	44.61
Mercury	0	0.096	0	0	841u	0.226	0	0	21.1u	0.323
Molybdenum	0	0.678	0	0	0.14	598.2	0	0	190u	599.1
Naphthalene	6.17m	0.865	0.241	0	0	0	0	0	0	1.112
Nickel	0	8.51	0	0	0.302	9.173	0	0	9.53m	18
Phosphorous	0	34.1	0	0	7.01m	7.378	0	0	1.69m	41.49
Selenium	0	1.521	0	0	0.035	2.526	0	0	50.6u	4.082
Toluene	0	0	0	0	0	0	0	0	0	0
Vanadium	0	16.3	0	0	1.261	13.29	0	0	498u	30.86
Total	6.17m	510.1	0.241	0	20.61	7509	0	0	1.124	8041

Return | Key m = 1.0e−3 u = 1.0e−6 n = 1.0e−9 p = 1.0e−12 K = 1.0e+3 M = 1.0e+6 B = 1.0e+9 T = 1.0e+12 | Print | Trace

A

Overall Trace Exiting | **Overall Trace Flow Rate** lbs/hr | Units

Species	Stack Gas	Bottom Ash Evaporation	Tower Drift & Evaporation	Ash Pond Residue	Ash Pond Leach	Flue Gas Waste Remain	Flue Gas Waste Leach	Net Cool Water Out	Domestic Water Dischg	Waste Water Dischg	Waste Water Evaporation	Waste Water Sludge	Remove Offsite	Total Out Plant
Arsenic	750u	0	0	0.079	37.8u	4.732	0	132.2	0.036	0.101	0	2u	0	137.1
Barium	3.98m	0	0	9.719	340u	44.93	0	1190	0.32	0.905	0	398u	0	1246
Benzo-pyrene	0.158	0	0	41.2u	83.2n	1.32m	0	0.291	78.3u	0	0	221u	0	0.451
Cadmium	146u	0	0	0.042	189n	0.25	0	0.661	178u	531u	0	872n	0	0.954
Chloride	107.4	0	0	1.058	1.53m	206	0	5354	1.441	3.297	0	1.783	0	5675
Chromium	86.4u	0	0	2.133	2.08u	6.698	0	7.27	1.96m	5.61m	0	1.12u	0	16.11
Cobalt	9.89m	0	0	0.423	18.9u	2.703	0	66.09	0.018	0	0	0.051	0	69.3
Copper	0.045	0	0	1.935	1.18u	7.857	0	4.131	1.11m	7.31m	0	3.66u	0	13.98
Fluoride	8.545	0	0	13.34	16.6u	42.85	0	58.16	0.016	0	0	0.048	0	123
Lead	104u	0	0	0.292	4.31u	4.976	0	15.07	4.06m	0.011	0	1.03u	0	20.35
Manganese	1.33m	0	0	7.441	3.37u	25.27	0	11.8	3.17m	0.097	0	36.1u	0	44.61
Mercury	0.032	0	0	2.97m	64.3n	0.062	0	0.225	60.5u	0	0	192u	0	0.323
Molybdenum	3.9m	0	0	0.148	170u	3.446	0	594.9	0.16	0.452	0	163u	0	599.1
Naphthalene	1.112	0	0	0	0	0	0	0	0	0	0	0	0	1.112
Nickel	1.2m	0	0	2.046	2.61u	6.807	0	9.121	2.45m	0.016	0	2.47u	0	18
Phosphorous	32.23	0	0	1.867	2.1u	0.04	0	7.337	1.97m	7.26m	0	1.23u	0	41.49
Selenium	0.606	0	0	0.039	718n	0.922	0	2.512	676u	1.96m	0	921n	0	4.082
Toluene	0	0	0	0	0	0	0	0	0	0	0	0	0	0
Vanadium	721u	0	0	2.994	3.78u	14.63	0	13.22	3.56m	0.011	0	1.69u	0	30.86
Total	150.1	0	0	43.56	2.14m	372.2	0	7466	2.009	4.914	0	1.883	0	8041

Return | Key m = 1.0e−3 u = 1.0e−6 n = 1.0e−9 p = 1.0e−12 K = 1.0e+3 M = 1.0e+6 B = 1.0e+9 T = 1.0e+12 | Print | Trace

B

Figure 13. Median values of trace species flow rates are summarized for streams (A) entering and (B) exiting the power plant.

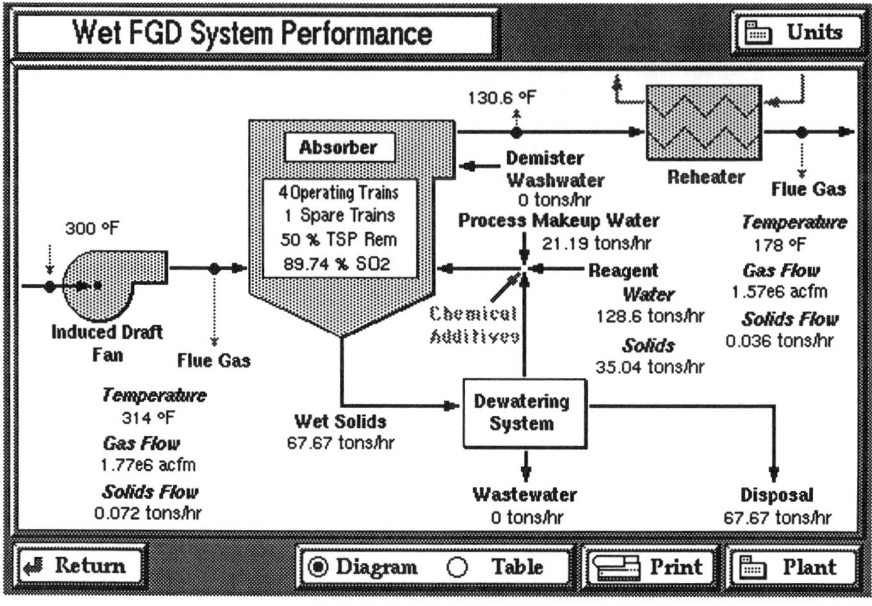

Figure 14. Median values of FGD flow rates, temperatures, and other data are shown in diagram form using the "Plant" button.

Wet FGD System — Trace Species Flow Rate (lbs/hr)

Species	Flue Gas In	Reagent Inject	Water Inject	Flue Gas Bypass	Total Remove	Dewater Waste	Waste to Dispose	Flue Gas Out	Flue Gas Bypass Remix
Arsenic	0.071	0.07	0.599	0	0.739	0	0.739	750u	750u
Barium	0.147	2.102	5.392	0	7.637	0	7.637	3.98m	3.98m
Benzo-pyrene	0.158	0	1.32m	0	1.32m	0	1.32m	0.158	0.158
Cadmium	0.024	0.067	3m	0	0.094	0	0.094	146u	146u
Chloride	204.6	3.504	24.26	0	125	0	125	107.4	107.4
Chromium	0.029	0.701	0.033	0	0.762	0	0.762	86.4u	86.4u
Cobalt	9.89m	0	0.3	0	0.3	0	0.3	9.89m	9.89m
Copper	0.045	0.224	0.019	0	0.243	0	0.243	0.045	0.045
Fluoride	29.5	8.199	0.264	0	29.42	0	29.42	8.545	8.545
Lead	0.022	0.911	0.068	0	1.001	0	1.001	104u	104u
Manganese	0.091	3.083	0.053	0	3.227	0	3.227	1.33m	1.33m
Mercury	0.081	841u	1.02m	0	0.051	0	0.051	0.032	0.032
Molybdenum	3.9m	0.14	2.696	0	2.836	0	2.836	3.9m	3.9m
Naphthalene	1.112	0	0	0	0	0	0	1.112	1.112
Nickel	0.031	0.302	0.041	0	0.373	0	0.373	1.2m	1.2m
Phosphorous	32.23	7.01m	0.033	0	0.04	0	0.04	32.23	32.23
Selenium	0.606	0.035	0.011	0	0.046	0	0.046	0.606	0.606
Toluene	0	0	0	0	0	0	0	0	0
Vanadium	0.058	1.261	0.06	0	1.378	0	1.378	721u	721u
Total	268.8	20.61	33.84	0	173.2	0	173.2	150.1	150.1

Units | Return | Diagram | Table | Print | Trace

Figure 15. Median values of FGD flow rates are show in table form for the chosen trace species using the "Trace" button.

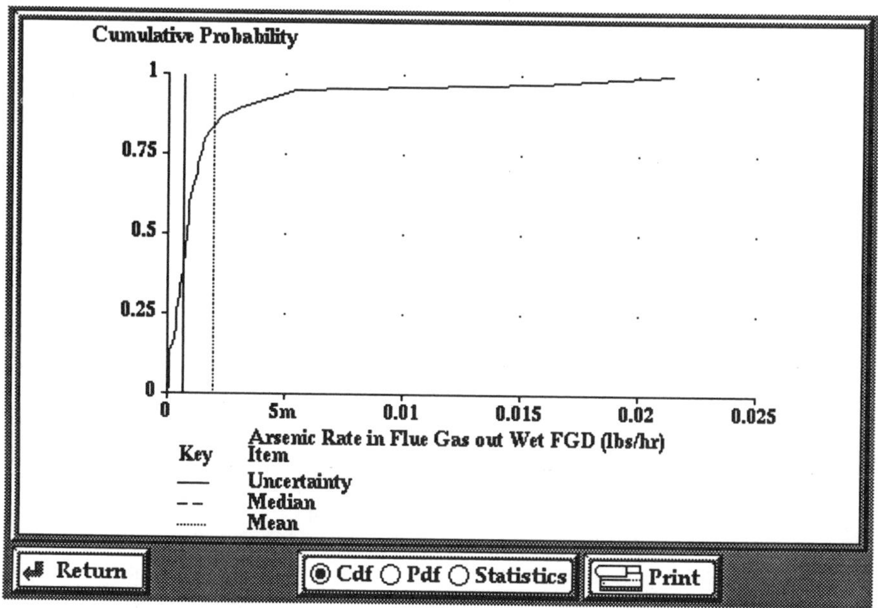

Figure 16. A cumulative probability distribution of the arsenic flow rate exiting the FGD system.

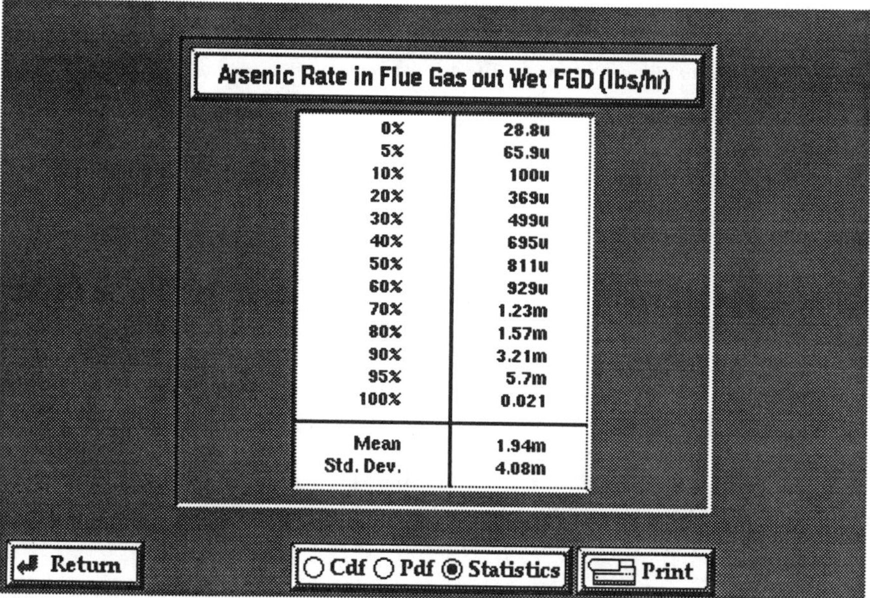

Arsenic Rate in Flue Gas out Wet FGD (lbs/hr)

0%	28.8u
5%	65.9u
10%	100u
20%	369u
30%	499u
40%	695u
50%	811u
60%	929u
70%	1.23m
80%	1.57m
90%	3.21m
95%	5.7m
100%	0.021
Mean	1.94m
Std. Dev.	4.08m

Figure 17. A tabular summary of the cumulative probability distribution for arsenic exiting the FGD system.

ACKNOWLEDGMENTS

The work described here was supported by the Electric Power Research Institute under Research Project RP2933-2.

REFERENCES

1. Public Law 101-549, November 15, 1990.
2. Boutacoff, D., New focus on air toxics, *EPRI J.,* 16, 4, 1991.
3. Chow, W., PICSES: Managing air toxics, *ECS Update,* 21, 2, 1991.
4. Chow, W., Miller, M. J., Fortune, J., Behrens, G., and Rubin, E. S., Managing air toxics under the new Clean Air Act Amendments, *Power Engineering,* January 1991, 30.
5. Rubin, E. S., Salmento, J. S., and Chow, W., A Probabilistic Approach to Multimedia Environmental Management, Paper No. 90-131.2, presented at the 1990 AWMA Annual Meeting, Air and Waste Management Association, Pittsburgh, PA, 1990.
6. Rubin, E. S., Salmento, J. S. and Chow, W., Chemical Characterization of Power Plant Waste Streams, Paper No. 90-37.1, presented at the 1990 AWMA Annual Meeting, Air and Waste Management Association, Pittsburgh, PA, 1990.
7. Wishbow, N. and Henrion, M., *An Introductory Tutorial for Demos,* Department of Engineering and Public Policy, Carnegie-Mellon University, Pittsburgh, PA, 1985.
8. Henrion, M. and Wishbow, N., *Demos User's Manual: Version Three,* Carnegie-Mellon University, Department of Engineering and Public Policy, Pittsburgh, PA, 1987.
9. Arnold, B. and Henrion, M., *Demos Interface Manual* (draft), Department of Engineering and Public Policy, Carnegie-Mellon University, November, 1990.
10. Berkenpas, M. B. and Rubin, E. S., *Power Plant Chemical Assessment Model: User Documentation*, prepared for the Electric Power Research Institute, Palo Alto, CA, by Center for Energy and Environmental Studies, Carnegie-Mellon University, Pittsburgh, PA, April 1991 (revised April 1993).
11. Balfour, W. D., Chow, W., and Rubin, E. S., PISCES: A Utility Database for Assessing the Pathways of Power Plant Chemical Substances, Paper No. 89-71.6, presented at the 1989 AWMA Annual Meeting, Air and Waste Management Association, Anaheim, CA, 1989.
12. Behrens, G. P. and Chow, W., Use of A Multi-Media Database for Chemical Emission Studies of Conventional Power Systems, Paper No. 90-131.1, presented at the 1990 AWMA Annual Meeting, Air and Waste Management Association, Pittsburgh, PA, 1990.
13. Berkenpas, M. B., Zalevsky, K., and Rubin, E. S., *Power Plant Chemical Assessment Model: PISCES Database Book*, prepared for the Electric Power Research Institute, Palo Alto, CA, by Center for Energy and Environmental Studies, Carnegie-Mellon University, Pittsburgh, PA, April 1991 (revised April 1993).
14. Bechtel Group, Inc., Power Plant Integrated Systems: Chemical Emission Studies (PISCES), prepared for Electric Power Research Institute, Palo Alto, CA, by Bechtel Group Inc., February 1989.
15. Rubin, E. S., Salmento, J. S., and Chow, W., Evaluating Power Plant Control Strategies for Air Toxics, Paper No. 91-103.20, presented at the 1991 AWMA Annual Meeting, Air and Waste Management Association, Vancouver, British Columbia, 1991.

Multimedia Health Risk Assessment for Power Plant Air Toxic Emissions

Elpida Constantinou
Christian Seigneur

I. INTRODUCTION

Under the sponsorship of the Electric Power Research Institute (EPRI), ENSR has developed a multimedia health risk assessment model for the prediction of the fate and transport of toxic pollutants in the different environmental media and the quantitative estimation of the associated health risks. The model combines a number of models to handle the transport and fate of contaminants in air, surface water, surface soil, vadose zone, groundwater, and the food chain. Concentrations calculated by the fate and transport models are used by exposure-dose models to calculate the individual exposure doses, which are then used to calculate health risks. A detailed description of the formulation of the multimedia health risk assessment model is presented by Constantinou and Seigneur.[1]

ENSR used the above-described model to perform a multimedia health risk assessment associated with the actual emissions of a coal-fired power plant unit. The results of the application are presented in this paper.

II. FACILITY DESCRIPTION

The subject power plant unit is about 700 MW. No liquid wastes are being produced at the site as all the water is recycled within the facility. Any significant amounts of solid wastes produced are carried offsite by trucks and disposed of at landfills. Consequently, the only type of emissions included in the present application is stack air emissions. The stack characteristics are as follows: stack height of 198 m, inner diameter of 9.3 m, flue gas temperature of 323 K, and flue gas velocity of 13 m/s.

A total number of 18 chemicals were sampled in the stack. Among those, only 14 were detected. The full list of these chemicals with their corresponding emission rate and nature (i.e., gaseous or particulate) is presented in Table 1.

III. MODEL STUDY AREA DESCRIPTION

The impact area examined in the present application is the area enclosed by a 50-km radius around the power plant. The model study area was divided into 40

Table 1. Stack Air Emissions

Chemical	Emissions (g/s)	Nature
Arsenic	4.44×10^{-4}	P
Beryllium	ND	P
Cadmium	5.39×10^{-4}	P
Chromium (Total)	1.08×10^{-2}	P
Copper	4.12×10^{-3}	P
Lead	4.76×10^{-3}	P
Manganese	2.40×10^{-3}	P
Mercury	1.08×10^{-4}	P or G
Molybdenum	2.09×10^{-2}	P
Nickel	1.01×10^{-2}	P
Selenium	1.46×10^{-2}	P
Vanadium	ND	P
Ammonia	NS	G or P
Chlorine compounds	1.24	G or P
Cyanides	NS	G
Fluorine compounds	1.90×10^{-2}	G or P
PAHs	ND	P or G
Benzene	1.40×10^{-2}	G
Formaldehyde	ND	G
Toluene	2.54×10^{-3}	G

Note: NS — not sampled; ND — not detected; P — particulate; G — gaseous.

subregions that are defined by 8 angular sectors of 45° each, and 5 radial divisions at distances of 10, 20, 30, 40, and 50 km from the facility (Figure 1).

IV. PHYSICAL CHARACTERISTICS OF ENVIRONMENTAL MEDIA

A. Local Meteorology and Climatology

Winds in the area blow primarily from the west and southwest to the east and northeast. Climatological data from a climatological station in the area indicated a variation of the ambient temperature throughout the year between –4.5°C (January) and 21.3°C (July), with an annual average temperature of 8.7°C. Monthly precipitation (including snowfall equivalent) ranges between 6.2 (February) and 10.3 cm (August).

B. Hydrologic System

From a total of 95.5 cm of annual precipitation reaching the ground, it was estimated that 65.5 cm returns to the atmosphere through evapotranspiration, 16 cm recharges the area's groundwater system through infiltration, and 14 cm recharges the area's surface water system through overland runoff. At places of surface water-groundwater interconnections, the groundwater may be locally recharging the sur-

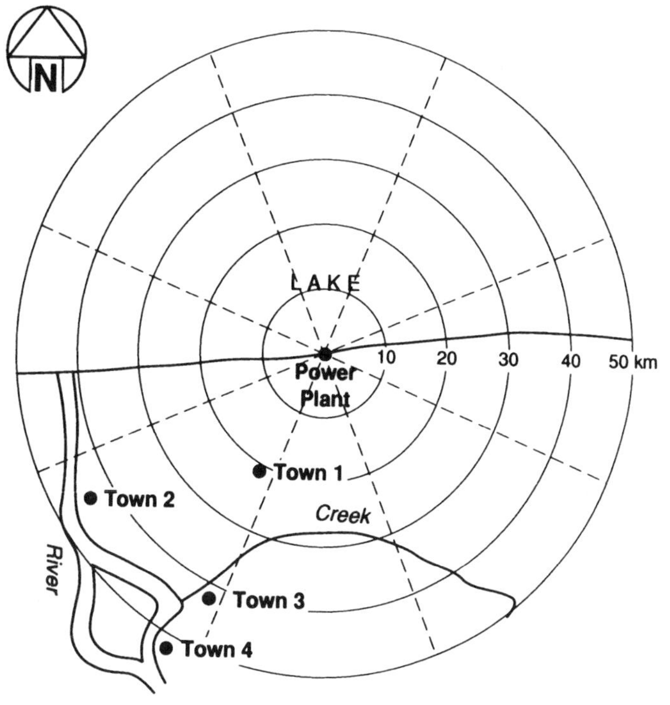

Figure 1. Model study area.

face water and vice versa, depending on the local relationship between surface water and groundwater.

The major surface water bodies in the area include a large lake, covering approximately the north half of the study area, a major river that stretches from south to north in the western part of the study area and drains into the lake, and a creek that stretches from east to west in the southern part of the study area and drains into the river.

The groundwater system consists of two parts: the unsaturated and the saturated zones. Based on the information collected, the unsaturated zone was estimated to extend to an average depth of 3 m below the ground surface. The values of porosity and intrinsic permeability of the unsaturated zone were estimated, through model calibration, to be 0.40 and 2.5×10^{-10} cm^2, respectively. The average saturated depth of the aquifer system was estimated to be approximately 7.5 m, with the groundwater flowing roughly toward the northwest, following the area's topography (from high elevation to low elevation). An average hydraulic gradient for the groundwater movement (i.e., the slope of the water table surface) was estimated to be 0.003. The values for porosity and hydraulic conductivity of the saturated zone (0.35 and 10^{-4} m/s, respectively) were taken from the literature to represent average properties of a sandy aquifer.

V. MATHEMATICAL MODELING OF THE FATE AND TRANSPORT OF CONTAMINANTS

A. Atmospheric Modeling

The modeling of the transport of chemicals in the atmosphere was performed with the help of the Industrial Source Complex-Long Term (ISC-LT) model. Inputs to the model consisted of source characteristics, emission data, and meteorological data in the form of statistical wind summaries. Ground-level pollutant concentrations in the air were calculated for a number of receptors placed on a polar grid with 10° increments, and at radial spacings ranging from 100 to 2000 m depending on the proximity to the power plant. The corresponding deposition rates were calculated through the use of deposition velocities for the cases of dry and wet deposition. It should be noted that deposition was only considered for particulate chemicals. Deposition of gases was not taken into account in this application because atmosphere/soil equilibrium is a complex phenomenon that cannot be simulated with a steady-state model.

A contour plot of the ground-level chromium (VI) air concentrations is presented as an example in Figure 2. The maximum zone of impact is located east of the power plant.

B. Overland Flow Modeling

The overland flow model used in this study is a subroutine incorporated into the Water Emission Risk Assessment model WTRISK. In this model, the total amount of pollutant deposited on the ground surface is distributed among different fates including interception on plant surfaces, overland runoff, and infiltration down the soil column, through the use of distribution coefficients that characterize the fraction of the contamination attributed to each of the above-mentioned fates. Pollutant interception on plant surfaces was considered insignificant, and the distribution coefficients for overland runoff and infiltration were chosen to reflect the same ratio as in the hydrologic balance (i.e., 0.47 and 0.53, respectively).

C. Surface Water Modeling

The modeling of the transport and fate of chemicals in surface water bodies was performed with the help of WTRISK. Input information to this model consisted of flow and geometric characteristics of the water bodies (i.e., point and distributed water discharges, flow velocities, and river cross-sectional areas), and pollutant loads.

The water bodies taken into account in the surface water analysis are a large lake, a river, and a creek. Even though the large lake was included in the analysis as a water body receiving runoff, and consequently contamination, from a large portion of the

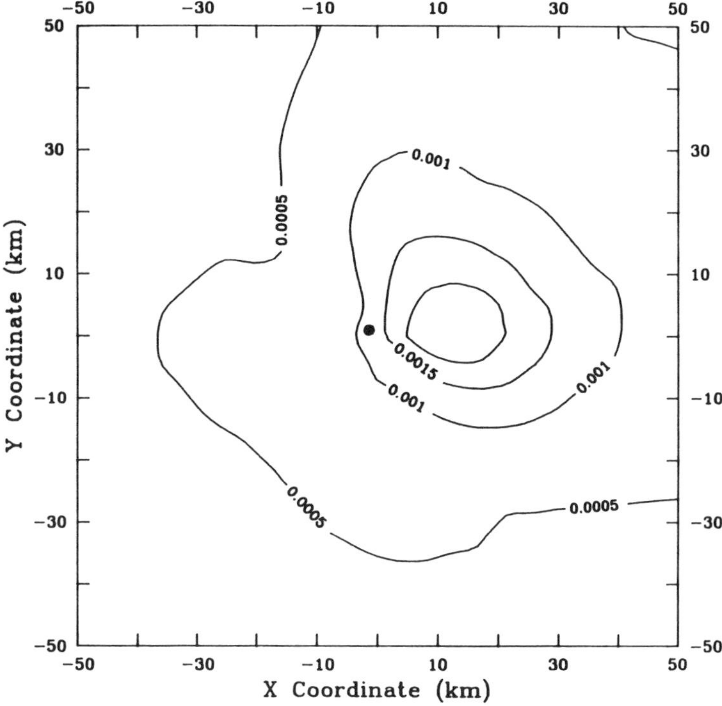

Figure 2. Chromium (VI) concentrations in air (ng/m^3).

study area, no actual fate and transport modeling was performed for the prediction of the resulting pollutant concentrations in the lake. Information acquired from telephone conversations with local county and city water resources departments indicated that the lake does not constitute a significant contributor to the area's water supply. A continuous river system composed of eight distinct river reaches (i.e., portions of the river with uniform flow and geometric characteristics) was considered in the analysis.

As an example, the resulting surface water concentrations of chromium (VI) ranged between 5×10^{-8} and 7.5×10^{-7} mg/L in the creek and between 2.2×10^{-9} and 4.3×10^{-9} mg/L in the river. For the health effect calculations, all the public water supply was considered to come from the last reach of the river.

D. Vadose Zone Modeling

The fate and transport modeling of pollutants in the vadose (unsaturated) zone was performed with the Seasonal Soil Compartment Model (SESOIL). Input information to the model consisted of climatological data, physical characteristics of the soil, chemical characteristics of the soil, and chemical characteristics of the 14 pollutants included in the risk assessment.

A 70-year simulation was performed, and an average surface soil concentration over the simulation period was calculated for the top 5-cm soil layer of each subregion. These soil concentrations were used by the food chain and health-effect models for the estimation of the pollutant concentrations in produce and the estimation of the human exposure dose due to ingestion of and dermal contact with the soil.

As an example, the average surface soil chromium (VI) concentrations in various subregions ranged between 9.5×10^{-5} and 9.0×10^{-4} mg/kg.

E. Groundwater Modeling

The fate and transport modeling of pollutants in the groundwater was performed with the Analytical, Transient, 1,2,3-Dimensional model (AT123D). Input information to the model consisted of pollutant waste release rates at the water table (calculated by SESOIL), and aquifer physical and flow characteristics. The resulting concentrations of all 14 chemicals in groundwater were very small [e.g., order of magnitude of 10^{-15} mg/L in the case of chromium (VI)]. Even though modeling of the fate and transport of pollutants in groundwater was performed, the resulting concentrations were not used anywhere in the risk assessment, since the information collected indicated that groundwater does not constitute a significant water supply source.

F. Food Chain Modeling

The fate and transport modeling of pollutants in the food chain was performed with the help of simplifying equations that are primarily based on the use of bioconcentration/bioaccumulation factors.

VI. EXPOSURE DOSE AND HEALTH EFFECT CALCULATIONS

For the calculation of the exposure dose, the EPA-recommended equations and parameter values were used.

The carcinogenic and noncarcinogenic health effect calculations were based on the use of cancer potency factors (CPFs) and reference doses (RfDs), respectively. For the selection of these values, priority was given to EPA established values. For some chemicals for which such values did not exist, the health effect parameter values were derived from Threshold Limit Values (TLVs) developed by the American Conference of Governmental Industrial Hygienists (ACGIH) or Permissible Exposure Limits (PELs) adopted by the Occupational Safety and Health Administration (OSHA). For benzene, the oral RfD was derived from the No Observed Adverse Effect Level (NOAEL) reported for a chronic inhalation study. For lead, the inhala-

Table 2. Maximum Carcinogenic Health Effects in the Study Area

	Risk (probability)						
	Ingestion		Inhalation		Dermal absorption		Total
Chemical	Absolute	%	Absolute	%	Absolute	%	absolute
Chromium (VI)	NA	NA	3.1×10^{-8}	100	NA	NA	3.1×10^{-8}
Arsenic	6.6×10^{-9}	41.8	9.0×10^{-9}	57.0	2.0×10^{-10}	1.2	1.6×10^{-8}
Cadmium	NA	NA	4.5×10^{-9}	100	NA	NA	4.5×10^{-9}
Benzene	2.6×10^{-13}	< 0.1	5.4×10^{-10}	100	0.0	0.0	5.4×10^{-10}
Total	6.6×10^{-9}	12.7	4.5×10^{-8}	86.9	2.0×10^{-10}	0.4	5.2×10^{-8}

Note: The chemicals in the above table are listed in the order of highest to lowest contribution to carcinogenic risk. %: Expresses the percentage of cumulative risk contributed by the specific pathway; NA: not applicable (i.e., chemical is not carcinogenic through specific pathway).

tion RfD was derived from the National Ambient Air Quality Standard (NAAQS). Dermal CPFs and RfDs were assigned the same values as the corresponding inhalation CPFs/RfDs, except where there was evidence that a chemical did not cause cancer via dermal exposure. The rationale behind this approach is that through the inhalation and dermal exposure pathways chemicals enter the blood stream directly, whereas through the ingestion pathway chemicals enter the gastrointestinal tract and the liver before entering the blood stream. In those organs, the chemicals could be transformed.

In the cases of chromium, nickel, chlorine compounds, and fluorine compounds, where chemical speciation is critical for the selection of dose-response values, the following compounds were considered: (1) total chromium: 5% as chromium (VI) and 95% as chromium (III) (i.e., not carcinogenic); (2) nickel: nickel salts or oxides (i.e., not carcinogenic); (3) chlorine compounds: hydrogen chloride; and (4) fluorine compounds: hydrogen fluoride.

The maximum chemical-specific and cumulative carcinogenic health effects in the study area with the contribution of the individual pathways to the total risk are presented in Table 2. The maximum chemical-specific and cumulative noncarcinogenic health effects in the study area with the contributions of the individual pathways to the total hazard quotient are presented in Table 3.

It should be emphasized that the model has not been fully tested yet; thus, the results presented in this report should be viewed as preliminary. Several alternative scenarios should be considered. For example, scenarios with different fractions of total chromium presented as hexavalent chromium and a scenario with nickel present as carbonyl or subsulfide compounds (i.e., carcinogenic) should be investigated.

VII. CONCLUSION

The application of a multimedia health risk assessment model to the actual emissions of a coal-fired power plant has been presented. It is clear from the results

Table 3. Maximum Noncarcinogenic Health Effects in the Study Area

	Hazard quotient						
	Ingestion		Inhalation		Dermal absorption		Total
Chemical	Absolute	%	Absolute	%	Absolute	%	absolute
Chromium (III)	2.9×10^{-7}	<0.1	2.4×10^{-2}	80.5	5.8×10^{-3}	19.5	3.0×10^{-2}
Nickel	1.1×10^{-5}	0.5	2.0×10^{-4}	10.0	1.8×10^{-3}	89.5	2.0×10^{-3}
Chromium (VI)	3.0×10^{-6}	0.2	1.3×10^{-3}	80.6	3.1×10^{-4}	19.2	1.61×10^{-3}
Hydrogen chloride	NA	NA	8.3×10^{-4}	100.0	0.0	0.0	8.3×10^{-4}
Molybdenum	4.2×10^{-4}	99.0	4.0×10^{-6}	1.0	1.6×10^{-7}	< 0.1	4.3×10^{-4}
Lead	1.0×10^{-4}	86.7	1.5×10^{-5}	13.0	3.2×10^{-7}	0.3	1.2×10^{-4}
Mercury	1.1×10^{-4}	98.5	1.7×10^{-6}	1.5	6.8×10^{-8}	< 0.1	1.1×10^{-4}
Cadmium	9.5×10^{-5}	99.2	7.2×10^{-7}	0.8	2.8×10^{-8}	< 0.1	9.6×10^{-5}
Selenium	4.9×10^{-5}	70.2	2.0×10^{-5}	28.7	7.6×10^{-7}	1.1	6.9×10^{-5}
Copper	3.9×10^{-5}	98.5	5.5×10^{-7}	1.4	5.9×10^{-8}	0.1	4.0×10^{-5}
Manganese	2.0×10^{-6}	6.4	2.8×10^{-5}	90.0	1.1×10^{-6}	3.6	3.1×10^{-5}
Hydrogen fluoride	1.6×10^{-9}	< 0.1	1.5×10^{-5}	100.0	0.0	0.0	1.5×10^{-5}
Arsenic	3.8×10^{-6}	46.9	4.2×10^{-6}	51.9	9.5×10^{-8}	1.2	8.1×10^{-6}
Benzene	1.1×10^{-9}	< 0.1	9.4×10^{-7}	100.0	0.0	0.0	9.4×10^{-7}
Toluene	3.1×10^{-11}	0.5	6.0×10^{-9}	99.5	0.0	0.0	6.0×10^{-9}
Total	8.4×10^{-4}	2.4	2.6×10^{-2}	74.9	7.9×10^{-3}	22.7	3.5×10^{-2}

Note: The chemicals in the above table are listed in the order of highest to lowest contribution to noncarcinogenic risk; %: expresses the percentage of cumulative risk contributed by the specific pathway.

of this application that, for some key chemicals such as arsenic and nickel, noninhalation pathways can contribute to a significant fraction of the health risk. These results suggest that a multimedia approach is needed for the analysis of the health effects of such chemicals.

REFERENCE

1. Constantinou, E. and Seigneur, C., Development of A Multimedia Health Risk Assessment Model, Proc. HAZMACON 91, Santa Clara, CA, April 16, 1991.

Chapter 6
Control Strategies and Applicable Technologies

Chair: Ian M. Torrens, EPRI
Co-Chair: Ramsay Chang, EPRI

The Behavior of Trace Elements During Coal Combustion and Gasification: An Overview

Lee B. Clarke

I. ABSTRACT

This paper provides an overview of the behavior of trace elements through coal combustion and gasification processes, and the influence of pollution control devices on trace element emissions. Trace elements are introduced into the combustion or gasification process primarily in coal, although some elements may be introduced in sorbents, additives, or fluxes.

Trace elements leave the combustor or gasifier either in the coal residues or in the flue gases. Elements in the flue gases may be present in the gas phase or bound on particulate aerosols. Gas cleaning devices, for SO_x and NO_x control, may have solid or liquid streams in which some elements from the flue gases are concentrated. Other constituents containing trace elements remain in the gas phase and are emitted into the atmosphere. The effective control of trace element emissions requires an understanding of their path through the various coal conversion processes and knowledge of the factors which control trace element distributions.

I. INTRODUCTION

Most naturally occurring elements have been detected in coal, at least in trace concentrations. Some of these elements are believed to be toxic to plant and animal life if present in sufficient quantities. Current environmental regulations (e.g., U.S. Clean Air Act and the FRG Air Quality Guidance — TA Luft 1986) already contain directives on the emission of trace elements to the atmosphere. More stringent regulations to be implemented in Europe, the U.S., and other countries will include strict limits for emissions of certain trace elements. It is important to assess the significance of each trace element in relation to coal combustion and gasification and the potential for emission of these trace elements to the environment.

The chemical and physical processes occurring during coal utilization are complex. Physical processes related to plant design, pollution control devices, turbulence, temperature profiles, and other parameters all affect the distribution of trace elements.

It is important to consider all relevant input and output streams from a plant. Input streams include the feed coal, fluxes or additive, sorbents for pollution control, and process water. Output streams consist of the coal residues from the combustor or gasifier, residues from pollution control systems, waste waters, and flue gases. Depending on the efficiency of particulate and pollution controls some trace elements may be emitted from the stack as particulates (aerosol bound) or as gases. This review concentrates on the path of trace elements through the plant, especially those emitted to the atmosphere via the stack, either as particulates or vapor phase elements. Controlling trace element emissions by increasing their concentration in other output streams may also cause environmental problems through, for example, increased leaching from residues, or additional waste water pollution.

A. Trace Element Partitioning

Trace elements are primarily introduced into the system in the feed coal as either inclusions in coal particles or discrete minerals and rock fragments. During combustion or gasification the particles undergo complex changes including the formation of char, agglomeration of melted inclusions, and vaporization of volatile elements. As the combustion or gasification products leave the combustor or gasifier and begin to cool, condensation of volatilized trace elements will occur. Figure 1 summarizes the formation of fly ash particles in coal combustion systems.

Studies have shown that trace elements can be classified into three broad groups according to their partitioning during coal combustion.[1,2] Trace elements show similar partitioning behavior between different gasifier streams.[3] Group 1 elements are concentrated in the coarse residues (combustor bottom ash or gasifier slag/ash) or are partitioned equally between coarse residues and particulates (combustor fly ash or gasifier particulates). They are typically lithophile elements. Group 2 elements are concentrated more in the particulates (such as fly ash) compared with coarse slag/ash. They are also enriched on the fine-grained particles which may escape particulate control systems. They are typically the chalcophile elements. Group 3 elements are concentrated in the vapor or gas phase and are depleted in all solid phases. Table 1 lists elements in each group for a coal-fired power station. Several elements, such as Cr, Ni, U, and V, often show partitioning behaviour intermediate between Group 1 and Group 2. Volatile elements, such as Se, may display partitioning behavior intermediate between Group 2 and Group 3.

Many authors have noted the enrichment of Group 2 elements with decreasing particle size.[1-5] During cooling volatile elements preferentially condense onto the surface of smaller particles in the flue gas stream. The enrichment occurs because these particles have a greater surface area to volume ratio. In both combustion and gasification systems the most volatile elements (such as Hg and the halogens) may remain in the gas phase during passage through the plant. In combustion systems oxides, such as As_2O_3, B_2O_3, and SeO_2, are important volatile species. In gasification systems trace elements which can form hydrides such as B_2H_4, SeH_2, AsH_3, and the transition elements which can form organometallic species such as Fe and Ni carbonyls may also volatilize readily.[6]

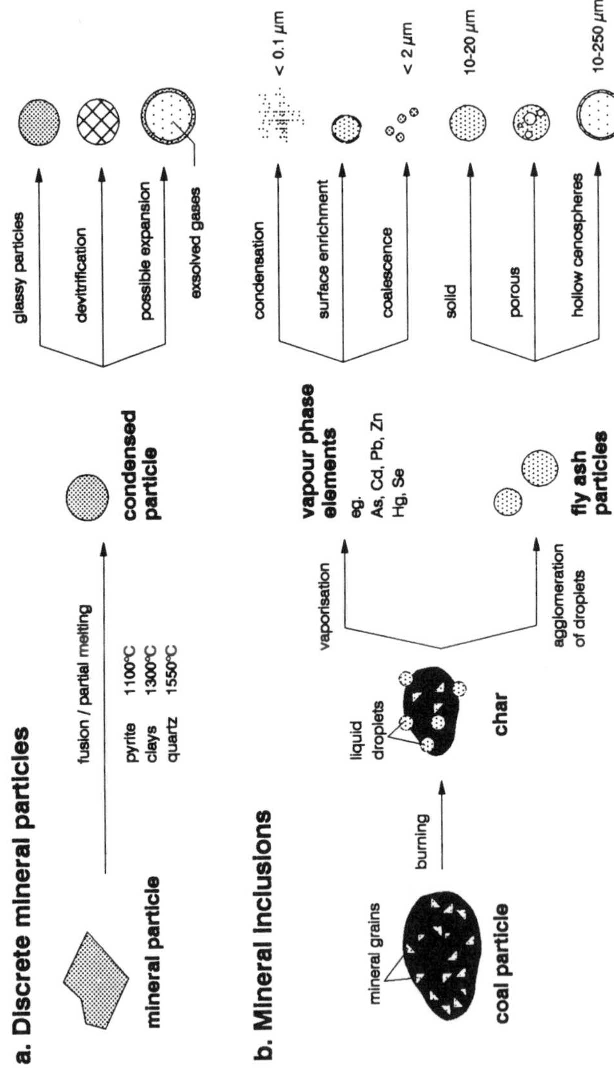

Figure 1. Mechanisms for the formation of fly ash.

Table 1. Partitioning of Trace Elements for the Gelderland Power Station, The Netherlands[16]

Group 1	Cs, Eu, Fe, Hf, K, La, Mg, Mn, Rb, Sc, Sm, Th, Ti
Group 2	Ba, Be, Ca, Co, Cr, Cu, Ge, Mo, Na, Ni, P, Pb, Sb, Sr, Tl, U, V, W, Zn
Group 3	B, Br, Cl, F, Hf, I, Se

II. TRACE ELEMENT BEHAVIOR IN COMBUSTORS

A. Pulverized Coal-Fired Power Plants

Coal combustion in utility power plants usually takes place in furnaces operating at temperatures above 1500°C. When the coal is burned, inorganic material is decomposed. A portion of the non-combustible material is retained in the furnace as either slag or bottom ash. The rest of the inorganic material exits in the flue gases as fly ash and vapor.

In a typical coal-fired power station, coal is injected into the furnace and ignited while in suspension. As the coal particles are heated, volatile matter, including pyrolyzed organic species, are vaporized and combustion occurs. A portion of the pulverized coal may be composed of free mineral matter leading to variable and quite different local conditions.[7] During combustion the mineral matter in the coal is exposed to rapid heating and high temperatures. Under these conditions minerals undergo thermal decomposition, fusion, disintegration, and agglomeration (Figure 1). It is possible that some larger mineral particles may only be partly melted, because of the short residence time in the high-temperature zone, resulting in some fly ash particles with nonspherical shapes. Mineral particles with high melting points may escape melting regardless of size. The main formation mechanism of the coarser particles (2 mm) is carryover of a proportion of the mineral material in the feed coal. Some involatile species may be transferred to the vapor phase because of an intimate association with organic material, while some potentially volatile trace element species may be trapped within melted material.[8]

As the combustion products leave the furnace and begin to cool, condensation of the volatilized trace elements will occur. Condensation will be complete for most elements before the flue gases reach the particulate control devices and the gas temperatures have decreased to about 350°C. Condensation may be delayed for volatile species with low boiling points or because of slower cooling of ash particles compared with the flue gas.[8] Elements that form chemical compounds with high vapor pressures may remain in the gas phase and pass through gas cleaning systems (e.g., Hg and Se). Figure 2 shows the relative distribution of trace elements between residues and flue gases.

B. Fluidized Bed Combustion (FBC)

The combustion characteristics in atmospheric FBC (AFBC) systems, operating at 750 to 900°C, are different from those in conventional pulverized coal-fired

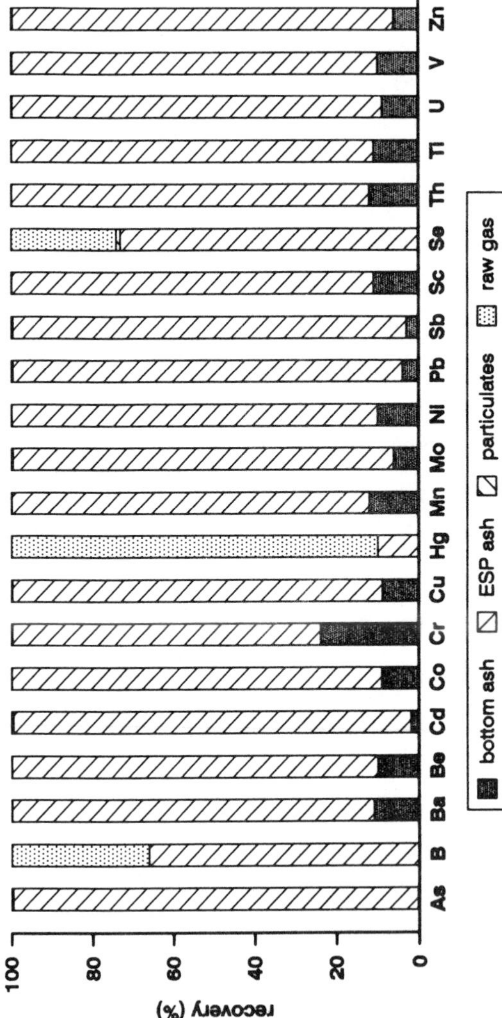

Figure 2. Relative distribution of trace elements between residues and flue gases (data from Gelderland coal-fired power station, The Netherlands).[16]

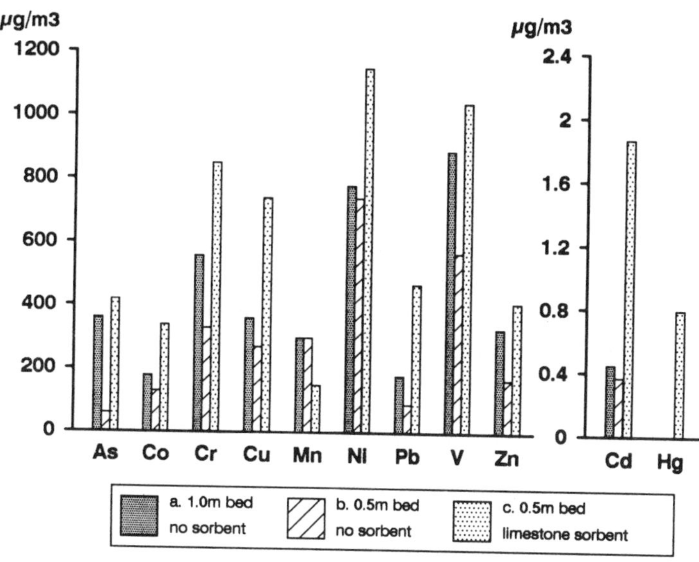

Figure 3. Trace element emissions from an experimental AFBC system.[10]

combustion. The emission of trace elements from AFBC units is expected to be reduced because volatilization of elements occurs to a lesser extent at the lower operating temperatures. This reduction could be partially offset by the longer residence times at a relatively high temperature in the fluidized bed, allowing more volatilization to occur.[9] Particulates that pass through flue gas cleaning systems in AFBC facilities are similarly expected to contain a lower concentration of trace elements compared with fines emitted from pulverized coal-fired systems.

A study carried out in the U.K. using an experimental scale 2.3-MW_t AFBC boiler has indicated that changing operating conditions can control trace element emissions (Figure 3).[10,11] It has been found that a reduction in the depth of the fluidized bed decreases emissions of selected trace elements by between about 5 and 50%. The higher rate of particulate emissions (and hence trace element emissions) for deeper beds is attributed to increased attrition of ash. Addition of limestone to the bed to control emissions of sulfur dioxide has the negative effect of increasing some trace element emissions by up to six times (e.g., Pb, Cd, and Mn) compared with earlier tests without sorbent. There was a corresponding increase in particulate emissions (by 3.5 times), and greater trace emissions may be explained by increased fine particulate loading when limestone is added, and by an increase in the bed height.[10] It has also been suggested that the limestone or dolomite sorbent may contribute to the trace element content.

In mass balance studies carried out on an experimental 4-MW_t AFBC system in The Netherlands, Hg was the only trace element detected in the gas phase in flue gases.[12] About 50% of the Hg present in the coal and limestone was emitted in the gaseous phase. Elements normally considered to be volatile (such as As, Sb, Se, and Pb) were not detected in the gas phase. The concentration and surface area of the fly ash particles in the colder parts of the system (convection system, cyclones, fabric filters) were sufficient to effect complete condensation of the more volatile elements.[12] However, for As, Cd, Cu, Hg, Pb, and Sb there were large enrichments in very fine particulates collected in sampling devices in the AFBC system stack downstream from the particulate control devices.

Most measurements of trace element emissions from fluidized bed systems have been from atmospheric plants, although some tests have been carried out on experimental pressurized FBC (PFBC) test rigs.[9] Recent studies have been carried out in Finland, using peat and coal as fuels, with sand or ash and limestone as bed materials, at operating temperatures of 800 to 900°C and pressures of up to 1 MPa. Trace element concentrations were generally found to be higher in ash streams consisting of finer particles, with levels in filter particulates > cyclone particulates > bed offtake. Enrichment is caused by the larger specific surface area (for a given volume) of smaller particles, which allows greater condensation or surface adsorption of volatile elements. Measurements of trace elements in the flue gases indicate that only a small proportion of the coal-bound trace elements left the PFBC system in the vapor phase: As <4%, Cd <5%, Cu 0.5 to 1.2%, Hg 7.5 to 16.0%, Pb 1 to 3%. Even at the lower temperatures in fluidized bed combustion (typically around 800°C) it is anticipated that most of the volatile elements would enter the vapor phase. The low concentrations of trace elements in the vapor phase have been attributed to adsorption of volatilized elements by PFBC ash particles.[9] The particles are crystalline and irregular in shape, compared with fly ash particles from conventional pulverized coal-fired combustion systems which are typically glassy and spherical, and thus the PFBC particles have a higher adsorption potential for vapor phase constituents. The PFBC tests also indicate that, for those elements investigated, there is little difference between trace element emissions with or without limestone addition. Changes in operating pressure have little effect on the distribution of the heavy metals studied.

III. TRACE ELEMENT BEHAVIOR IN GASIFICATION SYSTEMS

Trace elements are introduced into the gasification process in the fuel coal, apart from a few elements introduced in additives or fluxes (such as Mn and Mg), or from the degradation of gasifier lining refractories (e.g., Cr). The distribution and behavior of trace elements vary between individual gasification processes, but are broadly similar, and correspond to trace element behavior within conventional power plants. The trace elements may be distributed to one of three exit streams: they may be retained in the gasifier slag/ash, they may leave with particulates, or volatilized substances may exit with flue gases.

Results of analytical investigations carried out on gasifier slag and particulates also demonstrate the correlation of enrichment of certain trace elements, especially

As, B, Cd, Cu, Pb, Sb, and Zn, with decreasing particle size.[13] Volatile elements, such as As and Se, may become greatly enriched as the particle size decreases. The highly volatile elements will not condense out easily. Many gasifier systems currently under development feature wet scrubbers for final gas cleaning. These systems can control both particulates and halogen elements very effectively.

V. POST-COMBUSTION POLLUTION CONTROL

A. Particulate Control Systems

The fraction of the total ash reaching the pollution control devices depends on plant design. In some plants bottom ash may account for a significant fraction of the ash; in other plants it may be negligible. In most power stations >99% of the particulates are removed from the flue gases using electrostatic precipitators (ESP), fabric filters (baghouses), or wet scrubbers. This implies that most (>95%) of the trace elements are also removed. Although precipitators, baghouses, and scrubbers are all capable of high overall collection efficiencies, it is generally true that they are least effective for smaller particle size range (<5 mm).[5] Thus, a small proportion of the fly ash penetrates the particulate control systems and reaches the atmosphere with the stack gases.

The composition of the finer particles varies considerably, depending mainly on the properties of the mineral matter in the coal and on the combustion conditions. The submicron particles are of concern because they have been found to be enriched in certain trace elements, especially at the particle surface. There is a bimodal size distribution for the smaller fly ash particles.[8,14] In a study on particulate aerosols the mean diameters of the fine and coarse modes, collected after an ESP, were about 0.05 and 2 mm, respectively.[14] About 5% of the total mass of aerosol was concentrated in fine mode and about 95% in the coarse mode. The matrix-forming elements (Al, Fe, K, Mg, Na, Si, Ti) and the trace elements Mn and Zn were found to follow the distribution of total mass, with about 5% of these elements found in the fine-mode particles. The trace elements Cd, Cu, Pb, Sr, and V (together with Ca and S) were enriched in the fine-particle mode. About 34% of the Cd, 22% of the Cu, 9% of the Pb, 11% of the Sr, and 23% of the V were found in the submicron (<1 mm)-sized particles.

Fabric filter systems (baghouses) have a similar particulate removal efficiency to ESP. Both systems are more efficient at retaining the finer particles (<1 mm) compared with cyclone filters and venturi scrubbers. Some fluidized bed and gasification systems have been developed which use one or more cyclones for initial particulate removal (and sometimes recycle) followed by final gas cleaning using baghouses, wet scrubbers, or other filter methods. Experimental filter systems have been developed to remove fly ash and certain trace elements from hot flue gases. Granular-bed filter systems using sorbent particles (such as limestone) as the filter media are reported to effectively remove volatilized trace elements from hot flue gases.[15]

Particulate recycling has been employed in some gasification systems in order to deliver more residue as slag, and this may alter the trace element distribution. As the quantity of particulates decreases, more of the volatile trace elements are bound in the slag. Figure 4 shows the distribution of trace elements between residues and raw gas in a slagging gasifier with and without particulate recycling. As recycling increases Co, Cr, Cu, and Ni become enriched in the slag; Pb, Sn, and Zn become enriched on the surface of material captured as filter cake; and As, Cd, Hg, and Sb become enriched in the raw gas as volatile material. By increasing the production of slag some potentially environmentally harmful trace elements are bound into a glassy solid, which is more resistant to leaching;[3] however, increased trace element contents in the gas phase may present additional problems. By recycling particulates most of the trace elements end up in the gasifier slag, including many of the more volatile elements. Analyses of gasifier slags and feed coals have shown that most of the trace elements are passed to the slag; e.g., 95% of the Co, Cr, Mn, Se, Sr, Ti, and Th are retained in the slag.[6] Recoveries of Cd and Sb are lower, although concentrations are so low that mass balancing becomes inaccurate. Volatile trace elements may be locked into complex inorganic compounds, and thus concentrated in the slag.[3]

B. Flue Gas Desulfurization (FGD) Systems

The emission of trace elements from a coal-fired power station fitted with wet FGD plant has been studied in the Netherlands.[16] Trace elements are introduced into the FGD system partly by the flue gases, but primarily in the limestone used for desulfurization. The flue gases are the main source of volatile trace elements. Gaseous compounds condense on particulate matter in wet FGD systems. The particulate matter consists of fine fly ash particles (approximately 40%), gypsum particles (approximately 10%), and vapor droplets (approximately 50%).[16] The emissions depend on the behavior and operating conditions of the demisters. Trace metals leave the FGD system mainly in the gypsum and sludge. Most of the volatile elements are removed in the waste water effluent and only a few remain in the flue gases (Figure 5). The distribution between of trace elements between sludge and waste water depends on operating conditions, especially pH.

C. NO_x Control Systems

It has been noted that in certain combustion processes deactivation of selective catalytic reduction (SCR) systems occurs, caused by As poisoning of DeNOx catalysts.[17] Significant poisoning only occurs if As is in the gas phase, and is caused by it entering the catalyst pores, resulting in deactivation. Problems with As poisoning of catalysts are most pronounced in boilers operating with fly ash recirculation, designed to discharge residues as granulates. Whereas the catalysts in a dry bottom furnace plant are only exposed to the concentration contained in the fuel, ash recirculation and remelting in wet-bottom furnaces increase the arsenic oxide content

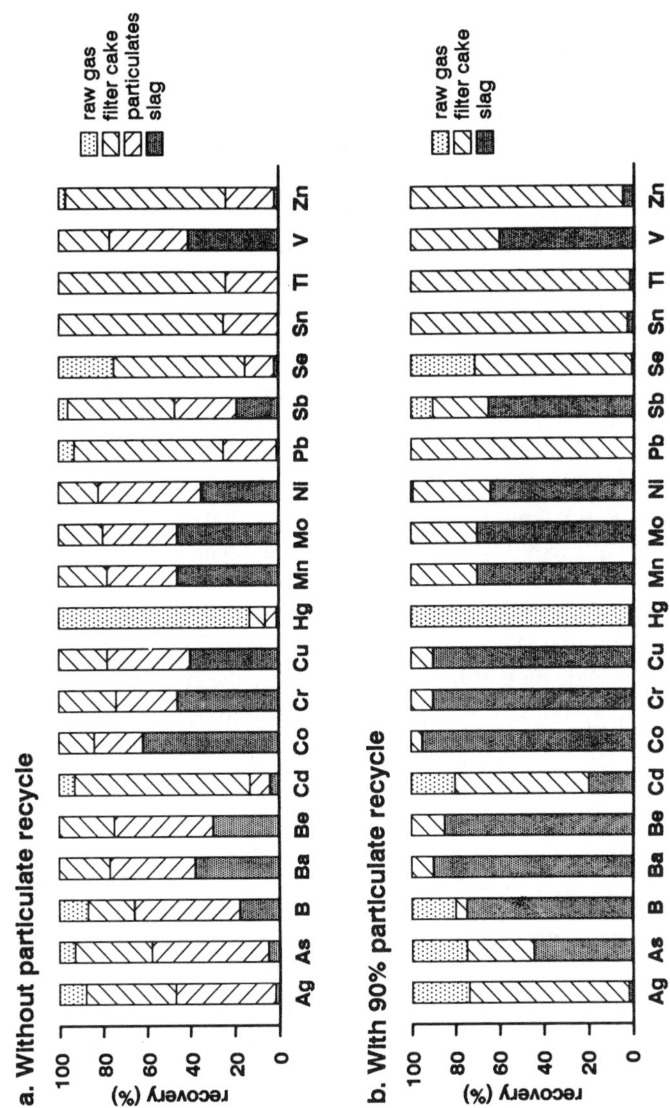

Figure 4. Relative distribution of trace elements between residues and flue gases in a slagging gasifier system (PRENFLO process).[13]

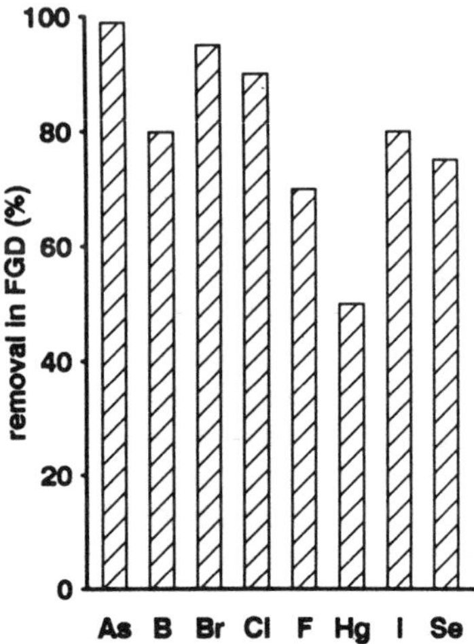

Figure 5. Removal of volatile elements in wet FGD systems (based on average values for coal-fired power stations in The Netherlands).[19]

in the flue gases by up to 15 times.[18] Even in dry-bottom boilers, As may still contribute to deactivation of catalysts if the conditions are unfavorable. Adsorption of gaseous As_2O_3 upstream from the DeNOx system will reduce poisoning. It has been demonstrated that As poisoning of catalysts can be effectively reduced by maintaining the CaO content of coal ash at >3%, and if possible at >5%, if necessary by adding limestone.[17] An admixture of limestone can reduce the As content of the flue gases from 1000 mg/m^3 to less than 100 mg/m^3, which although positive still results in premature deterioration of the catalysts. New As-resistant catalysts are under development.[18]

VI. CONCLUSIONS

Coal is the main source of trace elements in coal-fired combustion or gasification systems, although some elements are introduced in sorbents, additives, or fluxes. Trace elements are partitioned into several output streams, including solid coal residues, residues from pollution control systems, waste waters, and flue gases. Most trace elements are associated with the particulates. Volatile elements preferentially condense onto the surface of smaller particles in the flue gas stream. The emissions of these trace elements thus depend more on the efficiency of the gas cleaning system than upon the method of firing. In both combustion and gasification systems the most volatile elements may remain in the gas phase, although most of these (As, B, Se, halogens) are effectively removed from the flue gases in wet FGD systems. Only Hg passes through the systems in significant amounts (typically with about 50% removal in FGD systems).

REFERENCES

1. Smith, I. M., Trace Elements from Coal Combustion: Emissions, IEACR/01, IEA Coal Research, London, 1987.
2. Clarke, L. B. and Sloss, L. L., Trace Elements from Coal Combustion and Gasification, IEACR/49, IEA Coal Research, London, 1992.
3. Clarke, L. B., Management of By-Products from IGCC Power Generation, IEACR/38, IEA Coal Research, London, 1991.
4. Chadwick, M. J., Highton, N. H., and Lindman, N., The environmental significance of trace elements from coal combustion and conversion processes, in *Environmental Impacts of Coal Mining and Utilization*, Pergamon Press, Oxford, 1987, 171.
5. Germani, M. S. and Zoller, W. H., Vapor-phase concentrations of arsenic, selenium, bromine, iodine, and mercury in the stack of a coal-fired power plant, *Environ. Sci. Technol.*, 22, 1079, 1988.
6. Beishon, D. S., Hood, J., and Vierrath, H. E., The fate of trace elements in the BGL gasifier, 6th Annu. Int. Pittsburgh Coal Conference, Pittsburgh, PA, September 25–29, 1989; Greensburg, PA, Pittsburgh Coal Conference, MEMS, 1989, 539.
7. Wibberley, L. J. and Wall, T. F., An investigation of factors affecting the physical characteristics of fly ash formed in a laboratory scale combustor, *Combust. Sci. Technol.*, 48, 177, 1986.
8. Smith, R. D., The trace element chemistry of coal during combustion and the emissions from coal-fired plants, *Prog. Energy Combust. Sci.*, 6, 53, 1980.
9. Mojtahedi, W., Nieminen, M., Hulkkonen, S., and Jahkola, A., Partitioning of trace elements in pressurised fluidised bed combustion, *Fuel Proc. Technol.*, 26, 83, 1990.
10. British Coal, Coal Research Establishment, Trace element emissions from fluidised bed combustion units, Final Report, Technical Coal Research, EUR 11160, Commission of the European Communities, Luxembourg, 1987.
11. Hughes, I. S. C. and Littlejohn, R. F., Emissions of trace elements from coal-fired industrial boilers, *Int. J. Energy Res.*, 17, 473, 1988.
12. van Haasteren, A. W. M. B., *Fate of Trace Elements in Coal during Fluidised Bed combustion of Coal*, TNO-86-274, Netherlands Organisation for Applied Scientific Research (TNO), Apeldoorn, The Netherlands, 1987.
13. Hufen, K., Origin and properties of slag and fly ash obtained in the PRENFLO process, paper presented at International Energy Agency Expert Meeting on the Use of Coal Gasification Slag, Arnhem, The Netherlands, May 10–11, 1990.
14. Kauppinen, E. I. and Pakkanen, T. A., Coal combustion aerosols: a field study, *Environ. Sci. Technol.*, 24, 1811, 1990.
15. Mojtahedi, W. and Mrgueh, U. M., *Trace Elements Removal from Hot Flue Gases*, VTT-TUTK-663, Valtion teknillinen tutkimuskeskus, Espoo, Finland, 1989.
16. Meij, R., Tracking trace elements at a coal-fired power plant equipped with a wet flue-gas desulphurisation facility, *Kema Scientific & Technical Reports*, 7, N.V. KEMA, Arnhem, The Netherlands, 1989, 267.
17. Gutberlet, H., Influence of furnace type on poisoning of DENOX catalysts by arsenic, *VGB Kraftwerkstechnik*, 68, 264, 1988.
18. Balling, L. and Hein, D., DENOX catalytic converters for various types of furnace and fuels — development, testing, operation, in Joint Symposium on Stationary Combustion Nitrogen Oxide Control, EPRI-GS-6423-V2, San Francisco, March 6–9, 1989, Electric Power Research Institute, Palo Alto, CA, 1989.

19. Meij, R., The fate of trace elements at coal-fired plants, in Preprints of European Seminar on the Control of Emissions from the Combustion of Coal, London, February 18–29, 1992; Commission of the European Communities (Directorate General for Energy), Brussels, Belgium, 1992.

Controlling Trace Species in the Utility Industry

Frank B. Meserole
Winston Chow

I. INTRODUCTION

New controls for air toxics from industrial sources are called for as a result of the passage of the 1990 Clean Air Amendments. The targeted species are identified in a list of 189 hazardous air pollutants, which includes a number of heavy metals. Regulations regarding additional control of air toxics from electric utilities will be based on the results of studies to be conducted by the U.S. Environmental Protection Agency and the U.S. National Institute of Environmental Health Sciences. A 3-year period was specified for the evaluation of air toxics with the exception of mercury for which a 4-year time period was allocated.

The Electric Power Research Institute commissioned Radian Corporation to compile a database of information regarding the chemical composition of the various process streams associated with conventional fossil-fueled power plants. The resulting PISCES database contains nearly 80,000 individual records in which more than 390 chemical substances have been identified. This paper focuses on the properties that affect the removal of trace elements in the flue gas cleaning equipment including electrostatic precipitators (ESPs), wet scrubbers, and fabric filters.

Most of the trace elements that are associated with a coal- or oil-fired power plant enter with the fuel. For coal-fired power plants most of these elements are removed with the bottom ash or collected ash, e.g., precipitator ash. However, varying amounts of trace elements are released into the atmosphere with the flue gases. The amount that is emitted is dependent on several factors including the concentration in the fuel, the type of fuel, the size of the boiler, the efficiency of the control device, the distribution between vapor and condensed phases, and the enrichment characteristics with respect to ash particle size.

II. TRACE ELEMENT PARTITIONING

Many of the trace elements in coal are partially or totally vaporized during the combustion process. The degree of vaporization and condensation will determine how a particular species is partitioned between the bottom ash, fly ash, and flue gas. Vaporization in the boiler will tend to deplete the concentration of a species in the bottom ash relative to the coal concentration expressed as a fraction of the coal mineral matter.

Some of the trace elements are partitioned preferentially to the smaller size fractions of fly ash particles due to the condensation of the vaporized fraction onto the fly ash surfaces. Therefore, the removal efficiencies for these elements will, generally, be less than that of the overall particulate removal in an ESP or wet venturi scrubber. This results from the lower removal efficiencies of these devices for submicron-sized particles. This effect is less pronounced with fabric filters.

The trace element data from the PISCES database were analyzed to determine the degree of vaporization that was indicated for each of the reported sample sets. Because of the potential uncertainties in the reported coal and ash trace element data, a consistency check was made to screen the available data. This evaluation consisted of a comparison of the concentration of each trace element of interest in the coal to that of the bottom ash, collected ash, and emitted. This restricted the useable data sets to those where trace element data were reported for coal, bottom ash, and collected ash from samples collected over a common time period. Of the approximately 80 plants in the database that have stack data reported, the consistency requirement limited the number of data points per element to about 30. Of these 30, typically another 10% were rejected as outliers from the general set of data.

The coal equivalent values were calculated using the following equation:

$$CE_i = \left(X_{BA} \cdot [BA]_i + X_{CA} \cdot [CA]_i + X_{EA} \cdot [EA]_i\right) / X_{Ash} \tag{1}$$

where CE_i = coal equivalent concentration of element, i, associated with the combustion ashes; X_{BA}, X_{CA}, X_{BA} = the weight fraction of bottom ash, collected ash, and emitted ash, respectively; $[BA]_i$, $[CA]_i$, $[EA]_i$ = the concentration of element, i, in the bottom ash, collected ash, and emitted ash, respectively; and X_{Ash} = weight fraction of ash in the coal.

The data consistency check consisted of calculating the fraction of each trace element retained in the bottom and collected ashes and comparing that sum to the amount in the coal. The total amounts in the ashes was calculated and reported as an equivalent coal concentration which was correlated to the coal concentrations for all of the appropriate data sets. This comparison provided an evaluation method to select data sets of sufficient quality to perform subsequent analysis. Typical results of this type of analysis are shown graphically in Figures 1 and 2.

The mass balance results for chromium are shown in Figure 1. For most of the data sets the sum of the chromium in the ashes equals the amount in the coal within 25%. A linear regression analysis of these data gives a slope of 1.0 which indicates that essentially all of the chromium entering with the coal is distributed to the various ashes. This evaluation shows a couple of data sets that deviate substantially from the general population. These outliers were omitted from subsequent calculations. The other example shown here (see Figure 2) is the mercury results. The ash data are consistently low indicating that most of the mercury remains in the vapor phase across the ESPs with typically less than 10% of the mercury associated with the

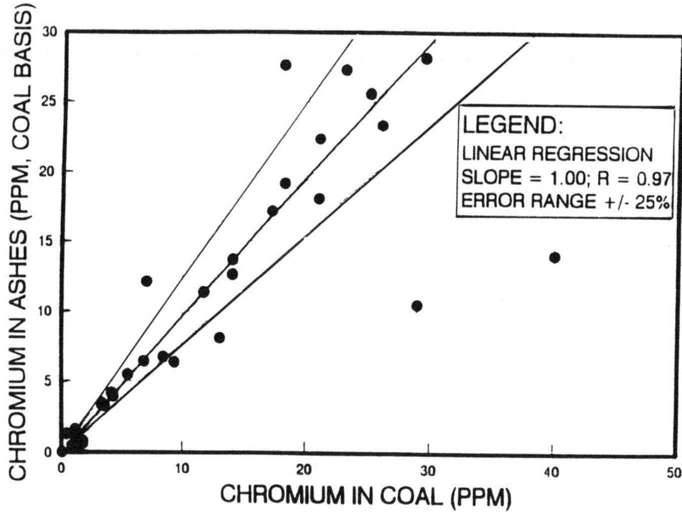

Figure 1. Determination of chromium phase distribution.

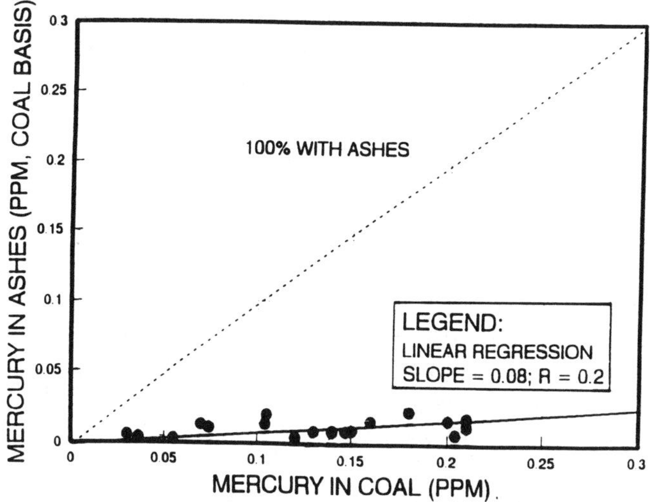

Figure 2. Determination of mercury phase distribution.

solids. It is interesting that, while it is recognized that mercury measurements in coal are difficult, there is a consistent pattern even though the data are from more than 15 different studies.

The fraction of an element vaporized in the boiler can be estimated by ratioing the concentration in the bottom ash to the concentration in the coal expressed on an

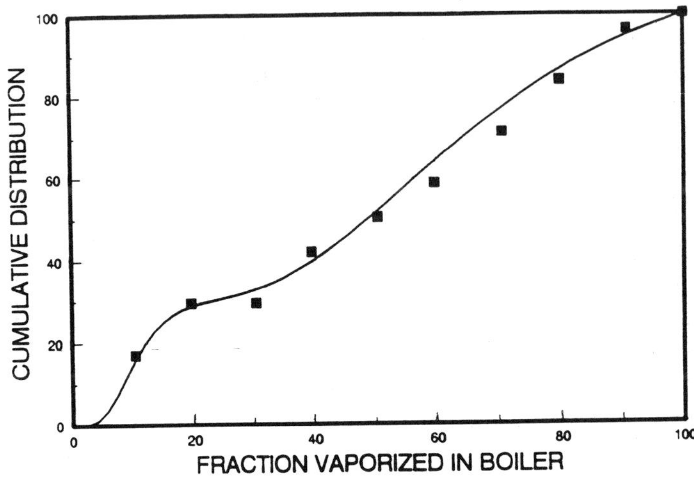

Figure 3. Population distribution of the fraction of cadmium vaporized.

ash equivalent basis. This approach is based upon the assumption that the bulk composition of the bottom ash and fly ash are the same as these ashes leave the boiler and that no significant amount of vaporized material is adsorbed or condensed onto the bottom ash. Using this method the fraction vaporized is calculated as follows:

$$FV_i = 1 - \frac{[BA]_i}{[C]_i / X_{Ash}} \tag{2}$$

where FV_i = the fraction of element, i, vaporized; $[BA]_i$ = the concentration of element, i, in the bottom ash; $[C]_i$ = the concentration of element, i, in the coal calculated based on the ash content; and X_{ash} = the fraction of ash in the coal.

The fraction vaporized in the boiler varies widely for the partially vaporized elements over the range of sample concentrations included in the PISCES database. Cumulative distributions of the different levels of vaporizations calculated for cadmium and nickel are shown in Figures 3 and 4.

Vaporization of a trace element, partial or complete, and the subsequent condensation or adsorption onto fly ash in the flue gas downstream of the boiler as cooling takes place results in enrichment on the surface. This condensation process also leads to an increase in concentration in the fly ash with decreasing particle size. Assuming that the condensation occurs exclusively on the available ash surface, the concentration as a function of size can be represented by the following equation:

$$C_T(D) = C_B + \frac{6C_s}{\rho D} \tag{3}$$

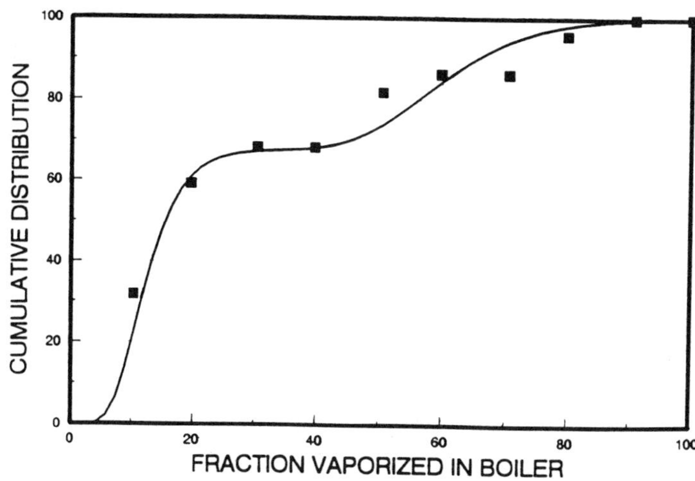

Figure 4. Population distribution of the fraction of nickel vaporized.

where $C_T(D)$ = the total concentration of a given species in fly ash as a function of particle size; C_B = the species concentration in the silicate matrix of fly ash; C_s = the species concentration on the ash surface; and D = the diameter of a given size fraction of ash.

This relation implies that a graph of concentration plotted against the reciprocal of the diameter will give a straight line if this model correctly describes the condensation mechanism. Furthermore, the slope is directly proportional to the surface concentration term and the intercept gives the concentration of that element in the bulk ash matrix. Figures 5 and 6 give two examples of size-dependent data obtained by chemically analyzing the various size fractions obtained by collecting fly ash downstream of an ESP using a multistage cascade impactor.

III. TRACE ELEMENT REMOVAL

Most of the particulate control devices presently in use on coal-fired power plants in the electric utility industry are ESPs. Other such devices include fabric filters and wet venturi scrubbers. The bulk of the trace element removal data available in the literature pertain to ESP. Therefore, the emphasis of this paper will be on the evaluation of the performance of ESPs with respect to trace element removal. However, some information concerning trace element removals of the other control systems will also be discussed.

The fly ash emitted from a coal-fired boiler will vary in particle size distribution depending primarily on the type of boiler. Three types that are most common are cyclone-fired boilers and tangentially fired and wall-fired pulverized coal (PC)

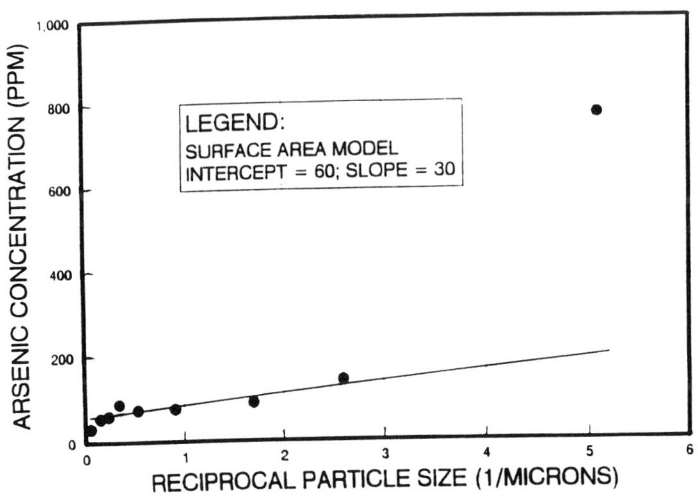

Figure 5. Concentration of arsenic by particle size in emitted ash.

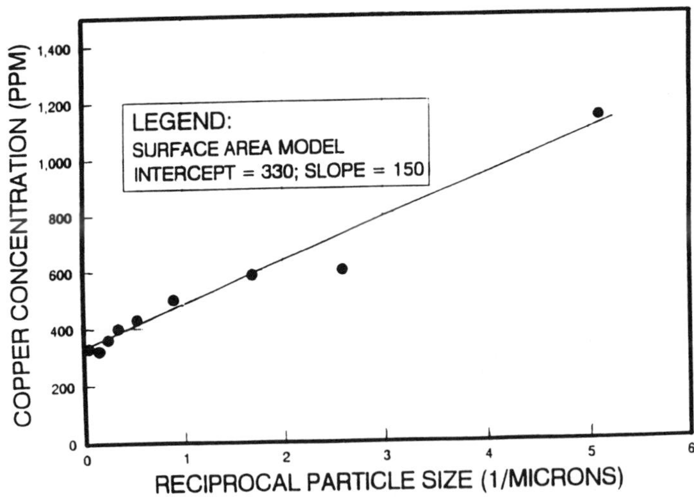

Figure 6. Concentration of copper by particle size in emitted ash.

boilers. The mean size fly ash particles from these boilers ranges from 5 to 10 μm for cyclones to 10 to 20 μm for PC boilers. The size distribution can be roughly represented using a log normal size distribution.

The size distribution of fly ash at the outlet of a collection device will depend on the removal characteristics of that device as a function of particle size. The removal efficiencies for both the ESPs and wet venturis decrease substantially for particles in the submicron-size ranges, while the removal efficiency for fabric filters is much more uniform over the size range.

The removal of trace elements across particulate control devices and SO_2 control systems will vary as a function of the distribution of the trace species between the vapor and condensed phases. For example, the particulate collection devices will be most effective on the condensed fractions while gas scrubbing system may be more effective in removing the vapor phase portion.

In order to predict the effect of the overall removal efficiency for control devices such as ESPs or wet venturis, the product of the inlet particle size distribution (PSD) and size-specific removal efficiency must be integrated over the particle size range. This can be represented mathematically as

$$P^o(D) = \int P^i(D) \cdot RE(D)dD \tag{4}$$

where $P^o(D)$ = the PSD at the control device outlet; $P^i(D)$ = the PSD at the inlet; and RE(D) = the particulate removal efficiency by size.

The removal efficiency, RE(ash), of the ESP towards the total ash is given by

$$RE(ash) = \frac{\int P^o(D)dD}{\int P^i(D)dD} \tag{5}$$

This calculation is equal to the total mass of ash entering the control device divided by mass of ash leaving. Combining the PSD data and the size dependence as shown in Equation 3, the removal efficiency by element can be calculated at various ESP performance levels using the following relation:

$$RE_i = \frac{\int P^o(D) \cdot C_T(D)_i dD}{\int P^i(D) \cdot C_T(D)_i dD} \tag{6}$$

where RE_i = the removal efficiency of element, i, and $C_T(D)_i$ = the size-dependent concentration of the fly ash for element, i.

The removal efficiencies across an ESP, wet venturi, and fabric filter were calculated for a model trace component as a function of the fraction vaporized in the boiler. It was assumed that all of the vapor phase fraction condensed on the fly ash prior to entering the control device. Removal efficiencies of the various control devices are shown in Figure 7 as a function of particle size. The coal concentration of this species was taken to be 20 mg/kg and the ash content of the coal to be 10 wt%. The results of these calculations for an ESP are shown in Table 1. The removal efficiencies as compared to the overall ash removal decrease with increasing levels

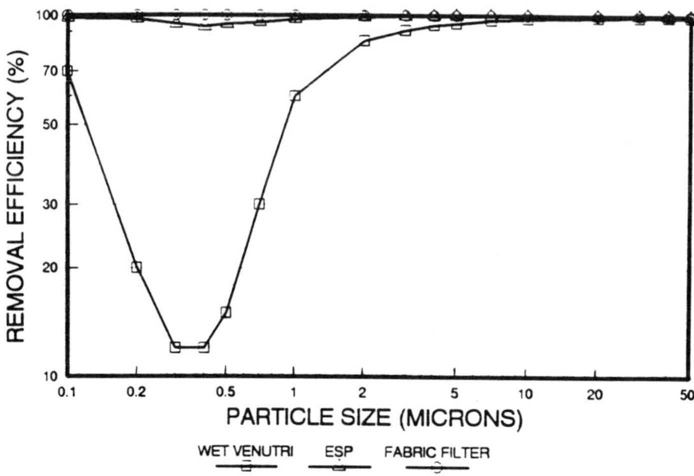

Figure 7. Particle removal performance as a function of particle size.

Table 1. Effects of Fraction Vaporized on Removal Efficiency across an ESP[a]

	FV (0%)[b]		FV (40%)		FV (90%)	
Ash removal efficiency (%)	**RE[c] (%)**	**Conc[d] (mg/kg)**	**RE (%)**	**Conc (mg/kg)**	**RE (%)**	**Conc (mg/kg)**
92	92	200[e]	89	270	85	370
98	98	200	97	370	94	610
99	99	200	98	430	96	760
99.6	99.6	200	98.9	560	97.9	1070
99.8	99.8	200	99.4	580	98.9	1110

[a] Calculations assume species is vaporized in boiler and fully condenses onto the fly ash prior to entering the ESP.
[b] Fraction vaporized in boiler.
[c] Species-specific removal efficiency across ESP.
[d] Species-specific concentration of ash emitted from ESP.
[e] Concentrations based on a coal of 10% ash and 20 mg/kg of species of interest.

of vaporization. Furthermore, the concentrations in the emitted ash increase with both the degree of vaporization and the removal efficiency. A comparison of the results of all three control devices are shown in Figure 8. The emission rates are given for a nominal 500-MW boiler for a compound at 20 mg/kg in the coal and vaporized at the 90% level.

IV. CONCLUSIONS

The PISCES database is a useful source of historical information regarding a range of data sets. In this paper trace element data for coal-fired power plants were

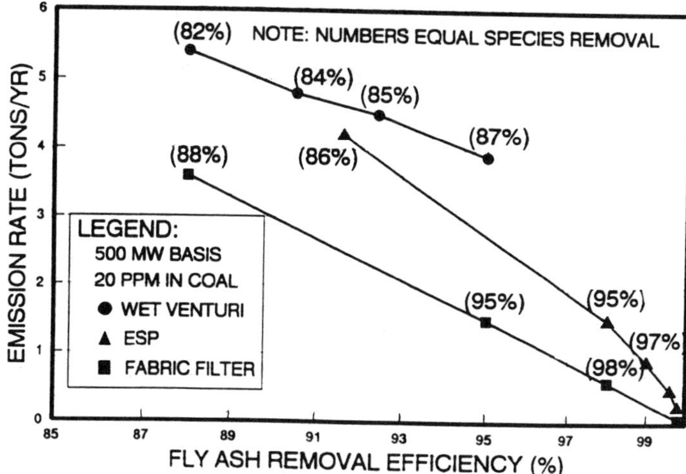

Figure 8. Comparison of emission rates of model compound from various control devices.

examined to determine the range of volatilities of several species in the boiler and the degree of condensation that occurs prior to the flue gas entering the particulate control device. Comparisons were made to identify possible cross correlations between the degree of vaporization and the coal composition. Calculations were made to model the effects of vaporization on the control efficiency using conventional particulate matter control devices.

The major conclusions of this paper are summarized below:

- Many trace elements such as arsenic, cadmium, mercury, and selenium are vaporized in the boiler.
- The degree of vaporization determined from historical data vary greatly for each of several species.
- Many vaporized elements are essentially totally condensed on to the available fly ash at the particulate control device inlet.
- Other elements such as mercury, chloride, and fluoride are primarily in the vapor phase at this point.
- The removal of surface enriched trace elements across ESPs and wet venturis is inversely related to the degree of vaporization.
- Many trace elements are enriched in the emitted particulate matter from ESPs and wet venturis, but theoretically, not from fabric filters.
- The fraction vaporized of some elements, such as arsenic, cadmium, and selenium, appear to correlate with the sulfur content of the coal.
- Additional field data measurements, which include more complete process characterization, are needed to more fully understand the factors that influence trace element emissions.

Control of Air Toxic Material by Novel Plasma Chemical Process — PPCP and SPCP

Senichi Masuda

I. INTRODUCTION

Most of the air-toxic and hazardous gaseous pollutants can effectively be decomposed by the novel plasma chemical processes, PPCP (pulse corona-induced plasma chemical process) or SPCP (surface discharge-induced plasma chemical process), both being produced in gases under ordinary temperature and pressure.

Copious active chemical species (radicals) are produced in a low-temperature, normal-pressure plasma region by collision of energy-enhanced electrons with gaseous molecules, and the molecules of pollutants are either oxidized or decomposed by reduction, depending upon the nature of the radicals: oxidizing radicals being O, O_2*, O_3, $\dot{O}H$, etc. and reducing radicals NH, NH_2, H, N, etc. Only electrons play an essential role in the radical generation, while ions play a role of raising the gas temperature to cause sparking and arcing. Hence, the plasma should be a highly nonequilibrium one in which the electron temperature is sufficiently high and ion temperature low enough.

One of the means for generating such nonequilibrium plasma under ordinary gaseous condition is to use corona discharge energized by a Nanosecond Pulse High Voltage with a very fast rise (rise time: 10 to 100 ns) and a very narrow width (half-tail: 100 to 500 ns). Use of positive corona is preferable since it produces streamers, much longer than negative ones, actually bridging across the entire electrode gap, and the space factor of the plasma reactor is about ten times higher. Electrons, very small in mass, can be effectively accelerated by an electric field even within such a short pulse duration time up to a sufficiently high energy level to cause satisfactory chemical effect, while ions, much larger in mass, remain almost unaffected. Because of such a short pulse duration time, no time is available for leaders — precursor for sparking ahd arcing — to be generated. As a result, a very high voltage can be applied in a corona gap without sparking, with a concurrently very high electric field. Hence, electrons can be very strongly heated. The heating of ions and molecules by electron collision is avoided by their cooling during a time interval between two successive intermittent pulses. This is PPCP. The generation rate of the radicals and thereby the end effect of the pollution control is proportional to the pulse frequency, which, however, is limited by the heating of ions and molecules to produce sparking and arcing.

Another alternative means is to use surface discharge occurring from the strip-like corona electrodes attached on a surface of a ceramic sheet having a film-like plane electrode attached on its opposite surface. An AC high voltage with 5 to 20 kHz is applied between these two electrodes, and the streamer discharges, interrupted by this ceramic sheet, develop along the surface within a very short time of several to tens of nanoseconds. The heating of ions and molecules, which is greatly enhanced by the use of a higher frequency, is avoided by an intense cooling of the opposite side surface of the ceramic sheet by enforced flow of air or water. The applicable frequency is limited by an over heat of the ceramic sheet to deteriorate the overall plasma chemical effect by thermal destruction of the radicals. This is SPCP.

Much work has been performed to apply PPCP in DeNOx and DeSOx of combustion gases by many researchers, including the author and co-workers, and its effectiveness has been confirmed.[1-3] Furthermore, an extensive study revealed that PPCP is likely to be the most cost-effective control means of NOx and SOx from coal-burning utility boilers among other means, e.g., calcium-gypsum process for SOx in combination with ammonia-catalytic process for NOx and electron beam process for DeSOx/DeNOx.[4]

The effectiveness of SPCP has also been confirmed by a number of works by the author and co-workers for DeNOx and DeSOx,[5] as well as for the surface treatment of plastic bumpers and other parts,[6] and also for the generation of ultrafine ceramic particles from the gas phase (SPCP-Plasma-CVD)[7] and generation of ozone.[8]

This paper reports the control of mercury vapor from combustion gas out of an incinerator plant by PPCP, and the control of freon gas from air by SPCP.

In view of the great effectiveness of PPCP and SPCP, it is expected that these two processes could be used for many other air-toxic materials.

II. CONTROL OF MERCURY VAPOR FROM COMBUSTION GAS BY PPCP

Tests are made using an experimental setup as shown in Figure 1 comprising a concentric cylinder-type corona electrode system (8) into which sample gases are introduced to be subjected to an intense streamer corona produced by the nanosecond pulse high voltage which is generated by a high-voltage pulse power supply shown in Figure 2. The corona reactor (8) used is either Cs, having a round wire and a smaller size (wire diameter: 0.26 mm; cylinder diameter: 56.5 mm ID; cylinder length: 210 mm effective), or C_1 having a square wire and a larger size (wire: 4×4 mm; cylinder diameter: 100 mm ID; cylinder length: 370 mm effective). The reactor (8) is placed in a thermostat (5) so that the reaction temperature can be altered up to 350°C. The carrier gas is either the combustion gas taken from the smoke duct of an incinerator [O_2: 10.5 to 11.7%; CO_2: 8.4 to 9.3%; NOx: 63 to 71 ppm; SOx: 39 to 41 ppm; HCl: 320 to 350 ppm; H_2O: 17 to 19%; total Hg: 0.11 to 0.24 mg/nm^3 (mostly water soluble)] or room air used for the purpose of baseline test. Hg, NO, and SO_2 are added to the carrier gases to make sample gases having the same initial concentrations: Hg = 0.5 mg/nm^3 (0.056 ppm); NO = 200 ppm; SO_2 = 200 ppm. The

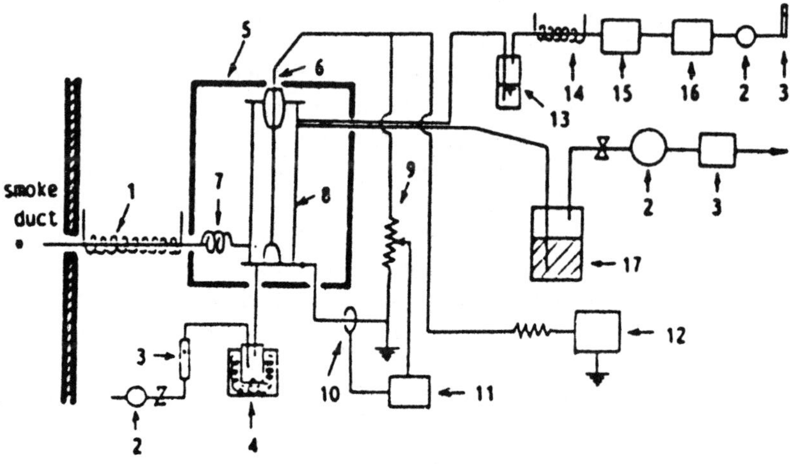

Figure 1. Experimental setup for PPCP control of $\overset{\circ}{Hg}$ vapor from incinerator combustion gas.

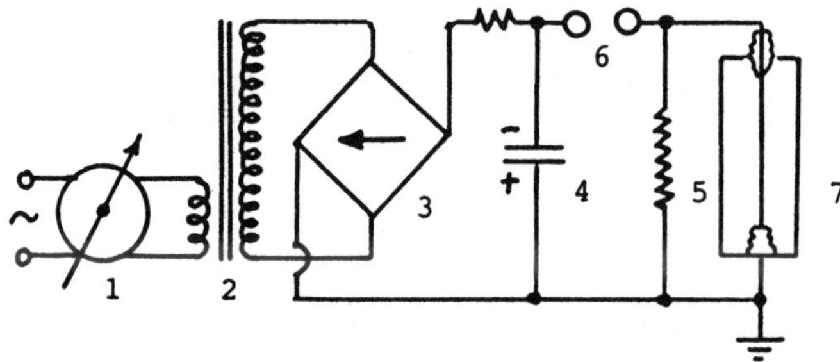

Figure 2. Circuit diagram of nanosecond high-voltage pulse power supply.

sample gas with combustion gas is referred to as E′ gas, while that with room air as A′ gas. The pulse power supply is a condensor-storage type with a stationary spark switch, where the pulse peak voltage, V_p, and pulse repetition frequency, f_p, are altered by changing the spark gap length and the condensor charging voltage. The nanosecond pulse high voltage obtained has pulse parameters: rise time $T_r = 100$ ns; half-tail $T_h = 300$ ns; peak voltage $V_p = 55$ kV (peak field intensity $E_p = 11$ kV/cm for C_1-reactor); frequency $f_p = 50$ to 250 Hz. The voltage and current wave forms of the nanosecond pulse are picked up with a pulse high-voltage probe (Pulse Denshi EP-50K) and a pulse current probe of Rogoski-Coil type (Sony-Techtronics P6021), monitored with a digital memory (Iwatsu DM-901), and stored in a microcomputer so that the energy consumption in the reactor is calculated as the time integration of voltage × current. The peak current amounts to 10 to 60 A, much higher than that of DC corona (several tens of mA). The gas temperature is varied from 150 to 350°C,

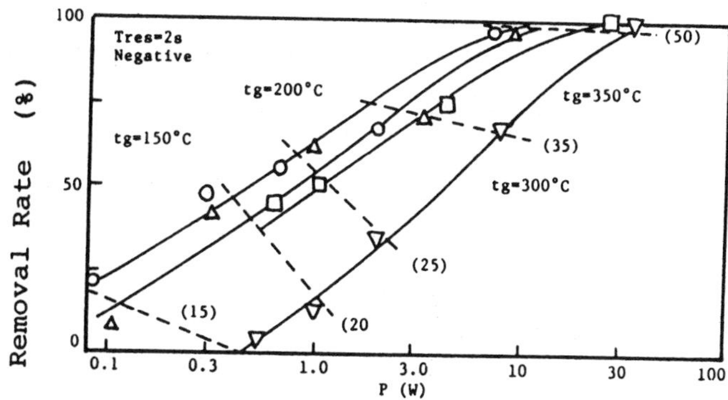

Figure 3. Power input vs. removal rate (T_{res} = 2 s).

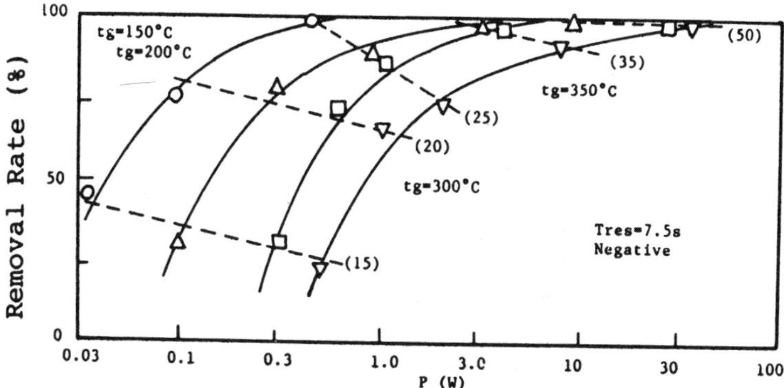

Figure 4. Power input vs. removal rate (T_{res} = 7.5 s).

while the gas residence time in the reactor is from 2 to 10 s (effective). The Hg concentration is measured after thermal decomposition of ozone produced in the reactor by using a flameless atomic absorption instrument (Toshiba-Beckman MV-253). This is because ozone has an absorption spectrum at the measuring line of Hg.

First, the effect of polarity is tested. The pulse power consumed in the reactor rises with increasing temperature and voltage, but it is always substantially higher in a positive corona than in a negative one. So far as the control performance for the mercury vapor is concerned, no difference can be observed in the corona polarity up to 200°C, while the negative polarity performed better at a temperature beyond 300°C. For this reason only the negative corona polarity is used throughout the present tests.

Second, the effect of pulse power input is investigated, using the reactor C_1. The results obtained for two different residence times, 2 and 7.5 s, are plotted in Figures 3 and 4, respectively. The Hg vapor is converted into solid particulate in a form of HgO and $HgCl_2$, easy to collect in the downstream precipitator or bag filter. The broken lines indicate the pulse peak voltage (V_p given in parenthesis in kV) applied

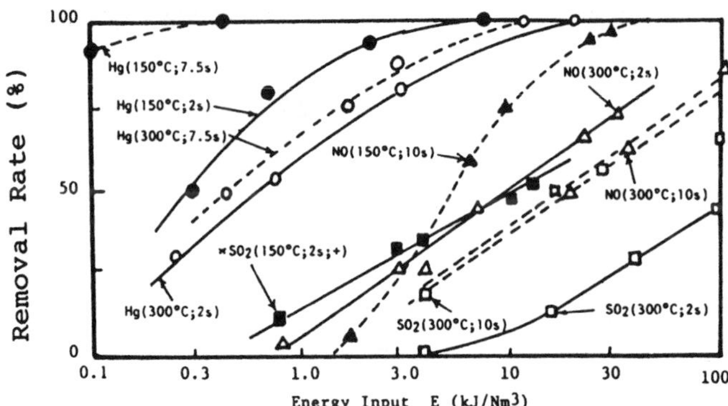

Figure 5. Energy input density vs. removal rate of $\mathring{Hg}$ (data for NO and SO_2 added for reference).

across the corona gap length of 5 cm. It can be seen that 100% removal of Hg vapor is achieved even at a short residence time, $T_{res} = 2$ s, and that the power consumption required rises greatly with increasing temperature for both shorter (2 s) and longer (7.5 s) residence times.

Since a shorter residence time means a higher gas volume treated, it is necessary to look at the control performance in terms of the pulse energy input per unit volume of gas, E (kJ/nm^3). Figure 5 indicates the results obtained, replotted from Figures 3 and 4. It is interesting to see that a great difference exists in the effect of gas residence time at a lower temperature ($T_g = 150°C$), while no difference appears at a higher temperature ($T_g = 300°C$). The temperature dependence of deterioration in performance is clearly observed in this plot.

Figures 6 and 7 show the effects of gas residence time, T_{res}, at a higher power input (P = 10 W), and the effect of the pulse frequency, f_p, at a shorter residence time (2 s) and a higher gas temperature (300°C). The saturation of performance occurs at $T_{res} = 2$ s for a higher power input (P = 10 W), and the logarithm of performance (removal rate) shows a linear rise with the pulse frequency, f_p.

From the foregoing figures the following empirical equation is derived for the Hg° removal performance, η_{Hg°, within a range where no performance saturation occurs:

$$\eta_{Hg^\circ} = 1 - \exp\left(-kPT_{res}f_p\right) \tag{1}$$

which can be explained by considering the continuity equation:

$$dN / dt = -k'nN \tag{2}$$

$$n = k''PT_{res}f_p \tag{3}$$

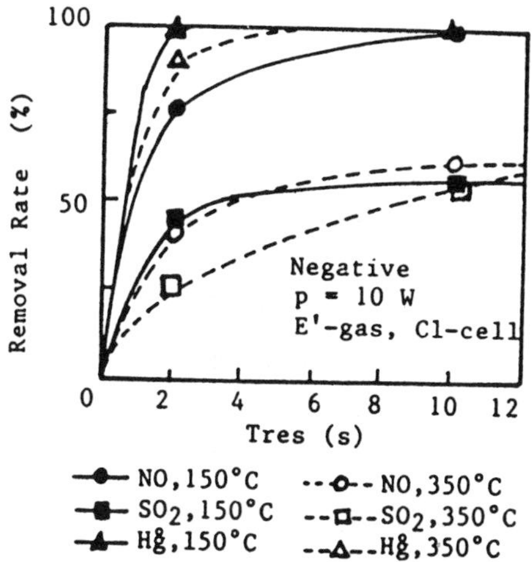

Figure 6. Gas residence time vs. removal rate.

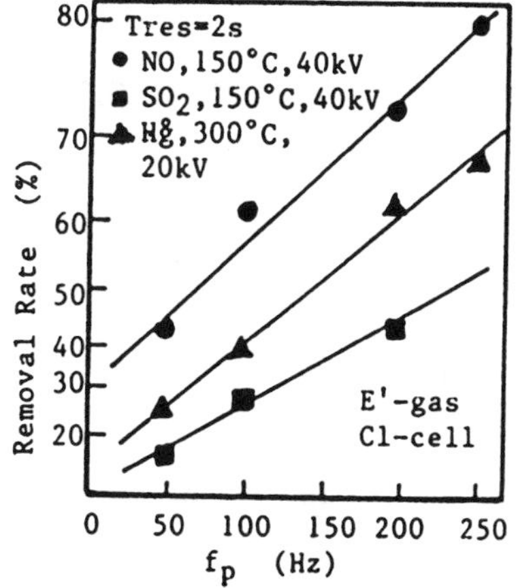

Figure 7. Pulse frequency vs. removal rate.

where N and n represent the number concentrations of Hg° molecule and radical, respectively, and k, k′, and k″ the proportional constants.

Figure 8 indicates for E′ gas and A′ gas the effect of the initial concentration of Hg° on its removal rate and total quantity removed per unit volume of gas measured at T_g = 300°C, f_p = 50 Hz, T_{res} = 2 s, P = 10 W, using the C_1 reactor. The increase in the initial concentration results in a comparatively smaller deterioration in performance for E′ gas with combustion gas, while it causes a large drop in performance for the air-based A′ gas. This may be attributed to the effects of HCl and CO existing in the combustion gas (see Figures 9 and 10).

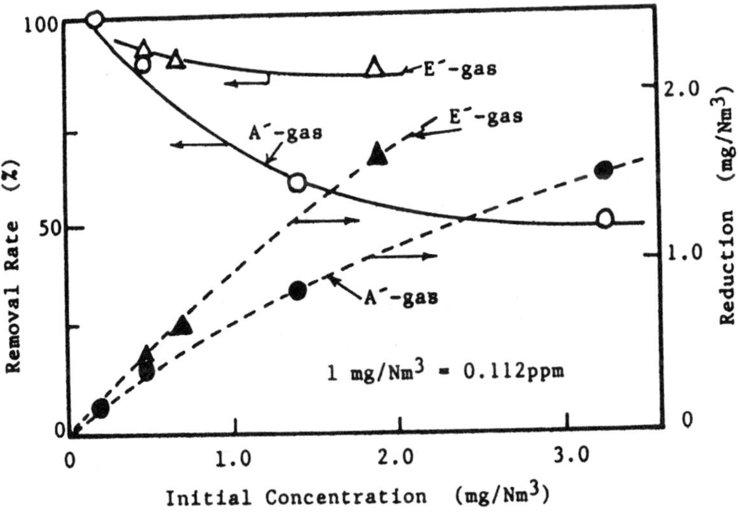

Figure 8. Initial concentration of Hg° vs. removal rate and reduction (300°C; 50 Hz; 2 s; (–); 10 W; C_1).

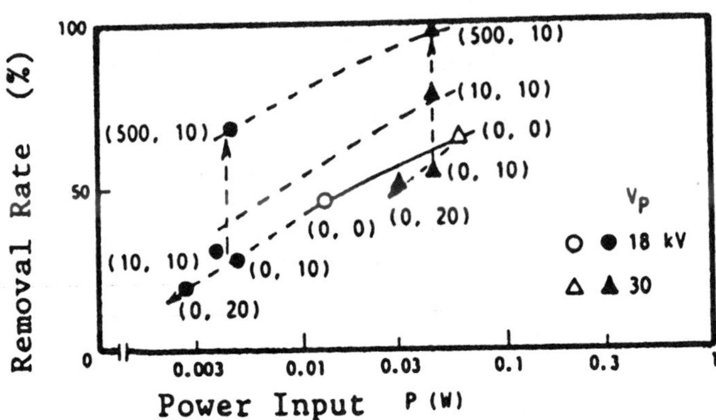

Figure 9. Effects of HCl and H_2O on removal rate of Hg° vapor [parentheses (a, b): a = HCl concentration ppm and b = H_2O concentration vol%; 100°C; 50 Hz; 18 s; Cs].

Figure 11 indicates the effect of gas temperature for E′ gas and A′ gas measured at a larger power input, P = 30 W, where no influence of the gas temperature is observed for the combustion-gas-based E′ gas. Again, in this figure a large performance deterioration is observed to occur with increasing gas temperature for the air-based A′ gas.

Finally, a comparative test is made between the PPCP and an ordinary DC corona in the control performance of Hg° vapor for both 1- and 10-s gas residence times. It has been known that Hg° vapor is removed by passing through a DC corona region

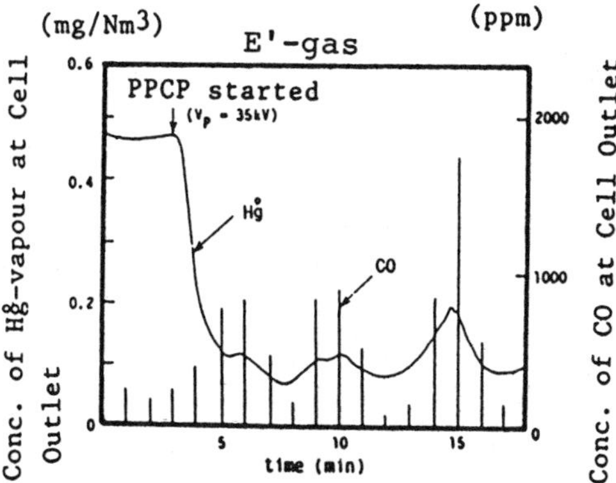

Figure 10. Correlation between Hg° and CO concentrations in time domain (300°C; 50 Hz; 2 s; Cl).

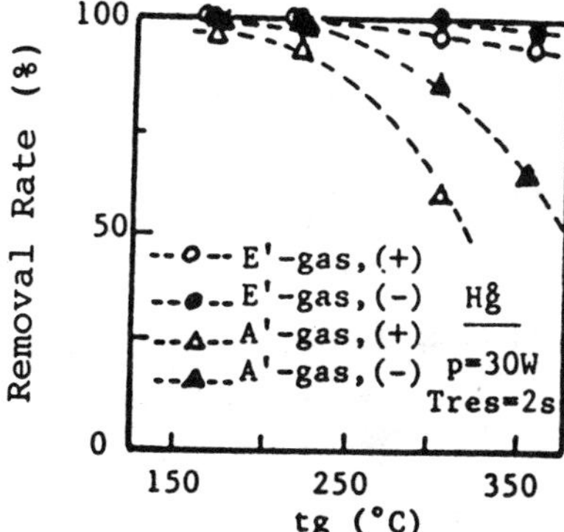

Figure 11. Effect of gas temperature on removal rate of Hg°.

of an electrostatic precipitator to some extent or another, so that the use of PPCP should be meaningful only when it has distinct merit in terms of power consumption. This advantage is clearly indicated in the results of the test, as shown in Figure 12.

III. DECOMPOSITION OF FREON GASES IN AIR BY SPCP

Figure 13 shows photographs of the SPCP units of planar type (Figure 13a) and cylinder type (Figure 13b), both in use for SPCP processing of toxic and hazardous gases. Figure 14 shows the cross-section of the cylinder-type units.

In the present test use is made of cylinder-type SPCP units in an experimental setup as shown in Figure 15. An AC high-voltage (2- to 6-kV peak; 5 to 10 kHz) is applied between the corona electrodes and film-like induction electrode (grounded); an AC surface discharge is then generated as shown in Figure 16. Three different

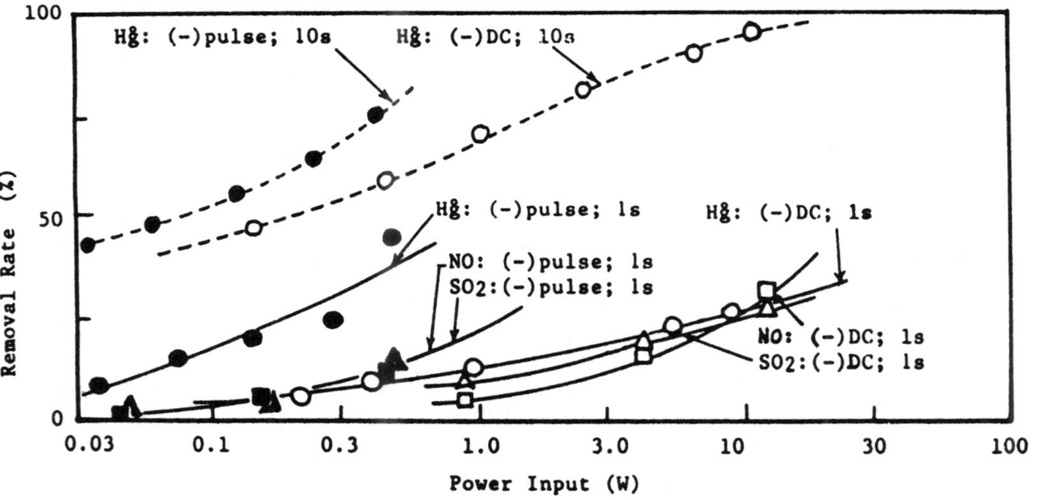

Figure 12. Comparison between PPCP and DC corona on removal of Hg° (150°C; 50 Hz; (–); initial concentration: Hg° = 0.5 mg/nm^3, NO = 200 ppm; SO_2 = 200 ppm; data for NO and SO_2 included for reference).

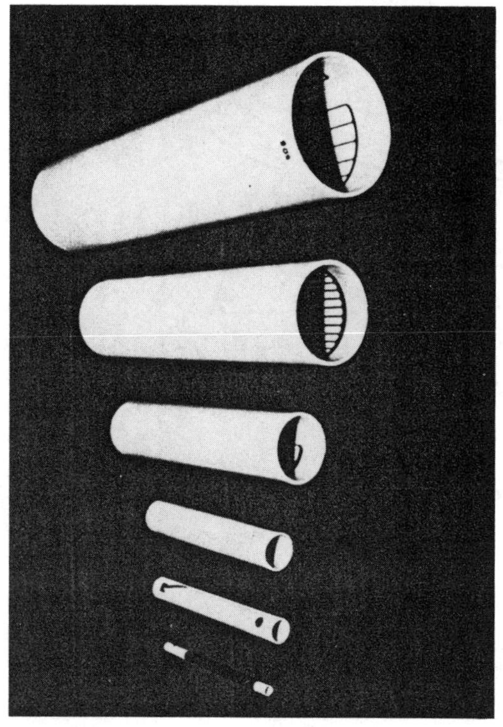

Figure 13. Photographs of SPCP unit.

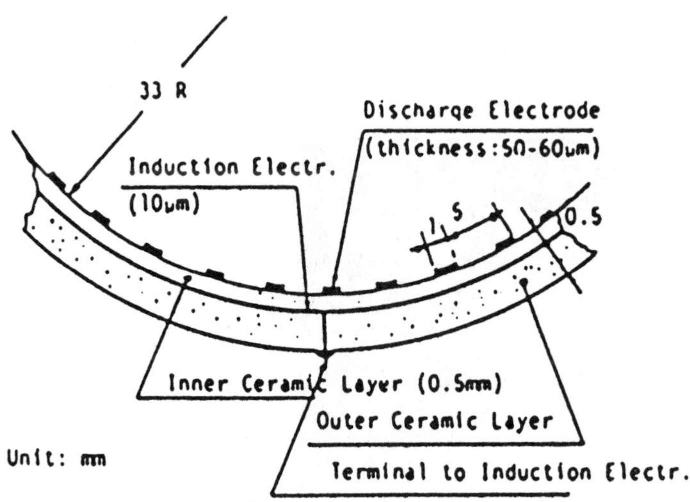

Figure 14. Cross-sectional view of cylinder-type SPCP unit (type: OC-70/5/1.0-A).

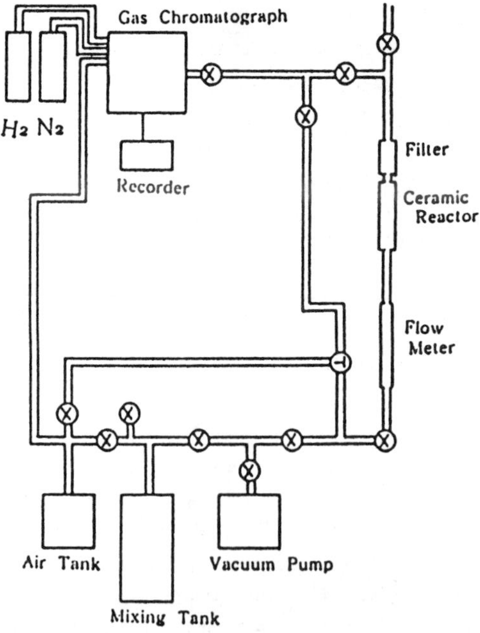

Figure 15. Experimental setup for SPCP decomposition of freon gases.

Figure 16. Photograph of AC surface discharge.

types of the SPCP unit are used in the present tests: OC-10-A (diameter: d = 10 mm ID; length: L = 150 mm; wire width: w = 1 mm; wire-to-wire pitch: p = 5 mm), OC-70/5/1.0-A (d = 70 mm; L = 300 mm; w = 1 mm; p = 5 mm), and OC-70/10/0.5-A (d = 70 mm: L = 300 mm; w = 0.5 mm; p = 10 mm), all attached with aluminum cooling fins and cooled by enforced air. Most of the data reported here are obtained with the smaller SPCP unit (OC-10-A) as a substantially higher power density (2.2 W/cm^2) with a concurrently greater performance of freon decomposition can be obtained than those of the other two larger SPCP units (OC-70/5/1.0-A and OC-70/10/0.5-A) (0.5 W/cm^2). This difference is not an essential one, but a result of the available capacity of the power supplies used.

Two kinds of fluorocarbon are tested: a substitution fluorocarbon, CFC-22 ($CHClF_2$), which is gas phase in room temperature with a melting temperature, Tm = –160°C, and boiling temperature, Tb = –40.8°C, and trichlorotrifluoroethane (CFC-113: $CF_2ClCFCl_2$), which is liquid phase in room temperature with Tm = –35°C and Tb = 47.6°C. CFC-113 is one of the most widely used freon gases (spray; cleaning liquid for semiconductor devices and electronic circuit boards; etc.), and CFC-22 is being used in place of CFC-113. CFC-113 represents the most difficult material to decompose; it is hardly decomposed by ozone gas in combination with UV-light irradiation. Even PPCP shows a comparatively poor decomposition performance. In the CFC-22 (gas-phase) decomposition test, a fixed amount of CFC-22 is taken by a gas-sampling syringe, and mixed with air in a pressurized mixing tank with 7-L volume at 5 atm pressure. The gas is sampled at the inlet and outlet of the SPCP reactor using a gas sampler (Shimazu: MGS-4) and analyzed by a gas chromatograph (Shimazu: GC-8). In the CFC-113 decomposition test, a fixed amount of liquid CFC-113 is taken by a liquid syringe, and injected into the mixing tank.

Figures 17 through 19 give the results obtained for CFC-22. It can be seen in Figure 17 that the doubling of the frequency at the same peak voltage (4 kV) results in an essential increase in the decomposition rate. Figure 18 indicates that higher than 90% decomposition rate can be achieved at a peak voltage beyond 6 kV even within 3 s residence time. Figure 18 shows a distinct saturation tendency to occur at a longer residence time (T_{res} = 8 s). In conclusion, a satisfactory decomposition rate can be

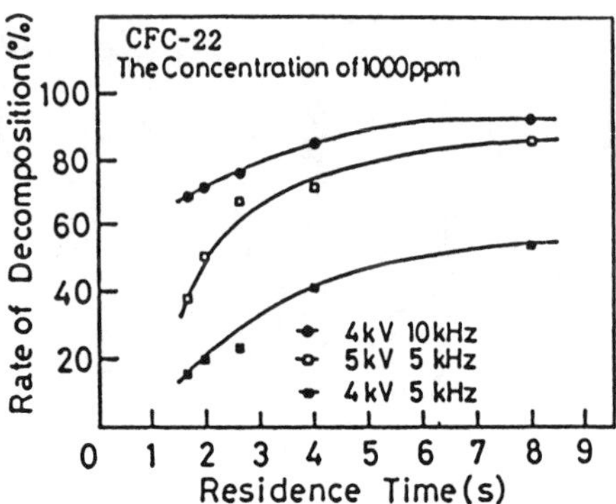

Figure 17. Decomposition rate of CFC-22 vs. residence time.

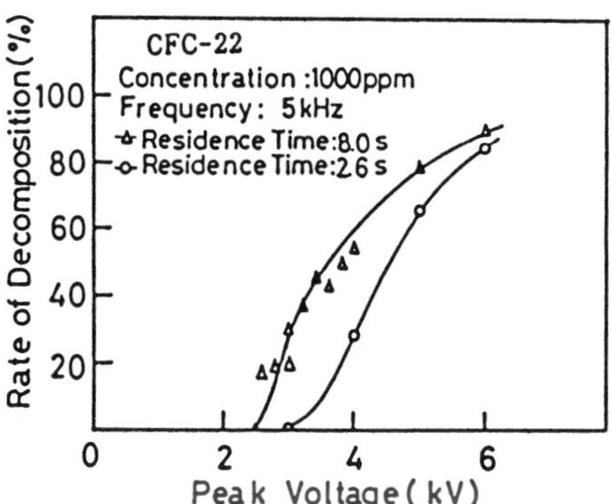

Figure 18. Decomposition rate of CFC-22 vs. peak voltage.

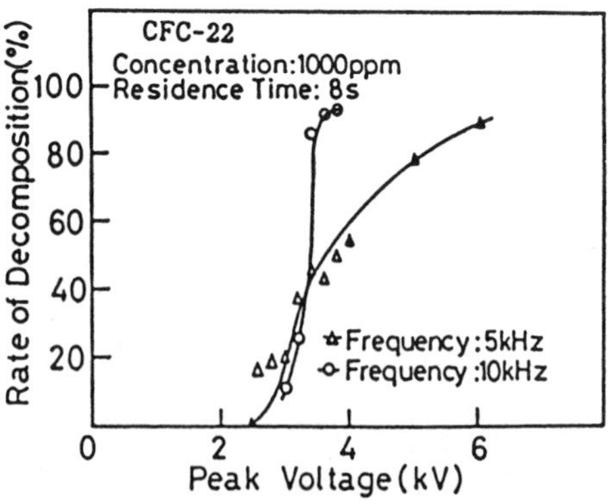

Figure 19. Decomposition rate of CFC-22 vs. peak voltage.

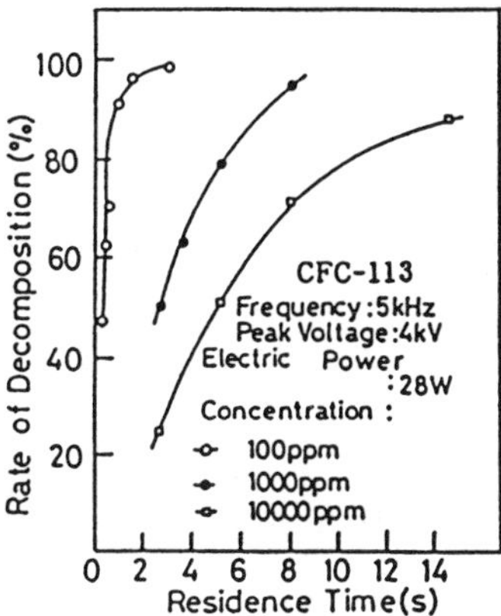

Figure 20. Decomposition rate of CFC-113 vs. residence time (P = 28 W).

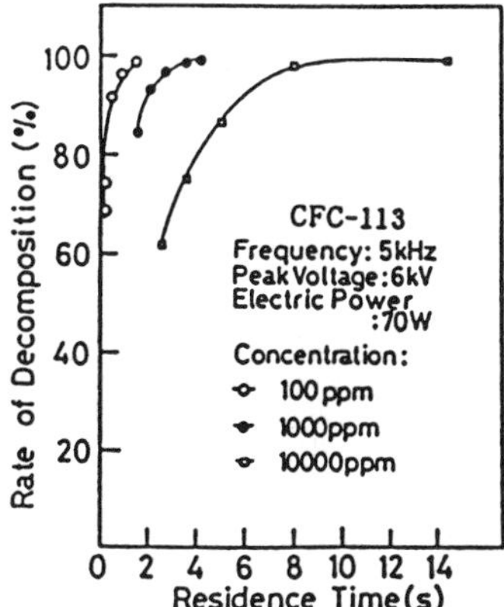

Figure 21. Decomposition rate of CFC-113 vs. residence time (P = 70 W).

achieved for CFC-22 with 1000-ppm concentration even under rather limited output parameters of the power supply (voltage: 5 kV peak max; frequency: 10 kHz max). A much better performance is expected by using a more powerful power supply.

Figures 20 through 22 indicate for CFC-113 test results obtained at three different inlet concentrations: 100, 1000, and 10,000 ppm. Figures 20 and 21 show a distinct effect of the voltage and a tendency of saturation to appear beyond a certain residence time. It can be seen that 100% decomposition can be obtained for a low concentration (100 ppm) even within 1 to 2 s residence time, and that even as high as 10,000 ppm inlet concentration it can be fully decomposed at a higher voltage (6 kV) and a longer residence time (7 s) with a low power consumption (70 W). Figure 22 illustrates the plots against the power input where the net power consumed in the SPCP unit is

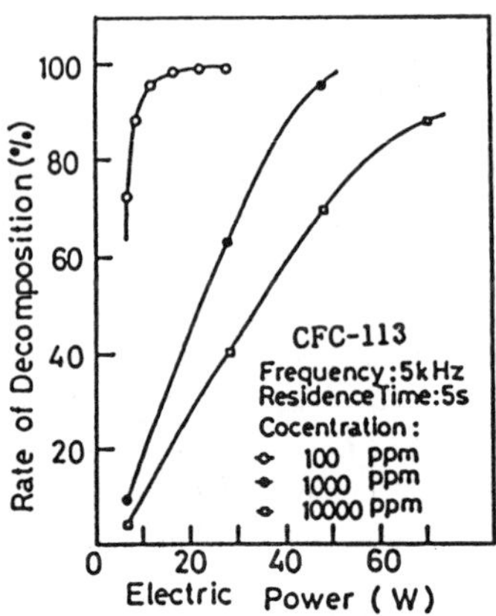

Figure 22. Decomposition rate of CFC-113 vs. power input.

estimated by Lissajous figure method. It can be seen that a proportionality exists between the power input and the decomposition performance within the unsaturated range.

In order to make clear the mechanism of freon decomposition by SPCP, two other carrier gases (N_2 and O_2) are used for the same tests, and their results compared with those obtained with air as the carrier gas. A substantial rise in the decomposition performance is observed with N_2 carrier gas, while a deterioration in the performance is observed with O_2 carrier gas. This suggests that nitrogen radicals play an essential role in the decomposition of freon gases.[9] The remarkable effect of the nitrogen radicals, strong in activity and long in life (several to several tens of seconds) is also observed in the SPCP control of NO, where the nitrogen radicals generate NO_x when mixed with air or combustion gas containing oxygen.[5]

Finally, gas-chromatographic mass-spectroscopy analysis is conducted to identify the decomposition products of CFC-113.[9] Because of complexity in the spectra obtained the identifications are yet to be completed. However, no poisonous gases such as phosgene ($COCl_2$) or fluorophosgene (COF_2O) are detected.

The use of a liquid or solid absorbent is necessary to finally remove the decomposition products from air.

IV. CONCLUSION

From the foregoing tests of PPCP and SPCP the following conclusions are derived:

- Hg°_{g} vapor in the combustion gas from incinerator plants, low in concentration (approximately 5 mg/nm^3 = 0.056 ppm), can be fully removed by PPCP within a short residence time (2 s) with the negative corona polarity with a substantially lower power input than DC corona discharge.

- The effectiveness of PPCP Hg^0_g control is confirmed within the temperature range between 150 and 350°C, where a higher voltage and concurrently greater power input become necessary with rising temperature.
- The reaction products of Hg^0_g vapor are either HgO or $HgCl_2$, both being solid particulates and easy to collect in the downstream ESP or bag filter.
- The energy consumption lies within an acceptable range (0.1 to 5 kJ/nm^3) which varies depending upon the gas temperature, initial concentration of Hg , and PPCP reactor design (residence time, peak voltage, pulse frequency, pulse rise time). The pulse rise time has an essential role for achieving a high performance.[3]
- Freon gases (CFC-22 and CFC-113) in air can be satisfactorily decomposed by SPCP within a short residence time (CFC-22: 3 s; CFC-113: 1 to 2 s) under ordinary temperature and pressure at a frequency higher than 10 kHz and peak voltage higher than 5 kV.
- The effectiveness of SPCP freon decomposition is confirmed at 1000 ppm initial concentration for CFC-22, and at 100, 1000, and 10,000 ppm for CFC-113.
- The power consumption in SPCP freon decomposition is within an acceptable range, and the decomposition products do not include toxic components.
- An absorbent, solid or liquid, must be used to remove the decomposition products in the downstream of the SPCP reactor.
- In general, PPCP is suitable for use if a large gas volume is to be treated, while SPCP can be used for a smaller gas volume.

V. ACKNOWLEDGMENTS

The author appreciates the dedicated works of his co-workers, Dr. T. Oda and Mr. X. Tu of the University of Tokyo, Mr. Y. Wu of Northeast Normal University, China, and Mr. S. Kito of Takuma Ltd. The author is also very much indebted to Mr. K. Sakakibara of Takuma Ltd. and JAMDA for their warm encouragement and generous financial support given to this work.

REFERENCES

1. Masuda, S. and Nakao, H., Control of NOx by Positive and Negative Pulsed Corona Discharges, *IEEE IA-Trans.*, Vol. 26, No. 2 (March/April 1990), pp. 374–383.
2. Masuda, S. and Wu, J., DeNOx and DeSOx by PPCP and SPCP, Proc. 8th EPA/EPRI Symposium on Transfer and Utilization of Particulate Control Technology, March 24–27, 1990, San Diego, CA.
3. Masuda, S. and Wu, Y., Removal of NOx by corona discharge induced by sharp rising nanosecond pulse voltages, *Inst. Phys. Conf. Ser. No. 85*, London, pp. 249–254 (Electrostatics '87, Oxford).
4. Japan Association of Machinery Industry/Research Institute of Energy Engineering: Study Report on Novel Dry Technology of Pulse Corona Induced Plasma Chemical Process (PPCP) for Control of SOx and NOx from Combustion Gas Out of Utility Boilers for Thermal Power Generation, Report No. 2EP-18, May 1991.
5. Masuda, S., u, X., Sakakibara, K., Kitoh, S., and Saito, S., Destruction of Gaseous Pollutants by Surface Corona Induced Plasma Chemical Process, *Proc. IEEE/IAS 1991 Annual Conference,* October 1–4, 1991, Dearborn, MI.

6. Masuda, S., Tochizawa, I., Kuwano, K., Akutsu, K., and Iwata, A., Surface Treatment of Plastic Material by Pulse Corona Induced Plasma Chemical Process — PPCP, *Proc. IEEE/IAS 1991 Annual Conference,* October 1–4, 1991, Dearborn, MI.
7. Yamamoto, H., Shioji, S., and Masuda, S., Synthesis of Ultra-Fine Particles by Surface Discharge-Induced Plasma Chemical Process (SPCP) and its Application, *Proc. IEEE/IAS 1990 Annual Conference,* pp. 282–285 (October 1990).
8. Masuda, S., Akutsu, K., Kuroda, M., Awatsu, Y., and Shibuya: A Ceramic-Based Ozonizer Using High-Frequency Discharge, *IEEE IA-Trans.,* Vol. 24, No. 2, pp. 223–231, 1988.
9. Oda, T., Takahashi, T., Nakano, H., and Masuda, S., Decomposition of Fluorocarbon Gaseous Contaminants by Surface Discharge Induced Plasma Chemical Processing, *Proc. IEEE/IAS 1991 Annual Meeting,* October 1–4, 1991, Dearborn, MI.

Suppression of PCDD/PCDF Formation in MSW Incinerator Flue Gas at Temperatures below 800°F by Ammonia Injection

L. Takacs
W. Gleason
G. L. Moilanen
A. McQueen

I. ABSTRACT

PCDD/PCDF emissions from MSW incinerators are a major source of concern, preventing MSW from reaching its full potential as a fuel for power-generating boilers. The majority of PCDD/PCDF found in MSW incinerator stack emissions is formed in low-temperature zones downstream of the boiler. This paper concerns a new technology for suppressing PCDD/PCDF formation in MSW incinerator flue gas at temperatures below 800°F using ammonia injection. The major advantage of the ammonia injection technology (AIT) is that it suppresses PCDD/PCDF formation rather than removing PCDD/PCDF after it is formed, and therefore does not lead to a pollution transfer problem. This technology was tested on a 4000-DSCFM pilot plant which draws flue gas from the boiler outlet of an existing RDF power plant. These experimental data can be used to infer a mechanism of ammonia suppression of PCDD/PCDF formation.

Based on the success of these pilot plant tests, consideration is being given to combining the AIT with a wet collector for ammonium salt, acid gas, and particulate matter removal. This paper will describe the features of the air pollution control device required to complete the AIT.

II. INTRODUCTION

PCDD/PCDF emissions from MSW incinerators have long been a concern. Recently, the new Clean Air Act has undertaken to regulate these emissions. This paper follows two previous papers which introduced a new technology for reducing PCDD/PCDF emission — AIT.[1,2] The successful pilot-scale demonstration of this technology is described here.

Background information on PCDD/PCDF emissions from MSW incinerators is presented in Section III. Section IV describes the development of the pilot project for

TABLE 1

TYPICAL PCDD/PCDF EMISSIONS FROM MSW INCINERATORS

COMBUSTOR TYPE	EXISTING FACILITY WITHOUT GCP	EXISTING/NEW FACILITY WITH GCP
Mass Burn,Refractory	4000	500/-
MB, Waterwall, Large	500	500/200
MB, Waterwall, Small	2000	200/200
RDF,Spreader Stoker	2000	1000/1000

REFERENCE: Kilgroe and Johnston, Solid Waste & Power Vol. 3 No. 6, 18-30, 1989

(ng/DSCM @ 7% O2)

evaluation of the AIT, and Section V presents the results. Future directions for the AIT are outlined in Section VI.

III. BACKGROUND ON PCDD/PCDF EMISSIONS FROM MSW INCINERATORS

Prior to discussing the suppression of PCDD/PCDF emissions from MSW incinerators using ammonia injection, it is useful to present data on typical PCDD/PCDF emissions from MSW incinerators (Section III.A) and review PCDD/PCDF emission limits for MSW incinerators under the new Clean Air Act (Section III.B). Current PCDD/PCDF control technologies focus on removal of PCDD/PCDF from the flue gas after it is formed. An alternate approach is to suppress PCDD/PCDF formation. This is the approach adopted in the AIT. Section III.C discusses the mechanisms of PCDD/PCDF formation, and introduces the concept behind the AIT.

A. Typical PCDD/PCDF Emissions from MSW Incinerators

Table 1 presents typical PCDD/PCDF emissions from MSW incinerators (estimated by the EPA), as a function of combustor type, facility age, and whether good combustion practice (GCP) is applied.[3] Table 1 indicates that GCP has more noticeable effects on mass burn (MB) than on refuse-derived fuel (RDF) facilities. Once good combustion practice is applied, emissions range from 200 to 1000 ng/DSCM at 7% O_2.

TABLE 2

NEW CLEAN AIR ACT REGULATIONS FOR PCDD/PCDF EMISSIONS FROM MSW INCINERATORS

COMBUSTOR TYPE	COMBUSTOR AGE	PCDD/PCDF <225 TON/D	EMISSION >225, <2000	LIMIT >2000
Non-RDF	New	75 (75%)	5-30 (90%)	5-30 (90%)
Non-RDF	Existing	500 (0%)	125 (75%)	5-30 (95%)
RDF	New	250 (75%)	250 (75%)	250 (75%)
RDF	Existing	1000 (0%)	250 (75%)	250 (75%)

REFERENCE: Brna and Kilgroe
Journal of Air & Waste Management Association
Vol. 40, 1324-1330, 1990

(ng/DSCM @ 7% O2)

B. New Clean Air Act Regulations for PCDD/PCDF Emissions from MSW Incinerators

New Clean Air Act PCDD/PCDF emission limits for various categories of MSW incinerators are outlined in Table 2.[4] Emission limits are stricter for both newer and larger facilities. The percent reduction required from GCP levels is shown in brackets. The emission limit for large RDF facilities is ten times higher than for large, non-RDF facilities: 250 vs. 30 ng/DSCM at 7% O_2. The limits chosen reflect the higher baseline emissions for RDF facilities than for non-RDF facilities (see Section III.A). The required reduction from GCP levels is 90% or better for non-RDF and 75% for RDF.

Current technologies for reducing PCDD/PCDF, for example, spray dryer/fabric filter systems chosen as BACT by the EPA, are designed to remove PCDD/PCDF from the flue gas after it is formed. An alternate approach is to suppress PCDD/PCDF formation. This approach requires some knowledge of PCDD/PCDF formation/suppression mechanisms.

C. Mechanisms of PCDD/PCDF Formation/Suppression

In a previous paper, the results of bench scale tests, which showed that NH_3 suppressed PCDD/PCDF formation, were published. These tests were performed at Occidental Chemical Corporation's (OxyChem's) Niagara Falls MSW incinerator, Energy from Waste (EFW). Other authors had established that NH_3 suppressed

TABLE 3

COMPARISON OF PROPOSED MECHANISMS FOR THE SUPPRESSION OF PCDD/PCDF FORMATION BY NH3

FEATURE OF MECHANISM	TAKACS AND MOILANEN, 1991	VOGG, METZGER AND STIEGLITZ, 1987
Type of reaction	Gas phase	Catalytic,on fly ash
Agent required	None	Copper catalyst
Effect of NH3	Competes for HCl	Poisons catalyst

TABLE 4

FACTORS AFFECTING PCDD/PCDF FORMATION RATES

FACTOR IN LAB. EXPERIMENT	REQUIRED FEATURES	FACTOR IN ACTUAL MSW INCIN.
Temperature	500-950oF	Boiler tube fouling
Ash residence time	>0.5 hrs.	Sootblowing freq.
Flue gas composition	O2, H20, HCl	Combustion practice
Fly ash composition	C, Cu	Waste composition

PCDD/PCDF formation in fly ash samples in the laboratory, and had proposed a catalytic mechanism.[5] An alternate mechanism for the suppression of PCDD/PCDF formation, involving competition of NH_3 with the PCDD/PCDF precursors for the available HCl, was outlined in a former publication.[2] These mechanisms are compared in Table 3. The issue of mechanism will be returned to in Section VI.B.

Table 4 presents other factors affecting PCDD/PCDF formation rates which were identified in laboratory and bench scale studies, together with the corresponding factors in actual MSW incinerators. The middle column of Table 4 indicates the required features for each factor.

Temperature is the main factor controlling PCDD/PCDF formation rates. Laboratory experiments by Vogg et al.[6] have demonstrated that the temperature range for PCDD formation was 500 to 750°F, while the temperature range for PCDF formation was 500 to 950°F. The optimal temperature for both PCDD and PCDF formation was 575°F. In an actual MSW incinerator train, these temperatures occur between the boiler outlet (economizer inlet) and the stack. The exact location of these temperatures varies as a function of time due to boiler tube fouling, which results in higher boiler outlet temperatures.

Ash residence time in the laboratory furnace was also found to be an important factor in PCDD/PCDF formation.[5] In a real boiler, ash residence time depends on the boiler soot-blowing frequency. Flue gas constituents whose concentration affected PCDD/PCDF formation rates included O_2, H_2O, and HCl (in addition to NH_3).[5] However, PCDD/PCDF formation rates were not as sensitive to these parameters as to temperature and ash residence time. In addition, it is likely that PCDD/PCDF formation will be sensitive to the precursor content in the flue gas, although this was not tested. In real boiler environments, flue gas composition is a function of combustion practice and varies with time. Both carbon and copper in the fly ash stimulated PCDD/PCDF formation.[7] Fly ash composition depends on waste composition, which changes appreciably in real systems.

The success of the bench scale study of the AIT prompted a decision to undertake a pilot scale study, as outlined previously.[5] The following section illustrates how the above information on factors affecting PCDD/PCDF formation/suppression was used in the design of the pilot scale test system.

IV. DEVELOPMENT OF THE AMMONIA INJECTION TECHNOLOGY PILOT PROJECT

The development of the AIT pilot project was described in a previous publication,[2] and will be briefly reviewed here. Pilot scale testing of the AIT for suppression of PCDD/PCDF formation was undertaken at OxyChem's EFW, who se features are summarized in Table 5. The objectives of the AIT pilot project are outlined in Section IV.A. Section IV.B describes the features of the pilot scale test system and Section IV.C presents the experiment design.

A. Objectives of the AIT Pilot Project

The objectives of OxyChem's AIT pilot project were

- To verify that the pilot plant can simulate the PCDD/PCDF-making properties of the full-scale plant.
- To demonstrate that ammonia injection suppresses this PCDD/PCDF formation and to determine the stoichiometry required

TABLE 5

OCCIDENTAL CHEMICAL CORPORATION'S ENERGY FROM WASTE

Location	Niagara Falls, N.Y.
Date of start-up	1980
Fuel type	RDF
Number of units	2
Fuel consumed	2200 ton/day
Flue gas flow rate	200,000 DSCFM (each)
Particulate control	Dry ESP
Acid gas control	Currently none

- To provide statistically sound data on PCDD/PCDF formation/suppression in order to make a decision concerning further tests

B. Description of the Pilot Scale Test System

Based on the optimal temperature for PCDD/PCDF formation determined in laboratory scale experiments, it was decided to draw flue gas from the boiler outlet, where the temperature was approximately 800°F, into the pilot plant, which consisted of a pilot-scale economizer and dry ESP simulator. The economizer cooled the flue gas to 575°F, while the dry ESP simulator, which contained a series of grates above an ash hopper, provided ash residence time with little temperature change.

Table 6 compares the pilot-scale and full-scale plants. In order to simulate the PCDD/PCDF formation properties of the full-scale plant, the pilot plant was designed to reproduce as many features of the full-scale plant as possible, including velocity, temperature, and ash hold-up characteristics.

A simplified process and instrumentation diagram for the bench scale test system is shown in Figure 1. Simultaneous PCDD/PCDF sampling was performed at two stations: pilot plant inlet/boiler outlet (background) and pilot plant outlet. In addition, some stack samples were collected simultaneously in order to compare the performance of the pilot plant and the full-scale plant.

C. Experiment Design for Pilot Scale Testing of the AIT

Table 7 lists the test conditions used, number of replicas, and number of simultaneous background samples. A total of 50 PCDD/PCDF samples were collected, the

TABLE 6

COMPARISON OF PILOT SCALE AND FULL SCALE PLANTS

FUNCTION OR PARAMETER	FULL-SCALE VERSION	PILOT-SCALE VERSION
Flue gas flow rate	200,000 DSCFM	4,000 DSCFM
Flue gas cooling	Economizer	Economizer
Ash removal/hold-up	Dry ESP	Simulator
Econ. inlet velocity	25 ft/s	25 ft/s
Econ. temp. profile	10 oF/ft	10 oF/ft
Cooling water temp.	275oF	275oF
Econ. outlet temp.	575oF	575oF
Sootblowing per day	1	1
Operation	Continuous	Continuous

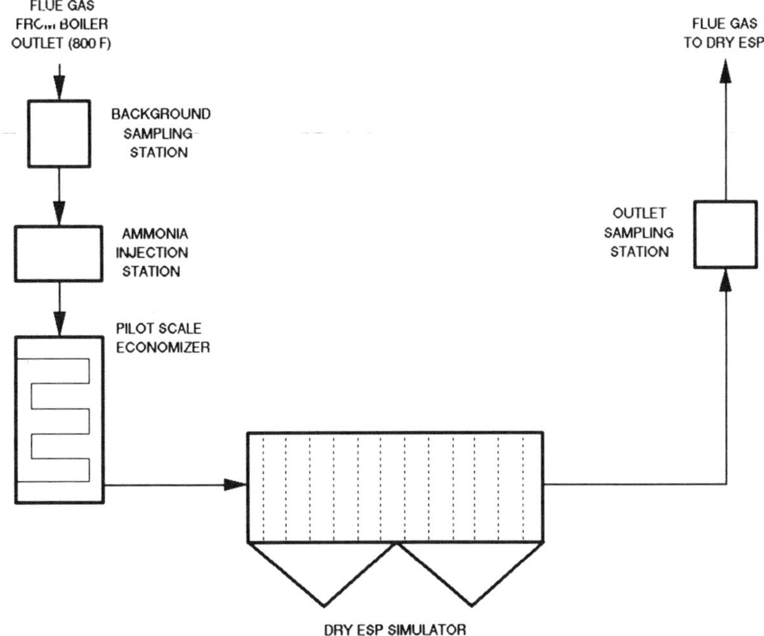

Figure 1. P & ID ammonia injection technology pilot project.

TABLE 7

TEST CONDITIONS FOR PILOT PLANT EXPERIMENTS ON PCDD/PCDF FORMATION/SUPPRESSION

TEMP.°F BOILER OUTLET	TEMP.°F ECON. OUTLET	NH3 STOICHIOMETRY (RATIO TO HCl)	# OF OUTLET SAMPLES	# OF BACKGROUND SAMPLES
858	575	0	2	1
		0.05	2	1
		0.72	2	1
	525	0	2	1
		0.72	1	0
778-805	575	0	6	2
		0.12	4	0
		0.18	4	0
		0.72	4	1
		1.3	5	1
	525	0	3	1
		0.72	4	0
		TOTAL	39	9

48 shown in the table and 2 stack samples. Two factors were varied in the experiments: economizer outlet temperature and ammonia stoichiometry. Two values of economizer outlet temperature were used: 575 and 525°F. Six values of NH_3 stoichiometry were tested: 0, 0.05, 0.12, 0.18, 0.72, and 1.3. In addition, a third factor was found to have a considerable effect on both background and outlet PCDD/PCDF levels, namely, the boiler tube fouling, as seen in the boiler outlet temperature. Initial tests were conducted with a relatively "dirty" boiler over a 2-week period, just prior to a scheduled tube water wash. A second set of experiments took place over a 6-week period after the water wash. These experiments demonstrated that the water wash cycle is an important factor to take into account in designing boiler PCDD/PCDF experiments.

The flue gas composition at EFW during the pilot scale experiments is shown in Table 8. The concentration of HCl in the flue gas was much higher than the other acid gases. The NH_3 flow rate was set based on a combination of HCl continuous emission monitoring and HCl source tests. The NH_3 flow rate was verified by NH_3 source tests. The highest NH_3 flow rate used corresponded approximately to a stoichiometric quantity of NH_3 for reacting with both HCl and SO_2.

V. RESULTS OF THE AMMONIA INJECTION TECHNOLOGY PILOT PROJECT

The results of the 50 flue gas samples collected under different conditions and analyzed for PCDD/PCDF are summarized in this section. All PCDD/PCDF results

TABLE 8

FLUE GAS COMPOSITION FOR PILOT EXPERIMENTS

SPECIES	UNITS	AVERAGE CONCENTRATION DURING TESTS
O2	Percent	12
CO2	Percent	8
H2O	Percent	12
HC1	ppmvd @ 7% O2	600
SO2	ppmvd @ 7% O2	200 (estim.)
NOx	ppmvd @ 7% O2	300 (estim.)
NH3	ppmvd @ 7% O2	30-780
	NH3/HCl molar ratio	0.05-1.3
Particulate Matter	gr/DSCF	1.5

quoted are in nanograms per DSCM at 7% O_2 and represent the average of two to six samples. The formation of PCDD/PCDF in the pilot plant, the suppression of PCDD/PCDF formation by NH_3, and the effect of NH_3 stoichiometry will be discussed in Sections V.A to C. In addition, the results of a statistical analysis of PCDD/PCDF results (analysis of variance) will be summarized in Section V.D. Finally, the outcome of ash PCDD/PCDF sampling will be presented in Section V.E.

A. Formation of PCDD/PCDF in the Pilot Plant

Table 9 compares average background total PCDD/PCDF results with average outlet results in the absence of NH_3. Table 9 indicates that background PCDD/PCDF levels were considerably higher in Phase IB than in Phase IA. Phase IA corresponded to a "dirty" boiler (average boiler outlet temperature 858°F), while Phase IB corresponded to a "clean" boiler (boiler outlet temperature increasing over the 6-week period from 778 to 805°F). The background PCDD/PCDF levels in Table 9 indicate that a cleaner (cooler) boiler produces more PCDD/PCDF. The implications for the mechanisms of PCDD/PCDF formation outlined earlier will be discussed in Section VI.B.

The outlet PCDD/PCDF data in Tables 9 and 10 are for an economizer outlet temperature of 575°F. Results at an economizer outlet temperature of 525°F were similar. Table 9 demonstrates that substantial PCDD formation occurred in the pilot plant (150 to 310% increase), but little PCDF formation took place. This is in agreement with the results of stack sampling conducted simultaneously (not shown), which indicated that substantial increases in PCDD levels took place between the boiler outlet and the stack, with considerably smaller increases in PCDF levels. The

TABLE 9

FORMATION OF PCDD/PCDF IN THE PILOT PLANT

	PHASE IA TOTAL PCDD	PHASE IA TOTAL PCDF	PHASE IB TOTAL PCDD	PHASE IB TOTAL PCDF
Background	16	121	71	272
Outlet w/o NH3	64	154	180	321
% Increase	310	27	150	18

TABLE 10

SUPPRESSION OF PCDD/PCDF FORMATION IN THE PILOT PLANT BY NH3 AT A STOICHIOMETRY CLOSE TO 1

	PHASE IA TOTAL PCDD	PHASE IA TOTAL PCDF	PHASE IB TOTAL PCDD	PHASE IB TOTAL PCDF
Background	16	121	71	272
Outlet without NH3	64	154	180	321
Outlet with NH3	24	122	97	287
% abs. suppression	63	21	46	10
Formed without NH3	48	33	109	48
Formed with NH3	8	1	26	15
% rel. suppression	83	96	76	69

optimal temperature level for PCDF formation must be above the boiler outlet/pilot plant inlet temperature.

B. Suppression of PCDD/PCDF Formation in the Pilot Plant by NH_3 at a Stoichiometry Close to 1

A comparison of outlet total PCDD/PCDF levels in the absence of NH_3 and in the presence of NH_3 at a stoichiometry of 0.72 is shown in Table 10. These outlet (absolute) values are used to calculate the percent absolute suppression as follows:

$$\%\ \text{absolute suppression} = [1 - (\text{outlet with } NH_3)/(\text{outlet without } NH_3)] \times 100\%$$

Formation values are calculated from outlet values by subtracting the average background value. The percent relative suppression is calculated from the formation values as follows:

$$\%\ \text{relative suppression} = [1 - (\text{formation with } NH_3)/(\text{formation without } NH_3)] \times 100\%$$

Table 10 demonstrates that the average absolute suppression of total PCDD by NH_3 at a stoichiometry of 0.72 was 50%. The average relative suppression of PCDD (suppression of PCDD formation in the pilot plant) was 80%. Since the extent of PCDF formation in the pilot plant was low, the percent absolute suppression of PCDF measured was also low. However, in the presence of NH_3 at a stoichiometry of 0.72, both PCDD and PCDF levels at the outlet of the pilot plant returned to essentially background levels. This is reflected by the uniformly high relative suppression levels. That NH_3 suppresses the majority of PCDD/PCDF *formation* suggests that, if NH_3 were injected at a location in the boiler where no formation had yet taken place (where background levels of PCDD/PCDF were zero), NH_3 would suppress the majority of PCDD/PCDF in *absolute* terms.

C. Effect of NH_3 Stoichiometry on the Suppression of PCDD/PCDF Formation by NH_3 in the Pilot Plant

Table 11 presents the effect of NH3/HCl molar ratio on the suppression of PCDD/PCDF by NH_3. These experiments were performed at an economizer outlet temperature of 575°F, in Phase IB only. Tests in Phase IA had shown that an NH_3 stoichiometry of 0.05 had no effect on PCDD/PCDF values. NH_3 stoichiometries up to 25% of the level which causes 50% absolute suppression of total PCDD (0.18 vs. 0.72) had little effect on PCDD values (less than 10% absolute suppression). An NH_3 stoichiometry 80% higher than the level found to be sufficient (1.3 vs. 0.72) did not increase the percent suppression of PCDD/PCDF. The implications of the dependence on NH_3

TABLE 11

EFFECT OF NH3 STOICHIOMETRY ON SUPPRESSION OF PCDD/PCDF FORMATION BY NH3 IN THE PILOT PLANT

	TOTAL PCDD	TOTAL PCDF
Outlet values		
Without NH3	180	321
With NH3 stoich=0.12	168	430
With NH3 stoich=0.18	165	330
With NH3 stoich=0.72	97	287
With NH3 stoich=1.3	95	290
% Absolute suppress.		
With NH3 stoich=0.12	7	0
With NH3 stoich=0.18	9	0
With NH3 stoich=0.72	46	10
With NH3 stoich=1.3	47	10

stoichiometry on the mechanism of suppression of PCDD/PCDF formation will be discussed in Section VI.B.

D. Statistical Analysis of PCDD Values from Pilot Plant Testing

An analysis of variance for total PCDD values from pilot plant testing (economizer outlet temperature = 575°F only) is summarized in Table 12. The mean square value is calculated by dividing the sum of squares for a factor, or for the error, by the degrees of freedom. In this analysis, two values were used for each factor: Phase IA or IB and NH_3 stoichiometry 0 or 0.72. The F value of a factor is equal to the mean square for the factor divided by the mean square for the error. A large F value means that the signal to noise ratio is high.

The results of the analysis of variance indicate that the experiment phase, which relates to the boiler condition (clean or dirty), is a significant factor in outlet PCDD levels. This means that the boiler condition affects not only background PCDD/PCDF levels, as discussed in Section V.A, but also PCDD/PCDF formation in the pilot plant. This may be because the boiler condition determines whether or not precursors are available downstream, as discussed in Section VI.B.

The statistical analysis of the outlet total PCDD values demonstrates that the suppression of PCDD/PCDF formation by NH_3 at a stoichiometry of 0.72 is significant at the 95% confidence limit.

TABLE 12

ANALYSIS OF VARIANCE FOR TOTAL PCDD VALUES FROM PILOT PLANT TESTING

SOURCE	MEAN SQUARE	F	F CRIT 95%	F CRIT 99%
Phase IA or IB	30,627	11.1644	4.9646	10.044
NH3 Stoichiometry	21,061	7.6774	4.9646	10.044
Error	2,743			

TABLE 13

RESULTS OF PILOT PLANT ASH PCDD/PCDF SAMPLING

	TOTAL PCDD	TOTAL PCDF
Ash without NH3	970	1,300
Ash with NH3=0.72	210	820
% abs. suppression	79	36

E. Results of Ash PCDD/PCDF Sampling in the Pilot Plant

Table 13 presents the results of PCDD/PCDF analysis on ash samples taken from the pilot plant when it was running either without NH_3 or with NH_3 at a stoichiometry of 0.72. Ash sampling was conducted in Phase IB, at an economizer outlet temperature of 575°F. Each value quoted is in picogram/gram ash and represents the average of two samples. The levels of PCDD/PCDF in the ash were low compared to those in the flue gas. If the PCDD/PCDF content of the ash in picogram/gram is multiplied by the ash content of the flue gas (2 to 4 g/DSCM at 7% O_2), the resulting PCDD/PCDF levels in ng/DSCM at 7% O_2 represent only 1 to 2% of those measured in the flue gas.

Ash sampling demonstrated higher levels of absolute suppression by NH_3 than flue gas sampling: 79 and 37%, respectively, for total PCDD and total PCDF. In agreement with flue gas results, ash results indicated greater suppression of PCDD than of PCDF.

VI. FUTURE DIRECTIONS OF THE AMMONIA INJECTION TECHNOLOGY

Based on the results of the AIT pilot project, the current status of the AIT will be evaluated in Section VI.A. In Section VI.B, the implications of the pilot scale results on the proposed mechanisms of PCDD/PCDF formation/suppression will be assessed. The requirements for the air pollution control device to be combined with the AIT are discussed in Section VI.C.

A. Summary of the Current Status of the AIT

The current status of the AIT, based on the results of the AIT pilot project, can be summarized as follows:

- The stoichiometry was found to be sufficient to suppress PCDD/PCDF formation was 0.72.
- The effect of NH_3 on total PCDD levels was found to be significant at the 95% confidence limit.
- The average absolute suppression of total PCDD was 50%.
- The average relative suppression of PCDD (suppression of PCDD formation in the pilot plant) was 80%.
- This represents the level of absolute suppression of PCDD which could be achieved if NH_3 were injected upstream of the boiler location where PCDD formation begins.
- The reduction of PCDD/PCDF emissions required for large RDF facilities under the new Clean Air Act is 75%.
- NH_3 at a stoichiometry of 0.72 also caused outlet total PCDF levels to return to background levels.
- However, since 80 to 85% of PCDF was formed prior to the pilot plant inlet, substantial reductions in absolute PCDF levels could not be demonstrated.
- In addition, 25 to 40% of PCDD was formed at temperatures above 800°F.
- In future work with the AIT, NH3 injection at temperatures above 800°F is recommended in order to verify the extent of suppression of PCDD/PCDF formation by NH_3.
- Based on the high relative suppression results obtained so far, it is likely that NH_3 injection at higher temperatures would result in 80% absolute suppression of both PCDD and PCDF.

B. Implications of Pilot Plant Results on the Mechanism of PCDD/PCDF Formation/Suppression

Certain features of the pilot plant results support the competitive mechanism of PCDD/PCDF formation/suppression proposed by Takacs and Moilanen and contradict the catalytic mechanism proposed by Vogg et al., both outlined in Table 3. The experiments on the effect of NH_3 stoichiometry (Section V.D) demonstrated that low levels of NH_3 did not suppress PCDD/PCDF formation, suggesting that NH_3 is unlikely to be acting as a catalyst poison in this case, since low levels of NH3 should be sufficient to inactivate all catalytic sites. Similarly, the low level of PCDD/PCDF in the ash relative to that in the flue gas implies that a gas phase reaction is likely, instead of a catalytic reaction between species adsorbed onto the ash. Furthermore, the reproducible results obtained from the pilot plant, under conditions where the refuse composition was likely to be changing, suggests that trace elements in the ash (e.g., copper) are not important.

The unexpected phenomenon of a clean boiler (lower boiler outlet temperature) giving higher background PCDD/PCDF levels, reflects the shift in the optimal temperature zone to a location inside the boiler (see below). That a clean boiler (lower boiler out let temperature) also results in more formation of PCDD/PCDF in the pilot plant may be explained by the higher precursor level at the boiler outlet in this case, due to the incomplete oxidation of organics to CO_2.

Both background PCDD and background PCDF levels were higher than expected relative to pilot plant outlet levels. Background PCDD re presented 25 to 40% of pilot plant outlet levels, implying that the optimal temperature range for PCDD formation extends above 800°F (the boiler outlet temperature where background samples are collected). Background PCDF represented 80 to 85% of pilot plant outlet levels, indicating that the optimal temperature for PCDF formation is higher than that for PCDD formation and much higher than the 575°F predicted.

C. Requirements for the Air Pollution Control Device to be Combined with the AIT

In the next phase of the AIT pilot project, the AIT must be combined with an air pollution control device (APCD) for removal of ammonium salts, acid gases, and particulate matter. Figure 2 presents a simplified P and ID for a complete pilot unit.

Table 14 outlines the requirements for the APCD. A wet collector with high efficiency on submicron particulate matter is required due to the low condensation temperature of NH_4^+ salts and due to the small particle size of the condensed solids. Some examples are listed in Table 14. An additional advantage of wet collectors is the higher efficiency of mercury and organics removal from the flue gas.

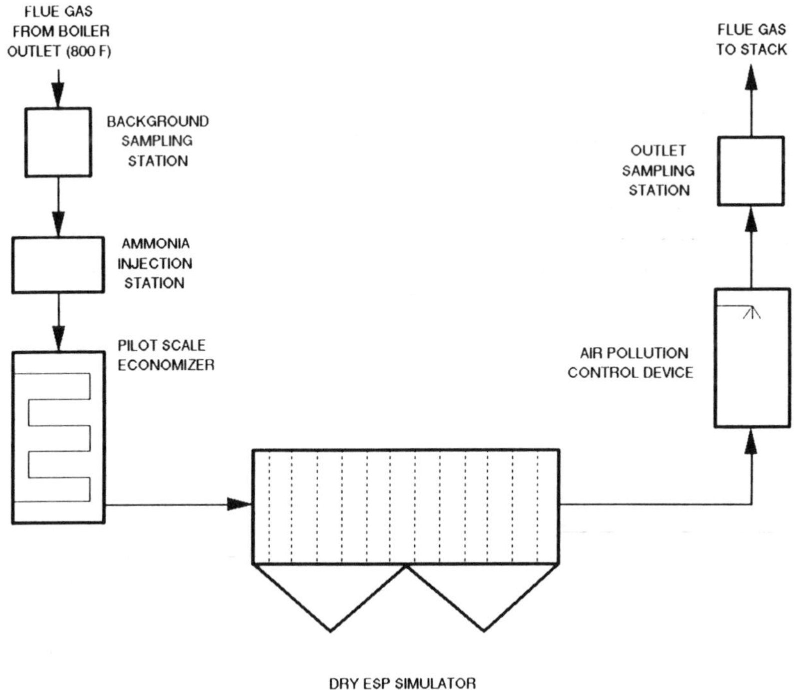

Figure 2. P & ID ammonia injection technology combined with APCD.

TABLE 14

REQUIREMENTS FOR AIR POLLUTION CONTROL DEVICE TO BE COMBINED WITH AMMONIA INJECTION TECHNOLOGY

- **Wet collector with high efficiency on submicron particulate matter, e.g.**
  - **Wet Electrostatic Precipitator**
  - **Ionizing Wet Scrubber**
  - **Electro-Dynamic Venturi**
  - **Frothing Wet Scrubber**
- **Solid-liquid separation**
- **Waste water treatment plant**
- **NH3 recovery and recycle or**
- **NH4+ salt separation and sale**

Any APCD considered for this application must include provisions for handling the liquid effluent produced, as shown in Table 14. The NH_3 recovery and recycle or NH_4 + salt separation and resale is necessary because NH_4^+ salts are soluble a nd there are limits on the NH_4^+ ion concentration in discharges to POTWs.

VII. CONCLUSIONS AND RECOMMENDATIONS

1. The principal advantage of the AIT is that it suppresses PCDD/PCDF formation, rather than removing PCDD/PCDF after it is formed.
2. A 4000 DSCFM pilot plant was used to test the suppression of PCDD/PCDF formation by NH_3 at temperatures below 800°F.
3. NH_3 at stoichiometry close to 1 suppressed TOTAL PCDD formation in the flue gas by 80%.
4. Ash total PCDD levels were also suppressed by 80% in the presence of NH_3 at a stoichiometry close to 1.
5. The majority of PCDF is formed at temperatures above 800°F. The extent of suppression of PCDF formation by NH_3 can only be determined by NH_3 injection at higher temperatures.

REFERENCES

1. Takacs, L. and Moilanen, G. L., Simultaneous control of PCDD/PCDF, HCl and NO_x emissions from municipal solid waste incinerators with ammonia injection, *J. Air Waste Manage. Assoc.*, 41, 716, 1991.
2. Takacs, L. and Moilanen, G. L., Pilot-scale testing of the ammonia injection technology for simultaneous control of PCDD/PCDF, HCl and NO_x emissions from municipal solid waste incinerators, 2nd Annu. Int. Conf. Municipal Waste Combustion, Tampa, FL, April 15–19, 1991.
3. Kilgroe, J. D. and Johnston, M. G., EPA assessment of technologies for controlling emissions from municipal waste combustion, *Solid Waste Power*, 3(6), 18, 1989.
4. Brna, T. G. and Kilgroe, J. D., The impact of particulate emissions control on the control of other MWC air emissions, *J. Air Waste Manage. Assoc.*, 40, 1324, 1990.
5. Vogg, H., Metzger, M., and Stieglitz, L., Recent findings on the formation and decomposition of PCDD/PCDF in municipal solid waste incineration, *Waste Manage. Res.*, 5, 285, 1987.
6. Vogg, H. and Stieglitz, L., Thermal behavior of PCDD/PCDF in fly ash from municipal waste incinerators, *Chemosphere*, 15, 1373, 1986.
7. Stieglitz, L. and Vogg, H., New aspects of PCDD/PCDF formation in incineration processes, Proc. Int. Workshop on Municipal Waste Incineration, Montreal, Quebec, October 1–2, 1987.

Enhanced Fine Particulate Control for Reduced Air-Toxic Emissions

Dennis L. Laudal
Stanley J. Miller
Ramsay Chang

I. BASIS FOR FINE PARTICULATE CONTROL

The primary basis for existing air pollution regulations is the potential for adverse effects of air pollution on public health. However, the quantitative effects of specific concentrations of individual pollutants are difficult to measure. Both gaseous contaminants such as SO_2 and NO_x, as well as particulate matter, may individually cause health problems, but when present together may produce an amplified, synergistic health effect.[1] The effect of particulate matter on health is highly dependent on the size of particulate matter present in the atmosphere, composition as a function of size, and the deposition rate of particles in the respiratory system. Added to the complexity of possible adverse health effects is the fact that certain hazardous trace elements such as arsenic and selenium are known to be highly enriched in respirable particulate matter.[2] Once fine particles are entrained in the atmosphere, they are subject to long-range transport. Therefore, the problem not only occurs near the pollution source, but is of national concern as well.

The main factor that determines where airborne particles will be deposited in the respiratory system is particle size. Pulmonary deposition is most significant for particles from 1 to 3 μm and for particles smaller than 0.1 μm. Above 3 μm, deposition in the pulmonary region falls off simply because few of the larger particles escape upper respiratory trapping. Particles from 1 to 3 μm are deposited in the pulmonary region by gravitational settling, and particles smaller than 0.1 μm are deposited primarily by diffusion. There is a minimum, at about 0.2 μm, for which pulmonary deposition is lowest.[3] Fine particles deposited in the pulmonary region may be retained for several hundred days, which clearly shows the potential for adverse health effects of fine particles in the atmosphere.

Fly ash from coal combustion is composed mainly of the earth's primary crustal elements such as silicon, aluminum, iron, and calcium. However, there are at least 58 elements which are present in coal.[2,4] Many of the trace elements in coal, such as arsenic, cadmium, lead, mercury, and selenium, are considered to be hazardous to human health, but are usually present in coal in very small quantities.[2] The concern over hazardous trace elements in coal is not simply due to the absolute concentration of the element in the coal, but also depends on the fate of the element during combustion and particle formation. Some of the trace metals are volatilized during combustion and condense on the surface of fine fly ash particles which subsequently become greatly enriched in trace metal concentration, compared to the bulk ash. Elements which have been shown to be significantly enriched with decreasing fly ash particle size include arsenic, antimony, barium, gallium, selenium, vanadium, zinc, mercury, nickel, cesium, cadmium, chromium, lead, and thallium.[5-7] The actual reported enrichment ratios, when comparing larger particle concentrations to fine particle concentrations, may be as high as 100. In addition, many of these elements are surface enriched which means that the elemental concentration in contact with respiratory tissue would be much greater than for the average bulk fine particle concentration. Thus, a basis for concern exists over fine particle emissions into the atmosphere because of enrichment of hazardous trace elements. Therefore, effective control of fine particulates would play an important role in controlling air-toxic emissions from coal combustion.

II. FABRIC FILTRATION COLLECTION AND PENETRATION MECHANISMS

Fabric filtration is one of the most effective method of removing fine particulate matter from coal combustion flue gas streams. Overall mass emissions from fabric filters are generally less than 20 mg/m^3, depending on the ash characteristics. However, once an adequate dust cake has been deposited, theoretically, a fabric filter should collect at very nearly 100% efficiency for all particles from 0.01 to 100 μm.[8] If this were the case, nearly all organic and inorganic air toxics emitted as particulate matter would be collected. The question then raised is, "What prevents a fabric filter from being a near-perfect collector?" There are two general mechanisms of ash penetration through a fabric filter. The first is direct, or straight-through, penetration in which suspended particles remain in the gas streamlines and pass directly through the filter along with the gas. Direct penetration implies that there must be large enough pores in the cake/fabric such that particles can avoid collection.[9] The second type of penetration is indirect, or delayed, penetration in which particles are initially collected and then reentrained into the gas exiting the clean side of fabric. Indirect penetration can occur by several mechanisms. One possibility occurs when large pores are initially bridged over, but with increased pressure drop and dust cake buildup the whole pore bridge breaks loose.[8] Indirect penetration may also occur if particles are not held rigidly in the dust cake, gradually seeping through the fabric.[8]

The degree of particulate penetration is dependent on the properties of the particles, the carrying gas, and the filter media. Particle properties which are important include size, density, cohesiveness, shape, electrical charge, and chemical composition. Collection efficiency of a fabric filter may be improved by strategically altering these properties. One such method is flue gas conditioning. By adding small amounts of ammonia and sulfur trioxide (SO_3) to the flue gas prior to the fabric filter, dramatic improvements in the collection efficiency of a fabric filter for fine particles have been observed. Conditioning improves collection efficiency by increasing ash cohesivity, thereby improving the pore-bridging ability of the ash and reducing direct penetration. In addition, conditioning also enhances the ability of the ash to remain in place (stick together) after initial collection, reducing indirect penetration.

III. FACILITIES AND EQUIPMENT

Fabric filter tests were completed with the EERC pilot furnace, known as the particulate test combustor (PTC), which is a 550,000-Btu/h pulverized coal-fired unit designed to generate fly ash representative of that produced in a full-scale utility boiler. The reverse-air baghouse contained a single, full-scale (30-ft long by 1-ft diameter) 10 oz/yd^2 woven fiberglass bag. A smaller baghouse located next to it was a source of cleaned flue gas during reverse-air cleaning. The pulse-jet fabric filter tests were also completed using the PTC, with the original reverse-air system modified to provide pulse-jet cleaning. The system was designed such that up to three bags, 5-inch diameter and up to 20 ft long, could be used to provide the desired air-to-cloth ratio.

Near real-time measurements were made for particles ranging in size from 0.01 to 30 μm. Particles from 0.5 to 30 μm were measured using an aerodynamic particle sizer (APS) manufactured by TSI, Inc. The particle-size distribution can be obtained as number concentration or mass concentration. In addition, another useful feature is the calculation of respirable mass. Respirable mass, as measured by the APS, is a sum of particle mass between 0.5 and 10 μm, including all particles from 0.5 to 1.5 μm, and a decreasing percentage of the particle mass from 1.5 to 10 μm. The American Council of Governmental and Industrial Hygienists' definition of respirable mass is used. This provides a convenient and effective method of plotting fine particle emissions as a function of time.

To measure the number concentration of submicron particles as a function of time (0.01 to 1.0 μm) a condensation nucleus counter (CNC) was used. An impactor with a cut point of about 1 μm was located prior to the CNC to remove larger particles. Detailed descriptions of each of the components were presented in a previous report.[10]

In addition to the CNC and APS, a Flow Sensor six-stage multicyclone provided an inlet particle-size distribution. EPA Method 5 was used to provide dust loading at the inlet and outlet of the baghouses to determine the overall particulate collection efficiency.

SO_3, injected into the flue gas just upstream of the baghouse, was generated by converting sulfur dioxide (SO_2) and air to SO_3 using a vanadium catalyst heated to 850°F. Ammonia was obtained from an anhydrous ammonia tank and was injected just upstream of the SO_3 injection point.

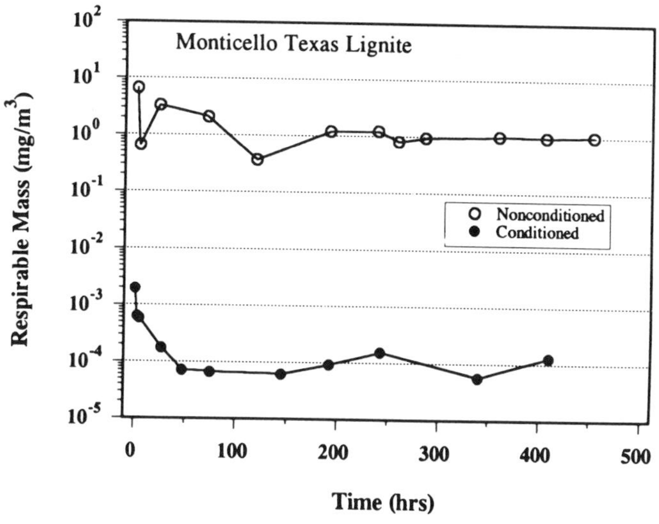

Figure 1. Respirable mass particulate emissions for reverse-air baghouse test with Monticello Coal. (Plotted values are integrated averages over time.)

IV. FLUE GAS CONDITIONING RESULTS

Tests have been conducted with a wide variety of coals, including Monticello and Big Brown Texas lignites and a Pittsburgh #8 bituminous coal. Although there was a substantial improvement in the collection efficiency for all the coals tested,[11,12] this paper will focus on the Monticello, TX, lignite for the reverse-air fabric filter test and the Big Brown, TX, lignite for the pulse-jet tests. These coals produce ashes that are difficult to collect in typical fabric filters. Complete analyses of these coals and their ashes are given in a previous report.[11]

A. Reverse-Air Fabric Filter Tests

The results of the APS measurements in 500-h tests with the Monticello, TX, lignite using the reverse-air system at an air-to-cloth ratio of 2 ft/min are shown in Figure 1. From Figure 1, it is apparent that emissions dropped somewhat during the first 75 h for both tests and then remained fairly steady for the duration of the test. Following the spike just after bag cleaning, for the baseline test, respirable mass emissions dropped to about 1 mg/m^3, compared to about 10^{-4} mg/m^3 for the conditioned test. This corresponds to a respirable mass collection efficiency of about 99.9% for the baseline test and 99.99999% for the conditioned test. A substantial portion of the emissions occurs in the first few minutes after bag cleaning. By integrating the respirable mass emissions over several cleaning cycles, the collection

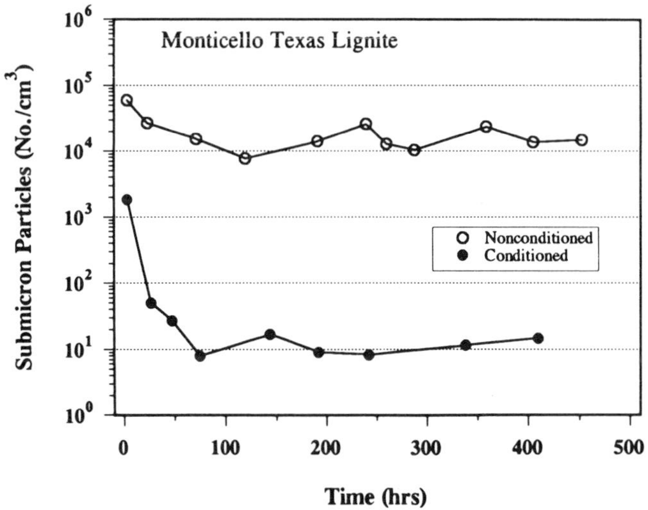

Figure 2. Submicron particulate emissions for reverse-air baghouse tests with Monticello coal as measured with a CNC. (Plotted values are integrated averages over time.)

efficiency based on an entire 2-h filtration period can be determined. With conditioning, the average respirable mass collection efficiency was 99.99996%, with about half of the emissions occurring in the first minute after bringing the bag back on line. This corresponds to a particulate emission concentration of 0.4 $\mu g/m^3$, which is substantially below the ambient particulate concentration in even the most pristine areas of the country.

Although both the EPA Method 5 and the APS data give strong evidence of the dramatic reduction in fine particulate emissions that can be achieved by using conditioning, the collection efficiency for submicron particles must also be considered. The most extensive submicron data were taken with the CNC because it provided a real-time determination of submicron particle emissions. However, several DMPS measurements were also taken to obtain the submicron particle-size distribution. Figure 2 shows the submicron particle emissions as measured by the CNC for both the baseline and conditioned tests. As can be seen, there is more than three orders of magnitude decrease in submicron particulate emissions with conditioning.

For any enhancement technology to be accepted, the fabric filter must operate at a reasonable pressure drop. The advantage of flue gas conditioning is that it not only improves the collection efficiency for fine particulate matter but also reduces the operating pressure drop. This is shown in Figure 3, where the peak baseline operating pressure drop was about 10 inches H_2O compared to 2 inches H_2O with conditioning. Conditioning with ammonia and SO_3 reduces pressure drop by changing the cohesive properties of the ash so that the dust cake porosity is significantly increased.[11,12]

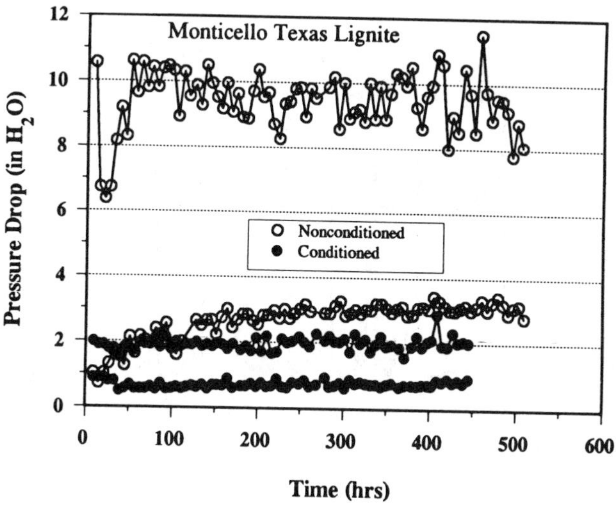

Figure 3. Baghouse pressure drop for reverse-air baghouse test with Monticello coal. Upper values are pressure drop before bag cleaning; lower values are after bag cleaning.

B. Pulse-Jet Fabric Filter Tests

Although reverse-air-cleaned fabric filters are primarily used in the electric utility industry, there has been increased interest in pulse-jet fabric filters.[13] Pulse-jet fabric filters can be operated at higher air-to-cloth ratios and, therefore, may be more economical. To evaluate the effects of conditioning on pulse-jet fabric filters, a series of 8-h tests using a Big Brown, TX, lignite was conducted. Using a woven glass fabric at an air-to-cloth ratio of 4 ft/min, after 6 h of testing, the respirable mass was almost three orders of magnitude lower with conditioning, compared to the baseline test, as shown in Figure 4. In addition, as is shown in Figure 5, the peak pressure drop is 70% lower. Submicron particle emissions, shown in Figure 6, reach a minimum that is five orders of magnitude lower than the baseline minimum. For comparison, ambient air typically has 10^3 to 10^4 submicron particles per cubic centimeter. Flue gas conditioning has been found to be an effective method of substantially reducing fine particulate emissions for both pulse-jet and reverse-air fabric filters. In both cases, particulate emissions are reduced to concentrations well below the levels typically observed in the ambient air of the most pristine areas of the country.

V. CONCLUSIONS

- Hazardous trace elements in coal such as arsenic and selenium are more associated with fine particulate emissions (<2.5 µm) due to surface enrichment.
- Fine particulate aerosols tend to be deposited in the pulmonary system where hazardous trace elements are likely to be absorbed.

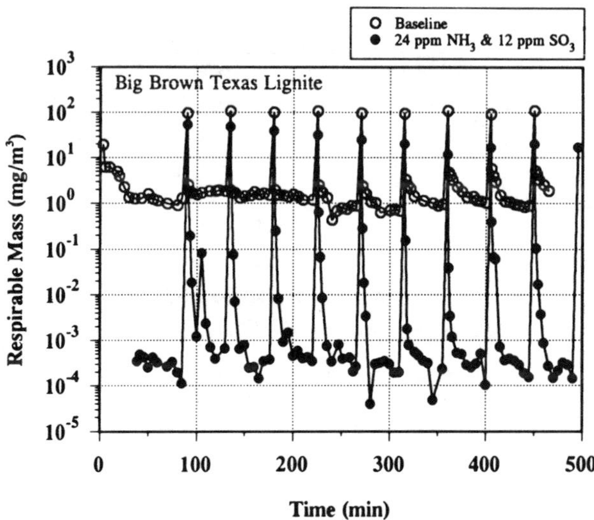

Figure 4. Respirable mass particulate emissions and pressure drop for pulse-jet baghouse test with woven glass fabric and firing Big Brown Coal.

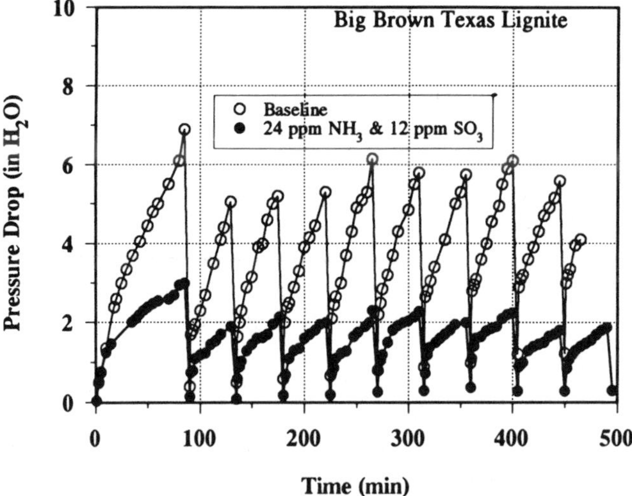

Figure 5. Baghouse pressure drop for pulse-jet baghouse test with woven glass fabric and firing Big Brown Coal.

- Flue gas conditioning with SO_3 and ammonia is an effective method for improving the fine particulate collection efficiency of a fabric filter and, at the same time, for reducing operating pressure drop.

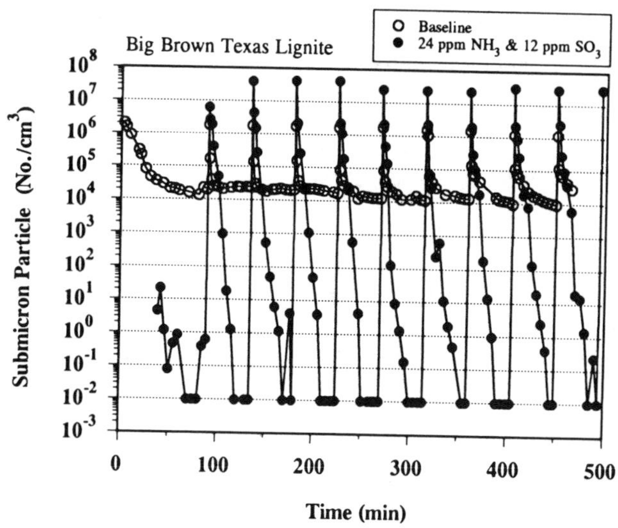

Figure 6. Submicron particulate emissions for pulse-jet baghouse test with woven glass fabric and firing Big Brown Coal.

- For either pulse-jet or reverse-air fabric filters, fine particulate emissions can be reduced to levels below ambient concentrations. Therefore, fabric filtration with flue gas conditioning is a superior method to control all particulate air-toxic emissions.

REFERENCES

1. U.S. EPA, Air Quality Criteria for Particulate Matter and Sulfur Oxides, Vol. 2, EPA-600/8-82-029bF, U.S. Environmental Protection Agency, Research Triangle Park, NC, 1982, 5.
2. Radian Corporation, Trace metals and stationary conventional combustion sources, Technical Report, Vol. 1, EPA-600/7-80-155a, August 1980.
3. Silverman, L., Billings, C. E., and First, M. W., *Particle-Size Analysis in Industrial Hygiene,* Academic Press, New York, 1971, 261.
4. Ray, S. and Parker, F., Characterization of Ash from Coal-Fired Power Plants, EPA-600/7-77-010, U.S. Environmental protection Agency, Research Triangle, NC, 1977.
5. Markowski, G. W. and Filby R., Trace element concentration as a function of particle size in fly ash from a pulverized coal utility boiler, *Environ. Sci. Technol.,* 19, 9, 1985.
6. McElroy, M. W., Carr, R. C., Ensor, D. S., and Markowski, G. R., Size distribution of fine particles from coal combustion, *Science,* 215, 4528, 1982.
7. Hansen, L. D. and Fisher, G. L., Elemental distribution on coal fly ash particles, *Environ. Sci. Technol.,* 14(9), 1111, 1980.
8. Leith, D., Rudnick, S. N., and First, M. W., High-velocity, high-efficiency aerosol filtration, EPA-600/2-76-020, U.S. Environmental Protection Agency, Research Triangle Park, NC, 1976.

9. Dennis, R. et al., Filtration model for coal fly ash with glass fabrics, EPA-600/7-77-084, U.S. Environmental Protection Agency, Washington, DC, 1977.
10. Miller, S. J. and Laudal, D. L., Flue gas conditioning for improved fine particle capture in fabric filters: comparative technical and economic assessment, in Low-Rank Coal Research Final Report, Vol. 2, Advanced Research and Technology Development, DOE/FC/10637-2424 (DE87006532), June 1987.
11. Miller, S. J. and Laudal, D. L., Flue Gas Conditioning for Fabric Filter Performance Improvement, Final Project Report for Contract No. DE-AC22-88PC88866 for Pittsburgh Energy Technology Center, December 1989.
12. Miller, S. J. and Laudal, D. L. Enhancing baghouse performance with conditioning agents: basis, developments, and economics, in Proc. 8th EPA/EPRI Symposium on the Transfer and Utilization of Particulate Control Technology, Vol. 2, EPRI GS-7050, 1990, 23.
13. Belba, V. H., Grubb, T., and Chang, R. L., A survey of the performance of pulse-jet baghouses for application to coal-fired boilers, worldwide, in Proc. 8th EPA/EPRI Symposium on the Transfer and Utilization of Particulate Control Technology, Vol. 2, EPRI GS-7050, 1990, 6.

Retention of Condensed/Solid Phase Trace Elements in an Electrostatic Precipitator

P. R. Tumati
M. S. DeVito

I. ABSTRACT

The performance of an electrostatic precipitator (ESP) in a pilot-scale combustion test facility was examined for retention of condensed/solid phase trace elements. Particulate samples were collected at the ESP inlet and outlet after an eastern bituminous coal — Pittsburgh seam — was burned in a pilot combustor. The samples collected, using U.S. EPA draft multimetal sampling method, were analyzed by inductively coupled plasma-mass spectrometry (ICPMS). The bulk concentrations of the trace metals were reproducible both at the ESP inlet and outlet. The ESP retention efficiency for most solid phase trace elements (except Ni and Se) was close to the overall particulate collection efficiency. Composition and size distributions for the elements at the ESP inlet were determined in the particle diameter range of 0.5 to 20 μm. In the particle size range studied, Cr, Mn, Zn and Se showed distinct bimodality with a fine peak around 1 μm and a coarse peak around 7 μm. The percent mass fraction in the fine mode correlated with the trace element removal efficiency in the pilot ESP. The trace element removal efficiencies and emission rates observed in the pilot-scale facility compared well with the reported data from large-scale commercial units.

II. INTRODUCTION

The term "trace elements" is applied to the constituents in coal which are usually present in concentrations less than 0.01%.[1] The concentrations of trace elements in coal vary depending on rank, geological surroundings, and geographical location.[1,2] Modes of trace element occurrences in U.S. coals vary considerably and may exhibit affinity for the inorganic and/or organic portions of the coal.[1-4]

During coal combustion, trace elements are released from both organic and mineral matter.[1] The physical and chemical transformations that the coal particle undergoes during combustion determine the fly ash particle size, composition, as well as the extent of mineral matter vaporization.[5] Trace element distributions from coal combustion arise from (1) elements not volatilized during combustion and (2) elements volatilized. Those elements that volatilize may exit as gases, condense as submicron particles, or condense on the surface of the particulate matter in the combustion gas stream. The elements not volatilized in the combustion zone form a melt of rather uniform composition that makes up the matrix of both fly ash and bottom ash.[6] The volatile elements are preferentially concentrated by condensation/absorption on the surfaces of the smallest fly ash particles as the flue gas cools. This preferential condensation or absorption on these particles can enrich trace element concentrations over those normally found in coal.[6,7]

For most pulverized-coal-fired boilers, about 80% of the coal ash is recovered as fly ash. An ESP is the most common control device used to remove particulate matter/fly ash from the combustion gases.[8] Current ESPs are highly efficient and can operate in +99.9% collection efficiency range. However, the overall ESP collection efficiency depends on various parameters such as particle size, ash resistivity, flue gas temperature, moisture content, and presence of conditioning agents such as SO_3, NH_3, etc.[9] The ESP efficiency is generally lowest in the 0.1 to 1.5 μm particle size range. A fraction of the particulate mass exiting the ESP falls in this size range. This material can be enriched in trace elements. It is, therefore, important to evaluate trace element concentrations as a function of particle size, ESP retention efficiency, and emission factors for various trace elements.

In the present study, ESP performance in controlling trace element emissions was studied in a pilot-scale test facility at Consol using a washed Pittsburgh seam coal.

III. TRACE ELEMENT CLASSIFICATION

During combustion, the elements originally present in the coal are partitioned between the bottom ash, the fly ash, and the flue gases that escape from the stack. Various classification schemes have been developed to describe the partitioning and enrichment behavior of trace elements during combustion. Four general classes of trace elements based on partitioning are[10]

- Class 1 — Elements that are approximately equally distributed between fly ash and bottom ash or show little or no particle enrichment
- Class 2 — Elements that are enriched in fly ash relative to bottom ash
- Class 3 — Elements that are intermediate between classes 1 and 2, e.g., several investigators associate these elements with class 1 or class 2
- Class 4 — Elements that are emitted in the gas phase

In the present study, 13 particulate bound trace elements in the class 2 and 3 categories were examined. The trace elements are arsenic (As), beryllium (Be), cadmium (Cd), chromium (Cr), copper (Cu), lithium (Li), manganese (Mn), nickel (Ni), lead (Pb), antimony (Sb), selenium (Se), tin (Sn), and zinc (Zn). Modes of occurrence of the subject elements have been studied and reported by several investigators.[1,2,11,12] In general, most of the trace elements in bituminous coals were associated with the mineral matter.[1,11]

Elements such as As, Cd, Pb, and Zn have been reported to have an affinity for the mineral portion of the coal and tend to associate with pyrite or accessory sulfide minerals.[1,12] Cadmium is commonly associated with sphalerite (ZnS). As was found to occur in solid solution with pyrite.[12] Other elements reported to be associated with pyrite include Mn and Se.[1,2] Mn commonly substitutes for Ca in calcite[12,13] while Zn appears to occur as sphalerite in most coals.

Be appears to have strong affinity towards the organic fraction of the coal.[2,12,13] Elements such as Cr, Cu, Ni, and Se can have strong affinities for either the mineral or the organic components.[2] However, most of the studies on organic/inorganic affinities of trace elements in coal were based on bulk chemical analyses of the float and sink fractions from conventional washability tests. A drawback to this method is that trace elements occurring in finely divided and widely disseminated minerals in the organic matrix are improperly classified as having an organic association.[2,14]

During combustion, trace elements that are trapped in the organic matrix or bonded in organic compounds have much higher probability of being transferred to the vapor state than a similar compound associated with the mineral fraction.[14] Therefore, the organic/inorganic affinity and volatility of trace elements play a significant role in the fly ash particle enrichment during combustion.

IV. FORMATION AND ENRICHMENT OF ASH PARTICLES IN COAL COMBUSTION

During pulverized coal combustion, several processes occur that determine the size and composition distributions of the resulting ash particles. Ash particle size distribution is influenced by various processes such as ash coalescence on the burning particle surface and fragmentation of char and possibly some minerals.[5,6,14,15] Ash composition is determined by the extent of ash coalescence, mineral oxidation and fragmentation, and vaporization and condensation of inorganic species during combustion.[14-16]

Many field and laboratory studies show that the ash particle size distributions that result from pulverized coal combustion are bimodal.[6,15,16] The fine particle mode is thought to be formed due to nucleation of vaporized ash components and growth via coagulation and heterogeneous condensation.[15,16] The coarse-particle formation mechanism involves combustion of char and fragmented char particles, and carryover of the incombustible mineral matter.[16]

If an element vaporizes completely during combustion, the form of the resulting composition distribution depends on the prevailing gas-to-particle conversion mechanisms. The particles that form via homogeneous condensation of inorganic vapors and coagulation result in a composition size distribution that is independent of particle diameter.[15,16] The heterogeneous condensation is thought to take place in two different regimes, i.e., free molecular and continuum. In the free molecular regime, where the Knudsen number (a ratio of mean gas free path to the particle diameter) is greater than 1, the heterogeneous condensation of vapors on existing particles may

Table 1. Coal Analysis

Proximate (dry) (wt %)		Ultimate (dry) (wt%)		Major ash elements (wt% of dry ash)	
Ash	(7.26)	Carbon	(77.75)	SiO_2	(41.76)
Volatile matter	(39.42)	Hydrogen	(5.22)	Al_2O_3	(21.87)
Fixed carbon	(53.32)	Nitrogen	(1.46)	TiO_2	(0.94)
Btu	(13,963)	Chlorine	(0.12)	Fe_2O_3	(22.04)
		Sulfur	(2.32)	CaO	(4.99)
		Oxygen (diff.)	(5.86)	MgO	(0.91)
				Na_2O	(0.89)
				K_2O	(1.35)
				P_2O5	(0.36)
				SO_3	(3.79)

Note: Washed Pittsburgh seam coal.

result in a composition distribution that is inversely proportional to the particle diameter.[15] In the continuum regime, where the Knudsen number is less than 1, the resulting composition size distribution is inversely proportional to the square of the particle diameter.[15] If an element vaporizes partially during the combustion process, heterogeneous condensation of vapor particles on the coarse particles may result in a composition size distribution that may differ from the inverse diameter square relationship.[16]

In this study, composition size distribution for the size range of (0.54 and 30 μm) was determined from measured differential size distributions. Particle diameter ranges and percent mass fraction in both the coarse and fine modes were determined for the combustor ash at the ESP inlet. ESP collection efficiency for the subject trace metals and the effect of percentage mass fraction in the fine mode on collection efficiency were determined. The collection efficiency and the emission factors for the individual trace elements were compared with the published data.

V. EXPERIMENTAL FACILITY

Combustion experiments were conducted in a 1.5-MM Btu pilot-scale combustor test facility, specifically designed to simulate the combustion environment of utility boilers. The test facility is also equipped with a pilot-scale electrostatic precipitator. Details of the pilot-scale test facility and operations procedures have been reported elsewhere.[9] During this test series, the combustor was operated in a tangential firing mode at a heat release rate typical of full-scale utility boilers. Washed Pittsburgh seam coal was pulverized to nominal 70% minus 200 mesh in a Williams Roll and Race mill prior to the test. The proximate and ultimate analyses and major ash elements of the as-fired coal are presented in Table 1.

A. Sampling and Analysis

Total gas flow rate, percentage moisture, and grain loading at the ESP inlet and outlet were measured using U.S. EPA Method 5. Vapor phase SO_3 in the flue gas was measured using an acid condensation method. Particle size distributions at the inlet and outlet, respectively, were measured using a five-stage cyclone and a seven-stage Anderson impactor.

Particulate matter for trace element analysis was withdrawn isokinetically from the ESP inlet and collected on a fiber glass filter using the most recent EPA draft multimetal sampling method. Samples for elemental analyses by particle size were collected using five-stage cyclone at the ESP inlet.

Coal samples were collected from each tote bin using a sample thief. The samples were mixed and riffled to form one composite sample for subsequent elemental analyses. Samples of bottom ash were taken when the combustor was cool, following the test.

All elements were determined by inductively coupled plasma-mass spectrometry (ICP-MS). This instrument ionizes the element in an argon plasma at 8000 to 10,000 K. The ions are then detected with a mass spectrometer.

Prior to the sampling tests, background filter concentrations were measured from several randomly selected quartz filters identical to those used in the actual sampling. For 10 of the 13 elements tested, the background concentrations (in the filters) were less than 2 ppm. However, the background concentrations for Zn, Cr, and Cu were relatively high: approximately 16, 4, and 3 ppm, respectively. Based on the trace element concentrations in the coal ash, sampling time for collecting the solids was chosen to yield large quantities (an order of magnitude higher than the background quantities) of various elements in the samples. Sampling times of approximately 6 h at the ESP outlet and 2 h at the ESP inlet were chosen to collect adequately large solid samples for the comprehensive analyses.

VI. TEST RESULTS

The ESP was operated in the current-limited mode at a current density of 40 nA/cm^2. Voltage and current data collected several times during the test series did not indicate any presence of back corona. The ESP operating data indicated normal operation at an overall collection efficiency of 99.54% on mass basis. The particle size data indicated that the average geometric mean particle size was 20 μm at the ESP inlet and 3.5 μm at the ESP outlet. This is consistent with the behavior of most ESPs, in that precipitators tend to act as classifiers; larger particles are removed near the inlet section and the average particle size diminishes as the gas proceeds through the unit.

In addition, the size distribution at the outlet was bimodal with peaks around 0.6 and 7 μm. The ESP fractional efficiency shown as a result was very low in the particle

Table 2. Average Trace Element Concentrations (ppm)

Element	Feed coal (ppm of ash)	Bottom ash	ESP inlet*	ESP outlet*	Retention efficiency
Cadmium (Cd)	1.2	0.3	1.9	5.2	98.57
Antimony (Sb)	4.0	2.3	9.2	12.9	99.26
Tin (Sn)	6.2	3.7	14.3	19.9	99.27
Beryllium (Be)	8.5	8.2	13.3	15.1	99.41
Arsenic (As)	49.5	14.6	141.5	181.4	99.33
Selenium (Se)	54.1	4.0	3.4	29.4	95.49
Lead (Pb)	52.6	39.7	167.9	277.2	99.13
Copper (Cu)	70.4	75.8	110.5	166.8	99.21
Nickel (Ni)	70.4	88.4	140.6	622.6	97.68
Zinc (Zn)	80.7	176.0	232.4	479.0	98.92
Lithium (Li)	127.0	127.0	168.3	181.8	99.43
Chromium (Cr)	143.3	210.0	361.7	406.1	99.41
Manganese (Mn)	249.5	249.0	311.6	411.0	99.31

size range of 0.2 to 1.0 μm. The collection efficiency was minimum (91%) for particles in the 0.65 μm size range. The d Mass/d Log Dp data computed from the measured differential size distributions indicated that approximately 15% of the total mass leaving the ESP is less than 0.65 μm in diameter.

The concentrations of various trace elements in the feed coal (microgram per gram of coal ash), bottom ash, fly ash at the ESP inlet and outlet, and the ESP retention efficiencies of the elements are shown in Table 2. Except for chromium and zinc, concentrations of most trace elements did not increase in the bottom ash. However, in all cases, the elemental concentrations increased significantly in the fly ash both at the inlet and outlet compared to coal ash. The ESP retention efficiency for individual elements was approximately the same as the overall collection efficiency except for nickel and selenium.

Elemental composition size distributions of the trace elements Cr, Mn, Li, and Ni in the inlet fly ash are shown Figure 1a. Due to difficulties in measuring small particle sizes, the composition size distribution data presented in the report are limited to particles larger than 0.5 μm. The composition size distributions for Li, Ni and Cr are nearly independent of the particle diameter. This indicates that a small fraction of these elements may vaporize and condense during the combustion process. However, the composition size distributions for Mn (Figure 1a), Cu, Zn, Pb, and As (Figure 1b), Be and Se (Figure 2a), and Sn, Sb, and Cd (Figure 2b) indicate that as the particle diameter decreased the elemental concentrations increased. Above 5 μm, the size distribution for most of the elements appear to be independent of particle diameter. The elemental concentrations when plotted against inverse particle diameter square (d^{-2}) appeared nearly linear in the size range of 0.5 to 3 μm. Heterogeneous condensation is the most probable reason for the observed particle dependence and the composition size distributions.

Elemental size distributions of all trace elements based on five-stage cyclone samples at the ESP inlet are shown in Figure 3. In the particle size range between 0.5

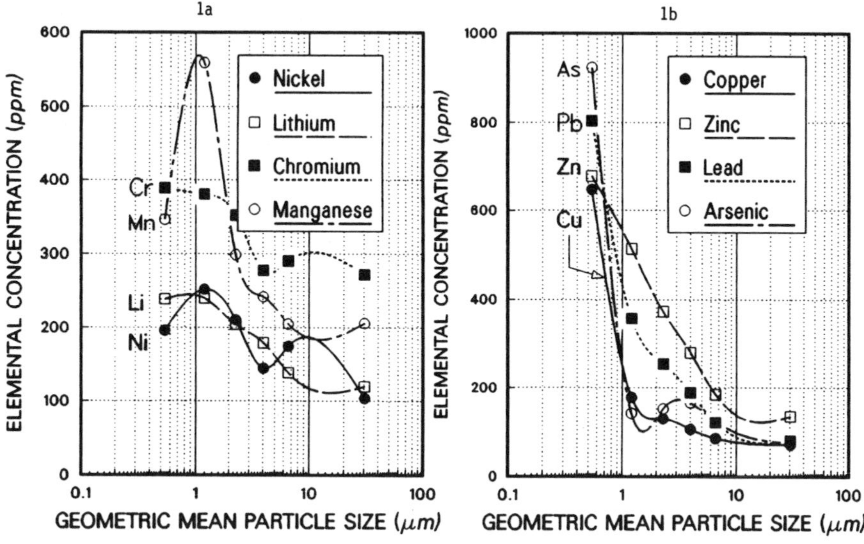

Figure 1. Elemental concentrations in size-classified fly ash.

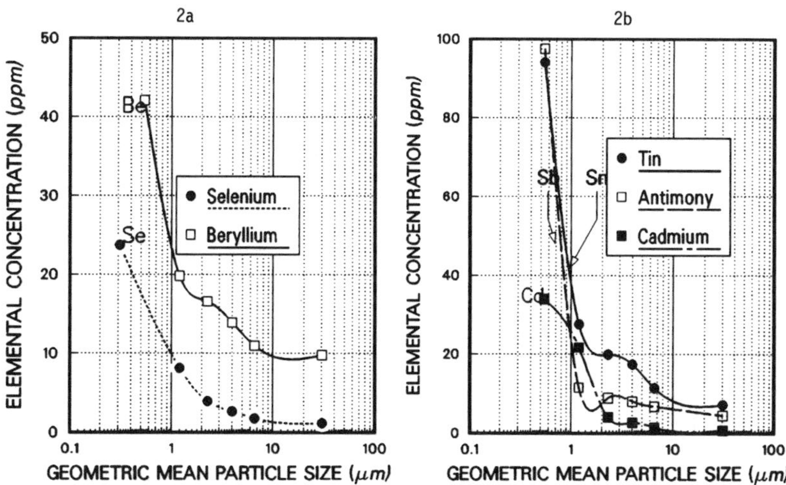

Figure 2. Elemental concentrations in size-classified fly ash.

to 20 μm, distinct bimodality can be observed for Cr, Mn, Zn, and Se. The coarse and fine particle modes peak at 7 and 1 μm. For the remaining elements, no distinct fine mode was observed in the data range presented. More data are probably needed in the submicron particle range (0.005 to 0.5 μm) to determine the fine mode peak for the remaining elements.

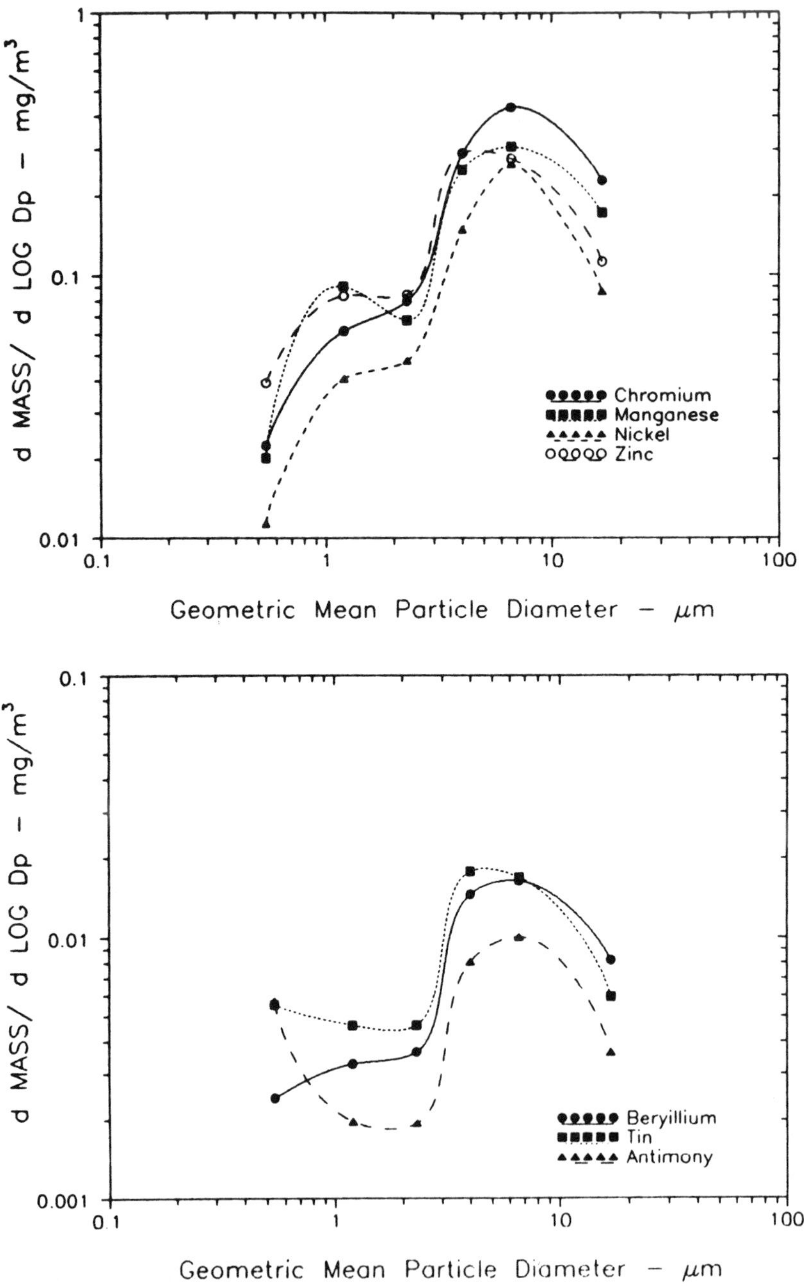

Figure 3. Composition size distributions.

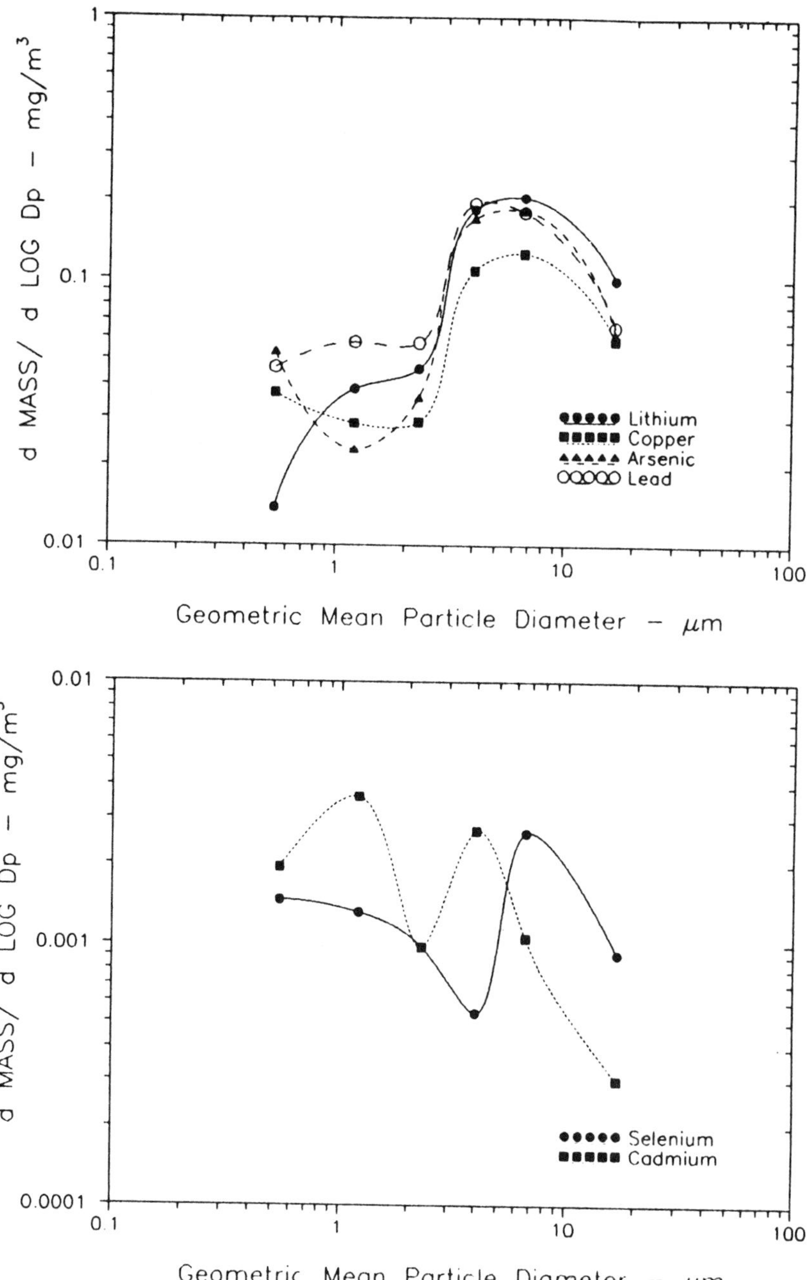
d MASS/ d LOG Dp – mg/m³
1
0.1
0.01
0.1
1
10
100
Geometric Mean Particle Diameter – μm
Lithium
Copper
Arsenic
Lead
d MASS/ d LOG Dp – mg/m³
0.01
0.001
0.0001
0.1
1
10
100
Geometric Mean Particle Diameter – μm
Selenium
Cadmium

Table 3. Elemental Mass Distribution between Fine and Coarse Modes

Element	Total mass flow rate (mg/h)	Coarse mode mass fraction	Fine mode mass fraction	Collection efficiency (%)
Chromium (Cr)	152.00	0.99	0.01	99.41
Manganese (Mn)	121.00	0.88	0.12	99.31
Zinc (Zn)	111.00	0.84	0.16	98.92
Arsenic (As)	69.32	0.80	0.20	99.33
Copper (Cu)	53.22	0.79	0.21	99.21
Antimony (Sb)	4.46	0.68	0.32	99.25
Selenium (Se)	1.19	0.58	0.42	95.49
Cadmium (Cd)	1.32	0.39	0.61	98.70

Total elemental mass flow rate distributions between the fine and coarse mode for various trace metals are shown in Table 3. For the elements shown, the percentage mass fraction in the fine mode correlated well with the trace element penetration. For example, Se and Cd with most of their mass (41 and 65%, respectively) in the fine mode have relatively lower collection efficiencies (95.5 and 98.7%). Similarly, chromium with only 1% of its mass in the fine mode had the highest collection efficiency.

A comparison of trace element removal efficiencies and emission rates with two other studies (Brooks,[10] EPA[17]) is shown in Table 4. The Radian study consisted of measured efficiencies and emission rates from several bituminous coal-fired dry bottom units equipped with coldside ESPs. The study presented range of values for each trace element studied. The EPA study consisted of actual operating data from large-scale commercial units as well. However, the number, size, and types of units involved are unknown. In general, there are very limited data available on the trace element removal efficiencies and emission rates from modern coal-fired power plants.

The trace element removal efficiencies observed in the present study compared reasonably well with the reported EPA data. Both the studies indicated relatively low collection efficiency for Se. A similar trend was observed for Ni as well. The retention efficiencies reported in the upper range of the Radian data agreed well with the pilot-scale data.

The emission rates for the pilot-scale unit were computed based on the particulate mass leaving the ESP and the measured exit concentrations. The emission rates in lb/10^{12} Btu agreed well with the lower data range in the reported Radian study. Large variations in the trace element emission rates exist between the reported sets of data. The variations are due to differences in total particulate flow rates resulting mainly from large variations in ESP collection efficiencies. In general, as the ESP efficiency increases the emission rates of trace elements decrease significantly. For example, an increase in collection efficiency from 90 to 99.9% can decrease the total particulate flow rate and as a result trace element emission rate by 100 times. Thus, an ESP can be an effective tool in controlling various trace element emissions. Other factors that

Table 4. A Comparison of Trace Element Removal Efficiencies and Emission Rates

	Collection efficiency (%)				Emission rate (lb/10^{-12} BTU)		
		Utility precipitators					
			Radian '89[b]				
Element	Pilot ESP	EPA '81[a]	Data range	Average[c]	Pilot ESP	EPA '81[a]	Radian '89[b] data range
Cadmium (Cd)	98.57	95.6	97.6–18.3	74.6	0.023	13.94	0.22–26.5
Beryllium (Be)	99.41	98.4	99.95–86.7	91.9	0.07		0.6–14.0
Arsenic (As)	99.33	95.3	97.6–50.5	87.5	0.80	116.00	0.35–138
Selenium (Se)	95.49	86.0			0.13		
Lead (Pb)	99.13	95.5			1.22	116.00	7.0–91.0
Copper (Cu)	99.21	99.2	99.2–28.6	85.0	0.73	46.47	34–974
Nickel (Ni)	97.68	52.5	99.5–48.8	79.1	2.73	125.00	700–2600
Zinc (Zn)	98.92	97.0			2.10		
Chromium (Cr)	99.41	95.1	98.6–46.7	71.5	1.78	185.00	1.6–7970
Manganese (Mn)	99.31	99.0	99.7–9.4	78.1	1.80	116.00	1.0–9240

[a] Larry Edwards et al.[17]
[b] 1989 Radian study includes ESPs of various sizes.[10]
[c] Pulverized dry bottom utility boilers fired with bituminous coal; two to eight different units and several tests with the same unit depending on element measured.

can influence the emission rates are trace element concentrations in the feed coal, boiler configuration and operating conditions.

VII. CONCLUSIONS

Retention efficiencies and emission rates from a pilot-scale electrostatic precipitator were determined for Cd, Be, As, Se, Pb, Cu, Ni, Zn, Cr, and Mn after burning an eastern bituminous washed Pittsburgh seam coal. The retention efficiencies for most elements were close to the overall ESP collection efficiency. The composition distribution at the ESP inlet was distinctly bimodal for several elements tested. The percent mass fraction in the fine mode correlated well with the trace element removal efficiencies for various elements. The measured retention efficiencies compared well with the reported data from large-scale commercial units.

REFERENCES

1. Raask, E., The mode of occurrence and concentration of trace elements in coal, *Prog. Energy Combust. Sci.,* 11,97, 1985.
2. Akers, D. J., Trace Elements in Coal and Coal Wastes, EPRI Report No. GS-6575, December 1989.
3. Swaine, D. J., *Trace Elements in Coal,* Butterworths, London, 1990.
4. Raask, E., *Mineral Impurities in Coal Combustion,* Hemisphere Publishing, New York, 1985.
5. Helble, J. J. and Sarofim, A. F., Influence of char fragmentation on ash particle size distribution, *Combustion Flame,* 76, 183, 1989.
6. Quann, R. J., Neville, M., Janghorbani, M., Mims, C. A., and Sarofim, A., Mineral matter and trace-element vaporization in a laboratory-pulverized coal combustion system, *Environ. Sci. Technol.,* 16, 776, 1982.
7. Smith, I., *Trace Elements from Coal Combustion: Emissions,* IEA Coal Research, London, 1987.
8. Trueblood, R. C., Wedig, C., and Gendreau, R. J., Efficiency of fabric filters and ESPs in controlling trace metal emissions from coal-burning facilities, 8th Symp. Transfer and Utilization of Particulate Control Technology, San Diego, March 1990.
9. Tumati, P. R. and Rees, D. P., Effect of flue gas conditioning on ESP performance, EPRI Symp. on Effects of Coal Quality on Power Plants, St. Louis, September 1990.
10. Brooks, G., Estimating Air Toxics Emission From Coal and Oil Combustion Sources, EPA-450/2-89-001, U.S. Environmental Protection Agency, Researach Triangale Park, NC, 1989.
11. Finkelman, R. B., Modes of Occurrence of Trace Elements in Coal, Ph.D. thesis, University of Maryland, Department of Chemistry, 1980.
12. Finkelman, R. B. and Gluskoter, H. J., Characterization of minerals in coal: problems and promises, Proc. Engineering Foundation Conference, 1981, 299.
13. Gluskoter, H. J., Ruch, R. R., Miller, W. G., Cahill, R. A., Dreher, G. B., and Kuhn, J. K., Trace Elements in Coal: Occurrence and Distribution, Illinois State Geological Survey, EPA-600/7-77-064, June 1977.

14. Smith, R. D., Campbell, J. A., and Felix, W. D., Atmospheric trace element pollutants from coal combustion, *Mining Engineering,* November 1980, 1603.
15. Flagan, R. C. and Friedlander, S. K. Particle formation in pulverized coal combustion — a review, in *Recent Developments in Aerosol Science,* John Wiley & Sons, New York, 1976, p. 48.
16. Kauppinen, E. I. and Pakkanen, T. A., Coal combustion aerosols: a field study, *Environ. Sci. Technol.,* 24, 1811, 1990.
17. Edwards, L. O., Muela, C. A., Sawyer, R. E., Thompson, C. M., Williams, D. H., and Delleney, R. D., Trace Metals and Stationary Conventional Combustion Processes, EPA Report 600/S7-80-155, U.S. Environmental Protection Agency, Research Triangle Park, NC, 1981.

Chapter 7
Poster Session

Moderator: Winston Chow, EPRI

The Clean Air Act Amendments of 1990: Hazardous Air Pollutant Requirements and the DOE Clean Coal Technology Program

P. D. Moskowitz
M. DePhillips
V. M. Fthenakis
A. Hemenway

I. INTRODUCTION

The purpose of the U.S. Department of Energy — Office of Fossil Energy (DOE FE) Clean Coal Technology Program (CCTP) is to provide the U.S. energy marketplace with advanced, efficient, and environmentally sound coal-fired technologies.[1] The design, construction, and operation of Clean Coal Technology Demonstration Projects (CCTDP) will generate data needed to make informed, confident decisions on the commercial readiness of these technologies. These data also will provide information needed to ensure a proactive response by DOE and its industrial partners to the establishment of new regulations or a reactive response to existing regulations promulgated by the U.S. Environmental Protection Agency (EPA). The objectives of this paper are to (1) present a preliminary examination of the potential implications of the Clean Air Act Amendments (CAAA) — Title III hazardous air pollutant requirements to the commercialization of CCTDP and (2) help define options available to DOE and its industrial partners to respond to this newly enacted legislation.

II. CAAA REQUIREMENTS

Figure 1 summarizes the issues and strategy being followed by the EPA to develop CAAA Title III hazardous air pollutant emission standards for routine releases from stationary sources. More specifically, Title III lists 189 hazardous air pollutants and directs EPA to promulgate maximum achievable control technology (MACT) standards for industrial sources emitting these contaminants. MACT standards may be achieved through process changes, installation of pollution controls, materials substitution, or operator training and certification. The failure of these controls to provide an ample margin of safety to public health, e.g., a residual cancer risk exceeding one in 10,000 to the most exposed person, would require the EPA administrator to develop more stringent emission limits.

Sources which may be regulated include "electric utility steam generating units" and "major sources". Section 301[(a)subsection 8] defines an electric utility steam

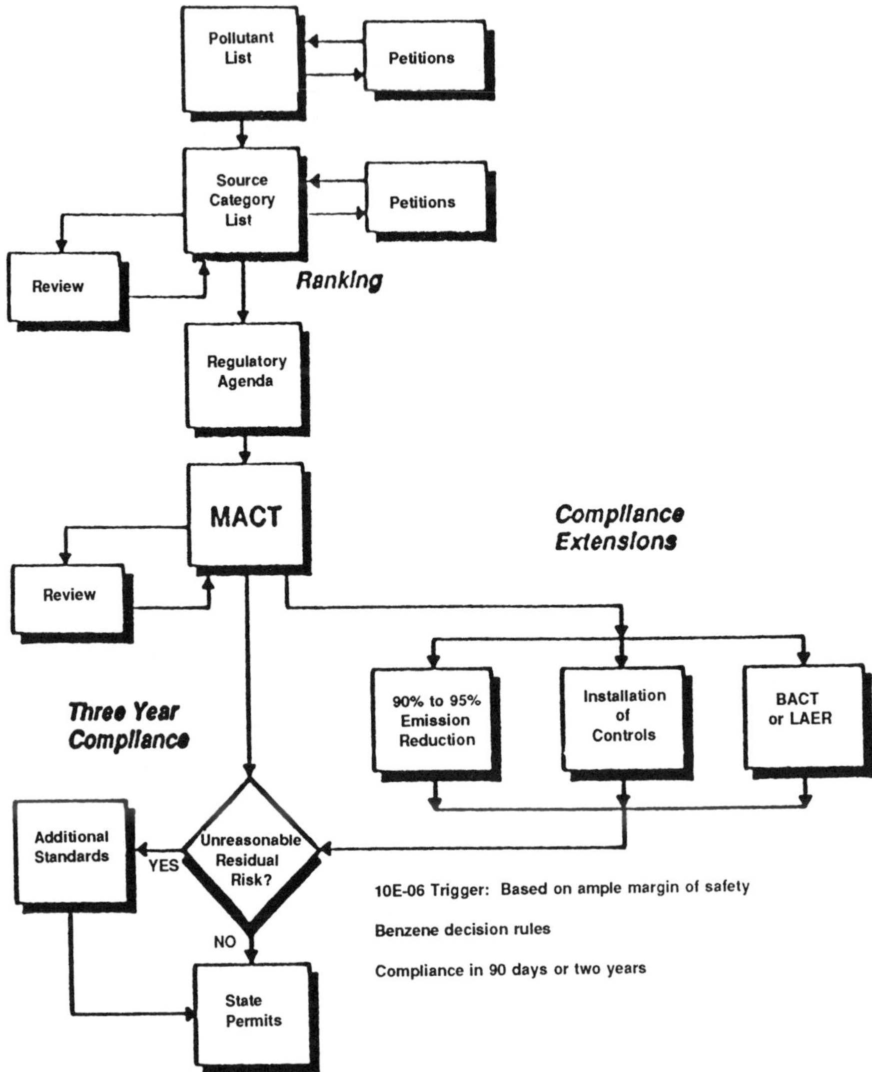

Figure 1. Title III air toxics strategy to be followed by the EPA.

generating unit as "...any fossil fuel fired combustion unit of more than 25 megawatts (MW_e) that serves a generator that produces electricity for sale..." A "major source" as any stationary source or group of stationary sources located within a contiguous area and can emit more than 10 tons per year (tpy) of any one listed pollutant and/or 25 tpy for any combination of listed pollutants.

Figure 2 presents the operational schedule for the CCTP and the statutory schedule for the Title III requirements. As shown, there is potential overlap in the

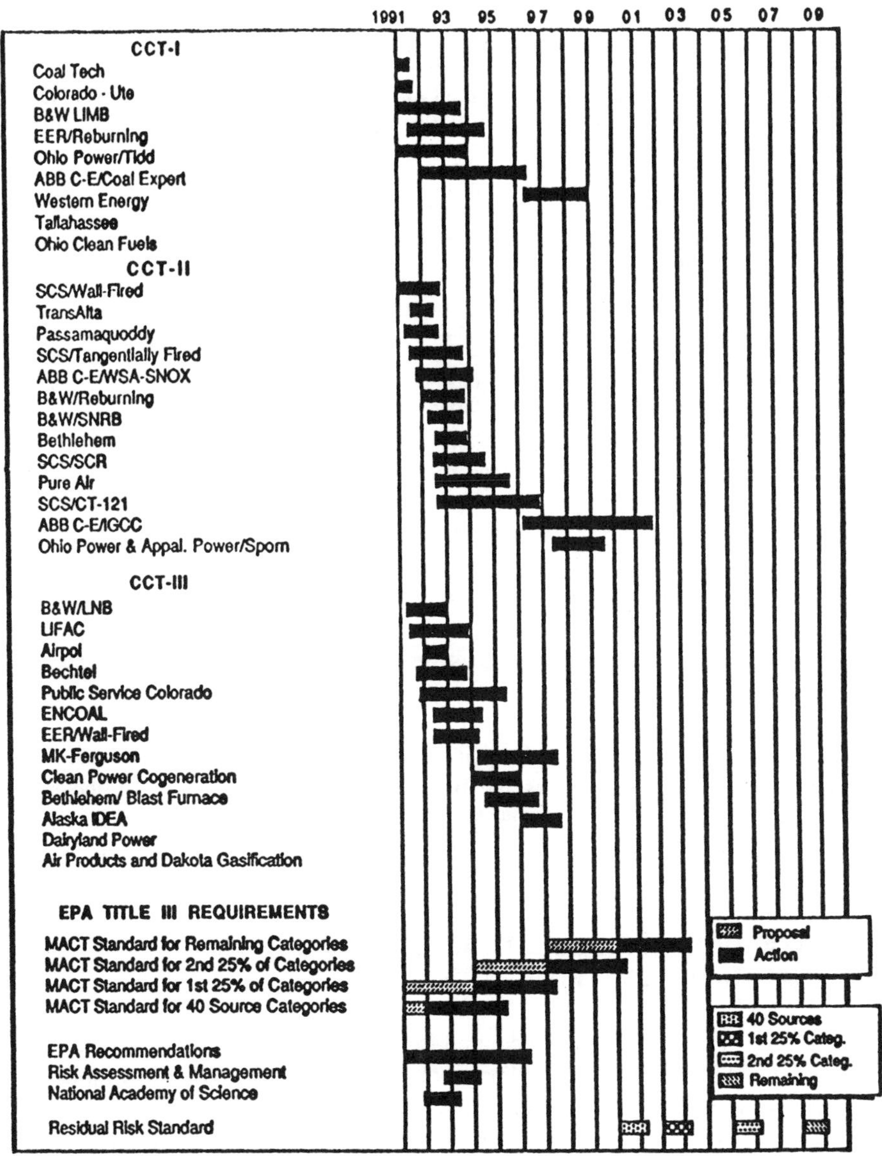

Figure 2. Clean coal technology demonstration project operating and Title III regulatory schedules.

schedules among these two programs. Thus there is a unique opportunity for the CCTP and its industrial partners to act proactively and collect data from commercial-scale fossil-based operations in time to contribute to the EPA rule-making process. These data can ensure the development of appropriate and defensible regulations for

Table 1. Base Case Estimate of Potential Trace Elements Discharged to Atmosphere without Scrubber (from EPA, 1980)

Included in Title III	Element	Emission >10 tpy	ppm in coal (dry basin)	Average % emitted	Emitted kg/day	Maximum tpy
Yes	Antimony		0.5	25	0.40	
Yes	Arsenic	x	8–45	25	11–36	13
Yes	Beryllium		0.6–7.6	25	0.5–6	2.2
No	Boron		13–198	25	10–160	
No	Bromine		14.2	100	46	
Yes	Cadmium		0.14	35	0.16	
Yes	Chlorine	x	400–1000	100	1300–3250	1170
No	Fluorine		50–167	100	160–550	
Yes	Lead	x	8–14	35	9–16	5.7
Yes	Mercury		0.04–0.49	90	0.1–1.4	0.5
No	Molybdenum		0.6–8.5	25	0.5–7	
Yes	Selenium		2.2	70	5.0	1.8
No	Vanadium		8.7–67	30	8.5–65	
No	Zinc		0–53	25	0–43	

Note: Based on a feed rate of 3445 tpd of Illinois No. 6 coal.

fossil-based technologies, that is, control of pollutants emitted in sufficient quantities that endanger public health. Development of regulations, to the extent needed, should be based on comprehensive sets of measurements, from representative technology, and processing options. Such measurements from fossil-based technologies do not exist today for most Title III contaminants. Without measurement data, engineering estimates can be prepared to guide monitoring and control efforts. However, these estimates should not be used as the basis for regulations because of the large variations that exist in fuel feedstock quality, combustion chemistry, and efficiency of existing pollution control systems. In this light, the following sections attempt to prioritize these needs.

III. HAZARDOUS AIR POLLUTANT EMISSIONS FROM COAL-BASED SYSTEMS

The air pollutants that may be emitted from coal-based technologies include but are not limited to the following: priority pollutants (e.g., SO_2, NO_x); low molecular weight hydrocarbons (e.g., CH_4, etc.); and trace emissions of metallic constituents (e.g., As, Se, Hg, Cd, Pb), polycyclic organics (e.g., benzo-a-pyrene), and fine particulate matter (0.1 to 0.6 μm).

Many trace elements are contained in coal (Table 1). Although their concentrations are low, the total potential mass of gaseous emissions from a coal-fired power plant may be relatively high because of the total amount of coal burned during a year. We calculate that a 400 MW_e coal-fired power plant will burn about 3500 tons per day (tpd) of coal. At this rate, trace elements in the coal at concentrations higher than 40 parts per million (ppm) and with 25% volatility have the potential to be in the effluent stream in quantities exceeding 10 tpy — the Title III threshold. Figure 3

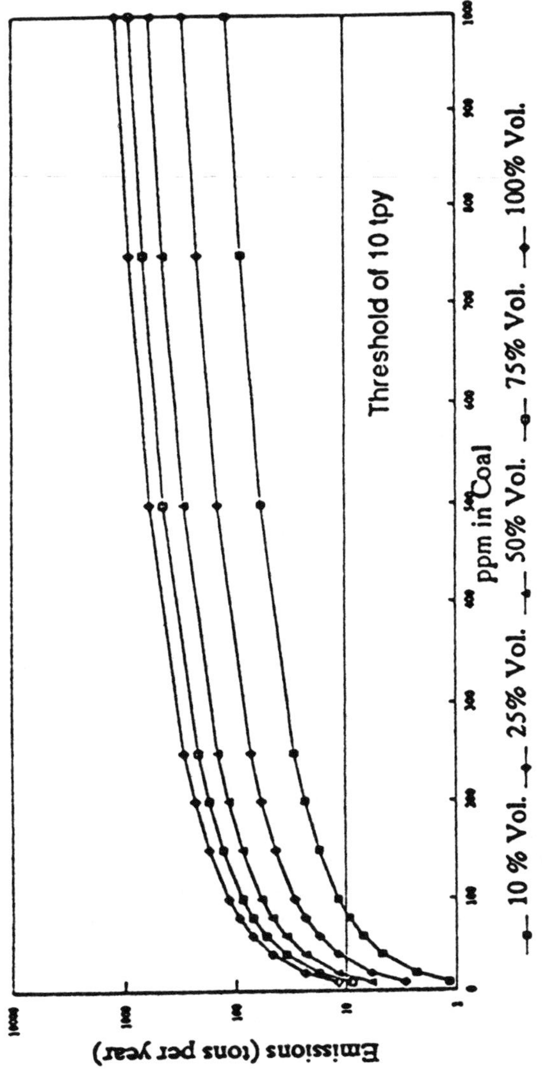

Figure 3. Estimated stack gas emissions as a function of trace element concentration and volatility.

graphically displays the relationship between volatility, concentration, and stack emissions for a plant burning 3445 tpd of coal. Estimates calculated from Figure 3 are based on the assumption that all the trace elements in the gas phase will be entrained within the stack gas. In reality, a fraction of these elements will condense on or be adsorbed by ash particles that will be removed by particulate pollution control equipment (e.g., electrostatic precipitators). However, further analysis seems warranted because some elements are preferably retained on very fine particles that can escape through the control equipment; some fraction also will be exhausted with the hot flue gases.

The composition of hydrocarbons in the effluent streams depends on several process and combustion conditions (e.g., temperature, pressure, steam/coal ratio, hydrogen/coal ratio, and residence times). Similarly, variations in coal type and reactivity will result in different gas emissions. Residence time affects the amount of carbon conversion and sulfur retention in the ash, and consequently the composition of the effluent stream. Thus, residence time is one factor that can be used to aggregate different types of coal-gasification and fluidized-bed technologies into different pollutant emission classes. An entrained-bed gasifier, for example, has low residence times (usually less than 1 s) whereas fluidized-bed gasifiers have much longer residence times (usually 3 to 7 min). Finally, the amount of water that enters a gasifier can affect the composition of the effluent stream. For example, in a coal gasifier water comes from three sources: steam injection, coal moisture, and feed of water slurry. In general, higher steam concentration result in higher hydrogen concentrations within the effluent gas.

Due to the paucity of measurement data of Title III-type emissions from coal-based facilities, we have attempted to identify the types of Title III compounds likely to be emitted. These characterizations are based on the fundamental assumption that the quantity and type of stack emissions will be a function of trace element concentration in the feedstock coal, and process and combustion chemistry.

Trace elements, if present in the feedstock coal, are sufficiently volatile that they would likely be present in all complete combustion systems. These elements include compounds of antimony, arsenic, beryllium, cadmium, chlorine, lead, mercury, radionuclides, and selenium. The chemical species present in the gaseous waste streams are likely to be different in oxidizing and reducing environments. In oxidizing environments, the metals would be mostly oxides, although some chlorides also would be present. In reducing environments, the metals would be mostly chlorides, hydrates, and sulfides. As shown in Table 1, the trace elements with the highest potential concentrations are arsenic, cadmium, chromium, lead, and selenium. In addition, some combustion by-products (e.g., polycyclic organic matter including primarily benzo-a-pyrene) are so highly refractory that they too could be present in complete combustion emission streams.

Although CCTDP generally have very low emission rates, certain Title III chemicals still require further investigation. For example, the following chemicals could form in reduced conditions and survive partial combustion: bis(2-

Table 2. Title III Pollutants that Are *Unlikely* to Exist in Effluents from CCTDP

Bromoform	Methyl methacrylate
Calcium cyanamide	4,4-Methylene bis(2-chloraniline)
Captan	Methylene diphenyl diisocyanate (MDI)
Carbaryl	4,4-Methylenedianiline
Chloramben	*N*-Nitrosomorpholine
Chlordane	Parathion
Chloroacetic acid	Pentachloronitrobenzene (quintobenzene)
Chloroform	Phosgene
Diazomethane	Phosphine
Dichloroethyl ether [bis(2-chloroethyl)ether]	Propionaldehyde
1,3-Dichloropropene	Styrene oxide
Diethanolamine	1,1,2,2-Tetrachloroethane
Dimethyl carbamoyl chloride	Titanium tetrachloride
Dimethyl formamide	Toxaphene (chlorinated camphene)
1,1-Dimethyl hydrazine	1,1,2-Trichloroethane
Hexachlorocyclopentadiene	Vinyl acetate
Hexachloroethane	Vinyl bromide
Hexamethylphosphoramide	Vinylidene chloride (1,1-dichloroethylene)
Hydrazine	

ethylhexyl)phthalate (DEHP), carbonyl sulfide, dimethyl sulfate, dibenzofurans, formaldehyde, hydrogen chloride, hydrogen fluoride, methyl chloride, pentachlorophenol, phenol, tetracloroethylene, trichloroethane, and 2,4,5-trichlorphenol. Similarly under highly reducing conditions the following substance also may be emitted: aniline, benzene, biphenyl, carbon disulfide, carbon tetrachloride, chlorobenzene, ethylbenzene, methanol, naphthalene, ortho-, meta-, and paracresols, quinoline, toluene, and xylenes. Trichlorethylene and nickel compounds (from some oils) also may be present in the gaseous waste streams from processes involving co-processing coal with other hydrocarbons. In Table 2, compounds unlikely to be formed or emitted from coal-fired facilities are listed.

IV. SUGGESTED INITIATIVES

The potential impacts that Title III regulations could have on projects supported by the CCTP are not known at this early date. Given the lack of data on the type and quantity of air pollutants emitted by coal-fired facilities, the time lag in collecting such data, and the intricacies of the EPA rule-making process, a proactive response by DOE and its industrial partners to the CAAA seems warranted. Among the range of options available to DOE and its industrial partners are the following: begin to develop independent critical estimates of the risks to health presented by hazardous air pollutants emitted by coal-based facilities; begin to examine critically the technical and economic efficiency of various pollution control strategies for Title III pollutants; sample existing coal-fired facilities for Title III contaminants; and sample newly emerging clean-coal processing options for Title III contaminants.

Of these options, serious, but significantly different, risks are assumed. The first two options reduce the short-term financial and contractual risks to the CCTP and the private sector, by simply letting the technological and regulatory processes move forward without collecting new data. The long-term risk to the commercial viability of clean-coal presented by these options, however, are large. The EPA administrator could determine that fossil-fuel-fired technologies need to be regulated through the application of controls that could be costly or not readily available. The other two options reduce the long-term risk, by collecting technology-specific data, but increase short-term administrative and financial costs.

DOE and its industrial partners have both a vested interest and opportunity to ensure the development of appropriate regulations for fossil-based technologies. Clearly, the foundation for this is the establishment of a measurements database. If sampling and analysis efforts begin, several important programmatic decisions must be made to increase the efficiency of the data collection and decrease the overall costs of these efforts. In addition, independent evaluations of health risks and pollutant control technologies are needed.

A. Analytical Protocols

There are no routine sampling and analytical protocols for many compounds listed in Title III. Table 3 cross-references analytical protocols identified by EPA[2] and the National Institute for Occupational Safety and Health — NIOSH[3] with the air pollutants identified by the CAAA. Only 31 of the 189 compounds listed have protocols identified by the EPA. NIOSH has developed protocols for 111 of the listed air toxics; however, these suggested protocols were developed solely for understanding threshold limits for worker exposure levels. Furthermore, because of the differences in the environments, i.e. "hot" stack gases as opposed to inhalable air, these protocols may be inappropriate. They may, however, be used to develop a comprehensive, uniform set of protocols that all contractors could follow when conducting chemical sampling and analysis.

B. Indicator Chemicals

In the collection of the supplemental monitoring data, contaminants that should be studied in greater detail must be identified. In this context, a formal definition for the term "considerable concern" is needed. In simple terms, the trigger for considerable concern can be based on concentration, mass, or risk. Triggers based on the first strategy can be as simple as a measured concentration in the stack that is equal to or greater than the Occupational Safety and Health Administration Permissible Exposure Level (OSHA PEL). The second trigger could be defined as Title III chemicals likely to be emitted in quantities greater than 10 tpy. Finally, the third trigger could be chemicals producing estimated lifetime cancer risks (or equivalent for noncarcinogens) exceeding one in 100,000 to maximally exposed individuals. The first approach may be sufficient for purposes of screening, the second for detailed sampling, and the third might be used for input into the EPA rule-making

Table 3. A Cross-Referencing of Hazardous Air Pollutants Listed in the CAAA with Analytical Protocols Identified by the U.S. Environmental Protection Agency and the National Institute for Occupational Safety and Health

Cas no.	Toxics	NIOSH analytical method	EPA	
			Method class	Ref.
75070	Acetaldehyde	—	—	—
60355	Acetamide	—	—	—
79061	Acylamide	—	—	—
75058	Acetonitrile	S165	—	—
98862	Actophenone	—	—	—
53963	2-Acetylaminofluorene	—	—	—
107028	Acrolein	118, 211	—	—
79107	Acrylic acid	—	—	—
107131	Acrylonitrile	202, S156	—	—
107051	Allyl chloride	S111	—	—
92671	4-Aminobiphenyl	269	—	—
62533	Aniline	168, S163	—	—
90040	O-Anisdine	—	—	—
1332214	Asbestos	239, 245, 309	—	—
71432	Benzene (including benzene from gasoline)	127, S311, 1008	T	12
92875	Benzidine	243, 315	—	—
98077	Benzotrichloride	—	—	—
100447	Benzyl chloride	S115	—	—
92524	Biphenyl	S24	—	—
117817	bis(2-Ethylhexyl)phthalate (DEHP)	—	—	—
542881	bis(Chloromethyl)ether	333	—	—
75252	Bromoform	S114, 1003	O	1[illegible]
106990	1,3-Butadiene	S91	D	3
156627	Cadmium cyanamide	—	—	—
105602	Caprolactam	—	—	—
133062	Captan	—	—	—
63252	Carbaryl	S273	—	—
75150	Carbon disulfide	179, S248	—	—
56235	Carbon tetrachloride	127, S314	T	15
463581	Carbonyl sulfide	—	—	—
120809	Catechol	—	—	—
133904	Chloramben	—	—	—
57749	Chlordane	115	—	—
7782505	Chlorine	209	—	—
79118	Chloroacetic acid	—	—	—
532274	2-Chloroacetophenone	291	—	—
108907	Chlorobenzene	133, 1003	O	11
510156	Chlorobenzilate	—	—	—
67663	Chlorform	127, S351	T	15
107302	Chloromethyl methyl ether	220	—	—
126998	Chloroprene	S112	—	—
1319773	Cresols/cresylic acid (isomers and mixture)	S167 R	—	—
95487	o-Cresol	S167	—	—
108394	M-Cresol	S167	—	—
106445	P-Cresol	S167	—	—

Table 3. A Cross-Referencing of Hazardous Air Pollutants Listed in the CAAA with Analytical Protocols Identified by the U.S. Environmental Protection Agency and the National Institute for Occupational Safety and Health (continued)

Cas no.	Toxics	NIOSH analytical method	EPA Method class	EPA Ref.
98828	Cumene	S23, 1501	O	6
94757	2,4-D, salts, and esters	S279	—	—
3547044	DDE	—	—	—
334883	Diazomethane	S137	—	—
132649	Dibenzofurans	—	—	—
96128	1,2-Dibromo-3-chloropropane	—	O	22
84742	Dibutylphthalate	S33	—	—
106467	1,4-Dichlorobenzene(P)	S281	—	—
91941	3,3-Dichlorobenzidene	246	—	—
111444	Dichloroethyl ether [bis(2-chloroethyl)ether]	S357 H	—	—
542756	Dichloropropen	—	—	—
62737	Dichlorvos	295	—	—
111422	Diethanolamine	221, S139	—	—
121697	N,N-Diethyl aniline (N,N-dimethylaniline)	—	—	—
64675	Diethyl sulfate	—	—	—
119904	3,3-Dimethoxybenzidine	—	—	—
60117	Dimethyl aminoazobenzene	—	—	—
119937	3,3-Dimethyl benzidine	—	—	—
79447	Dimethyl carbamoyl chloride	—	—	—
68122	Dimethyl formamide	S255	—	—
57147	1,1-Dimethyl hydrazine	248, S143	—	—
131113	Dimethyl phthalate	—	—	—
77781	Dimethyl sulfate	—	—	—
534521	4,6-Dinitro-o-cresol and salts	S166	—	—
51285	2,4-Dinitrophenol	—	—	—
121142	2,4-Dinitrotoluene	S215	—	—
123911	1,4-Dioxane (1,4-diethyleneoxide)	S215	—	—
122667	1,2-Diphenylhydrazine	—	—	—
106898	Epichlorohydrin (1-chloro-2,3-epoxpropane)	—	—	—
106887	1,2-Epoxybutane	—	—	—
140885	Ethyl acrylate	S105, 2519	—	—
100414	Ethyl benzene	1501	O	6
51796	Ethyl carbamate (urethane)	—	—	—
75003	Ethyl chloride (chloroethane)	2519	O	20
106934	Ethylene dibromide (dibromoethane)	1008	O	12
107062	Ethylene dichloride (1,2-dichloroethane)	S118	T	13

Table 3. A Cross-Referencing of Hazardous Air Pollutants Listed in the CAAA with Analytical Protocols Identified by the U.S. Environmental Protection Agency and the National Institute for Occupational Safety and Health (continued)

Cas no.	Toxics	NIOSH analytical method	EPA	
			Method class	Ref.
107211	Ethylene glycol	338	—	—
151564	Ethylene oxide	286, 1607	O	9
96457	Ethylene thiourea	281	—	—
75343	Ethylidene dichloride (1,1-dichloroethane)	—	—	—
50000	Formaldehyde	125, 235, 318, S327, 354	—	—
76448	Heptachlor	287	—	—
118741	Hexachlorobenzene	—	—	—
87683	Hexachlorbutadiene	307	—	—
77474	Hexachlorocyclo-pentadiene	308, 2518	O	4
67721	Hexachloroethane	S101	—	—
822060	Hexamethylene-1,6-diisocyanate	—	—	—
680319	Hexamethylphos-phoramide	—	—	—
110543	Hexane	S90, 1500	O	2
302012	Hydrazine	248, S237	—	—
7647010	Hydrochloric acid	—	—	—
7664393	Hydrogen fluoride (hydrofluoric acid)	117, 262, S176	—	—
123319	Hydroquinone	S57	—	—
78591	Isophorone	S367	—	—
58899	Lindane (all isomers)	S290	—	—
108316	Maleic anhydride	302	—	—
67561	Methanol	247, 559, 2000	O	1
72435	Methoxychlor	S371	—	—
74839	Methyl bromide (bromomethane)	S372, 2520	O	16
74873	Methyl chloride (chloromethane)	201, S99	O	17
71556	Methyl chloroform (1,1,1-trichloroethane)	127, S328	T	13
78933	Methyl ethyl ketone (2-butanone)	127, S3, 2500	O	8
60344	Methyl hydrazine	—	—	—
74884	Methyl iodide (iodomethane)	S98	—	—
108101	Methyl isobutyl ketone (hexone)	S18, 1300	O	7
624839	Methyl isocyanate	—	—	—
80626	Methyl methacrylate	S43	—	—
1634044	Methyl-tert-butyl-ether	—	—	—
101144	4,4-Methylene bis(2-chloroaniline)	236, 342	—	—
75092	Methylene chloride (dichloromethane)	121, S329	T	18
101688	Methylene diphenyl diisocyanate (MDI)	—	—	—

Table 3. A Cross-Referencing of Hazardous Air Pollutants Listed in the CAAA with Analytical Protocols Identified by the U.S. Environmental Protection Agency and the National Institute for Occupational Safety and Health (continued)

Cas no.	Toxics	NIOSH analytical method	EPA	
			Method class	Ref.
101779	4,4′-Methylenedianiline	—	—	—
91203	Naphthalene	264	—	—
98953	Nitrobenzene	S217	—	—
92933	4-Nitrobiphenyl	213	—	—
100027	4-Nitrophenol	—	—	—
79469	2-Nitropropane	272	—	—
684935	N-Nitroso-N-methylurea	—	—	—
62759	N-Nitrosodimethylamine	252, 299	—	—
59892	N-Nitrosomorpholine	—	—	—
56382	Parathion	244, 253, 329, S120, S121	—	—
82688	Pentachloronitrobenzene (quintobenzene)	—	—	—
87865	Pentachlorophenol	230, S29722	—	—
108952	Phenol	330, S330	—	—
106503	P-Phenylenediamine	—	—	—
75445	Phosgene	219	—	—
7803512	Phosphine	S332	—	—
7723140	Phosphorus	242, 257, 351 S334	—	—
85449	Phthalic anhydride	S179	—	—
1336363	Polychlorinated biphenyls (aroclors)	244, 253, 329 200	—	—
1220714	1,3-Propane sultone	—	—	—
57578	β-Propiolactone	—	—	—
123386	Propionaldehyde	—	—	—
114261	Propoxur (baygon)	—	—	—
78875	Propylene dichloride (1,2-dichloropropane)	S95, 1013	O	14
75569	Propylene oxide	S75, 1612	O	10
75558	1,2-Propylenimine 92-methyl (aziridine)	—	—	—
91225	Qinoline	—	—	—
106514	Quinone	S181	—	—
100425	Styrene	121, S30, 1501	O	6
96093	Styrene oxide	303	—	—
1746016	2,3,7,8-Tetrachloro-dibenzo-p-dioxin	—	—	—
79345	1,1,2,2-Tetrachloro-ethane	S124	—	—
127184	Tetrachloro-ethylene (perchlor-ethylene)	127, S335	T	13
7550450	Titanium tetrachloride	—	—	—
108883	Toluene	127, S343, 1500 1501	O	2, 6
95807	2,4-Toluene diamine	—	—	—
584849	2,4-Toluene diisocyanate	141, 168	—	—

Table 3. A Cross-Referencing of Hazardous Air Pollutants Listed in the CAAA with Analytical Protocols Identified by the U.S. Environmental Protection Agency and the National Institute for Occupational Safety and Health (continued)

Cas no.	Toxics	NIOSH analytical method	EPA Method class	EPA Ref.
95534	O-Toluidine	141, 326	—	—
8001352	Toxaphene (chlorinated camphene)	S672	—	—
120821	1,2,4-Trichlorobenzene	343	—	—
79005	1,1,2-Trichloroethane	127, S134	—	—
79016	Trichloroethylene	127, S336	T	13
95954	2,4,5-Trichlorophenol	—	—	—
88062	2,4,6,-Trichlorophenol	—	—	—
121448	Triethylamine	221, S152	—	—
1582098	Trifluralin	—	—	—
540841	2,2,4-Trimethylpentane	—	—	—
108054	Vinyl acetate	278	—	—
593602	Vinyl bromide	349	—	—
75014	Vinyl chloride	178	R	21
75354	Vinylidene chloride (1,1-dichloroethylene)	266	O	19
1330207	Xylenes (isomers and mixture)	127, S318, 1501	O	6
95476	O-Xylenes	127, S318	—	—
108383	M-Xylenes	127, S318	—	—
106423	P-Xylenes	127, S318	—	—
0	Antimony compounds	—	—	—
0	Arsenic compounds (inorganic including arsine)	—	—	—
0	Beryllium compounds	—	—	—
0	Cadmium compounds	—	—	—
0	Chromium compounds	—	—	—
0	Cobalt compounds	—	—	—
0	Coke oven emissions	—	—	—
0	Cyanide compounds	—	—	—
0	Glycol ethers	—	—	—
0	Lead compounds	—	—	—
0	Manganese compounds	—	—	—
0	Mercury compounds	—	—	—
0	Fine mineral fibers	—	—	—
0	Nickel compounds	—	—	—
0	Polycyclic organic matter	—	—	—
0	Radionuclides (including radon)	—	—	—
0	Selenium compounds	—	—	—

Note: R = reference — EPA promulgated method. T = tentative — EPA method development complete; EPA reference available. D = development — EPA method currently under development. O = other — method development completed by organization other than EPA. N = none — no reference available.

process, including the evaluation the effectiveness of different control strategies. These triggers should be identified early in any program so that monitoring schedules can be quickly adjusted to eliminate unnecessary tests and implement more useful ones.

C. Quality Assurance/Quality Control

To ensure the credibility of all data collected, it would be important to integrate a quality assurance/quality control program with the sampling and analysis program. The appropriate methods should be clearly defined in the early stages of any monitoring effort.

D. Evaluation of Health Risks

Preliminary estimates from coal-conversion facilities suggest that these operations might emit hazardous air pollutants in excesss of the 10- or 25-tpy guidelines. Given the strong possibility that EPA will regulate these sources, efforts are needed to develop independent, realistic estimates of the health hazards from these releases. These estimates could be presented to EPA rule makers as they evaluate the need to control emissions from coal-based facilities.

E. Pollutant Control Technologies

Independent efforts should begin to evaluate the technical and economic efficiency of different control strategies. Initial efforts should focus on the trace element. With the collection of more data, the program can be refined to include analyses of other Title III materials.

V. CONCLUSION

At present only limited data exist to characterize the types and quantities of Title III that could be emitted to the atmosphere from fossil-based technologies. The operation of CCTDP provides an opportunity to collect and sample data from a wide range of new coal-based technologies. Collection of these data will ensure the development of appropriate regulations, without a bias resulting from a lack of data, for fossil-based technologies. This will help ensure equitable treatment for all clean coal projects regarding any Title III regulations promulgated by EPA rule makers. In this context, DOE is now evaluating its options to contribute to this process.

REFERENCES

1. U.S. Department of Energy, Clean Coal Technology Demonstration Program, DOE/FE-021p, Washington, DC, 1991.
2. U.S. Environmental Protection Agency, Quality Assurance Handbook for Air Pollution Measurement Systems, Vol. 3, Stationary Source Methods, EPA-600/4-77-027b, Research Triangle Park, NC, 1977.
3. U.S. Department of Health and Human Services, NIOSH Manual of Analytical Methods, Vol. 7, Washington, DC, 1988.

In-Plume Measurement of Fugitive and Point Source Emissions Using Airborne Instrumentation

Ken Lionarons

I. ABSTRACT

Instrumentation and delivery equipment has been developed for determination of contaminent isopleths and to measure the species, concentration, or other physical and chemical properties of time-dependent emissions from point or fugitive sources. Data may be obtained by *in situ* analysis and recording of selected species for immediate or subsequent transmission and display at a ground receiving station. In addition, the capability has been developed to isolate individual or composite samples at altitude or to retrieve individual or continuous samples to a ground station for laboratory examination.

Equipment developed for dispersion model verification at any convenient time, or determination of gaseous and particulate isopleths following accidents at nuclear power plants, may be used to provide data on containment transport between the source and selected downwind targets in industrial environments. The methods described are useful as a real-time release tracking tool or to verify dispersion models prior to construction utilizing a variety of tracers in all atmospheric stability classes.

Mission-specific instrumentation, sampling, and communications equipment carried aloft by tethered airfoils, lighter-that-air balloons, or internally powered devices may remain at altitude for extended periods to retrieve data or samples for the duration of emission events or atmospheric conditions of interest.

II. BACKGROUND

Following the accident at Three Mile Island in Middletown, PA, during March and April of 1979, the Nuclear Regulatory Commission felt the need to improve disaster recovery at commercial nuclear reactors throughout the U.S.[1] During the accident, radionuclide concentrations at the discharge stack were reported to be site boundary determinations, resulting in an unprecedented evacuation of the local population. Since the effects of ionizing radiation from proximity (ground shine), as well as direct immersion in a plume containing particulate-borne radionuclides, noble gases ($krypton_{85}$ and $xenon_{133}$), and various condensables present such a potential damaging effect on the public, extraordinary precautions were implemented for protection of public safety. In this case, exact location of the discharge plume was significant since exposure to the shine associated with the decay of radioactive matter may have deposited damaging levels of ionizing energy on ground-level population targets.

The release concentrations reported at the site boundary were not subjected to a "sanity check", which would have cast doubt on their relationship to the point source values, and this conservative estimate of dispersion which assumed members of the public to be exposed as if in direct contact with the discharge stack. The actual concentrations were many orders of magnitude lower than those used as a basis for an evacuation decision.

The investigation and corrective actions taken as a result of the accident yielded programs to improve source instrumentation, point of discharge hardware, dose dispersion models, meteorological monitoring, and emergency response coordination. Annual emergency drills which simulate accidents and elevated release rates are held to evaluate the effectiveness of field survey teams, effluent monitoring equipment, and dispersion/dose models used to protect the public.

During the emergency drills simulating an actual release, event teams of technicians are dispatched to locate the plume by taking ground level shine data using hand-held ionizing radiation and particulate collection instruments. This method provides data reflected from a discharge path which may be interpreted as a line or plane source at an estimated height, so isopleths and the actual plume centerline may be determined. Vertical dispersion, an important component of the data, may only be inferred. Decisions on the location of the plume centerline, both horizontally and vertically, are then based on computer-generated models which received data from plant discharge equipment, meteorological monitors, and plant operations personnel.

The computer models were originally developed using historical meteorological data and local topography. The great number of variables and changes to topography due to construction and changing land use, as well as the methods used to report background deposited energy information from field survey teams may cast doubt on the utility of these programs for some predicted discharge conditions.[2] Equipment that maps the dispersion of discharges by verification and modification of models, using actual samples and plume characteristic data, would therefore simplify the task of predicting the exact path of radionuclides from the discharge point to affected individuals in the downwind area of risk.

To aid this effort, special sampling equipment was developed to determine the concentrations of radionuclides out to a radius of 10 miles throughout the vertical and lateral extent of the discharge path. The equipment includes a reconfigurable instrument package containing all measurement and communication instrumentation and a delivery device to locate the instrument payload directly in the path of airborne contaminants.

Instrumentation that serves to gather data for at-altitude evaluations includes ionizing radiation instruments (β/γ), humidity, temperature, pressure altitude, fixed flow particulate filters, and a data recorder/transmitter. Remotely triggered sample bottles, or a line to transport discreet or continuous samples to a ground station, are used to obtain representative radionuclide samples for decay or attenuation (dispersion) counting as well as individual radionuclide and physical identification.

Delivery equipment, developed to locate sampling and analysis instruments in the plume interior, includes tethered, faired helium balloons, specially modified airfoils with messenger capability, and modified gasoline-powered helicopters for use in close quarters and where contrary air currents or topography limitations rule out the lighter-than-air delivery devices. Detailed aeronautical sectional charts, surveying and navigation instruments, and pressure altitude encoders complete the list of equipment used to place instruments in the plant discharge for precise location and measurement capability. The use of kites, kytoons, parafoils, and similar airfoils is not new. Kites with the capability of lifting soldiers (usually *not* volunteers) were utilized as early as 2000 B.C. The use of kites to lift instruments, weapons, and observation cameras as well as men was not uncommon in the 19th century or during the first and second world wars. The use of these devices in scientific endeavors for the measurement of atmospheric contaminants or meteorological data has a long history of success, and holds considerable economic and common-sense advantage over alternative methods.

III. DESCRIPTION AND CAPABILITIES

The modern application of atmospheric sampling, as applied to these effluent monitoring programs, would not be possible without recent advances in the miniaturization of electronics, especially memory and power devices. Measurement instrumentation used for this application must withstand large shocks, temperature and pressure variation, and be capable of running for long periods of time without attention or evidence of changes to data quality. Equipment available off the shelf with only minor modification has seen service in extended flight for more than 20 days without failure or loss of data. The use of lightweight lithium and mercury cells, more than any other factor, has provided designers with a dependable and lightweight instrument package capable of extended operation.

A. Instrumentation

Instrument packages consisting of a central power supply, three data recorders, radiation, temperature, humidity, and wind velocity sensors, an altitude sensor, miniature gyrostabilizer and control circuits, and an electronic (visual) strobe assembly were assembled for a 9-day study at a total weight of under 2 kg.

All instruments may be mounted in a single sealed unit for a specific mission and sent aloft for data gathering with either an on-board recording capability, with data downloaded post-mission on a personal computer, or for immediate transmission to a ground data station. In addition, air samplers capable of programmed filter changes using sealed cartridges may be employed for solid-phase sampling. These same pumps are capable of drawing air samples through prepackaged reagents, although temperature and other ambient conditions limit the number of different reactions possible on board. The use of cameras and image sensors may be accommodated.

The fitting of instrumentation for *in situ* analysis is limited by the size and weight of the instrumentation. In high or variable wind conditions and in some populated locations, payloads in excess of 7 kg may cause problems for high altitude delivery. Various "permissive" devices may be fitted to limit sampling to certain atmospheric or other predetermined conditions. Ultraviolet, infrared, chromatography, or similar instruments may be sent aloft, but in cases where the sample may be pumped a short distance remote sample transport is preferable. For example, specific altitude, ambient light, humidity, or wind speed criteria may alone or in combination be required for analysis or retrieval of samples as determined by the investigator's needs.

In some cases, samples must be taken at altitude and retrieved for immediate analysis. The sample may be withdrawn through a length of lightweight tubing for ground station analysis, if the limited intersample blending between contaminants is not a factor and the sample tubing available is inert. The alternative is to fit a multiple sampling sequencer to the instrument package for archiving of individual samples or to send a small vial or evacuated sample tube up a messenger line for quick sample retrieval.

The number and types of remote chemical determinations and sample retrieval protocols are limited only by the imagination of the air contaminant test engineer or scientist.

B. Delivery Devices

The delivery devices used to place the instrument package or sample vessel in the desired sample location are sensitive to a number of factors beyond the control of the investigator. Meteorological conditions, terrain, and risk to local population or property play a large part in the planning of atmospheric sampling downwind of industrial facilities.

A variety of sizes, configurations, and control schemes are needed to ensure that sampling may be performed in a large variety of wind, precipitation, and visibility conditions. Often, payloads must be shared between two lifting devices if safety or stability requirements cannot be met. Areas near airports or their approaches are often restricted for high altitude flight (all require an FAA permit) and the use of contrasting colors, on-board strobes, and use of *notices to airmen* announcements enhances safety.

The possibility of the failure of an airfoil or instrument package and resultant free fall is addressed by the use of autodeployable parachutes and loss-of-stability operator warnings generated by on-board sensors and gyro controls.

The initial choice of a delivery system is made after evaluating payload, altitude, terrain, population density, man-made obstructions, air traffic, and anticipated meteorological conditions. In light to variable winds, in locations where ground access is reasonable beneath the sampling site(s), helium-filled balloons are the best choice. In heavier breezes, a variety of parafoils and other specialized airfoils are available for a wide range of wind and terrain conditions. Some, with on-board gyrostabilizers, variable camber surfaces, and moveable vents are capable in winds from 3 to 22 knots for a wide range of loads. Both of the above devices may be used with messengers and tubing for sample retrieval.

The use of radio-controlled helicopters is limited to loads of 1.7 kg, in most cases, and only in light winds up to 14 knots. They are, however, quite useful for retrieving samples with little preparation and from locations with difficult access, especially at low altitudes. The noise, short flight duration, and limited load make the use of helicopters or other heavier-than-air powered devices a highly specialized endeavor.

C. Communications Systems

Bidirectional radio links may be used to change protocols, initiate analysis or sampling routines, or modify the flight characteristics of delivery systems from ground-based stations. The use of frequency phase-shift communications simplifies the verification of required operations on the instrument communications board, by signaling acknowledgment that required operations were completed successfully.

D. Safety and Precautions

Some precautions must be followed when operating sampling equipment around or over personnel, the public, or around flight operations.

1. Local Flight Safety Officer of the local Federal Aviation Administration region should be notified of flight operations for all devices. Most operations require a permit, warning devices, and pilot notification.
2. Operation over populated areas requires failure contingency and use of fail-safe designs.
3. Dangerous meteorological conditions involving high winds or lightning must be avoided. In addition the local lapse rate should be reviewed for signs of icing.
4. Power lines, roads, and other manmade hazards require prudent examination of flight paths.
5. All airborne delivery devices should be equipped with radar reflectors, contrasting highly visible colors, and an electronic strobe.

IV. SUMMARY

The available instrumentation and delivery devices create an opportunity for engineers and scientists to investigate the actual path of contaminants from point or line sources with or without the use of ground-based field sampling teams.

The data collected by sending analysis and sampling equipment into the plume of fugitive or point source emissions may be used to verify assumptions based on models or may be used as raw data for the creation of new models. Some advantages include

1. Synergistic or unpredicted reactions that take place between the time reagents leave the source and the time they reach ambient monitoring stations may be uncovered. Reactions with sunlight, moisture, or emissions from other sources may change the properties of emitted contaminants.
2. The effects of topography, local differential heating/cooling, or other changes to the environment may be assessed for their effects on contaminant transport.
3. Transport at return of contaminants to ground level for both dry and wet deposition conditions may be closely approximated for improved location of ambient monitoring stations.
4. Existing models of contaminant transport may be more closely evaluated to understand their worth for discharge rates beyond design or unusual meteorological conditions. Still conditions, plume trapping, inversions, and storms may yield unpredicted dispersion and transport characteristics. Model sensitivities may be uncovered and evaluated.
5. Low cost/flexible deployment.

It is not often that a hobby or hobby equipment can be so readily adapted for use in scientific testing. The marriage of lightweight instrumentation and well-designed airfoils provides a significant advantage on the tracking, analysis, and sampling of airborne contaminants and protection of target populations.

REFERENCES

1. NUREG 0654 Rev 1, U.S. Nuclear Regulatory Commission Criteria for Preparation and Evaluation of Radiological Emergency Response Plans in Support of Nuclear Power Plants.
2. NUMARC/NESP-008, Nuclear Management and Resources Council Analysis of Dose Models for Accidental Airborne Radioactive Releases.

An Integrated Approach for Ambient Air Toxics Impact Analysis at Industrial Complexes

Robert Wells
Ted Palma
Sarah Osborne

I. INTRODUCTION

Air toxics is an area of continuing environmental concern. The Clean Air Act (CAA) Amendments of 1990, state air toxics regulations, and public concern over air emissions reported under SARA Title III are creating an urgent need to deal with air toxics issues. Environmental managers must take stock of existing air toxics emissions. In many cases, industry must relate these emissions to public exposures and use this information to find creative ways to comply with new regulations. The air toxics management challenge is compounded by the increased amount of data that must be assembled, analyzed, and maintained. Many different sources of hundreds of pollutants must be incorporated in compliance planning.

The Galson Corporation (Galson) has created a management tool to support this effort. The AIR-1© Air Toxics Management System was originally developed to combine database management, dispersion modeling, and environmental management tools to provide a comprehensive data summary and impact analysis of complex facilities. AIR-1 has recently evolved to the pcAIR-1© system, which is designed for personal computers (PCs). pcAIR-1 provides both access to a popular, cost-effective computer platform and the enhanced data management capabilities available in a Geographic Information System (GIS) graphical computing environment.

This paper relates air toxics management issues raised by both state programs and the CAA Amendments to the data management needs they create, and shows how the pcAIR-1 system can automate both database management and impact assessment. Increased automation and graphical computing streamline the mundane data management chores, and allows environmental managers to focus on critical issues.

II. AIR TOXICS MANAGEMENT ISSUES

The CAA Amendments of 1990 have altered the regulation of air toxics (hazardous air pollutants, or HAPs) drastically. The list of HAPs will include at least 189 chemicals and chemical categories, most of which have not been regulated at the federal level before. These chemicals will initially be regulated by technology-based emission standards (Maximum Achievable Control Technology, or MACT). Some of these HAPs will be considered for regulation under residual risk standards based on health risk resulting from ambient impacts as well.

Federal New Source Review (NSR) under Prevention of Significant Deterioration (PSD) regulations also can require air toxics analysis. Following the North County Remand, BACT decisions should consider air toxics emissions; such a BACT analysis is developed on a case-by-case basis, but can include consideration of air toxics impacts, and those large projects that require a Federal Environmental Impact Statement (EIS) may need to consider toxics impacts and/or health risk.

In addition to federal regulations, many states have developed their own regulatory programs for air toxics. These programs differ widely as to the pollutants regulated, the philosophy of regulations (source control vs. ambient criteria), and the level of control required and/or impacts allowed. Some states regulate emissions, some require "best" control, some establish ambient guidelines or standards, and many states combine these approaches. As of 1990 there were 41 state and local regulatory programs in place.[1] Of these, 39 rely on ambient guidelines and/or standards; 25 programs conduct risk assessments, primarily for carcinogens. State permitting for large industrial projects can require an EIS or health risk assessment (HRA) as well.

Many industries have received a great deal of largely negative publicity from the public distribution of emission reports required under SARA Title III. SARA reports are routinely reprinted in local newspapers, often as front page stories. Industries have had to respond to this publicity, either voluntarily, to maintain good public relations, or in adversarial and emotional confrontations with the public.

With the exception of SARA emissions reporting, most of these issues are of concern to the electric utility industry. The public perception of power plants, particularly coal-fired plants, is often as prominent sources of air pollution, including air toxics. Analysis of toxic air pollutants is often a required component of state-level environmental permitting. Power plants are a good example of sources that must consider air toxics in BACT analysis under the North County Remand, and energy facilities may ultimately be regulated for specific HAPs under Title III of the 1990 CAA Amendments. (Two different studies will likely be conducted of utility air toxics emissions under the CAA Amendments before that decision is made.)

III. THE pcAIR-1© SYSTEM

The original AIR-1 system was developed by Galson in cooperation with the Eastman Kodak Company[2] for Digital Equipment's VAX line of minicomputers. It was designed to assist environmental managers in compiling, maintaining, analyzing, and reporting the data needed to satisfy concerns for air toxics. AIR-1 used customized software to create a relational database linked with a dispersion model and analysis modules for specific air toxics management issues. The system provided a mechanism to comprehensively assess the emissions and impacts of a complex facility and to revise that assessment as both the facility and the regulations change.

pcAIR-1 is an enhanced Air Toxics Management System for PCs, which has been designed to extend the capabilities of the minicomputer AIR-1. The computing power and storage space requirements that initially dictated a minicomputer implementation can now be provided on cost-effective PC platforms. The pcAIR-1 user has improved access to the system for editing data, conducting dispersion modeling, and analyzing the results.

pcAIR-1 is implemented as an integral part of the Arc/Info® GIS. GIS was chosen as the platform for several reasons. First, dispersion modeling analyses are dependent upon spatial data such as source locations, receptors, and terrain. Second, GIS is adept at handling large data sets and integrating data from many different sources. Third, many advanced graphical and analytical functions, such as layered storage of data, zooming and panning for interactive graphical analysis, creating custom graphics, and on-screen point-and-click queries, are fundamental GIS operations[3] that do not require extensive custom programming.

The pcAIR-1 design is based on linked databases, which store information needed to identify sources of many different pollutants, facilitate maintenance of source and emission records, and support permitting. In addition, the system maintains the input data and current results for dispersion modeling. The system also maintains compliance records, emission inventories, and other output from the pcAIR-1 analysis modules. Internal database structures are essentially similar for AIR-1 and pcAIR-1. These have been presented in detail elsewhere.[2] pcAIR-1 provides important new features, in that relations between databases can be based on direct graphical point-and-click access, as well as through menus.

Access and updates to all the pcAIR-1 databases can be made by users through a series of easy-to-use menu commands. The GIS foundation of the system also provides a point-and-click menu interface to operating functions such as database management (querying, linking, and editing), air dispersion modeling and analysis, report generation, and software access (to external software and tools, such as word processors, spreadsheets, and other database managers). Users can view the data on screen and generate tabular or graphical reports. Hard copy and screen display are available for both reports and plots, and the reporting system is menu driven, with appropriate prompts for data queries. Because data security is an important issue in environmental management, pcAIR-1 includes password-controlled access to the system.

IV. DISPERSION MODELING

Considerable effort has been expended in defining dispersion modeling procedures to evaluate new sources of air pollutants in a regulatory context.[4-6] In principle, these modeling procedures, or combinations of otherwise recognized procedures, can account for most of the situations expected from even complex facilities. However, modeling generally requires a significant effort for each individual pollutant and individual source in question.

The prospect of analyzing impacts from up to 189 (or more) pollutants from many different sources requires special consideration. The pcAIR-1 dispersion model was designed to conduct a general analysis of each source, then interactively determine impacts for a pollutant of concern with the same precision as pollutant-specific analyses. In this way, impacts of many different pollutants — or simultaneous impacts from multiple pollutants — can be readily analyzed.

The primary dispersion model currently recommended for most pcAIR-1 installations is the U.S. EPA's Industrial Source Complex — Short Term (ISCST) model. This model is generally accepted by state and federal agencies that will be evaluating impacts of HAPs. It includes provisions both for point sources and for fugitive emissions that may emanate from building vents, open storage areas, tanks, or other locations.

The dispersion modeling module of the pcAIR-1 system can be adapted to suit the individual location and source characteristics. This may be necessary, as the appropriate model for a given situation may differ. Models such as COMPLEX I or SHORTZ can be used where complex terrain (ground elevation above stack height) is important. Galson's CAVITY model is available to account for impacts near buildings. CAVITY is based upon an approach by Halitsky,[7] and implements currently recommended U.S. EPA procedures.[4-6] Specialized models such as FDM (for fugitive dust) or deposition models incorporating the California Air Resources Board (CARB) SEHMEL algorithm[8] can also be made available. In addition to accessing dispersion models within pcAIR-1 for routine analysis, the system can be used to generate input data for specialized modeling studies, which can be conducted off line.

V. AIR TOXICS MANAGEMENT ANALYSES

pcAIR-1 provides the capability to analyze and summarize the emissions, impacts, and projected air pathway health risk from a large number of sources of many different pollutants. Comprehensive multisource dispersion modeling studies can be done interactively, with both numerical and graphical results for even the most complex analysis generally available in real time.

Data in two different internally maintained databases are updated each time an impact analysis is made. This maintains a record of the facility's compliance status with respect to guidelines or standards (if applicable), the maximum annual average concentration, location of the maximum concentration, and the date of analysis. pcAIR-1's "what-if" analysis feature also allows the user to evaluate emissions and impacts under various source and control scenarios (adding or modifying sources, adding control equipment, changing emissions estimates, etc.) without disturbing the permanent facility databases.

Tabular reports can be obtained at any point in an analysis via menu options. Reports include comprehensive summaries of complete results, analogous to standard EPA dispersion model results, and specialized reports designed to focus on common problems in interactive modeling analysis. For example, results can readily be obtained to show sources contributing to predicted impacts of each pollutant under study. These can be obtained for selected impact locations (receptors) or for the entire analysis. pcAIR-1 also provides the option to create special purpose reports for site-specific needs, and data can be extracted from the databases in formats compatible with spreadsheets, word processing systems, or other software. Reports can be generated from either the graphics screens (without interrupting an interactive analysis) or from the database management screens.

Graphical output to support interactive analysis can also be selected from menus. A plot can be generated simultaneously with any tabular report through on-screen graphical query. The graphical output can be modified and copied, saved for later retrieval, and/or printed on a laser printer or plotter. Specific graphical analyses that are used repeatedly, such as impact isopleths or source location plots, can be assigned a menu selection or function key that will automatically create the standard output format.

The pcAIR-1 system can also analyze impacts based on toxicity due to the air-exposure pathway. It can be used to estimate both maximum health risk (the hypothetical Maximally Exposed Individual, or MEI risk) and impacts to areas of concern, such as nearby residential areas. Ambient impacts for each pollutant can also be presented in the form of isopleths of predicted health risk. Cumulative health risks, assuming additive impacts from user-selected pollutants, is also possible.

Additional applications may require combining impact analysis with other data. These extensions to customary impact analysis can be readily accommodated with off-line GIS analysis. For example, quantitative health risk assessment can require comparison of detailed concentration and deposition predictions with census data in order to estimate exposures due to air toxics emissions. Census data are readily available in a format compatible with Arc/Info. Impact and census data can be combined in Arc/Info, then analyzed further with all of the spatial analysis capabilities of a sophisticated GIS. Similar extensions can be accommodated to integrate pcAIR-1 data and analyses with other data sets. Databases are maintained in dBase® format, and can be converted to text files or spreadsheet formats as needed. Graphics are stored in the proprietary format developed for Arc/Info, which is the most popular GIS system. Graphical data files can also be converted to a variety of formats, including Autocad® DXF and others.

VI. APPLICATION

A hypothetical situation has been defined to illustrate pcAIR-1 features. The analysis involves a 500-MW coal-fired power plant. The plant is assumed to be located adjacent to an industrial facility that has both process sources and an oil-fired reciprocating internal combustion (IC) engine, which cogenerates electricity and thermal energy. Concern has been raised over emissions of chromium and formaldehyde from the coal-fired plant. The coal-fired plant is the only nearby source of chromium; the coal-fired plant, process sources, and IC engine each emit formaldehyde.

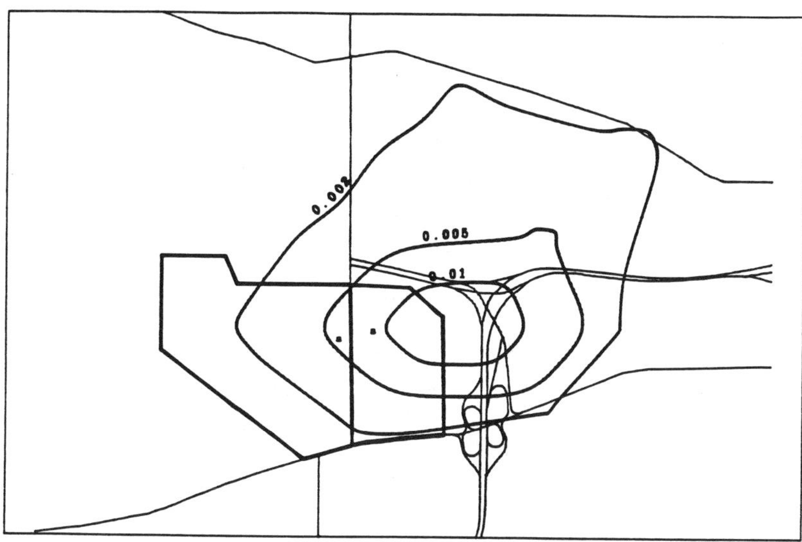

Figure 1. Predicted chromium concentrations.

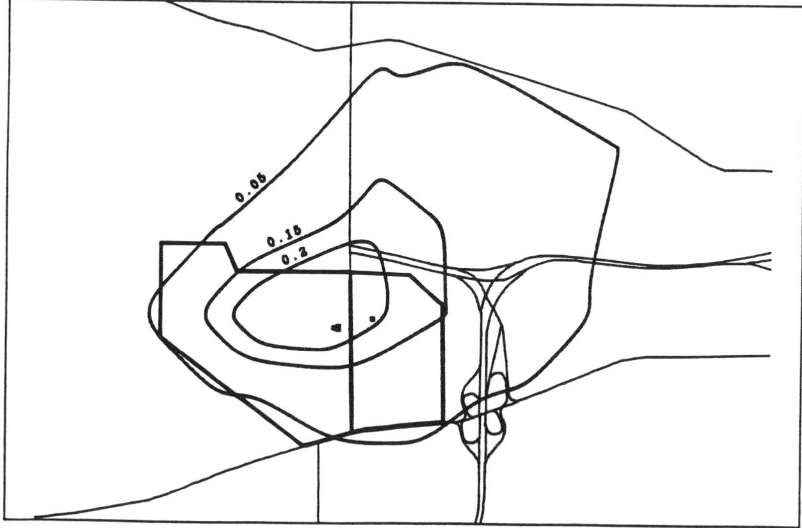

Figure 2. Predicted formaldehyde concentrations.

The layout of the two facilities is illustrated in Figure 1 and 2 along with the impact analysis for chromium and formaldehyde. Source data for the coal-fired plant are detailed in Figure 3, in pcAIR-1 report format. Figure 4 shows the analysis of source contributions to maximum impacts. While the coal-fired plant is the only source of chromium, the IC engine appears to be the primary factor in formaldehyde impacts.

SOURCE ANALYSIS FOR pcAIR-1 SOURCE: BOLR01 NUMBER OF EMISSIONS: 27

SOURCE PARAMETERS	
X COORD (STATE PLANE)	728248
Y COORD (STATE PLANE)	1149528
ELEVATION (FEET)	405
STACK HEIGHT (FEET)	185.00
STACK TEMP (F)	350
STACK VELOCITY (FPS)	32.80
STACK DIAMETER (IN)	61 0

BUILDING PARAMETERS	
BUILDING	
HEIGHT (FEET)	95
N-S LENGTH (FEET)	158
E-W WIDTH (FEET)	227
X COORD (STATE PLANE)	728227
Y COORD (STATE PLANE)	1149475

EMISSIONS					
CAS NUMBER	ANNUAL (TONS/YR)	AVG HOUR (LBS/HR)	PEAK HOUR (LBS/HR)	OPERATING HOURS DAYS	HOURS
7446-09-5	9150000.00	1830.000000	2013.000000	350	24
10102-43-9	7380000.00	1476.000000	1623.600000	350	24
NYS-00-0	203000.00	40.600000	44.660000	350	24
630-08-0	2585000.00	517.000000	568.700000	350	24
7439-92-1	2628.00	0.300000	0.330000	350	24
7440-41-7	210.00	0.024000	0.026000	350	24
7439-97-6	3504.00	0.400000	0.440000	350	24
7429-90-5	67468.00	7.702000	8.472000	350	24
7440-21-3	116620.00	13.313000	14.644000	350	24
7723-14-0	356.00	0.041000	0.045000	350	24
7782-50-5	352.00	0.040000	0.044000	350	24
7740-70-2	519.00	0.059000	0.065000	350	24
7440-47-3	2628.00	0.300000	0.033000	350	24
7740-09-7	6580.00	0.751000	0.826000	350	24
7439-89-6	11189.00	1.277000	1.405000	350	24
74-84-0	16221.00	1.852000	2.037000	350	24
71-43-2	2978.00	0.340000	0.374000	350	24
50-00-0	876.00	0.100000	0.110000	350	24
74-98-6	13815.00	1.577000	1.735000	350	24
106-97-8	7606.00	0.868000	0.955000	350	24
106-98-9	6131.00	0.700000	0.770000	350	24
78-78-4	9469.00	1.081000	1.189000	350	24
627-20-3	8149.00	0.930300	1.023330	350	24
110-54-3	73345.00	8.372700	9.209970	350	24
142-82-5	11642.00	1.329000	1.461900	350	24
592-41-6	28174.00	3.216180	3.537800	350	24
108-08-7	49595.00	5.661540	6.227690	350	24

Figure 3. Source analysis for pcAIR-1.

A real-world analysis would require considerable additional complexity. Fugitive emissions from both coal and ash handling at the power plant (for chromium) and building ventilation at the industrial facility (for formaldehyde) may be significant. Hexavalent chromium impacts may be initially estimated based on total chromium, then revised to more reasonable levels if the preliminary risk assessment shows it to be a significant factor. Several other pollutants may be of concern. Exposure pathways other than inhalation may be a factor for air toxics in the particulate phase. Actual human exposure may be small if maximum cumulative impacts do not coincide with populated areas.

The range of potential air toxics concerns is considerable. pcAIR-1 helps provide timely answers to these concerns by facilitating database development, maintenance, and revision, and by providing the analytical tools to answer the complex questions raised.

SOURCE CONTRIBUTION REPORT

SITE: STUDY AREA #1 DATE = 09/05/91

CAS#: 007440-47-3 CHROMIUM % OF IMPACT: 90.00

RANK: 1 RECEPTOR: 62 CONC (ug/m**3): 1.57E-02
TOP SOURCES CONTRIBUTING TO 90.00 OF IMPACT

SOURCE	CONC (ug/m**3)	% OF IMPACT
BOLR01	1.56E-02	99.36

CAS#: 000050-00-0 FORMALDEHYDE % OF IMPACT: 90.00

RANK: 1 RECEPTOR: 61 CONC (ug/m**3): 2.963E-01
TOP SOURCES CONTRIBUTING TO 90.00 OF IMPACT

SOURCE	CONC (ug/m**3)	% OF IMPACT
SORC01	2.483E-01	83.80
SORC02	1.56E-02	5.26
SORC03	1.56E-02	5.26
SORC04	1.56E-02	5.26

Figure 4. Source contribution report.

VII. CONCLUSIONS

The growing regulatory concern for air toxics, as reflected in the CAA Amendments of 1990, has created management challenges that require both extensive data and extensive analysis. Regulatory plans must be formulated to respond to these new federal initiatives. There will be a need to consider initial MACT requirements together with possible further restrictions under residual risk standards. In addition, many states either have or are developing regulations for air toxics. A significant number of these states are concerned with ambient impacts as well as identifying sources, emissions, and controls. Public sensitivity to air toxics emissions and impacts is being raised by SARA reports that are readily available and widely publicized by the news media. For most facilities, this will ultimately translate into more questions, more analysis, and more permitting issues, both state and federal, for more different sources.

Analysis of air toxics emissions presents problems on a larger scale than analysis of traditional air pollutants. The major constraint in air toxics emissions and impact analysis is often data management. pcAIR-1 provides the tools to simplify data management and assist with the elaborate analysis frequently required for major industrial and utility facilities. In addition, pcAIR-1 was designed to be modular. This modularity allows for adding or modifying existing capabilities without significantly altering the system. pcAIR-1 is based on data structures that can readily be extended or can communicate with other database management systems and general-purpose software.

pcAIR-1 offers a range of access to data and software for air toxics management, which gives environmental managers the freedom to consolidate data management tasks and focus on interpretation and problem solving.

ACKNOWLEDGMENTS

We would like to thank Michael Stoogenke, Geosource Operations Manager, for contributing Figures 1, 2, 3, and 4 included in this paper.

REFERENCES

1. U.S. EPA, National Air Toxics Information Clearinghouse: NATICH Data Base Report on State, Local, and EPA Air Toxics Activities, EPA-450/3-90-012, U.S. Environmental Protection Agency, OAQPS, June 1990.
2. Mathews, J., Wells, R. C., Palma, T., and Muschet, D., Modeling Air Toxics Exposure from a Complex Industrial Site, at the 83rd Annual Meeting and Exhibition, Air and Waste Management Association, Paper No. 90-73.8, Pittsburgh, 1990.
3. Burrough, P. A., *Principles of Geographical Information Systems for Land Resource Assessment,* Clarendon Press, Oxford, 1986.
4. U.S. EPA, Regional Workshops on Air Quality Modeling: A Summary Report (Revised), U.S. Environmental Protection Agency, OAQPS, October 1983.
5. U.S. EPA, *Guideline on Air Quality Models* (revised), EPA-450/2-78-027R, U.S. Environmental Protection Agency, OAQPS, July 1987.
6. U.S. EPA, Supplement B to the Guideline on Air Quality Models (revised), U.S. Environmental Protection Agency, OAQPS, September 1990.
7. Halitsky, J., Gas diffusion near buildings, *ASHRAE Trans.,* 69, 464, 1963.
8. ARB, SEHMEL — FORTRAN 77 Program, Technical Support Division, California Air Resources Board, P.O. Box 2815, Sacramento, CA, 95812, 1987.

Removal of Pyrite from Coal before Combustion

J. D. Bignell

I. ABSTRACT

Worldwide, utilities are facing tighter controls on sulfidic emissions from coal-fired plants and elimination of at least part of the pyrite in the coal before combustion would ease the burden put upon permit sharing schemes or other, costly, abatement systems. For pyrite to be removable, the coal has to be ground fine enough for liberation, but still be coarse enough to process. Conventionally, grinding is done dry at the site of combustion. To avoid expense and plant disruption, a dry separation process is essential. Computer models based on dry sluice tests at 630 kg/h have demonstrated that 30% of the toal sulfur (~40% of the pyrite) in the classifier returns of two U.K. utilities was removable for the loss of only 3% of the calorific value. As discussed in the paper, the potential for pyrite removal is entirely coal and grind dependent but refinement and application of sluice technology could permit the consideration of several alternative abatement strategies even where control is strict.

II. INTRODUCTION

Throughout the world, particularly in North America, there is a growing public and political concern about the effects that acid depositions are having on the environment. Coal-burning power plants are perceived as prime sources of the acidic pollutants nitrogen and sulfur oxides, and are coming under increasing legislative pressure to reduce the emissions. The nitrogen oxides can be reduced by modified burning practices; however, the sulfur is a part of the coal itself and will inevitably go to the oxide form on combustion.

The sulfur oxides can be captured in the bed of fluidized bed plants by the addition of lime or, for existing plants, by the fitting of flue gas desulfurization equipment. Both methods are inevitably expensive on capital, can be complex to run, and require the consumption and disposal of much lime.

An underestimated alternative approach is physical removal of pyrite at the power plant. This cannot be as comprehensive a method for sulfur removal as the above but it still offers the possibility of removing significant sulfur relatively inexpensively.

At present many, but by no means all, coals are washed for the removal of ashy materials after crushing no finer than 50 mm; some pyrite is removed but this is

Table 1. Power Plant Classifier Return — Size-Assay Analysis

Size fraction (mm)	Density product (g/mL)	Wt %	Total sulfur Assay (%)	Total sulfur Distribution (%)
1.0–0.3	<1.4	18.02	0.87	5.73
	1.4–2.0	1.24	2.43	1.10
	2.0–3.3	0.84	6.62	2.03
	>3.3	0.24	40.8	3.58
		20.34	1.68	12.44
0.3–0.15	<1.4	36.40	0.78	10.30
	1.4–2.0	2.20	2.32	1.86
	2.0–3.3	1.30	7.31	3.47
	>3.3	0.91	32.6	10.81
		40.81	1.78	26.44
0.15–0.075	<1.4	19.01	1.03	7.15
	1.4–2.0	4.56	1.65	2.74
	2.0–3.3	1.54	6.21	3.48
	>3.3	1.61	38.3	22.48
		26.72	3.68	35.85
0.075–0.053	<1.4	5.14	1.00	1.87
	1.4–2.0	2.86	1.36	1.42
	2.0–3.3	0.74	9.15	2.47
	>3.3	0.66	48.5	11.68
		9.40	5.08	17.44
<0.053		2.73	7.86	7.83
Total		100.00	2.74	100.00

purely incidental. To be removable the pyrite has to be effectively released, or liberated, from the coal and be at a size suitable for treatment. Pyrite crystals or crystal aggregates are frequently finer than 1 mm and are hence inseparable at the washery. However, at the power plant the coal is usually dry pulverized to about 0.075 mm; during the pulverization a last opportunity to separate the pyrite from the coal presents itself.

During pulverization, the coal passes through a size range suitable for treatment on the dry pinched sluice (0.5 to 0.1 mm). The dry sluice is an inclined tapered chute similar to the conventional wet sluice except the deck is made of a porous plastic through which air is gently passed.[1,2] This air fluidizes the coal and any liberated pyrite gravitates through the coal to the deck and a separation can be effected at the lower narrow end.

The effectiveness of the method is dependent upon the specific coal but the suitability of any coal for treatment can be readily assessed by a size-density analysis on a sample ground to about 1 to 0.5 mm (e.g., Table 1). Any sulfur that passes to

Table 2. Actual and Predicted Dry Sluice Test Results

		Total sulfur (%)		Ash (%)		CV (MJ/kg)	
	Wt %	Assay	Distribution	Assay	Distribution	Assay	Distribution
				Actual Single Pass			
Heavies	10.1	8.7	31.6	27.5	26.6	21.5	7.9
Lights	89.9	2.1	68.4	8.5	73.4	28.2	92.1
Feed	100.0	2.8	100.0	10.4	100.0	27.5	100.0
			Predicted Rougher Pass with Repass of Rougher Heavies				
Heavies	5.7	15.1	31.1	33.6	18.5	15.4	3.2
Lights	94.3	2.0	68.9	8.9	81.5	28.2	96.8
Feed	100.0	2.7	100.0	10.3	100.0	28.2	100.0

Table 3. Recovery of Dense Grains to Heavies Product

		Recovered sulfur in grains	
Size (mm)	Recovered total wt in grains SG >3.3	SG >3.3	SG 2.0–3.3
1.0–0.5	84	85	87
0.5–0.3	94	94	84
0.3–0.15	90	91	48
0.15–0.075	51	52	38
0.075–0.053	6	6	1

a SG >3.3 density product must be present in the form of effectively liberated pyrite and will be separable from the coal if in a suitable size range. Similarly, sulfur in the SG 2.0 to 3.3 product will also be mostly in the form of pyrite locked with some coal and much of it will also be recoverable. Any sulfur present in SG products <2.0 must be present as organic sulfur or as relatively small pyrite crystals thinly dispersed through the coal; this will not be separable from the coal.

The classifier returns from a U.K. power plant have been analyzed in detail (Table 1) and tested on a 1.5 m long pilot scale sluice operated at 630 kg/h (Table 2).[3] The products from five test runs were all analyzed in detail and a computer model of the operation of the sluice was constructed.

Overall, 30% of the total sulfur (40% of the pyrite) was separated from the coal tested. However, the pyrite recovery was highly dependent on particle size, ranging from over 90% for grains of SG >3.3 in the >0.15 mm fraction to 6% in the 0.075 to 0.053 mm fraction (Table 3). The pyrite in the SG 2.0 to 3.3 product (along with much of the silicate ashy material) behaved similarly. Most of the pyrite not recovered into the sluice heavies product was lost solely because of its fineness.

The heavies product also contained about 10% of the calorific value of the feed. The detailed analysis revealed that most of this loss was due to misplaced coarse

(>0.3 mm), low-density coal grains. The model predicted that this coal should be largely recovered to the coal product with no loss of pyrite from the final heavies product by repassing the first heavies product (Table 2).

III. DISCUSSION

The test work has proved that it is possbile to recover pyrite to a low-weight heavies product by using the dry-pinched sluice on partially ground coal at a power plant. The potential for pyrite removal is dependent upon the individual coal concerned. In the detailed test work, 30% of the total sulfur was separated from the classifier return from two U.K. power plants. This sulfur was virtually all pyritic and represented 40% of the pyrite present in the return. Most of the pyrite not separated had already been ground too fine for recovery. Since the recovery of dense grains >0.15 mm was 90%, had the coal been ground specifically for sluice separation a total sulfur rejection of around 50% might have been anticipated.

Plant contractors have said that the sluice system should not be too difficult or expensive to incorporate into a power plant operation.

Grinding high pyrite coals to a size specifically suited to the individual coal and for separation on the sluice may be a cheap option for rejecting much of the sulfur in the coal. Depending on circumstances, this could make high pyrite coals more competitive with lower sulfur coals, permit their more widespread use, or at the very least, reduce the cost of running flue gas desulfurization plants and limiting the consumption of lime and its subsequent disposal.

The heavies product still must be disposed. After removal of the silicate ashy minerals it may be possible to burn the residue in small plants for the recovery of sulfur and heat.

REFERENCES

1. Douglas, E. and Sayles, C. P., Dry sorting using pneumatically fluidised powders, *Am. Inst. Chem. Eng.*, November 29–December 3, 1970.
2. Muller, L. D. and Sayles, C. P., Processing dry granular materials, *Min. Eng.*, 23(3), 54, 1971.
3. Bignell, J. D. and Gooriah, B., Physical removal of pyrite at the power plant, in *Processing and Utilisation of High-Sulfur Coals,* Vol. 3, Markuszewski, R. and Wheelock, T. D., Eds., Elsevier, Amsterdam, 1990, 321.

The PISCES Database: Structure and Uses

Robert G. Wetherold
Douglas A. Orr
Kevin J. Williams
Winston Chow

I. INTRODUCTION

The Electric Power Research Institute (EPRI) is conducting a project to assist utilities with managing and controlling chemical substances in the process streams of power-generating systems in light of current regulatory and health effects issues. This project, entitled Power Plant Integrated Systems: Chemical Emissions Studies (PISCES), involves the collection and validation of information on chemical substances in process and discharge streams from power-generating systems. One of the objectives of the EPRI research has been the development of a relational database to store this information. The database development represents one element in the overall EPRI effort to (1) define the source, pathways, and concentration of potentially hazardous pollutants; (2) establish the health and environmental risks associated with these chemical substances; (3) determine the performance of available technologies for their measurement and control, if warranted; and (4) develop a model for estimating their distribution and discharge.

The PISCES database contains information from selected public and confidential literature sources on the chemical composition of process streams present in conventional power-generating systems. Related information, such as plant operating characteristics, sample collection methods, analytical measurement techniques, and relevant regulations are also compiled in the database. Currently, over 80,000 individual records have been validated, qualified, and entered into the PISCES database.

This paper discusses the structure of the database, the type of information available in the database, and the distribution of data within the database. Typical ways in which the PISCES database has been and can be used are illustrated. Finally, current methods of accessing the existing database are described, and methods of making the database more accessible in the future are discussed.

II. DATABASE DESCRIPTION

The PISCES database is a relational database; it is composed of several separate tables which may be linked together. Each table contains a large number of records,

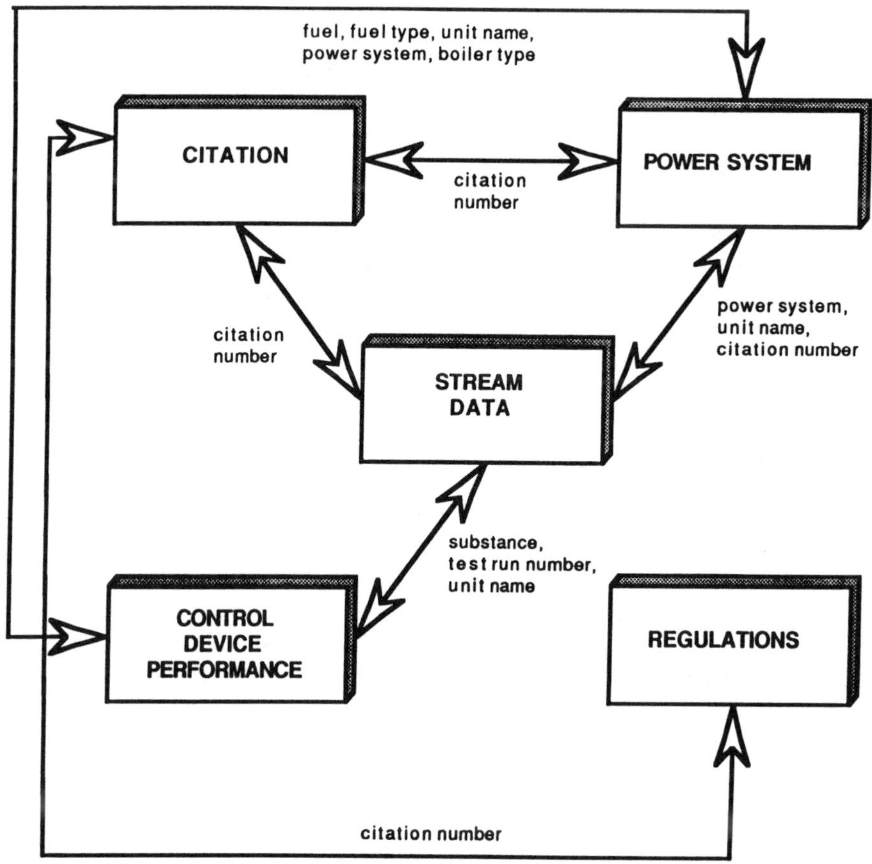

Figure 1. Relationship between tables in the PISCES database.

and each record consists of several fields called attributes. Each attribute contains a specific type of information pertaining to the record, such as the name of the unit or the sampling method. Attributes which are common to more than one table may be used to link these tables together when information is retrieved from the database.

Figure 1 illustrates the structure of the PISCES database, showing the five most important tables. Arrows drawn between the tables show which tables may be linked together; the common attributes are listed next to the arrows. The PISCES tables include

- The **Citation** table, which contains information pertaining to each of the literature sources
- The **Power Systems** table, which contains a description of each unit from which stream data were obtained

- The **Stream Data** table, which contains measured concentration data for specific chemical substances in power plant streams
- The **Control Device Performance** table, which contains paired inlet and outlet data for air pollution control devices
- The **Regulations** table, which is a compilation of regulatory limits for individual substances

The attributes for each of the PISCES tables are summarized in Table 1.

The PISCES database contains stream data for both organic and inorganic substances in a wide array of power plant streams. Table 2 lists 59 of the 71 individual stream names for which data are available. These streams include gas, solid, and liquid phase streams from major groups such as fuels, boiler additives, reagents, intermediate streams, and discharges.

III. DATA SOURCES

Currently, the PISCES database contains information from 259 individual citations. Much of the data contained in the PISCES database was obtained from studies available through the public literature, including journal articles, conference proceedings, and reports. Process stream composition data are currently available for 425 unique power-generating systems; utilities in the U.S., Canada, and Europe are represented. In addition, the database also contains information from confidential sources. Each citation is given an index designation which distinguishes confidential data.

IV. DISTRIBUTION OF DATA

The existing PISCES database contains approximately 80,000 records for individual process streams. Of these 80,000 records, roughly 60% are associated with solid streams, 28% are associated with liquid streams, and 12% are associated with gaseous streams. The relatively small amount of data for gaseous streams is not surprising since the measurement of trace species in gas streams has historically been a costly, labor-intensive effort. As a result, power plants have not routinely measured trace substances in gaseous streams. The PISCES database contains stack gas (i.e., emitted gas) composition data for about 80 individual units which represent only a fraction of the approximately 2000 steam-electric units listed in the Utility Data Institute Power Directory.

The distribution of data records among selected electric utility process streams in the database is depicted in Figure 2. The records for the streams shown comprise about 80% of all records in the database. Among solid streams, the majority of the data are for coal, bottom ash, and collected fly ash. Smaller quantities of data exist for emitted particulate, solid streams in flue gas desulfurization (FGD) systems, and solids from water treatment systems. Liquid stream data are primarily associated with

Table 1. PISCES Attribute Summaries

Citation (18)	Power systems (41)	Stream data (31)	Control device performance (18)	Regulations (11)
Document information, such as title and author	Power system category	Power system category	Power system category	Chemical substance
Citation number	Unit name	Unit name	Unit name	CAS number
Comments	Citation number	Citation number	Citation number	Citation number
	Fuel type, source, and specifications	Stream name	Control device	Document type
	Boiler type and additives	Run number	Fuel type	Media
	Capacity	Chemical substance	Boiler type	Limit
	Control device configuration	Sampling and analytical methods	Chemical substance	Comments
	Solid waste disposal information	Concentration and detection flag	Inlet/outlet concentrations and detection flags	
	Cooling system information	Variability information	Removal efficiency	
	Wastewater treatment information	Credibility		
	Comments	Comments		

Note: Total number of attributes for each table is shown in parentheses.

Table 2. Selected Power Plant Streams Contained in the PISCES Database

Solid streams	Liquid stream	Gas streams
Bottom ash	Ash pond water	Air
Coal	Boiler chemical cleaning waste	Boiler outlet gas
Coal pile solids	Boiler blowdown	Boiler outlet gas, gas phase
Coal, ash phase	Bottom ash makeup water	Boiler outlet gas, solid
Collected economizer ash	Bottom ash water	Emitted gas
Collected fly ash	Clean coal waste	Emitted gas, FGD pond
Collected fly ash, sized	Coal pile runoff	Emitted gas, gas phase
Collector inlet ash	Cooling tower blowdown	Emitted gas, sized
Collector inlet ash, sized	Cooling water	Emitted gas, solid phase
Cooling tower basin sludge	Evaporation pond water	Emitted particulate
Economizer ash	FGD lime, liquid	Emitted particulate, sized
FGD lime	FGD liquid	FGD inlet gas
FGD lime grit	FGD slurry	Low NOx gas
FGD limestone	Floor drain	Spray dryer outlet gas
FGD solids	Landfill runoff	
Fixed fly ash/FGD solids	Makeup water	
Municipal solid waste	Oil	
Oil, ash	Vapor compression	
Sodium injection waste solids	evaporation brine	
Spray dryer solids	Vapor compression	
Superheater ash	evaporation feed water	
Vapor compression	Vapor compression evaporation	
evaporation solids	liquid waste	
Wastewater treatment	Vapor compression	
sludge	evaporation product water	
	Wastewater treatment	
	effluent	

ash pond water, FGD liquid, and makeup water. Data for gaseous streams are primarily associated with the emitted gas (i.e., stack gas); data for intermediate gas process streams are quite sparse.

The PISCES database contains data for approximately 390 unique chemical substances, 156 inorganic substances, and 234 organic substances. Table 3 lists the hazardous air pollutants (HAPs) for which stack gas data are available; 46 of the 189 HAPs specified in the 1990 Clean Air Act Amendments are represented. The distribution of records for selected substances among solid, liquid, and gas streams is shown in Figure 3. The number of records for important elements and organic compounds are provided. Generally, only 5 to 10% of the total records for a given element are associated with gas streams (i.e., stack gas, boiler outlet gas, FGD inlet gas). For example, the database contains 2591 records for arsenic in all solid, liquid, and gas streams; 213 of these records (8%) are associated with gas streams. Relatively large data sets are available for most elements in solid and liquid streams. Very little data (less than 100 records) are available in the literature for organic compounds such as benzene, dioxins, and furans in any of the utility process streams.

One goal of the PISCES program has been to use the PISCES database to identify data gaps for particular substances, streams, and power system configurations. This

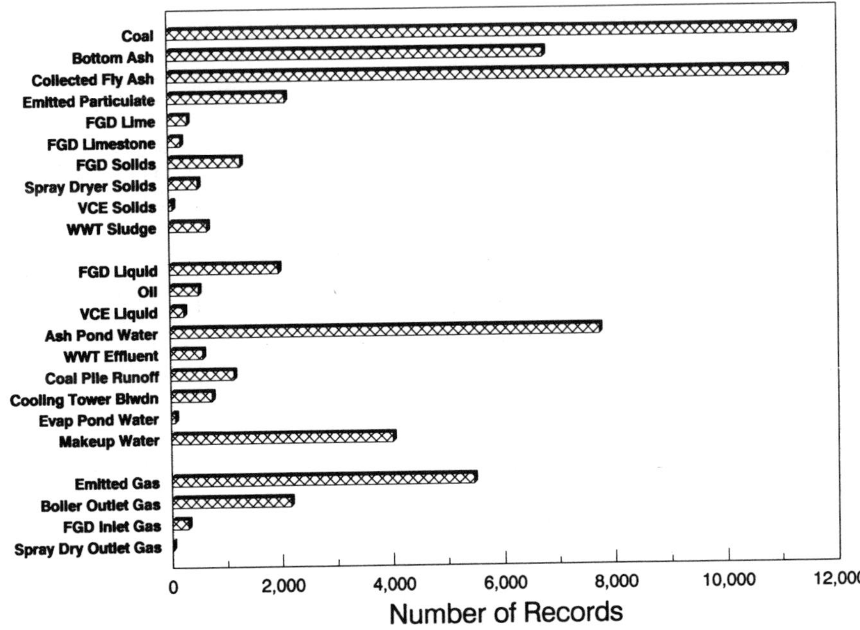

Figure 2. Number of records for selected streams in the PISCES database.

type of information can then be used to direct future literature review and field sampling efforts to fill gaps in emission information in the utility industry. Figure 4 shows the distribution of stack gas records among the different plant configurations (i.e., different combinations of fuel type and emission control device). The number of records for organic and inorganic substances are shown separately. A substantial amount of stack gas data (approximately 74% of the total stack gas records) exists for coal-fired boilers equipped with electrostatic precipitators (ESPs); however, the volume of data decreases dramatically for other control device configurations. A relatively large data set for coal-fired boilers equipped with ESPs is expected since this configuration represents approximately 40% of the total megawatt generating capacity in the U.S. To date, very little information on emissions from systems using fabric filters has been identified. Data sets for oil- and gas-fired units are also small. Figure 4 illustrates the need for additional testing and literature review.

V. USES OF THE DATABASE

The PISCES database can be used to retrieve data sets for particular streams, chemical substances, and plant configurations. Data queries can be tailored to include all desired information pertaining to an individual data value, such as the sampling method, the analytical method, and comments, to assist in evaluating the quality of the data. Thus, the database can be readily applied to the development of emission

Table 3. Hazardous Air Pollutants for which Stack Gas Data Are Available

Organic substances	Inorganic substances
Acetaldehyde	Antimony
Acrylonitrile	Arsenic
Benzene	Beryllium
Biphenyl	Cadmium
Carbon disulfide	Chlorine
Carbon tetrachloride	Chromium
Chlorobenzene	Cobalt
Chloroform	Hydrochloric acid
Dibenzofurans	Hydrogen fluoride (hydrofluoric acid)
1,4-Dichlorobenzene(p)	Lead
Ethylene dichloride (1,2-dichloroethane)	Manganese
Formaldehyde	Mercury
Hexachlorobenzene	Nickel
Hexane	Phosphorus
Methyl bromide (bromomethane)	
Methyl chloride (chloromethane)	
Methyl chloroform (1,1,1-trichloroethane)	
Methyl ethyl ketone (2-butanone)	
Methyl iodide (iodomethane)	
Methylene chloride (dichloromethane)	
Naphthalene	
N-Nitrosodimethylamine	
N-Nitrosomorpholine	
Pentachlorophenol	
Phenol	
Polychlorinated biphenyls (arochlors)	
2,3,7,8-Tetrachlorodibenzo-*p*-dioxin	
Tetrachloroethylene (perchloroethylene)	
Toluene	
1,2,4-Trichlorobenzene	
Trichloroethylene	
Polycyclic organic matter	

factors and the evaluation of control technology performance. Additionally, subset databases can be generated for use in alternate formats for any specific application. Also, the database can be used to develop partitioning coefficients for species traversing the various power plant subsystems. This type of information, plus the variability of substance composition data, is useful for the operation of the PISCES Plant Chemical Assessment Model being developed by Carnegie-Mellon University under a parallel PISCES effort.

Two figures are included to illustrate the type of information which can be obtained from the PISCES database. Figure 5 shows the relationship between the concentrations of arsenic in the stack gas and in the coal from bituminous coal-fired power plants employing either an ESP or an ESP/FGD combination for air pollution control. Each point represents a single stack gas/coal data pair sampled during the same test run. The lines drawn on the figure indicate the stack gas concentration that would be expected if the indicated percentage of the inlet arsenic is emitted from the stack. This serves as an indicator of the relative removal efficiency. Figure 6 shows

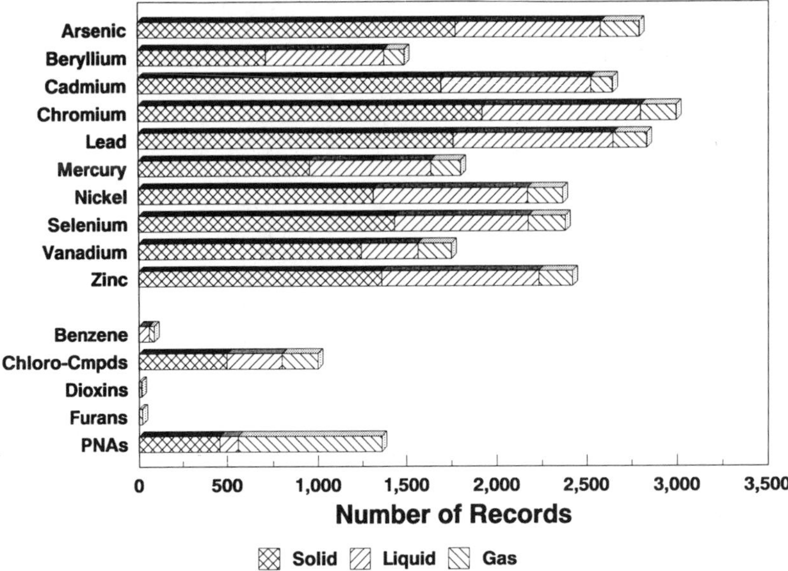

Figure 3. Number of records for selected substances in the PISCES database.

the mean concentrations of selected trace elements in bottom ash and collected fly ash from subbituminous coal-fired power plants employing an ESP. As can be seen, for each element the concentrations in the two ash streams are similar, with the fly ash concentration somewhat higher in the majority of the elements shown.

VI. ACCESSING THE DATABASE

The PISCES database uses a relational database management software program. It is currently maintained on a network of SUN Microsystems computers. Access to the database system is achieved via telephone modem, personal computers, and terminal emulator software. The current system for accessing the database allows users to browse through tables, link to information in more than one table, and view records one at a time. The system also provides a flexible mechanism by which the user may select a subset of data to satisfy a specific data need. These customized data retrievals are performed using the native standard query language (SQL) provided with the database software.

In addition, a standardized data retrieval program has been developed which allows the user to select stream composition data and supporting information for different combinations of fuel type and power system configuration.

Future plans for accessing and distributing the PISCES database include development of a CD ROM version of the database and development of subset "mini-

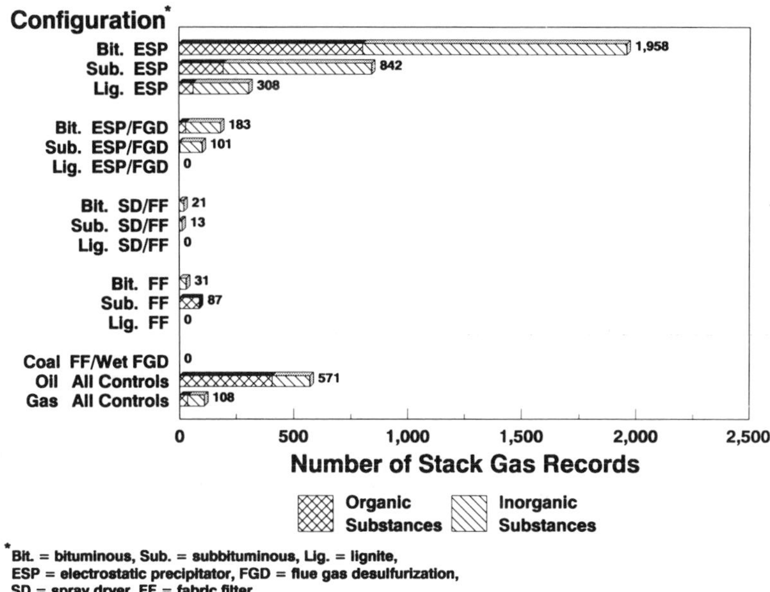

Figure 4. Number of stack gas records for selected power system configurations in the PISCES database.

databases". PC software suitable for accessing the CD ROM version will also be developed. Current plans are to produce a series of mini-databases on 3.5" diskettes for use with Lotus® 123™ spreadsheets. Each diskette will contain information and data associated with categories such as those described below.

Category	Contents
Power systems	Unit configuration, plant location, control system type, plant size, etc.
Streams	Trace substance levels in candidate streams, such as coal, oil, ash, emitted gas, water steams, etc.
Performance	Matched sets of inlet and outlet concentrations along with calculated removal efficiencies of various control systems.

These mini-databases will be available in the winter 1991/spring 1992 period.

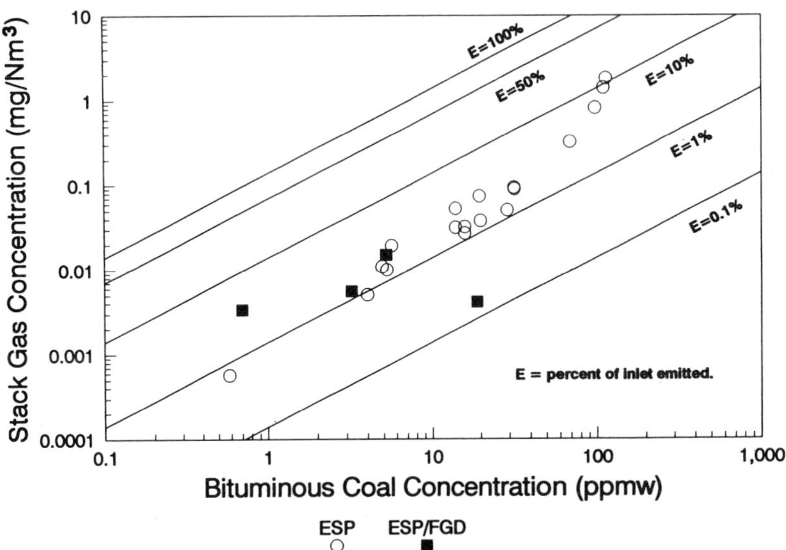

Figure 5. Comparison of stack gas arsenic concentrations with inlet coal concentrations for bituminous coal-fired boilers equipped with ESP and ESP/FGD controls.

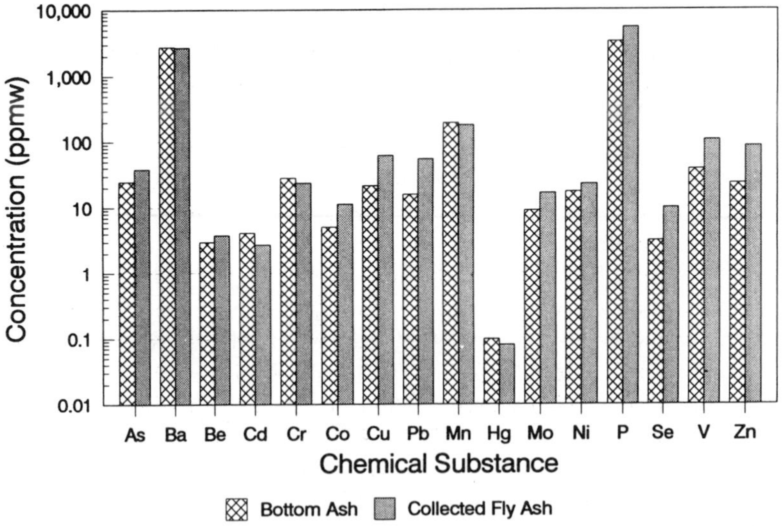

Figure 6. Comparison of bottom ash and fly ash mean concentrations for subbituminous coal-fired boilers equipped with ESP controls.

Coal Cleaning: A Trace Element Control Option

David J. Akers

I. INTRODUCTION

Most naturally occurring elements have been detected in coal, at least in trace concentration. In sufficient quantity, many of these trace elements are potentially toxic to plant and animal life. Although no new constraints on trace element emissions were placed on the power generation industry under the 1990 Clean Air Act, it is expected that new regulations will be forthcoming following the federally mandated 3-year study of air toxics. Despite the uncertainty of the full effects of any new laws, utility compliance could be very costly. Because coal contains potentially toxic trace elements, reducing the emission of these elements may be a large part of any new emissions regulations.

Many trace elements in coal are associated with mineral matter. For example, arsenic is commonly associated with pyrite; cadmium with sphalerite; chromium with clay minerals; mercury with pyrite and cinnabar; nickel with millerite, pyrite, and other sulfides; and selenium with lead selenide, pyrite, and other sulfides.[1] There are also cases in which some of these elements are organically bound. Just as both organic and pyritic sulfur can be found in the same coal, the same trace element may be both organically bound and present as part of a mineral in the same coal.

Physical coal-cleaning techniques are effective in removing mineral matter from coal and can, therefore, potentially remove at least some of the trace elements associated with specific minerals if the mineral is physically separate from the coal (liberated) and if the cleaning process is designed to remove the mineral. Problems encountered in removing pyrite by froth flotation provide a classic example of the interplay between the cleaning process and removal of a specific mineral. Froth flotation, the most frequently used method of cleaning fine-sized coal, sorts by differences in surface properties. Unfortunately, the surface properties of coal and pyrite are so similar that froth flotation sometimes concentrates pyrite (an iron/sulfur mineral) in the clean coal rather than sorting it to refuse.

In spite of these limitations, physical coal cleaning can be very effective in trace element removal, guaranteeing that the element will not be released into the atmosphere. Cleaning processes such as heavy-media cyclones that sort by differences in density are commonly very effective in pyrite removal because of the large density difference between pyrite and coal. Also, physical coal cleaning reduces the ash

Table 1. Conventional Coal Cleaning: Upper Freeport Seam Coal

Test 1 (80% Energy Recovery)

	Raw		Clean		
Elements	**ppm**	**grams/BBtu**	**ppm**	**grams/BBtu**	**Reduction (%)**
Arsenic	66	3,619	18	600	83
Barium	254	13,926	62	2,068	85
Cadmium	5	274	3	100	63
Chromium	50	2,741	25	834	69
Fluorine	670	36,735	136	4,536	87
Lead	12	658	11	367	44
Mercury	0.7	38	0.25	8	78
Nickel	27	1,480	23	767	48
Selenium	1.1	60	1.89	63	–4
Silver	2	110	0.4	13	87
Zinc	73	4,002	35	1,167	71

Test 2 (56% Energy Recovery)

	Raw		Clean		
Elements	**ppm**	**grams/BBtu**	**ppm**	**grams/BBtu**	**Reduction (%)**
Arsenic	73	4,002	19	618	84
Barium	276	15,133	54	1,757	88
Cadmium	6	329	4	114	65
Chromium	51	2,796	28	911	67
Fluorine	550	30,155	137	4,457	85
Lead	18	987	7	228	76
Mercury	1	38	0.28	9	76
Nickel	30	1,645	21	683	58
Selenium	2	88	2	54	38
Silver	2	110	0	13	88
Zinc	75	4,112	29	943	77

Note: ppm — parts per million; grams/BBtu — grams per billion Btu.

content of coal and increases the heating value, reducing transportation costs and increasing boiler efficiency. Finally, coal cleaning provides other environmental benefits by reducing the sulfur dioxide emissions potential of the coal and the amount of ash for collection and disposal.

II. CONVENTIONAL COAL CLEANING TO REMOVE TRACE ELEMENTS

As part of a project funded by the Electric Power Research Institute, CQ Inc. — a wholly owned EPRI subsidiary located in western Pennsylvania — has demonstrated that large reductions in the concentration of many trace elements are possible if conventional coal-cleaning techniques are properly applied. Four examples are given in Tables 1 to 4. In each example, the results shown were generated by cleaning the coal at CQ Inc.'s commercial-scale cleaning test facility.

Table 2. Conventional Coal Cleaning: Rosebud/McKay Seam Coal

Test 1 (99% Energy Recovery)

	Raw		Clean		
Elements	**ppm**	**grams/BBtu**	**ppm**	**grams/BBtu**	**Reduction (%)**
Arsenic	2	117	2	77	34
Barium	284	16,676	118	4,567	73
Cadmium	0.8	47	0.5	19	60
Chromium	5	294	11	426	–45
Fluorine	50	2,936	30	1,161	60
Lead	0.7	41	4	155	–277
Mercury	0.12	7	0.1	4	45
Nickel	15	881	11	426	52
Selenium	0.51	30	0.67	26	13
Zinc	6	352	5	194	45

Test 2 (90% Energy Recovery)

	Raw		Clean		
Elements	**ppm**	**grams/BBtu**	**ppm**	**grams/BBtu**	**Reduction (%)**
Arsenic	3	165	2	78	53
Barium	320	17,581	117	4,560	74
Cadmium	0.8	44	0.8	31	29
Chromium	6	330	10	390	–18
Fluorine	50	2,747	50	1,949	29
Lead	4	220	4	156	29
Mercury	0.14	8	0.12	5	39
Nickel	14	769	10	390	49
Selenium	1.87	103	0.78	30	70
Zinc	7	385	6	234	39

Note: ppm — parts per million; grams/BBtu — grams per billion Btu.

Cleaning results for Upper Freeport Seam coal from Pennsylvania are provided in Table 1. Data are presented in the table in two ways, as a weight-based concentration (parts per million) and as a concentration per heat unit (grams per billion Btu). Grams per billion Btu is analogous to pounds per million Btu, but avoids the use of numbers with many decimal places. The heat-based concentration provides a better measure of boiler impacts because the increased heating value obtained through coal cleaning reduces the number of tons that must be burned to produce a given thermal output. Reducing the quantity of coal burned reduces the quantity of trace elements entering the boiler.

This raw coal is relatively high in several trace elements of environmental concern, including arsenic, cadmium, and chromium. Cleaning provided large reductions in the quantity of arsenic, barium, cadmium, chromium, fluorine, lead, mercury, nickel, silver, and zinc. The concentration of selenium was essentially unchanged by cleaning.

The results for tests with a Powder River Basin coal, Rosebud/McKay, are presented in Table 2. Large reductions in arsenic, barium, cadmium, fluorine, mer-

Table 3. Conventional Coal Cleaning: Croweburg Seam Coal

Test 1 (91% Energy Recovery)

	Raw		Clean		
Elements	**ppm**	**grams/BBtu**	**ppm**	**grams/BBtu**	**Reduction (%)**
Arsenic	7	243	4	115	53
Barium	69	2,484	10	315	87
Cadmium	0.4	14	0.1	1	91
Chromium	29	1,037	8	252	76
Fluorine	35	1,267	50	1,650	–30
Lead	10	358	8	270	25
Mercury	—	—	—	—	—
Nickel	33	1,180	27	899	24
Selenium	1	36	1.6	51	–44
Zinc	80	2,861	20	638	78

Test 2 (90% Energy Recovery)

	Raw		Clean		
Elements	**ppm**	**grams/BBtu**	**ppm**	**grams/BBtu**	**Reduction (%)**
Arsenic	4.8	172	3.6	116	32
Barium	29.5	1,055	3.4	109	90
Cadmium	0.3	11	0.0	1	94
Chromium	—	—	—	—	—
Fluorine	36.4	1,303	46.2	1,479	–14
Lead	7.7	274	7.1	227	17
Mercury	0.1	3	0.1	4	–26
Nickel	26.5	950	28.4	910	4
Selenium	0.6	21	1.3	41	–96
Zinc	70.8	2,532	17.3	555	78

Test 3 (95% Energy Recovery)

	Raw		Clean		
Elements	**ppm**	**grams/BBtu**	**ppm**	**grams/BBtu**	**Reduction (%)**
Arsenic	5.5	197	2.5	83	58
Barium	47.1	1,687	13.3	441	74
Cadmium	0.3	12	0.1	3	78
Chromium	28.9	1,035	9.9	328	68
Fluorine	—	—	—	—	—
Lead	10.1	360	6.2	205	43
Mercury	0.1	4	0.2	6	–46
Nickel	28.9	1,035	27.9	923	11
Selenium	1.0	36	0.8	28	22
Zinc	75.4	2,700	25.8	851	68

Note: ppm = parts per million; grams/BBtu = grams per billion Btu; — = sample undergoing analytical rechecks.

Table 4. Conventional Coal Cleaning: Kentucky No. 11 Seam Coal

Test 2 (73% Energy Recovery)

	Raw		Clean		
Elements	**ppm**	**grams/BBtu**	**ppm**	**grams/BBtu**	**Reduction (%)**
Arsenic	1.8	88	1.5	51	42
Barium	99	4,929	12	392	92
Chromium	41	2,060	13	425	79
Fluorine	717	35,851	59	1,956	94
Lead	21	1,070	4	116	89
Mercury	0.151	7,573	0.119	3,925	48
Nickel	43	2,144	12	385	82
Selenium	1.5	77	0.9	29	61
Zinc	97	4,834	13	431	91

Note: ppm — parts per million; grams/BBtu — grams per billion Btu. Data from Test 1 are not yet available.

cury, nickel, selenium, and zinc were observed with cleaning. The concentration of chromium increased with cleaning, while lead concentration increased in one test and decreased in another.

Table 3 presents test results for Croweburg Seam coal from Oklahoma. Large reductions in arsenic, barium, cadmium, chromium, and zinc were obtained with cleaning while fluorine, mercury, and selenium increased. Small reductions were obtained with the other elements measured.

Table 4 presents cleaning test data for Kentucky No. 11 Seam coal. In this case, large reductions were obtained with all elements measured.

In general, these data indicate that physical coal cleaning is effective in reducing the concentration of many trace elements, especially if they are present in the coal at relatively high concentrations. The degree of reduction achieved relates to the degree of mineral association of the specific trace element and the degree of liberation of the trace element-bearing mineral.

The impact of degree of mineral association is illustrated by the behavior of mercury and selenium. These two elements can be organically bound or associated with minerals such as pyrite. In some cases, these minerals are increased by coal cleaning, indicating an organic association and, in some cases, they are decreased, a certain sign of mineral association. Increasing concentration with cleaning is not a sure sign of organic association because some trace element-bearing minerals are not sufficiently liberated for removal by a physical process.

The extent of trace element removal also depends on the method of cleaning the coal. Figure 1 is a washability plot, by size fraction, of mercury vs. ash content for Upper Freeport Seam coal. Notice that the relationship between ash and mercury is fairly similar for each size fraction except the -200 mesh size and that the mercury concentration is highest in this size fraction. The mercury concentrations at the two levels of cleaning given in Table 1 are, therefore, a function not only of total ash

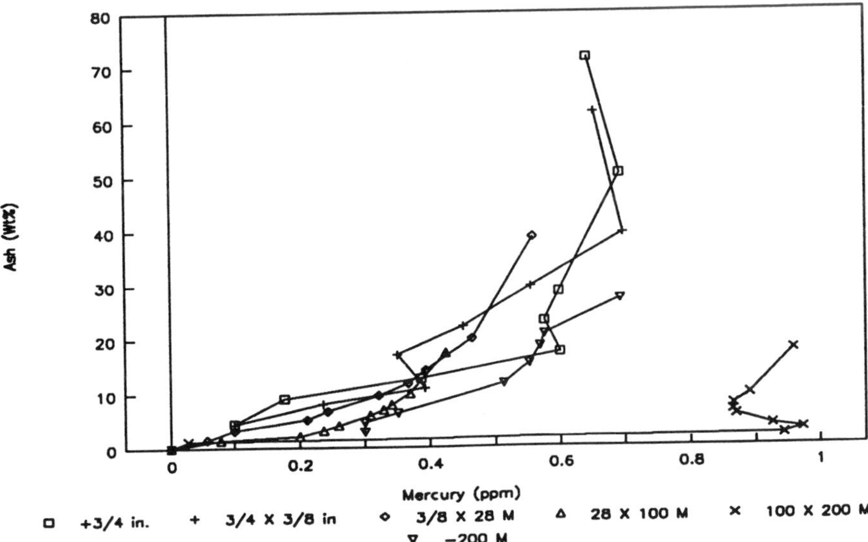

Figure 1. Upper Freeport seam coal — ash vs. mercury concentration.

reduction, but also of the way the various size fractions are cleaned. Cleaning the fine-size fraction of this coal to an ash level lower than the coarse fraction reduces the proportion of this high mercury fraction in the clean coal, potentially providing a greater reduction in mercury concentration in the clean coal than the reverse case, even though both clean coals have a similar ash content.

Figure 2 presents Upper Freeport Seam washability data for selenium. As with mercury, the fine-sized coal contains the highest concentration of selenium. Also, the relationship between ash and selenium is very different for the size fractions. For example, ash reduction provides no reduction in selenium concentration for the two coarsest size fractions until a low ash content is reached, indicating that the element is evenly disseminated throughout the coal and the ash-forming mineral matter, possibly as fine-grained mineral matter.

These examples illustrate that the relationship between ash content and trace element concentration is not constant, even within a single coal. An understanding of the interplay between traditionally measured cleaning performance parameters such as ash removal and trace element reduction is therefore essential in designing cleaning plants better able to remove potentially hazardous elements.

III. ADVANCED COAL CLEANING TO REMOVE TRACE ELEMENTS

Advanced coal-cleaning technologies may offer even further advantages in reducing trace elements. First, advanced processes typically crush or grind coal to very small particles to increase the chance of liberating sulfur-bearing and ash-forming mineral matter, possibly also liberating trace element-bearing mineral matter. Sec-

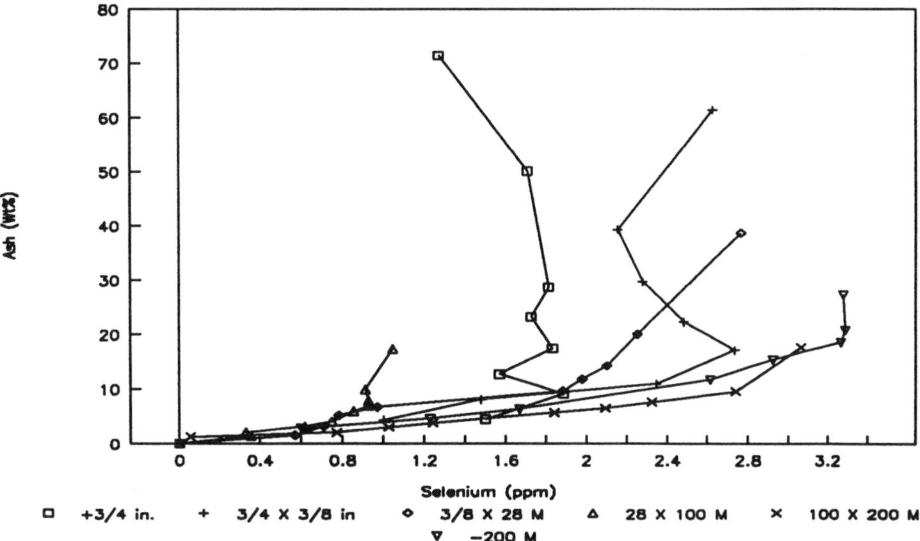

Figure 2. Upper Freeport seam coal — ash vs. selenium concentration.

ond, advanced processes are designed to clean fine-sized coal, making them more efficient than conventional processes in removing mineral matter from this material. As many trace elements are concentrated in fine-sized coal, efficient methods of removal are especially important.

CQ Inc. engineers compared an advanced coal-cleaning process developed by Custom Coals International to conventional coal-cleaning techniques. As part of this evaluation, extensive washability and liberation tests were performed on the coal. CQ Inc. engineers developed a model of a conventional coal-cleaning plant and a plant using the advanced process along with middlings crushing for liberation. Both plants were designed to clean Sewickley Seam coal. The output of the model includes a projection of the size and density fractions of the raw coal that would report to clean coal during the operation of each plant. This information was used to produce a laboratory-simulated clean coal by combining the appropriate size and density fractions of the raw coal in the proportions predicted by the models to produce both the conventional and the advanced clean coal.

The results of this evaluation are presented in Table 5. Conventional cleaning techniques reduced the concentration of antimony, arsenic, chromium, cobalt, lead, mercury, and nickel and the advanced technology provided a further reduction in all cases except mercury. For example, conventional cleaning reduced the arsenic concentration of the coal from 14 to 7 ppm, while advanced cleaning provided a further reduction to 4 ppm.

Data from pilot-scale tests of a combined chemical and physical cleaning process (The Midwest Ore Process) is presented in Table 6. While the details of this process are confidential, analysis of feed and clean coal samples indicates high removals of

Table 5. Conventional and Advanced Physical Cleaning: Sewickley Seam Coal

	Raw	Conventional cleaning	Custom coal advanced process
Ash content (%)	28.7	14.9	13.7
Antimony	0.80	0.48	0.26
Arsenic	14.0	7.2	3.5
Cadmium	0.20	0.63	0.34
Chromium	16.07	8.35	8.22
Cobalt	0.27	0.24	0.22
Lead	14.73	6.96	6.16
Mercury	0.16	0.14	0.14
Nickel	13.39	9.13	8.21
Selenium	1.14	1.54	1.24

Note: ppm except where noted.

Table 6. Chemical/Physical Coal Cleaning: Pittsburgh Seam Coal

	Raw coal	Midwest ore process
Antimony	<0.5	<0.5
Arsenic	7.0	2.0
Beryllium	1.1	0.7
Cadmium	<0.1	<0.1
Chromium	37.8	11.7
Lead	13.1	1.8
Mercury (ppb)	311	106
Nickel	32.2	9.3
Selenium	1.8	0.5

Note: ppm except where noted.

many trace elements. The observed reduction in beryllium is especially interesting because this element has such a strong organic affinity[1] that removal by a physical process is unlikely.

IV. WASTE DISPOSAL IMPACTS

While the removal of toxic trace elements by coal cleaning assures that these elements will not be released into the atmosphere, the cleaning operation does produce a waste enriched in these same elements. Coal-cleaning waste typically contains sufficient pyrite and other acid-forming compounds to produce highly acidic leachate if proper disposal procedures are not followed. Worse, many trace elements are mobile in an acid environment.[2] CQ Inc. has performed field studies of the leaching characteristics of the waste produced by cleaning various coal seams. These studies involved placing approximately 100 lb of cleaning refuse in plastic drums exposed to the atmosphere and monitoring leachate quality. It is important to note that this procedure does not simulate current refuse disposal practice, which involves compaction of the refuse and sealing of the disposal site to control air and water entry, greatly reducing or eliminating acid formation.

These field studies have demonstrated that the concentrations of some trace elements can be significant if an acid environment is allowed to form. For example, Table 7 contains the range of water quality parameters observed during a 21-month leaching period using refuse generated by cleaning Pittsburgh Seam coal. At low acidity, insignificant quantities of the monitored elements are present; however, at high acidity measurable concentrations of aluminum, arsenic, barium, cadmium, chromium, lead, mercury, manganese, nickel, selenium, and zinc were found in the leachate. The observed concentrations of aluminum, manganese, and zinc were greater than 100 ppm.

As a second example, Table 8 contains the range of water quality parameters observed during a 21-month leaching period using refuse generated by cleaning a Texas lignite. Again, at low acidity, insignificant quantities of the monitored elements are present; however, at high acidity measurable concentrations of aluminum, arsenic, cadmium, chromium, lead, manganese, nickel, selenium, and zinc were found in the leachate. The observed concentrations of aluminum and manganese were greater than 100 ppm.

V. SUMMARY

Physical coal cleaning can provide large reductions in the concentration of many trace elements that, if released into the atmosphere, are considered to be potentially toxic or environmentally harmful. Trace elements that are commonly reduced by cleaning include arsenic, barium, cadmium, chromium, lead, manganese, nickel, silver, and zinc. In some cases, antimony, fluorine, mercury, and selenium can be reduced by cleaning. In addition to reductions in concentration, coal cleaning increases the heating value of coal and can improve boiler efficiency, further reducing the quantity of the various trace elements entering the boiler.

Coal-cleaning technologies appear especially effective in removing high percentages of trace elements if the trace elements are present in relatively high concentrations. Also, trace element removal is not always directly proportional to ash removal, potentially allowing changes in cleaning plant flowsheets and operational parameters that increase the removal of these potential air toxics without changing the ash content of the clean coal.

Advanced physical cleaning technologies may provide greater reductions than conventional technologies because these processes are normally used to clean a coal that has been ground to increase mineral matter liberation, including trace element-bearing mineral matter, and they are very efficient at cleaning fine-sized coal, generally the size fraction highest in trace element concentration. Chemical cleaning processes can also remove many trace elements, possibly including organically bound elements such as beryllium.

Table 7. Water Quality Parameters: Pittsburgh Seam

	Acidity	Al	As	Ba	Cd	Cr	Hg	Mn	Ni	Pb	Se	Zn
Low	<1	T	T	T	T	T	T	0.07	T	T	T	T
High	199,000	8,160	1	3.5	1.34	52.5	0.01	462	150	1.3	8.2	550

Note: T: trace, concentration too low to measure; measurements in parts per million.

Table 8. Water Quality Parameters: Texas Lignite

	Acidity	Al	As	Ba	Cd	Cr	Hg	Mn	Ni	Pb	Se	Zn
Low	13	1	T	T	T	T	T	1	T	T	T	T
High	7200	1750	1	<1	1	54	0	150	39	1	5	71

Note: T: trace, concentration too low to measure; measurements in parts per million.

Care must be taken in the disposal of coal-cleaning plant refuse because oxidation and leaching of this material can create highly acidic leachate and many trace elements are mobile in an acid environment.

REFERENCES

1. Finkelman, R. B., Modes of Occurrence of Trace Elements in Coal, Ph.D. dissertation, University of Maryland, College Park, MD, 1980.
2. Akers, D. J., A Geochemical Study of Headwater Streams of West Run, M.Sc. thesis, West Virginia University, Morgantown, WV, 1976.

Rain Scavenging of Toxic Air Pollutants: The Rain Scavenging Ratio and Research Needs

Wangteng Tsai
Yoram Cohen

I. INTRODUCTION

Rain scavenging is an important intermedia transport process responsible for the removal of toxic air pollutants from the atmosphere. Rain scavenging is also an important pathway for chemical exchange between the atmosphere and the terrestrial and aquatic ecosystems. Given the complexity of rain scavenging, it has been common practice in most mass balance studies to apply empirical approaches that make use of an overall rain scavenging ratio ($W_{overall}$) for the calculation of wet deposition fluxes,[1] as defined below,

$$W_{overall} = W_g(1-\phi) + W_p\phi \tag{1}$$

in which

$$W_g = C_w^{(d)} / C_a^{(g)} \tag{2}$$

$$W_p = C_w^{(p)} / C_a^{(p)} \tag{3}$$

where W_g and W_p are the gas and particle scavenging ratios, respectively, and ϕ is the fraction of organic compound adsorbed on the atmospheric particle phase. $C_w^{(d)}$ is the dissolved pollutant concentration in rainwater at ground level, $C_w^{(p)}$ is the pollutant concentration in the particle-bound form in rainwater, and $C_a^{(g)}$ and $C_a^{(p)}$ are the atmospheric concentrations of the chemical in the gaseous and particle phases, respectively. It is important to note that often W_g and W_p have been taken as time-invariant parameters independent of the particle size distribution or the rate of rainfall.

In this paper, we explore the process of below-cloud rain scavenging for toxic air pollutants, with a special focus on semivolatile organics, through the use of a simple rain scavenging model that accounts for the dynamic partitioning of semivolatile organics in the gas/particle/rain phases, as well as the dynamic variation of particle size distribution during the rain event. Finally, a discussion is presented regarding the uncertainty in estimating the scavenging ratio in relation to the interpretation of field data for rain scavenging of semivolatile organics.

II. FORMULATION OF THE RAIN SCAVENGING MODEL

The formulation of the present model for the rain scavenging of semivolatile organics (RSSVO) is based on a chemical mass balance which includes the gas, particle, and rain phases in the below-cloud region. We approximate the complex below-cloud rain-scavenging process by taking the atmospheric air phase (below cloud) to be uniform and where convective winds are neglected. Accordingly, the overall chemical mass balance on the below-cloud air phase (including the gas and particle phases) can be expressed by

$$(1-q)\frac{d(V_aC_a)}{dt} = -C_a^{(g)}H_{wa}J_{rain}A\,\Lambda_g{}^* - C_a^{(p)}J_{rain}A\,\Lambda_r - kV_aC_a + S_a \qquad (4)$$

where

$$C_a = C_a^{(g)} + C_a^{(p)} \qquad (5)$$

$$C_a^{(p)} = C_a\phi \qquad (6)$$

$$\Lambda_g{}^* = W_g / H_{wa} \qquad (7)$$

$$\Lambda_p = \overline{C_{wf}^{(p)}} / C_a^{(p)} \qquad (8)$$

in which, $C_a^{(g)}$ and $C_a^{(p)}$ are the concentrations of the gaseous and particle-bound chemical in the atmospheric phase (ng/m^3 air), respectively, and C_a is the total chemical concentration in the atmosphere (ng/m^3 air). $\overline{C_{wf}^{(p)}}$ is the chemical concentration, in the particle-bound form, in rainwater (averaged over raindrop size) at ground level (ng/m^3 water), and q is the volume fraction of the atmosphere occupied by raindrops. The volume of the atmosphere is denoted by V_a (m^3), H_{wa} is the dimensionless water to gas partition coefficient for the given chemical, J_{rain} is the precipitation rate (m/s), and A is the interfacial area (m^2) between the atmosphere and the land (or water) surface below in the region under consideration. It is noted that the left-hand side of Equation 4 represents the rate of accumulation of the chemical in the air phase. The first term on the right hand side of Equation 4 accounts for the rate of chemical mass scavenged by raindrops via gas absorption, where $\Lambda_g{}^*$ is the normalized gas scavenging ratio (Equation 7) which varies between 0 and 1. The second term represents the rate of removal of the particle-bound chemical through particle scavenging by raindrops where Λ_p is the particle scavenging ratio (Equation

8). The third term represents the degradation by chemical reaction. The last term is the net input of the chemical into the atmosphere from source emissions (ng/s).

In order to determine the variation in the total atmospheric concentration of the chemical, C_a, during rain (using Equation 4), one must first realize that the particle rain scavenging ratio Λ_p is a function of the particle size distribution. The particle size distribution, however, changes during the course of a rain event since the rain scavenging removal efficiency is a function of particle size. Also, the fraction of adsorbed chemical is a function of the particle surface area. As the particle size distribution changes, during rain, the available particle surface area for adsorption will change and thus ϕ will also vary during the course of a rain event. Therefore, the change in the adsorbed fraction ϕ is also reflected in a variation in the vapor phase concentration of the chemical during rain. In the following section, the various model parameters in Equation 4 and the solution algorithm are described.

A. Gas Scavenging Ratio

The normalized gas scavenging ratio (Λ_g^*) for chemicals which are nonreactive in the aqueous phase can be derived theoretically based on a chemical mass balance on a raindrop as it falls from cloud base to ground level, and subsequently integrating over the spectrum of raindrop sizes to obtain the average concentration of the dissolved chemical (in rain water) at ground level, $C_w^{(d)}$

$$\frac{\overline{C_w^{(d)} / C_a^{(g)}}}{H_{wa}} = \Lambda_g^* =$$

$$\frac{1}{V_r} \int_0^\infty \left[1 - \left(1 - \frac{C_{wo}^{(d)} H_{aw}}{C_a^{(g)}} \right) \exp\left(\frac{-3K_{OL} L_c}{V_t R_d} \right) \right] \frac{4\pi R_d^3}{3} N_{R_d} dR_d \qquad (9)$$

in which $C_{wo}^{(d)}$ is the dissolved pollutant concentration in a raindrop of radius R_d at the cloud base, K_{OL} is the overall liquid phase mass transfer coefficient of chemicals from air to water, L_c is the height of cloud base, and V_t is the terminal velocity of a given size raindrop. The size distribution of raindrops N_{Rd} is defined such that $N_{Rd}dR_d$ is the number of falling raindrops in the size range R_d to R_d+dR_d in a unit volume of air and V_r is the volume of rain per unit volume of air.

B. Particle Scavenging Ratio

The particle scavenging ratio of Λ_p can be derived theoretically based on a mass balance on the particle-bound chemical in a raindrop as it falls from cloud base to ground level. The final expression for particle scavenging ratio (Λ_p) is given by[2]

$$\Lambda_p = \frac{1}{\phi C_a V_r} \sum_{j=1}^{m} \int_0^{\infty} \left(C^{(p)}{}_{woj} + \frac{3C^{(p)}{}_s L_c}{4R_d} \int_{a_j}^{a_{j+1}} E_j(a, R_d) \pi a^2 n_j(a) da \right) \frac{4\pi}{3} R_d^3 N_{R_d} dR_d \tag{10}$$

in which $C_{wo}^{(p)}$ is the initial concentration at the cloud base of the particle-bound chemical in a given size raindrop, and $C_s^{(p)}$ is the mass of particle-bound chemical adsorbed per unit surface area of particle. The collection efficiency of a particle with diameter a by a raindrop of radius R_d is denoted by $E(a, R_d)$, and m is the total number of particle size intervals. Finally, $n(a)$ is the number distribution function of particle in the atmosphere (cm^{-3} cm^{-1}), which is expected to vary during rainfall due to primarily rain scavenging.

In this study, the Junge correlation[3] is applied to estimate the fraction of chemical adsorbed onto the particle phase, ϕ, and the empirical correlation of Ryan and Cohen[4] is used to approximate the collection efficiency.

Given the gas and particle scavenging ratios, the instantaneous total pollutant concentration in rainwater (i.e., dissolved plus particle-bound), $\overline{C_w}$ (units of ng/L), at ground level, is given by

$$C_w = C_a^{(g)} H_{wa} \Lambda_g^* + C_a^{(p)} \Lambda_p \tag{11}$$

In comparing model results with field data it is important to note that pollutant rain concentrations reported in most field studies are the average pollutant rain concentrations in each sequential sample of rainwater collected during a rain event. The average pollutant rain concentration in each sequential sample, $\overline{C_w}$, can be easily obtained by performing a simple mass balance on the pollutant in a rain sampler,

$$\overline{C_w} = \frac{A_s}{V_s} \int_{t_o}^{t} C_w J_{rain} dt \tag{12}$$

in which V_s is the accumulated volume of rainwater in the sampler, A_s is the cross-sectional area of the sampler, t_o represents the beginning of the sampling period, and C_w is the instantaneous rainwater concentration obtained from Equation 11.

It is convenient to quantify the degree of rain scavenging using the following overall rain scavenging ratios

$$W = C_w / C_a \tag{13}$$

$$\overline{W} = \overline{C_w} / \overline{C_a} \tag{14}$$

Table 1. Meteorological Data Applied in the Simulation of the November 9 and 10, 1982 (Los Angeles) Rain Event

Sample	Time duration	Precipitation[a] rate (mm/h)	Cloud[b] base (m)	Temperature[b] (°C)
A	1:30–4:30	1.3	1830	13.3
Break	4:30–9:30	—	2134	11.1
B	9:30–12:45	2.0	400	13.3
C	12:45–16:00	1.1	850	14.4
Break	16:00–19:45	—	1520	11.7
D	19:45–23:25	0.75	1220	11.1
E	23:25–9:40	1.4	1160	10.6

[a] From Kawamura and Kaplan.[5]
[b] From the observations (on 3-h interval) at Los Angeles International Airport, 10 km south of UCLA.

where W and $\overline{W}$ are the instantaneous and average overall scavenging ratios, respectively, and $\overline{C_a}$ is the cumulative average air phase concentration of the chemical (during each sampling period of rain event).

The governing equations of the RSSVO model (Equations 4 to 14), subject to the appropriate initial conditions, were solved simultaneously using the predictor and corrector method. The basic model results include the instantaneous concentrations of the dissolved and particle-bound chemical in rainwater (at ground level) and the instantaneous gaseous and particle-bound concentrations in the atmospheric phase. Using these results, the various averaged concentrations can be determined as described above.

III. CASE STUDIES

Sequential sampling data of rain rates and PAHs concentrations in rainwater were previously reported by Kawamura and Kaplan[5] for a rain event on November 9 and 10, 1982 monitored in Los Angeles. In particular, the field data for pyrene and fluoranthene in that event were selected for this case study. The meteorological data required for running the rain scavenging model include rain rate, height of cloud base, and temperature, as shown in Table 1. The compartmental data, emission rates of particles and PAHs, and physicochemical properties of PAHs have been described elsewhere. With the above information, simulations of rain scavenging of pyrene and fluoranthene during the November 9 and 10, 1982 rain event in Los Angeles were performed.

Figure 1 illustrates the comparison of actual and predicted pyrene and fluoranthene rain concentrations ($\overline{C_w}$) for each of the sequential samples taken during the November 9 and 10, 1982 rain event. From Figure 1, it is seen that the highest pyrene and fluoranthene rain concentrations were obtained for the second sample which are consistent with the morning rush hour traffic, and the low cloud base (Table 1). Clearly, the rush hour emissions and height of cloud base are significant in determining the levels of pollutants in the collected rain water.

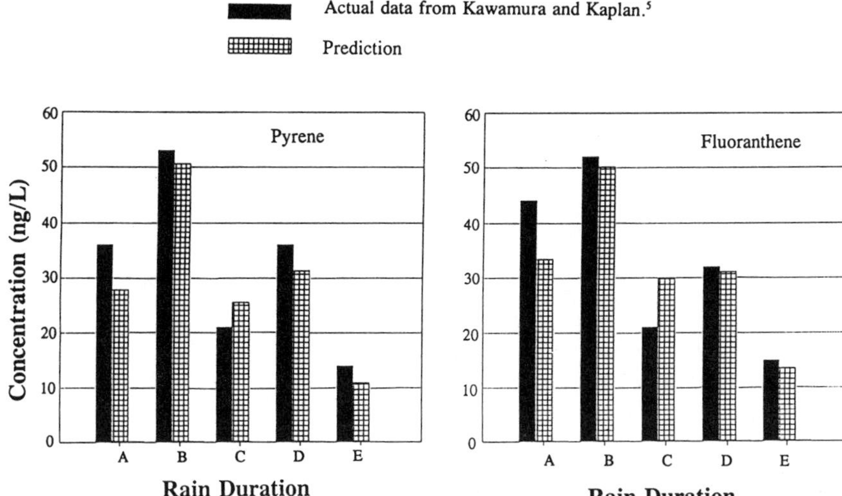

Figure 1. Predicted and measured pyrene and fluoranthene sampler rain concentrations ($\overline{C_w}$) at ground level for the November 9 and 10, 1982 Los Angeles rain event. The time duration for each sequential rain sample is A (1:30 to 4:30), B (9:30 to 12:45), C (12:45 to 16:00), D (19:45 to 23:25), and E (23:25 to 9:40).

It is important to note that in most field studies the reported rain scavenging ratio (Equation 1) is determined based on a constant atmospheric concentration (C_a). Due to sampling difficulties, reported field measurements of the rain scavenging ratio are generally values averaged over different sampling periods. In reality, pollutant concentrations in both the air and rain water are time dependent during rain events. Thus, the rain scavenging ratio is expected to vary with time during a given rain event. In the present model, the temporal variations in the atmospheric and rainwater concentrations are determined and thus one can evaluate the corresponding change in the rain scavenging ratio (Equations 13 and 14) during the rain event. As an illustration, the predicted instantaneous total fluoranthene concentrations in rainwater (C_w) and in the air phase (C_a) at ground level for the November 9 and 10, 1982 event are illustrated in Figure 2.

Given the chemical concentration in rain water and in the atmospheric phase, it is possible to determine the temporal variations of the overall instantaneous and average scavenging ratios as illustrated in Figure 3 for the rain scavenging of fluoranthene during the November 9 and 10, 1982 rain event. The predicted overall scavenging ratios W and $\overline{W}$ vary in the range of 1.82×10^4 to 9.87×10^4 and 2.15×10^4 to 9.87×10^4, respectively. The nearly order of magnitude variation in the scavenging ratios during rainfall suggests that the application of average experimental values of rain scavenging ratios for mass balance calculations is questionable. It is also important to note that the reporting basis for field measured scavenging ratios is not uniform. Some studies report the scavenging ratios, as calculated from Equation 1, based on atmospheric concentration measured before, during, or even after the

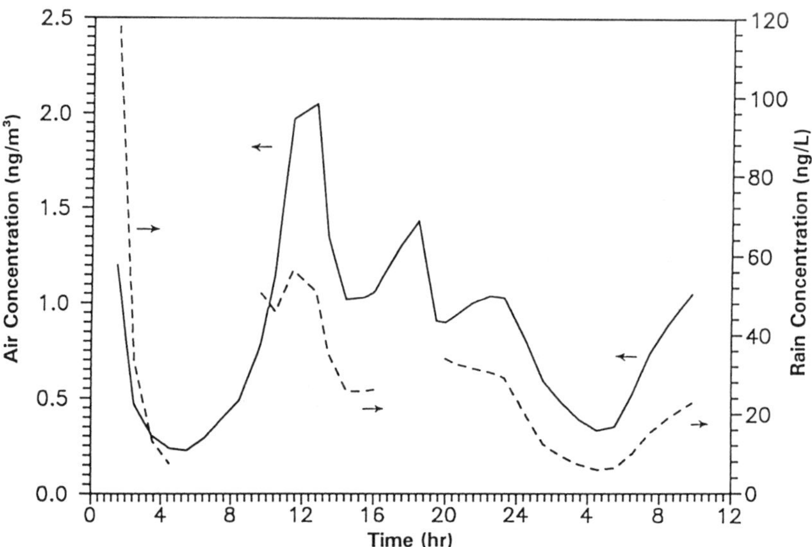

Figure 2. Predicted instantaneous fluoranthene concentration in rainwater (C_w) and air phase ($\overline{C_a}$) for the November 9 and 10, 1982 Los Angeles rain event.

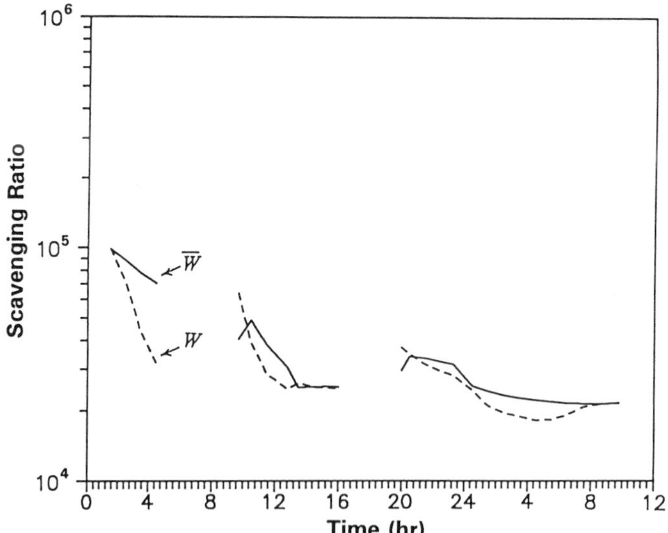

Figure 3. Predicted overall scavenging ratio of fluoranthene for the November 9 and 10, 1982 rain event.

specific rain event. The inconsistency in the reporting basis for rain scavenging ratios ultimately results in an uncertainty as to the extrapolation of reported values to other regions, different rain conditions, or atmospheric concentrations.

The present study suggests that the rain scavenging ratio is a function of numerous variables including the chemical type, rate, and duration of rainfall, height of cloud base, source emissions, and particle size distribution. Consequently, in order to properly interpret rain scavenging data, field studies should at the minimum report the above information. Also, a consistent basis should be used for reporting the concentrations in the atmospheric and rainwater phases.

IV. SUMMARY AND CONCLUSION

Below-cloud rain scavenging of semivolatile organics was investigated by a mass balance model that accounts for the dynamic partitioning of atmospheric organics in gas/particle/rain phases during a rain event. Results of case studies for pyrene and fluoranthene illustrated that the atmospheric concentrations of gaseous and particle-bound chemical can vary significantly during a rain event, and thus the use of a time-invariant scavenging ratio in multimedia mass balance calculations and the associated health and environmental risk analysis should be revisited.

ACKNOWLEDGMENT

Although the information in this document has been funded in part by the U.S. Environmental Protection Agency under assistant agreement CR-812271-03 to the National Center for Intermedia Transport Research at UCLA, it does not necessarily reflect the views of the Agency and no official endorsement should be inferred. This work was also partially funded by the University of California Toxic Substances Research and Teaching Program.

REFERENCES

1. Baker, M. P. and Eisenreich, S. J., Concentration and fluxes of polycyclic aromatic hydrocarbons and polychlorinated across the air-water interface of Lake Superior, *Environ. Sci. Technol.*, 24, 342, 1990.
2. Tsai, W., Cohen, Y., Sakugawa, H., and Kaplan, I. R., Dynamic partitioning of semivolatile organics in gas/particle/rain phases during rain scavenging, *Environ. Sci. Technol.*, 25, 2012, 1991,
3. Junge, C. E., Basic considerations about trace constituents in the atmosphere as related to the fate of global pollutants, in *Fate of Pollutant in the Air and Water Environments*, Suffett, I. H., Ed., Advances in Environ. Sci. Technol., John Wiley & Sons, New York, 1977, 5.
4. Ryan, P. A. and Cohen, Y., Multimedia transport of particle-bound organics: benzo(a)Pyrene test case, *Chemosphere*, 15, 21, 1986.
5. Kawamura, K. and Kaplan, I. R., Compositional change of organic matter in rainwater during precipitation events, *Atmos. Environ.*, 20, 527, 1986.

Determining Emissions and Assessing Risks: A Case Study of Large Industrial Facilities

William R. Oliver
Michael T. Alberts
Ronald J. Dickson

I. ASSESSING HEALTH RISKS

Many in the air pollution control field view health risk assessment as a valuable, if not required, technique for examining the effects of emissions from stationary point source operations, but this has not long been the case. In 5 short years, health risk assessments of air toxics from individual industrial facilities have proceeded from virtually unknown to commonplace. Although the risk assessment process itself is not new, its application to air emissions from single point sources has expanded significantly. This increased use of health risk assessments has coincided with the expanding interest in evaluating toxic and potentially toxic air contaminant emissions.

Additional weight was given to the health risk assessment process in the 1990 Clean Air Act Amendments. For instance, under Title III, risk assessments will be used to evaluate the residual risk after controlling categories of major point sources. Furthermore, several states have begun using a risk assessment process in the last few years to review certain impacts from modernizing or expanding point sources.

II. AIR POLLUTANT RISK ASSESSMENT — A CASE STUDY

Because of the use of chlorine at many Kraft pulp mills, dioxin emissions have been considered a potential issue for the pulp and paper industry. In 1987 and 1988, a study of dioxin and other emissions from two pulp mills near Eureka, CA, was conducted. This paper presents a synopsis of that study.[1]

Local citizens initially pressed for a health assessment of emissions from the pulp mills operated by Louisiana-Pacific Corporation and Simpson Paper Company, primarily due to concerns over potential dioxin air emissions. Thus, in mid-1987, the North Coast Unified Air Quality Management District requested Louisiana-Pacific and Simpson Paper to conduct a health risk assessment for air emissions from their Humboldt Bay pulp mills. The study was conducted jointly for both mills by Radian Corporation and included participation by Rifkin and Associates. In addition, a Citizens Advisory Committee was established to oversee and guide the effort.

III. ACTIVITIES OF THE AIR EMISSIONS STUDY

The study consisted of four phases: program planning, on-site testing of air emissions, evaluation of potential exposure to the substances emitted by the mills, and characterization of environmental health risks. In the first phase, Radian performed a technical evaluation to identify substances most likely to be emitted in significant quantities. Three Kraft pulping processes were reviewed: wood chip digestion, recovery furnace operations, and pulp bleaching. An extensive literature search was then performed which led to the following compounds for source characterization:

Volatile inorganics	Volatile hydrocarbons	Semivolatile hydrocarbons
Chlorine	Benzene	Chlorinated dioxins
Chlorine dioxide	Carbon tetrachloride	Chlorinated dibenzofurans
	Chloroform	Polycyclic aromatic hydrocarbons
	Ethylene dichloride	
	Methylene chloride	
	Perchloroethylene	
	Trichloroethylene	
	Vinyl chloride	

IV. SOURCE TESTING

Little published information existed that described the emission characteristics of toxic substances from Kraft pulp mills. Consequently, we designed and conducted a source sampling effort to gather data about pulp mill emissions for use in the health risk assessment. The following sampling and analytical methods were used to determine the concentrations of the selected substances in the vent and stack gases:

- The Volatile Organic Sampling Train (VOST) was used to collect samples of volatile organic hydrocarbons from all sources considered in the study. The samples were analyzed by gas chromatography/mass spectrometry (GC/MS).
- The modified EPA Method 5 (SW 846, Method 0010) sampling procedure was used to collect samples of chlorinated dioxins and furans and polycyclic aromatic hydrocarbons (PAHs) from the recovery furnace stacks. The samples were analyzed by GC/MS with high-resolution GC/MS used for the dioxin and furan samples.
- The National Council of the Paper Industry for Air and Stream Improvement (NCASI) procedure (Method 521) was used to determine the concentrations of chlorine and chlorine dioxide.

An extensive literature search was conducted to gather information for the development of a sampling and analytical plan. The Quality Assurance Project Plan described the sampling and analytical methods to be used to characterize emissions.[2] The sampling effort was then carried out at each mill in July 1988.

Chloroform was found to be the substance emitted in the most environmentally significant amount from the Kraft pulping process. This substance is formed during the bleaching of pulp with chlorinated substances. In addition, chlorinated dioxins and furans were detected in the recovery furnace stacks at each mill. Finally, most PAHs, including benzo(a)pyrene, were not detected in the recovery furnace emissions. Fluoranthene was the only carcinogenic PAH detected, and it was detected in only one of the six samples collected at the recovery furnace stacks.

V. EMISSION ESTIMATES AND ATMOSPHERIC MODELING

Data generated during the source characterization phase were used to estimate average emission rates. Emission estimates were calculated by multiplying the stack gas concentration (micrograms per cubic meter) of each substance by the stack gas flow rate (cubic meters per second) from that source. This value was then adjusted to account for annual plant downtime. Table 1 presents the annual emission estimates for both mills.

Ambient concentrations of the emissions were estimated for the populated areas of Eureka and southern Arcata. Both the U.S. Environmental Protection Agency (EPA) and the California Air Resources Board (ARB) approved Industrial Source Complex Short Term (ISCST) and COMPLEX I dispersion models were used to predict annual average concentrations. The dispersion modeling was performed using 4 years of meteorological data collected at the Arcata Airport. The modeling calculated concentrations for 138 receptor locations spaced 500 m apart.

VI. RISK ASSESSMENT RESULTS

The health risk assessment utilized cancer potency estimates developed by the California Department of Health Services (DHS) and the EPA. Three different scenarios were evaluated to provide a range of potential risk estimates. One scenario relied on the potency estimates developed by the DHS. Where DHS values did not exist, EPA values were used. The other two scenarios relied on potency estimates developed by the EPA. These latter two scenarios differed only in the potency assumed for chlorinated dioxins and furans. One scenario used the existing EPA estimate of potency for 2,3,7,8-TCDD (referred to as the EPA scenario), while the other utilized a 1987 proposed revision to the potency estimate (referred to as the EPA Draft scenario).

Risk can be characterized in a variety of ways. In this assessment, health risk was characterized using three different techniques:

- Maximum risk: This is the risk estimate for a continuous 70-year exposure at the point of highest pollutant concentration in air. The risk to the maximally exposed individual (MEI) represents a theoretical worst-case estimate.
- Cancer burden: The increased incidence of cancer, or cancer burden, is defined as the estimate of the increased number of cancer cases that could potentially develop in the exposed population.

Table 1. Estimated Annual Emission Rates

Substance	Average emission rate (lb/year)	
	Louisiana-Pacific	Simpson paper
Chlorine	4,500	2,900
Chlorine dioxide	440,000	520,000
TCDD — DHS weighting[a]	0.019	0.014
TCDD — EPA weighting[b]	0.0030	0.0025
Vinyl chloride[c]	11	11
Methylene chloride	96	170
Chloroform[d]	8,300	8,900
Ethylene dichloride[c]	9.8	10
Carbon tetrachloride	16	47
Trichloroethylene	16	64
Benzene	120	4,800
Perchloroethylene	8.2	11
Polycyclic aromatic hydrocarbons[e]	2.0[c]	16

[a] Calculated using DHS weighting scheme.
[b] Calculated using EPA weighting scheme.
[c] This substance was not detected in any of the samples taken. The emission rate is based on one half the method detection limit.
[d] See the text for a discussion of the uncertainty in these estimates.
[e] Includes anthracene, pyrene, fluoranthene, benzo(a)anthracene, chrysene, benzo(b)fluoranthene, benzo(k)fluoranthene, benzo(a)pyrene, indeno(1,2,3-cd)pyrene, dibenzo(a,h)anthracene, and benzo(g,h,i)perylene. Of these substances, only fluoranthene was detected, and it was detected in only one sample. The emission rate assumes the other compounds were present at one half the method detection limit.

- Population average risk: The population average risk is calculated by dividing the cancer burden by the study area population.

Table 2 summarizes the risk estimates resulting from inhalation of the pulp mill emissions. The maximum risk calculated for all three scenarios was 2 in 1 million (2×10^{-6}). Furthermore, the increased incidence of cancer (also referred to as cancer burden) ranged from 0.02 to 0.03 cases for the 3 scenarios considered. This rate of excess cancer was determined using the standard assumption of a 70-year exposure period. Since 3/100 of a cancer case provides little meaning, it is helpful to present these estimates in an alternate form. For example, the burden estimate of 0.03 is equivalent to 1 cancer case in the Eureka/Arcata area if the population were exposed continuously for 2333 years.

The population weighted risk, the average risk experienced by the population, is another way to characterize the risk to Eureka/Arcata residents. Population weighted risk estimates ranged from 4 to 6 in 10 million (4×10^{-7} to 6×10^{-7}).

VII. DISCUSSION OF THE RISK ASSESSMENT

This risk assessment was prepared using commonly accepted guidelines and practices. These guidelines and practices include several health-conservative assumptions and methods that tend to overestimate rather than underestimate the actual

Table 2. **Cancer Risk Estimates**

Scenario	Maximum risk[a]	Cancer burden[b]	Population average risk[c]
DHS	2×10^{-6}	0.03	6×10^{-7}
EPA draft	2×10^{-6}	0.02	4×10^{-7}
EPA	2×10^{-6}	0.03	5×10^{-7}

Note: These estimates are based on the inhalation of the substances considered in this study. See the text for a discussion of the uncertainty in these estimates.

[a] This is the risk estimate for a continuous 70-year exposure at the point of highest pollutant concentration.

[b] The increased incidence of cancer, or cancer burden, is defined as the estimate of the increased number of cancer cases that could potentially develop in the exposed population from a 70-year exposure to the pulp mill emissions.

[c] The population average risk is calculated by dividing the cancer burden by the study area population.

risk. For example, it was assumed in the exposure assessment that all inhaled pollutants are completely absorbed in the lungs and that all residents are exposed continuously for 70 years. Cancer potency estimates used in estimating risk are also health conservative. According to EPA, risk estimates calculated using published cancer potency values represent upper-bound or worst-case risks. The actual risk is unlikely to be higher than the risk calculated with the cancer potency values and could be as low as zero.

There are, however, three specific instances where the risks presented in this study may be underestimated:

- There is uncertainty in the chloroform emissions due to the very large concentrations encountered during the sampling of the bleach plant vents and sewer line vents.
- There is uncertainty in the PAH analytical results because of low recovery efficiencies for the PAH samples.
- Exposure to chlorinated dioxins and furans, as well as PAH, resulting from noninhalation pathways (i.e., dermal contact and ingestion) was not rigorously calculated. Since the results of the inhalation assessment for solid phase substances showed risk levels much less than 1×10^{-6}, a comprehensive multiple pathway risk assessment was not performed.

Each of the above factors was evaluated carefully to determine its effect on the health risk calculations. The chloroform sampling results are judged to be low. It is not possible from the data generated in this study to determine how low the chloroform emission estimates might be. Other chloroform emission data were made available to Radian after the sampling effort was completed. Using the initial chloroform emission data reported by the pulp mills as a part of the Superfund Amendments and Reauthorization Act (SARA) Title III requirements, an estimate was made of the uncertainty in the chloroform emission estimates developed in this study. On the basis of the chloroform emissions reported under SARA Title III, the risk of 2 in 1 million (2×10^{-6}) for the maximally exposed individual would increase to 35 in 1 million (35×10^{-6}) or 3.5 in 100,000 (3.5×10^{-5}). In addition, the cancer burden estimate of 0.03 would be revised to 0.5, or one excess cancer case in the area assuming exposure for 140 years.

Consideration of noninhalation pathways also increases the risk estimates. For the maximally exposed individual, the estimated risk of 35 in 1 million (35×10^{-6}) would increase to 36 in 1 million (36×10^{-6}) as a result of exposure through noninhalation and inhalation pathways, using a ratio of noninhalation to inhalation risk calculated from other studies.

In summary, the risk at the point of maximum impact is estimated to be 2 in 1 million (2×10^{-6}). Accounting for the above uncertainties, the risk estimate ranges from 36 in 1 million (36×10^{-6}) to zero.

VIII. RISKS FROM SPECIFIC POLLUTANTS

The fraction of total risk that results from each of the individual pollutants is shown in Figure 1 for each scenario. As seen in the figure, chloroform dominates the risk in all three scenarios. Benzene and 2,3,7,8-TCDD equivalents have a minor contribution to risk under the DHS scenario due to differences in the DHS and EPA cancer potency estimates for these compounds.

From the data gathered in this study, it is apparent that chloroform is the dominating factor in the health risk assessment. A range of risks from zero to 36 in 1 million (0 to 36×10^{-6}) was calculated for a hypothetical, maximally exposed individual. A new sampling program with a focus on chloroform could be conducted to reduce this range.

However, at the beginning of the study, the real impetus for the risk assessment was dioxin and furan emissions. Rather than being based on technical information (which was lacking at the time), this driving force was more the result of a general concern over dioxin in air on the part of the public in the Humboldt Bay region. Nevertheless, the results of this study showed dioxin emissions to contribute less than 10% of the total risk from air emissions. This outcome emphasizes the need to maintain an inquisitive attitude and to be cautious about preconceptions early in such studies.

REFERENCES

1. Radian Corporation, Air Emissions Study for Humboldt Bay, California, Pulp Mills: Final Report, Sacramento, CA, 1988.
2. Radian Corporation, Air Emissions Study for Humboldt Bay, California, Pulp Mills: Source Testing Quality Assurance Project Plan, Sacramento, CA, 1988.

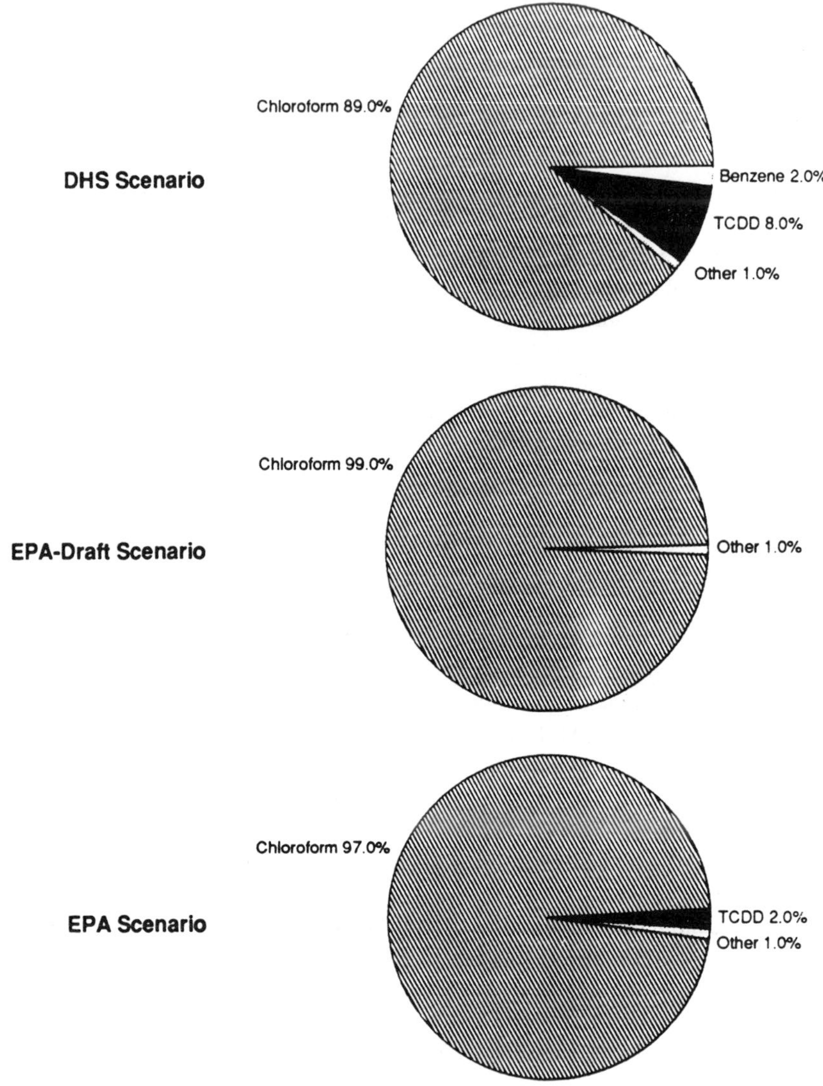

Figure 1. Pollutant contributions to maximum risk.

Clean Power from Coal: The British Coal Topping Cycle

M. St. J. Arnold
A. J. Minchener
S. G. Dawes

I. INTRODUCTION

Increasing environmental concern has focused attention on the development of cleaner and more efficient power stations. The long-term availability of coal, together with the desirability of preserving premium fuels for their most appropriate applications, has led to a resurgence of interest in advanced coal-fired technologies. Such systems will be required to offer a number of advantages compared with conventional technology:

- Low capital cost and short construction periods
- Low cost of electricity
- Good environmental performance
- High efficiency with a consequent reduction in CO_2 emissions
- Good availability, turndown, and response characteristics
- Fuel flexibility with respect to coal and ash characteristics

These requirements have been taken into account by British Coal in the development of an alternative advanced coal fired system known as the British Coal Topping Cycle. This has the potential for a 20% or more reduction in specific capital cost and cost of electricity compared with conventional pulverized fuel stations equipped with flue gas desulfurization. Improved efficiency will result in a 20% or more reduction in coal burned per unit of electricity, with corresponding reduction in CO_2 emissions. Inherent features of the process will ensure low emissions of dust, sulfur, and nitrogen oxides. The near-term potential is for a 46 to 47% station efficiency (lower heating value basis) with further developments able to achieve over 50%.

Because of these potential advantages, British Coal with financial and technical support from Power Gen plc and GEC ALSTHOM have embarked upon the initial phase of a substantial development program. Further financial support is being provided by the U.K. Department of Energy, the European Economic Community (EEC), the European Coal and Steel Community (ECSC), and the Electric Power Research Institute (EPRI). There is also some input from the U.S. Department of Energy (U.S. DOE) for specific work on coal injection into a pressurized combustor.

The proposed development program comprises three phases to permit the design, construction, and operation of a commercial prototype to proceed with confidence.

- Proof of concept by the mid 1990s
- Scale up and cycle component development before year 2000
- Demonstration of a commercial prototype early in the next century

II. THE BRITISH COAL TOPPING CYCLE

A. System Description

Figure 1 shows a schematic diagram of the Topping Cycle selected by British Coal for immediate development. Coal is gasified in a pressurized air-blown spouted bed gasifier. Sorbent is injected into the gasifier to retain sulfur compounds. The fuel gas is cleaned in a cyclone and cooled to 600°C or below. The remaining dust is removed using ceramic candle filters. At this temperature, almost all the volatile alkali salts condense onto the particulate matter collected by the filter. The clean fuel gas is burned in a high-temperature (1260°C) gas turbine with an exhaust waste heat boiler.

In the gasifier 70 to 80% of the coal is converted into a low calorific value gas. The 20 to 30% which remains unconverted is burned in a circulating fluidized bed combustor (CFBC) where the sulfided sorbent is also converted to calcium sulfate. Additional sorbent can be fed to the combustor if required.

B. Rationale for System Choice

A range of new options is available for coal firing.[1] British Coal has examined these options, assessing factors such as efficiency, cost, environmental factors, and development risk. It has concluded that a Topping Cycle based on a partial fluidized bed gasifier with fluidized bed combustion of the residues offers a highly effective development route toward improved efficiency and reduced cost consistent with environmental and other objectives set out in the introduction.

Many coal gasification concepts have been tested throughout the world over a number of decades. They fall into three main generic types: fixed bed, entrained flow, and fluidized bed. Fixed bed gasifiers generally require sized coal; additional coal processing is required to use the total output from mechanized mines. Substantial quantities of tars and oils are produced which are generally recycled to the gasifier. Entrained gasifiers operate at very high temperatures which exacerbates heat recovery problems and results in significant vaporization of alkali metal compounds. In addition, variation in ash composition can cause slag viscosity problems. Fluidized bed gasifiers operate at moderate temperatures and have the advantage that limestone may be added to the fluidized bed for sulfur removal. This property of fluidized beds may then be used to advantage by utilizing dry dust removal systems.

Particular advantages of the British Coal spouted fluidized bed gasifier are that it can operate using a wide range of hard coals, including high swelling coals, and operates in a nonagglomerating mode, thereby avoiding problems due to variation in

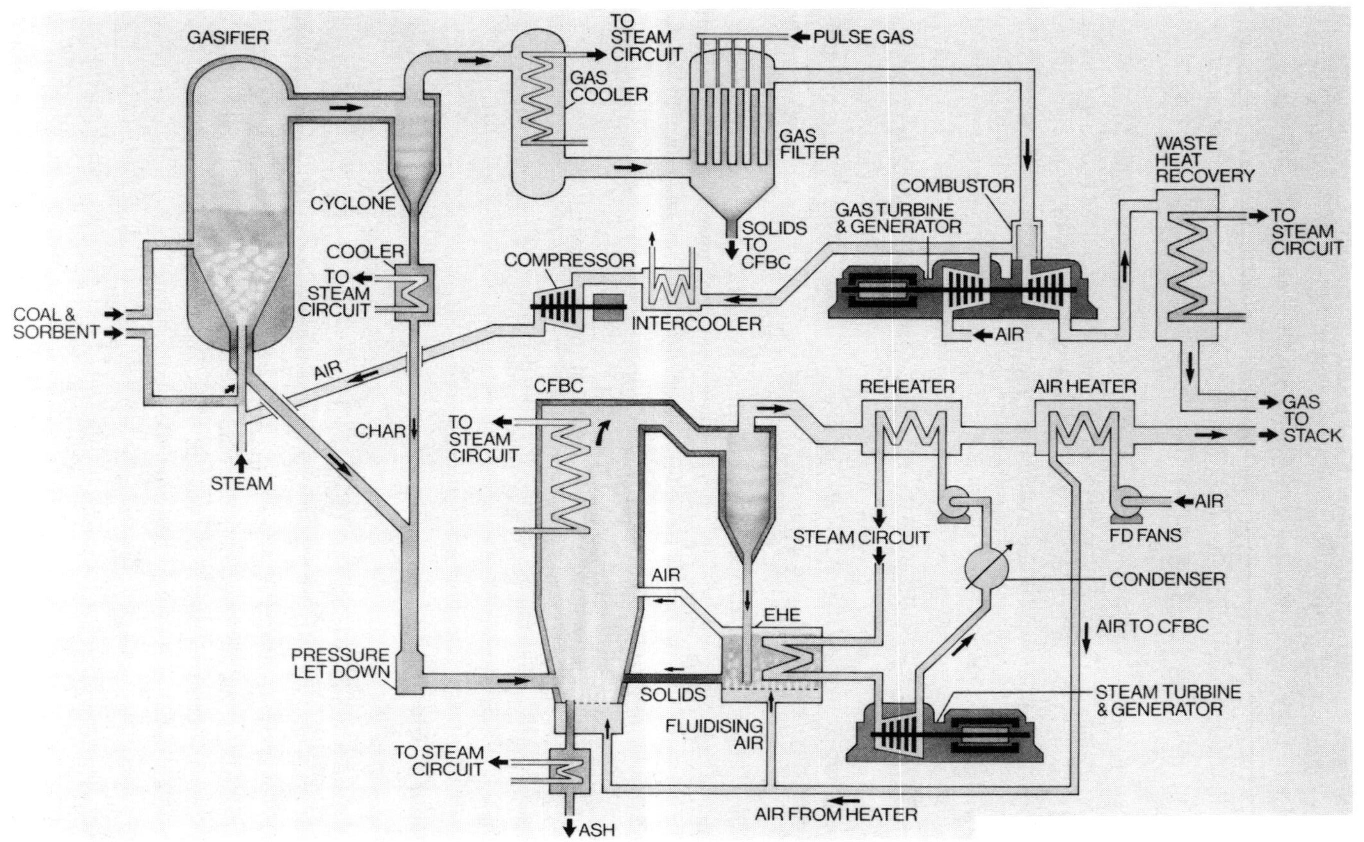

Figure 1. The British Coal Topping cycle.

ash properties. Further advantages of the British Coal Topping Cycle lie in the use of an air-blown system. By contrast to the oxygen blown total gasification system required for IGCC, extremes of process conditions are not required, and the cost and efficiency loss arising from oxygen production are avoided. The use of a separate char combustor means that high-temperature heat is available in an oxidizing atmosphere to ensure that state of the art steam turbines can be used, including supercritical conditions.

There are a number of Topping Cycle variants which can employ different gas cleaning options and use either pressurized or circulating fluidized bed char combustion. The system shown in Figure 1 was selected on the basis of minimum risk and development requirements while maintaining high efficiency.

C. Efficiency

The efficiency of the British Coal Topping cycle has been calculated using the British Coal ARACHNE[2] network analysis program which is able to take into account power station efficiencies and losses. The database was developed during extended discussions with the Central Electricity Board. All efficiency calculations have been made on a strictly comparable basis ensuring meaningful comparisons.

The efficiencies of a number of coal-based power generation systems have been calculated[3] and are shown in Table 1. The figures were calculated for state of the art gas turbines and subcritical steam conditions of 538°C, 160 bar, and reheat to 538°C. There will be scope for significant improvement on these efficiencies in due course. Thus the development of advanced supercritical steam turbines (for which the CFBC system is particularly well suited) will mean that they can readily be introduced to the Topping Cycle configuration. There should also be benefit from future developments in gas turbines. Such advances are predicted to raise the Topping Cycle efficiency to around 52%.

D. Cost of Electricity

In a study with Bechtel Ltd. (who produced the necessary capital cost data), the cost of electricity has been calculated for a number of coal-fired technologies.[4] Estimated electricity generation costs are shown in Figure 2 in terms of contributions from capital charges, operation and maintenance (O & M) charges, and fuel costs. The capital charges are based on a discount rate of 10% and an assumed plant life of 25 years with an average load factor of 85%. The fuel costs are based on a coal price of £1.70/GJ. The O & M costs include labor, maintenance, local taxes, insurance, water, chemicals, ash disposal, and, where relevant, limestone costs. It is evident that the lower specific capital charges and fuel costs for technologies incorporating gas turbines result in considerably cheaper electricity costs compared with the steam-only plants. This is especially true for the Topping Cycle, with its high generation efficiency.

Table 1. Comparison of Generating Efficiency

Technology	Net efficiency (% LHV)
PF+FGD	38.3
CFBC	39.1
PFBC	40.8
IGCC	41.7
BCTC (CFBC)	46.4

Note: PF + FGD: pulverized fuel with flue gas desulfurization, CFBC: circulating fluidized bed combustion, PFBC: pressurized fluidized bed combsution, IGCC: integrated gasification combined cycle, and BCTC (CFBC): British Coal Topping cycle — CFBC option.

E. Environmental Effects

Emission standards have been set for particulates, SOx, and NOx in a number of countries. Concern over the potential effect of enhanced global warming has led to interest in CO_2 emissions. The British Coal Topping Cycle is able to take advantage of the inherent characteristics of the process to ensure low emissions levels. Estimates set against current EEC standards are given in Table 2.

- Particulates — The requirement to protect the gas turbine ensures that gas cleaning will be highly efficient resulting in very low levels of particulate emissions.
- Sulfur oxides — Control is achieved by the use of sorbent in the fluidized beds. Retention levels of greater than 90% have been shown to be possible in both gasification and combustion systems. Should further reductions be necessary, polishing systems such as the use of zinc ferrite are under development.
- Nitrogen oxides — The use of low calorific value gas in a suitably designed gas turbine combustion system results in a low level of NOx emissions.
- Carbon dioxide — The improved efficiency of the Topping Cycle compared with other power generation systems will result in a proportionate decrease in the amount of CO_2 emitted per unit of electricity generated.
- Trace elements and hydrocarbons — The system of coal conversion together with the high-efficiency gas cleaning and combustion system is expected to result in very low emissions of trace elements and hydrocarbons.

III. DEVELOPMENT PROGRAM

The proof of concept phase is being carried out at the Coal Research Establishment and at the Grimethorpe PFBC Establishment. It covers all the main processes, i.e., pressurized coal gasification, char combustion, gas cleaning, gas combustion, and utilization in a gas turbine.

A. Pressurized Gasification

The gasifer is based on an atmospheric pressure spouted bed design developed previously[5] for industrial fuel gas applications to handle a wide range of bituminous

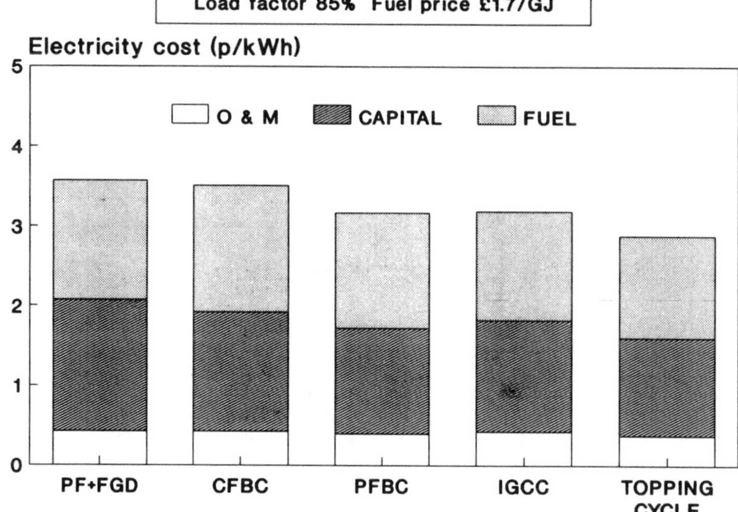

Figure 2. Cost of electricity.

coals. The pressurized pilot plant which is being used for this development has a coal throughput of 12 tonnes per day and can be operated at pressures up to 20 bar.

Initial gasification tests were completed in July 1991. Preliminary evaluation has provided confidence that gasification performance in terms of coal conversion (70 to 80%) and gas calorific value (3.6 to 4.0 MJ/m^3) will be suitable for the requirements of the Topping Cycle.

In addition to pressurized gasifier operation, a substantial supporting program is being carried out aimed at understanding the chemical and physical process in the gasification and desulfurization processes, and at providing a basis for future scale-up.

B. Char Combustion

Atmospheric pressure CFBC was selected for the initial development of the Topping Cycle because it is proven in many commercial plants worldwide. Also, the relatively low degree of cycle integration in the CFBC Topping Cycle should result in simpler control and higher operating availability, both of which are important considerations in a utility application.

The main issues in CFBC char combustion are the efficiency of low volatile fuel combustion and conversion of calcium sulfide produced in the gasifier to sulfate. There is also a need to establish whether formation of nitrous oxide (N_2O) is an issue and if so what control measures can be applied. Characterization of other emissions, e.g., trace elements and hydrocarbons, will also be carried out. Combustion tests will be undertaken on a 2-MW CFBC test facility at CRE using residual char and fines

Table 2. Environmental Impact Data

Technology	CO_2 (kg/kWh)	SOx retention (%)	NOx (mg/m³ at 6% O_2)	Particulates (mg/m³ at 6% O_2)
PF+FGD	0.87	90	500–650	50[a]
CFBC	0.86	90	100–300	~30[b]
PFBC	0.82	90	150–300	~10[c]
IGCC	0.78	99	120–300	Negligible emission
BCTC (CFBC)	0.72	90	200–300	~30[b]
BCTC (PFBC)	0.73	90	200–300	~5[c]
EEC emission standard		90[d]	650	50

[a] UK HMIP inspectorate, 1986.
[b] Filter bag house.
[c] Ceramic filters.
[d] Based on a 500 MW_{th} plant burning a typical U.K. sulfur coal.

produced from the gasification tests. Initial tests on this unit started in September 1991.

C. Gas Filtration

Critical to the success of the system is the production of a reliable flow of hot clean fuel gas to the gas turbine combustion system. The hot gas entering the gas turbine must contain low levels of dust and alkali metals if acceptable lifetimes of the turbine rotor and stator blades are to be achieved. A detailed technical and economic assessment of gas cleaning options has resulted in the above of dry gas cleaning at temperature of 400 to 600°C. The fuel gas will be cooled before the filter system, which uses ceramic candle filter elements. This represents a lower technical risk compared with gas cleaning at higher temperatures and lower capital costs and higher cycle efficiency than using a conventional scrubber. Cooling has the additional advantage that condensation of alkali metal compounds onto the particulates is expected to occur.

British Coal has considerable experience in the filtration of hot gas in both a combustion environment[6,7] and in a fuel gas environment.[8] Because of concerns about the life of ceramic media under the conditions of high temperature and repeated pulsing, a program of work has been initiated to study their durability. A filter test using over 100 elements to establish their durability when filtering PFBC flue gas at around 830°C and 10.5 bar is under way as part of the Grimethorpe Topping Cycle Project (see Section III.E, Gas Turbine Integrity). In addition, a 2- to 3-year program of accelerated thermal cycling of candle filter elements in a specially constructed high pressure rig (see Figure 3) is under way. Stresses produced in elements due to cleaning will be determined with the aid of mathematical modeling techniques. Corrosion effects will also be studied.[9]

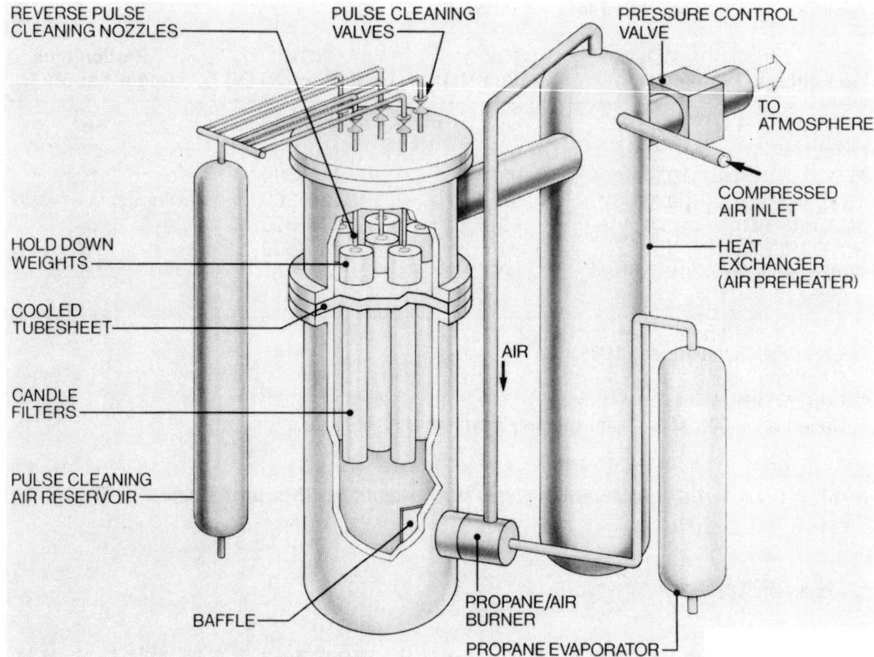

Figure 3. The candle durability rig.

D. Gas Combustion

The duty for a low cv gas combustion system operating at a combustor outlet temperature of 1380°C (corresponding to a TET of 1260°C) and a fuel gas temperature of 400 to 600°C are sufficiently different from conventional technology to require separate development as part of British Coal's Topping Cycle development strategy. The particular features of the fuel gas which the combustion system must burn are

- Calorific value of 3.5 to 4.0 MJ/m^3 wet gross (approximately 100 Btu/ft^3)
- This low calorific value means that to achieve the required temperature air:fuel mass ratios, typically 2:1 or less, will be required, thereby restricting the amount of air available for combustor wall cooling
- The fuel gas contains small amounts of nitrogen-containing species, principally ammonia; when burned, a portion of this ammonia will be converted to NOx

To achieve the development of a suitable combustion system, a multiphase development program is in progress, initially using a 1 MWth tubo-annular combustor[10] and ultimately as part of engine-specific tests in collaboration with the appropriate turbine manufacturer. This work will be backed by modeling techniques.

In tests so far carried out on the 1 MWth unit,[10] the NOx emission in the absence of ammonia addition to the fuel gas, i.e., thermal NOx, was below 5 ppmv (dry, 15% O_2) with COT in the range 970 to 1370°C.

When ammonia was added to the fuel gas, higher levels of NOx were produced (Figure 4). However, the efficiency of ammonia conversion decreased as the ammonia level increased. At the 300-ppmv level of ammonia (that predicted for an air-blown gasifier in the Topping Cycle application), the overall NOx emission at 7 bar with a COT of 1300°C was about 20 ppmv (dry, 15% O_2). This corresponds to an ammonia to NOx conversion rate of about 40%.

In addition to measurement of thermal and fuel NOx, measurements were also made of N_2O. The concentration of N_2O was less than 1 ppmv for all test conditions studied. This indicates that, although the adiabatic flame temperature is lower when burning a low calorific value fuel gas compared with conventional fuels, it is still above that at which significant levels of N_2O are produced.

Integrated tests of gasification, gas cooling, filtration, and combustion are being set up. This will provide data on contaminants such as dust and alkali vapor reaching the gas turbine. Emissions of trace elements and hydrocarbons will also be measured.

E. Gas Turbine Integrity

As part of the development program for the Topping Cycle, the pressurized fluidized bed combustor at Grimethorpe is being run in conjunction with a small gas turbine. The PFBC acts as a source of coal-derived gas which, after passing through an existing high-temperature, high-pressure ceramic filter, is fed to propane-fired topping combustors. These increase its temperature before being expanded through a modified Ruston TB 5000 gas turbine and a high-temperature sidestream blade cascade. This part of the Topping Cycle development is known as the Grimethorpe Topping Cycle Project. The principal objective is to appraise the feasibility of operating a state of the art gas turbines on coal-derived gas.

The TB5000 turbine blades will be made of a variety of different materials, with overlay and diffusional coatings typical of those currently used in industrial machines. Exposure will be for various intervals up to about 1500 h. Following exposure, metal loss and deposition rates will be measured. A set of models will be used to predict the lifetime of components. Aerodynamic models will predict gas flows, pressure, and temperature distributions through the turbine. A transport/deposition model will use these data together with gas composition and thermochemical predictions to determine the nature and arrival rate of contaminants at the blade surface. The deposition model will also predict the proportion of ash retained at the surface. An erosion/corrosion model will predict the degradation rate of the blade surface. These models will be used to predict the blade life of utility machines.

The TB 5000 machine was delivered in January 1991. Hot commissioning, leading to the full experimental program, started in March 1991. Materials performance assessment, modeling, and validation will continue until around August 1993. Lifetime prediction and the final assessments of technical and economic validity of a utility scale Topping Cycle are expected to be completed by November 1993.

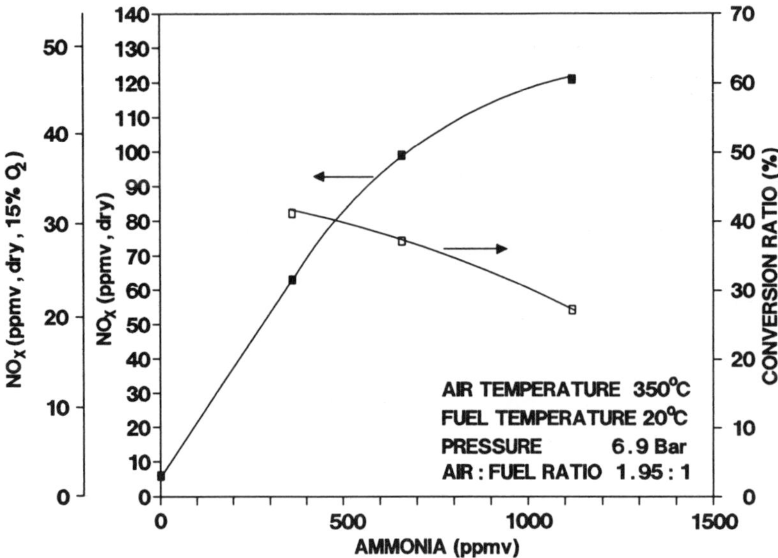

Figure 4. NOx formation from ammonia.

F. Process Control

In addition to the main component developments, a further requirement is for process control. Part load and dynamic performance is being modeled, since the overall cycle is relatively complex and the performance of some of the subsystems is unknown. Load loss protection methods are being developed.

IV. SUMMARY

The British Coal Topping Cycle has the potential to provide a cost-effective, high-efficiency combined cycle system for power generation with low environmental impact. Such a system should offer significant economic benefit over both conventional coal-fired systems and emerging advanced technologies such as CFBC, PFBC, and IGCC. Of the Topping Cycle options available, studies have identified that the preferred initial development choice is the CFBC Topping Cycle with dry gas cleaning at 400 to 600°C, since this has a reasonable technical risk while maintaining high cycle efficiency. The proof of concept program established by British Coal has been directed toward the development of this system. It is intended that this will be followed by scale-up and cycle integration, leading to the demonstration of a commercial prototype.

The environmental performance of the system has been designed to conform to the current U.K. and EEC emissions standards. Due regard has been given to possible future tightening of these standards and to the possible future need to limit carbon dioxide. The program includes establishing the environmental impact of trace elements and hydrocarbons.

ACKNOWLEDGMENTS

The work described in this paper form part of the British Coal Topping Cycle program. The contributions of colleagues at both British Coal and other participating organizations are gratefully acknowledged. Any views expressed are those of the authors and not necessarily those of the supporting organizations.

REFERENCES

1. Dawes, S. G., Brown, D., Hyde, J. A. C., Bower, C. J., and Henderson, C., Options for advanced power generation from coal, Inst. Mech. Eng. Conference, Power Generation and the Environment, London, 1990.
2. Topper, J. M., Cross, P. J. I., Davison, J. E., and Goldthorpe, S. H., Process assessment with the Arachne process simulation package, *I. Chem. E Symp. Ser. No. 114,* EFCE Publication No. 77, 1989.
3. Dawes, S. G., Arnold, M. St. J., Cross, P. J. I., and Holmes, J., British Coal development of advanced power generation technologies, Inst. Mech. Eng. Conference, Steam Plant for the 1990's, London, 1990.
4. Cross, P. J. I. and Sheikh, K. A., Development of advanced coal-fired power generation technologies in the UK, Power-Gen '90 Conference, Orlando, FL, 1990.
5. Gale, J. J., Steel, J. G., Laughlin, K. M., and Reed, G. P., Development of a pressurised fluidised bed gasifier for use in an advanced coal fired power generation system, VGB Coal Gasification Conference, Dortmund, Germany, 1991.
6. Morrel, R., Butterfield, D. M., Clinton, D. J., Barrat, P. G., Oakey, J. E. O., Durst, M., and Burnard, G. K., The mechanical performance of ceramic dust filter of the Grimethorpe pressurised fluidised bed combustor (PFBC), Proc. 1st Int. Conf. Ceramics in Energy Applications, Sheffield, England, 1990.
7. Tassicker, O. J., Burnard, G. K., Leitch, A. J., and Reed, G. P., Performance of a large filter module utilising porous ceramics on a pressurised fluidised bed combustor, 10th Int. Conf. Fluidised Bed Combustion, San Francisco, 1989.
8. Bower, C. J., Arnold, M. St. J., Oakey, J. E., and Cross, P. J. I., Hot particulate removal for coal gasification systems, Proc. Filtech 1989, Karlsruhe, W. Germany, 1989.
9. Hudson, D. M., Twigg, A. N., Clarke, R. K., Holbrow, P., and Leitch, A. J., Durability of ceramic systems for particulate removal at high temperature, 11th Int. Conf. Fluidised Bed Combustion, Montreal, 1991.
10. Kelsall, G. J., Smith, M. A., Todd, H., and Burrows, M. J., Combustion of LCV coal derived fuel gas for high temperature, low emissions gas turbines in the British Coal topping cycle, 36th Gas Turbine and Aeroengine Conference and Exposition, ASME 91-GT-384, Orlando, FL, 1991.

Pilot-Scale Evaluation of Sorbent Injection to Remove SO_3 and HCl

Joseph R. Peterson
Gordon Maller
Andrew Burnette
Richard G. Rhudy

I. ABSTRACT

This paper presents the results of a pilot plant test program conducted at the Electric Power Research Institute's (EPRI) High Sulfur Test Center (HSTC) as part of EPRI Research Project 2250-3 to investigate the feasibility of injecting dry alkaline materials into flue gas upstream of the electrostatic precipitator (ESP) for removal of gaseous SO_3 and HCl.

Four sorbents were tested: commercial hydrated lime, high-surface-area hydrated lime, commercial-grade sodium bicarbonate ($NaHCO_3$), and activated alumina. Conditions which were varied during the test program included the sorbent injection rates, flue gas flow rate, temperature, ESP specific collection area, and SO_3 and HCl concentrations.

Test results showed that the SO_3 removal was greater than the HCl removal for all sorbents and process conditions evaluated. For a given sorbent, the most important parameter for SO_3 removal was the sorbent injection rate, which agrees well with the predictions from a simple mathematical model. For SO_3 removal, the commercial-grade $NaHCO_3$ and the regular and high-surface-area hydrated limes performed about the same when compared on a weight basis. However, at high injection rates, the hydrated limes degraded the operation of the ESP, causing both the outlet opacity and outlet mass loading to increase. The operation of the ESP improved when $NaHCO_3$ was injected compared to baseline operation. The injection of activated alumina did not appear to affect the operation of the ESP, but the sorbent was relatively unreactive toward SO_3 and HCl.

II. INTRODUCTION

The results of a test program to evaluate the technical feasibility of removing sulfuric acid vapor (H_2SO_4) and hydrochloric acid (HCl) vapor from flue gas by injection of dry sorbents are presented in this paper. The testing was performed at the EPRI's HSTC located at the New York State Electric and Gas' Kintigh Station near

Barker, NY. The testing was sponsored under EPRI Research Project 2250-3. Additional funding for the sorbent injection study presented in this paper was provided by Kansas City Power and Light Co. and Louisville Gas and Electric.

Sorbent injection technology involves the injection of a dry alkaline sorbent into a flue gas duct upstream of a particulate control device (e.g., ESP, baghouse, or particulate scrubber). The application of this technology has the potential to reduce stack plume opacity resulting from condensed sulfuric acid droplets in the stack exit gas. The presence of these very fine droplets (ranging from about 0.1 to 0.5 μm) can significantly affect visual opacity.

The application of sorbent injection technology can also remove vapor-phase HCl. For utilities operating wet flue gas desulfurization (FGD) systems, removing HCl upstream of the scrubber can reduce the soluble chloride concentration of the scrubber recirculation liquor. This has the potential to improve scrubber performance (e.g., removal efficiency, limestone utilization) and to reduce the corrosion tendencies of scrubber materials of construction. If this technology could be successfully applied to new FGD systems, the use of less-expensive materials of construction may be possible.

The current study was a follow-up to an earlier evaluation of this technology by EPRI, performed at the HSTC. Although very limited in scope, the earlier study suggested that removal of H_2SO_4 and HCl was feasible and that some of the operating variables that may affect removal efficiency included the type of sorbent, the addition rate of the sorbent, the sorbent duct residence time, and the gas temperature at the point of injection. The primary objective of the current study was to perform a more-exhaustive evaluation of sorbent injection technology in an attempt to more fully characterize the process.

III. CHEMICAL REACTIONS

Although the SO_3 in the flue gas at the process conditions existing at the HSTC is actually present primarily as sulfuric acid vapor (H_2SO_4), it is more convenient to discuss the chemical reactions as if the SO_3 is the true chemical species. Therefore, the term SO_3 is used throughout this paper in place of H_2SO_4.

$$SO_3(g) + Ca(OH)_2(s) \rightarrow CaSO_4(s) + H_2O(g) \quad (1)$$

$$2HCl(g) + Ca(OH)_2(s) \rightarrow CaCl_2(s) + 2H_2O(g) \quad (2)$$

$$SO_3(g) + 2NaHCO_3(s) \rightarrow Na_2SO_4(s) + H_2O(g) + 2CO_2(g) \quad (3)$$

$$HCl(g) + NaHCO_3(s) \dashrightarrow NaCl(s) + H_2O(g) + CO_2(g) \quad (4)$$

$$3SO_3(g) + Al_2O_3(s) \rightarrow Al_2(SO_4)_3(s) \quad (5)$$

$$6HCl(g) + Al_2O_3(s) \rightarrow 2AlCl_3(s) + 3H_2O(g) \quad (6)$$

In addition to the above reactions, all of the sorbents have the potential to react with SO_2 and CO_2, both of which are present in flue gas. These reactions are not expected to have a significant effect on the tests at the temperatures evaluated. Therefore, they are not addressed in this paper.

While the above reactions are known to proceed and produce relatively stable products, it was not known whether the overall rates would be sufficient to remove SO_3 and/or HCl in a cost-effective manner at typical flue gas conditions.

IV. TEST METHODOLOGY

A. Process Diagram/Description

A simplified process flow diagram for the pilot unit showing the configuration used for the current test program is presented in Figure 1. Flue gas was extracted approximately isokinetically from the outlet duct on the Kintigh Station boiler for use at the HSTC. This test program was conducted on the HSTC 4-MW spray dryer/ESP pilot unit flow path. The flue gas passed through the spray dryer vessel (which was not in operation during this program) and then proceeded to the outlet duct where sorbent injection occurred. The flue gas then passed through a five-field ESP for sorbent and fly ash removal. For most of the tests, only the first three fields were energized for an SCA of about 300 ft^2/kacfm. A few tests were also conducted with two fields for an SCA of 200 ft^2/kacfm. After the ESP, the flue gas was returned to the Kintigh Station ductwork.

The normal sulfur content of the coal fired at Kintigh (2.8%) produces a flue gas SO_2 concentration of about 1600 to 1800 ppmv and a SO_3 concentration of about 10 to 15 ppmv.

The chloride content of the Kintigh Station's coal (0.1%) produces a flue gas HCl concentration of about 50 to 55 ppmv. For most of the current tests, the inlet HCl concentration remained at the baseline level, but for a few tests it was increased to approximately 100 ppmv by spiking the flue gas with anhydrous HCl.

The flue gas SO_3 concentration was varied for many of the tests by spiking with SO_3. The SO_3 was produced by passing an SO_2/air mixture over a vanadium catalyst at 800°F. The SO_2 content of the SO_2/air mixture was changed to alter the amount of SO_3 that was injected into the gas stream. The SO_3 was injected into the flue gas just upstream of the spray dryer vessel.

The gas flow rate and temperature at the outlet of the spray dryer vessel were controlled to their desired setpoints using a variable-speed fan and an electric heater. The flue gas SO_2 and O_2 concentrations were measured at the spray dryer inlet, the spray dryer outlet, and the outlet of the ESP to determine if measurable SO_2 removal occurred and to correct the measured concentrations and calculated removals for air inleakage into the system.

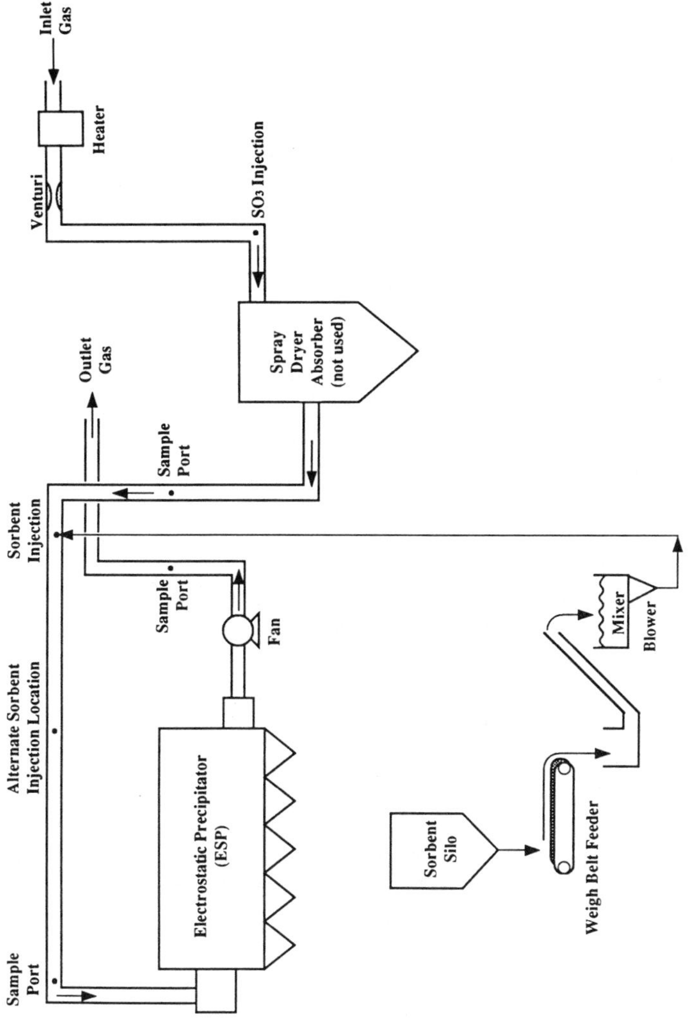

Figure 1. Pilot unit configuration for sorbent injection experiments.

Sorbent was gravimetrically fed into small hoppers and then pneumatically conveyed into the flue gas downstream of the spray dryer vessel. A small weight loss feeder with a self-contained hopper was used for the low sorbent flow rates (3 to 50 lb/h). A 4-inch weigh belt and a sorbent silo were used for the higher flow rates (32 to 160 lb/h) and for overnight tests.

B. Gas Sampling

The SO_3 concentration in the flue gas was determined by a controlled condensation technique. This technique involved pulling a sample of flue gas through a heated filter, then through a glass condenser which was maintained at 140°F. This temperature was below the SO_3 dewpoint but above the water dewpoint. As a result, SO_3 and not water condensed on the walls of the glass condenser. Condensation appeared as a visible "fog" in the condenser. The gas sample then entered a set of impingers designed to remove gaseous HCl and water vapor. The sample then exited through a pump and a dry gas meter.

For this study, the process inlet and outlet flue gas streams were sampled simultaneously for at least 30 min, which was more than adequate to observe the condensation of SO_3 in the condenser. At the end of the sampling time, condensed SO_3 was recovered by rinsing the condenser with about 60 mL of distilled water into previously weighed sample bottles. The SO_3 concentrations in the flue gas streams were determined by analyzing the samples for sulfate (by ion chromatography) and by recording the amount of gas sampled (i.e., from the dry gas meter readings).

The HCl concentrations in the inlet and outlet flue gas streams were determined by two methods. For most tests, the impinger solutions from the inlet and outlet flue gas samples were analyzed for chloride by ion chromatography. An infrared HCl monitor was also used to continuously measure the HCl concentration in the flue gas at the outlet of the ESP. This monitor was checked with span gases and found to be quite accurate over the concentration range of interest (less than 150 ppmv). The HCl concentrations measured by the impingers did not agree well with those measured by the monitor. However, from past experience with the monitor on the spray dryer system, it is believed that the HCl concentrations determined by the monitor better represent the true HCl concentrations in the flue gas. This monitor has shown that the HCl concentration in the flue gas at the inlet to the HSTC is normally about 52 ppmv when the Kintigh power plant is near full load. Since all of the current tests were performed when the power plant was close to full load, the removal of HCl across the pilot system was determined using the ESP outlet concentration measured by the HCl monitor and an assumed inlet concentration of 52 ppmv.

The spiked flue gas was sampled for SO_3 and HCl just downstream of the spray dryer vessel but upstream of any sorbent injection. Sampling at this location supplied the inlet SO_3 and HCl concentrations (i.e., before any sorbent injection). The outlet concentrations were measured by sampling at two locations: immediately upstream of the ESP, and downstream of the ESP and induced draft fan. By sampling simultaneously at all three locations, which was done for a selected number of tests, one could determine the SO_3 and HCl removal occurring across the flue gas ductwork and across the combination of the ductwork and the ESP.

C. Reagent Properties

Four reagents were tested: commercial hydrated lime, a special high-surface-area hydrated lime, commercial-grade $NaHCO_3$, and activated alumina. Samples of each reagent were taken twice each day when that particular sorbent was being injected into the ductwork. Selected samples were analyzed for specific surface area (using a one-point BET method) and for sorbent particle size. A summary of the reagent properties is presented in Table 1.

D. Experimental Conditions

The experimental conditions for the current study are summarized in Table 2. Most of the tests were conducted at an ESP inlet temperature of 315°F and an inlet flue gas flow rate of 13,600 acfm. For almost all of the tests, only the first three fields of the five-field ESP were energized. At that flue gas rate, operation with three fields yielded a specific collection area of about 300 ft^2/kacfm. Throughout the program, the first field was rapped every 5 min, the second every 10 min, and the third every 20 min. The last two fields, which were not energized, were rapped every 20 min.

Most of the tests lasted less than 2 h. For these tests, the system was allowed to equilibrate for about 15 minutes after the sorbent flow was initiated prior to beginning data collection. The equilibration time period was chosen based on data from the continuous HCl analyzer which sampled the gas exiting the ESP. These data showed that the HCl concentration stabilized about 15 min after the sorbent injection rate was changed.

V. DISCUSSION OF RESULTS

This section discusses the results from the current program. First, a theoretically based model developed to aid in data interpretation is discussed; the measured results are then discussed in light of the model.

A. Mathematical Model Development

The results of this program are best interpreted by a theoretically based model which was developed for predicting SO_3 removal as a function of operating conditions. With this model, SO_3 removal can be predicted for other locations and other operating conditions.

At the high reagent ratios tested in this program [e.g., $Ca(OH)_2$ to SO_3 molar ratios ranging from 2 to over 40], it is likely that the rate-controlling step for SO_3

Table 1. Summary of Reagent Properties

Sorbent	$Ca(OH)_2$	$Ca(OH)_2$	$NaHCO_3$	Al_2O_3
Grade	Commercial	High surface	Commercial	Activated
Source	Chemical lime	Chemical lime	Kerr McGee	Alcoa
Average surface area (m^2/g)	20	35	3	170
Average particle diameter (μm)	14	12	11	7

Table 2. Summary of Experimental Conditions

	Lowest value	Base-case value	Highest value
Gas flow rate (acfm)	7,700	13,600	13,600
Gas temperature (°F)	305	315	350
Inlet SO_3 (ppmv)	10	24	40
Inlet HCl (ppmv)	52	52	100
Duct residence time (s)	1.0	2.0	3.5
Total residence time (s)[a]	2.3	3.3	5.7

[a] Assuming particles are collected in the first half of the first ESP field.

removal was the diffusion of SO_3 from the bulk gas to the sorbent particles. Therefore, a gas diffusion model was developed to compare the measured SO_3 removal to that predicted by the model and to see if any knowledge could be gained by exercising the model for a variety of conditions.

The development of the model assumed

- A large excess of reagent was present, relative to the amount of SO_3 removal.
- All of the resistance to mass transfer occurred in a thin film surrounding the particle.
- The competing reactions of HCl, SO_2, and CO_2 with the sorbent particles were not important.
- The sorbent particles were spherical with smooth external surfaces (i.e., internal or pore surface areas did not contribute to the overall reaction rates at the relatively low sorbent conversion efficiencies).
- The average particle diameter accurately approximated the true distribution of sorbent particle diameters.
- The particles were well dispersed in the flue gas at all times.
- There was no net velocity between the particles and the flue gas.
- Constant temperature and pressure were maintained.

Since the particles were assumed to be well dispersed in the flue gas, the problem could be reduced to a single particle associated with some amount of flue gas. Therefore, the modeling process involved calculating the volume of flue gas per particle, then calculating the rate of diffusion, or flux, of SO_3 to that particle.

In its general form, the flux of SO_3 to the sorbent particle is given by

$$N_{SO_3} = k_g \cdot A \cdot C_t \cdot (y_{SO_3\,bulk} - y_{SO_3\,surf}) \tag{7}$$

where N_{SO_3} = the flux of SO_3 to the sorbent particle (gram moles SO_3/s); kg = gas-phase mass transfer coefficient (cm/s); A = external surface area of the particle (cm^2); C_t = concentration of flue gas (gram moles total gas/cm^3); $y_{SO_3\,bulk}$ = mole fraction of SO_3 in the bulk gas (moles SO_3/total moles gas); and $y_{SO_3\,surf}$ = mole fraction of SO_3 at the sorbent's surface.

Since the model assumes that the rate-controlling step for the SO_3 removal process is the diffusion of SO_3 through a thin film surrounding the sorbent particle, the concentration of SO_3 at the surface of the particle must be zero and the flux expression is reduced to

$$N_{SO3} = k_g \cdot A \cdot C_{SO_3bulk} = k_g \cdot \pi \cdot d_p^2 \cdot C_{SO_3bulk} \tag{8}$$

where d_p = diameter of sorbent particle (cm) and $C_{SO_3\,bulk}$ = concentration of SO_3 in the bulk flue gas (gram moles SO_3/cm^3).

The above equation can be rearranged and solved analytically to give

$$\text{Percent SO}_3\text{ removal} = 100 \cdot \left[1 - \exp\left(\frac{-2\pi \cdot D_{SO_3} \cdot M}{p_p \cdot d_p^2 \cdot G} t\right)\right] \tag{9}$$

where D_{SO_3} = H_2SO_4 (v) diffusion coefficient (cm^2/s); M = sorbent injection rate (g/s); p_p = particle density (g/cm^3); G = total gas flow rate (actual cm^3/s); and t = reaction time (s).

A very important parameter in this equation is the diameter of the sorbent particle. The model assumes a single particle size, but all of the sorbents showed a distribution of particle sizes. To correctly model the SO_3 removal data, the model would have to integrate the removal occurring for each of the particle sizes. Since this was beyond the scope of this study, an average value for the particle diameter was used. Furthermore, the aerodynamic particle size (i.e., the actual agglomerated particle size in the ductwork) is more important for modeling the SO_3 removal process. The SO_3 removal data seemed to closely fit the diffusion model if a particle diameter of 10 μm was assumed. As shown in Table 1, this assumed diameter does not differ greatly from the average diameter determined from the particle size distribution data.

The time for the reaction between the sorbent and the flue gas is also an important parameter in the modeling equation. Since the flue gas flow rate, duct length, and duct diameter were well known for the current study, it was possible to accurately determine the reaction time for the sorbent in the ductwork. However, since most of the particles were removed in the first field of the ESP, it is difficult to predict the total reaction time of the particles with the flue gas. For the modeling results presented in this study, it was assumed that the particles continued to react with the flue gas in the ESP for a time equal to one half of the flue gas residence time in the first field of the ESP (i.e., 1.32 or 2.31 s, depending upon the flue gas flow rate).

Another important parameter in the above equation is the diffusion coefficient for SO_3. Since the SO_3 is present as gaseous H_2SO_4 under typical flue gas temperatures and flue gas moisture levels, the diffusion coefficient was estimated by calculating the diffusion coefficient of H_2SO_4 in nitrogen at the temperature and pressure of the flue gas stream. The diffusion coefficient was calculated to be 0.169 cm^2/s at 315°F using the method of Fuller as described in Reid et al.[4] The diffusion coefficient would increase with the absolute temperature of the flue gas (e.g., the diffusion coefficient was calculated to be 0.183 cm^2/s at 350°F), so the model predicts higher SO_3 removal at higher temperatures.

B. Sorbent Addition Rate — SO_3 Removal

For a given particle size, flue gas flow rate, and reaction time, the diffusion model predicted that the only other important parameter for SO_3 removal is the sorbent addition rate. This rate determines the number of particles injected into the flue gas and, therefore, the volume of total gas surrounding each particle. The model assumed that there was no effect of sorbent type. The experimental results from the current test program agree fairly well with this model as shown in Figure 2.

The data for the commercial-grade $NaHCO_3$ and the regular and high-surface-area hydrated limes (Figure 2) show that the observed SO_3 removal agrees fairly well with the model predictions at moderate sorbent injection rates. However, the model tends to overpredict SO_3 removal at very high sorbent injection rates and to underpredict SO_3 removal at low sorbent injection rates. For activated alumina, the model seems to overpredict SO_3 removal at nearly all injection rates. There are plausible explanations for the deviations from the model.

At very high sorbent injection rates, it is likely that the assumption of well-dispersed sorbent particles is much less valid than at low injection rates. Therefore, the model will tend to overpredict SO_3 removal. This is also supported by the fact that the measured SO_3 removal never reached 100%, even at very high sorbent injection rates.

At lower sorbent injection rates, the diffusion model tends to underpredict SO_3 removal across the system. We speculate that this may result from some SO_3 removal being caused by condensation of H_2SO_4 at cold spots in the ESP. Since the ESP is a pilot-scale unit, it has more external surface area per unit volume than a full-sized unit. As a result, cold spots in the ESP are much more important on a pilot-scale unit than on a full-scale unit. The background SO_3 removal across the pilot ESP was

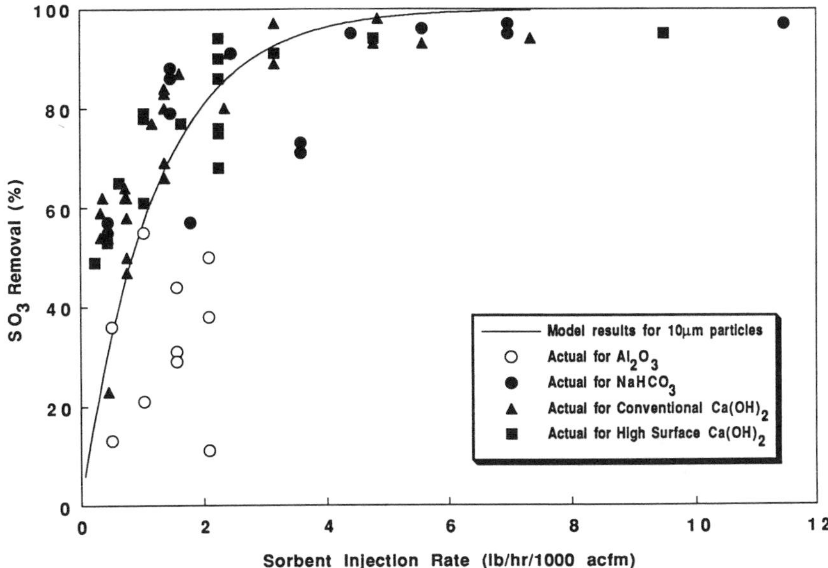

Figure 2. Experimental and model results for SO_3 removal by injection of $NaHCO_3$, conventional hydrated lime, high-surface-area hydrated lime, and activated alumina. The model assumes a 3.28-s reaction time. All data are at a gas temperature of 315°F.

quantified by simultaneously measuring the SO_3 concentrations at the system inlet and outlet when no sorbent was injected into the ductwork. The background removal ranged from 10 to 30% and seemed to increase with the inlet SO_3 concentration. This observation is consistent with the background SO_3 removal since higher inlet SO_3 concentrations create higher dew point temperatures, resulting in higher SO_3 removal.

Activated alumina is known to readily agglomerate, which, as for high injection rates with the other sorbents, tends to increase the effective particle diameters and lower the actual SO_3 removal. Thus, the model's overpredictions of the SO_3 removal for activated alumina could be rationalized but not proven.

C. Sorbent Addition Rate — HCl Removal

Figures 3 and 4 show the HCl removal data for the four sorbents at a gas temperature of 315°F. These data are quite different from the SO_3 removal data for several reasons:

- The magnitude of the HCl removal was less than that for the SO_3 removal.
- The effect of sorbent type on HCl removal was more pronounced than for SO_3 removal when the sorbents were compared on a mass basis (Figure 3). Compar-

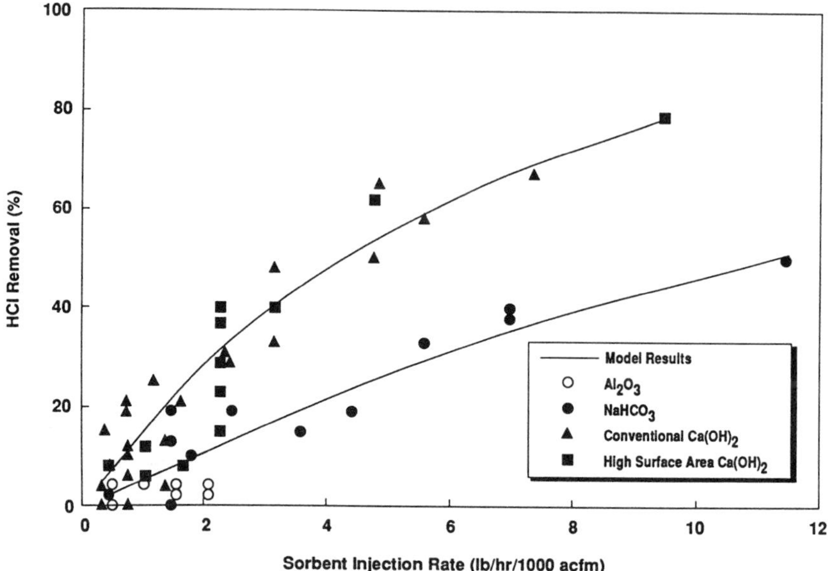

Figure 3. Experimental and model results for HCl removal by injection of $NaHCO_3$, conventional hydrated lime, high-surface-area hydrated lime, and activated alumina. Sorbents are compared on a mass basis. All data are at a gas temperature of 315°F.

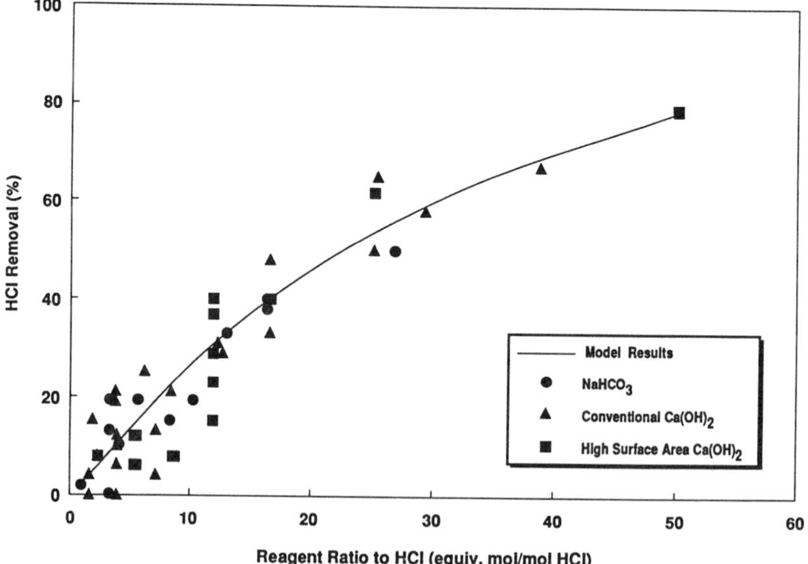

Figure 4. Experimental and model results for HCl removal by injection of $NaHCO_3$, conventional hydrated lime, high-surface-area hydrated lime, and activated alumina. Sorbents are compared on a reagent ratio basis. All data are at a gas temperature of 315°F.

ing the data on a reagent ratio basis caused most of the data to collapse onto one curve (Figure 4).

- The shape of the removal vs. sorbent injection rate curve was more linear.

These data suggest that HCl removal was not limited by gas-phase diffusion. Some other mechanism evidently controlled HCl removal. The data were not sufficient to prove which mechanism controlled the overall reaction, but it is easy to fit the data if one assumes that the overall reaction was controlled by the kinetics of a first-order reaction between the sorbent and HCl. The curves shown in Figures 3 and 4 were the results of fitting a first-order reaction rate expression to the data.

D. Inlet SO_3 Concentration

For a given sorbent *mass* injection rate, the diffusion model predicted that there was no effect of inlet SO_3 concentration on the percent SO_3 removal for the sorbent injection process. The data taken in this test program appear to agree with this prediction. However, for a given *reagent ratio* (i.e., moles sorbent per moles SO_3), the data show that the percent SO_3 removal was higher for higher inlet SO_3 concentrations (Figure 5). This trend was also predicted by the diffusion model and can be explained by noting that, for the same reagent ratio, more particles must be injected into the flue gas for the higher inlet SO_3 concentrations than for the lower concentrations. Therefore, less gas volume is associated with each particle at high inlet SO_3 concentrations, and the distance that the SO_3 has to diffuse to reach the sorbent particles is reduced.

E. Flue Gas Temperature

Another objective of this test program was to evaluate the effect of flue gas temperature on SO_3 and HCl removals for the sorbent injection process. Most of the experiments were completed at a flue gas temperature of 315°F (ESP inlet temperature). Additional tests were performed at 350°F at the same sorbent residence time.

The data from the tests showed no significant effect of flue gas temperature on SO_3 and HCl removal levels. The diffusion model predicted a slight increase in the SO_3 removal when the temperature was increased to 350°F because the diffusion coefficient of H_2SO_4 increases with temperature. It is likely, however, that in this pilot-scale system the predicted increase in SO_3 removal was negated by the decrease in SO_3 removal due to cold spots in the ESP. The cold spots become less effective as the gas temperature is increased.

F. Sorbent Residence Time

The effect of sorbent residence time on the SO_3 removal level was investigated by injecting the sorbent into the flue gas at a location closer to the ESP inlet (Figure 1). Injecting at this point decreased the duct residence time from 2.0 to 1.3 s and the

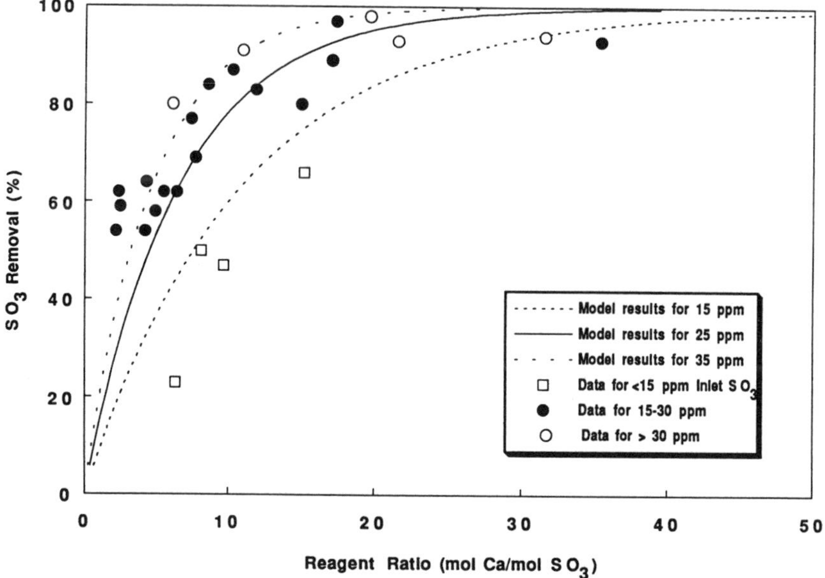

Figure 5. Effect of inlet SO_3 concentration on SO_3 removal for conventional hydrated lime. The model assumes 10-μm particles and a 3.28-s reaction time. All data are at a gas temperature of 315°F.

total (duct plus ESP) estimated residence time from 3.3 to 2.6 s. Figure 6 shows the effect of changing the sorbent injection location on the SO_3 removal obtained with the commercial hydrated lime. No significant effect of changing the total residence time from 3.3 to 2.6 s was observed. Any change in SO_3 removal with residence time was apparently within the ability to measure the SO3 removal, which was estimated to be ±5%.

G. Sorbent Surface Area

The data from the current study show that the high-surface-area hydrated lime performed no better than the commercial-grade hydrated lime, even though the high-surface-area hydrated lime had almost twice the surface area (35 vs. 20 m^2/g). In addition, the activated alumina had a very high surface area (170 m^2/g) but was much less reactive toward SO_3 and HCl. Sodium bicarbonate had the lowest surface area of all the sorbents tested (3 m^2/g), but it performed as well as the hydrated limes. However, the sodium bicarbonate thermally decomposes in the flue gas to form higher-surface-area sodium carbonate; the measured value of about 3 m^2/g probably understates the actual reactive surface area of the reagent after it is injected into the flue gas. Even so, data from the literature[1] indicate that the surface area of the thermally decomposed $NaHCO_3$ is probably much less than those of the other sorbents.

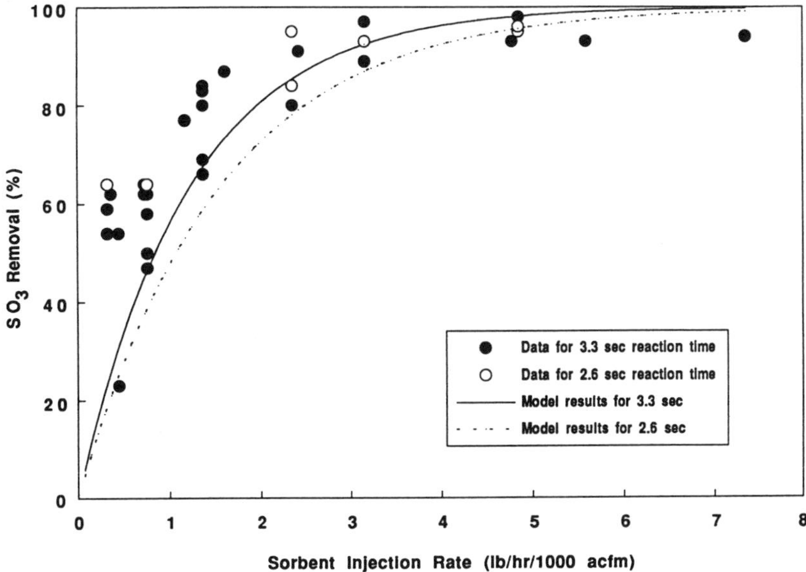

Figure 6. Effect of sorbent residence time on SO_3 removal for commercial-grade hydrated lime. The model assumes 10-µm particles. All data are at a gas temperature of 315°F.

The apparent lack of dependence upon sorbent-specific surface area agrees with the predictions of the diffusion model. The model states that the internal surface area of a sorbent particle is not important since the SO_3 removal process is assumed to be limited by the diffusion of SO_3 from the bulk gas to the external surface of a sorbent particle.

VI. EFFECT ON ESP OPERATION

One goal of this test program was to determine the effects of sorbent injection on ESP operation. The injection of alkaline sorbents into the flue gas upstream of the ESP can affect the operation of the ESP due to the increased mass loading, changes in the overall particle size and resistivity, and the removal of SO_3 which is a known ESP conditioning agent.

The following measurements were made for baseline (i.e., fly ash only) and sorbent injection conditions:

- Voltage-current relationships for each field of the ESP
- Continuous flue gas opacity measurements at the outlet of the ESP
- Flue gas mass loadings at the outlet of the ESP

Most of these measurements were conducted when three of the ESP fields were energized, corresponding to an SCA of 300 ft^2/kacfm. Some measurements were performed when only two ESP fields were energized (an SCA of 200 ft^2/kacfm) to simulate a smaller ESP. The results of these measurements are discussed below.

A. Results of ESP Testing

Early in the program, it was observed that injecting hydrated lime at high flow rates (greater than 2.2 lb/h/1000 acfm) had adverse effects on the operation of the ESP. This was first evidenced by strong sparking in the first field of the ESP. If the power to the first field was turned off to stop particle collection in the first field, sparking immediately started in the second field of the ESP.

At lower hydrated lime injection rates (less than 2.2 lb/h/1000 acfm), the severity of the sparking was diminished, but the voltage-current relationships in the first field were still altered. The corona current in the first field was much lower when the hydrated lime was injected than the during fly-ash-only conditions. Since the lower corona current probably indicated a low particulate collection efficiency in the first field, it was speculated that the second and third fields would exhibit the same behavior as the first field if the sorbent injection continued for an extended period of time.

Several overnight tests were performed to investigate the effect of hydrated lime injection on the ESP for a longer time period. The ESP outlet opacity and outlet mass loading are summarized in Table 3 and illustrated in Figure 7. Figure 7 shows that the ESP outlet opacity increased when hydrated lime was injected at high flow rates (greater than 2.2 lb/h/1000 acfm). Figure 8 compares the voltage-current relationships in the absence of sorbent injection with those for the injection of a large amount of hydrated lime. These data were taken after the hydrated lime had been injected continuously for about 36 h. The data show that the hydrated lime drastically reduced the operating current in the first two fields and produced back corona in the third field. The current may be reduced to very low levels in all fields if the sorbent were to be injected for a longer time.

The results from the outlet mass loading tests (Table 3) tend to agree with the outlet opacity measurements and appear to support the observation that ESP performance deteriorates over time when large amounts of hydrated lime are injected into the flue gas. For example, the outlet mass loading increased from 0.025 lb/MBtu under fly-ash-only conditions to 0.068 lb/MBtu after hydrated lime was injected at 50 lb/h/1000 acfm for approximately 48 h. However, for the same injection rate of hydrated lime, another outlet mass loading test showed a lower-than-baseline outlet mass loading of 0.019 lb/MBtu. Data collection for this test began approximately 1 h after the start of the hydrated lime injection and lasted for approximately 10 h. As shown in Figure 7, the ESP outlet opacity during this time period was relatively low until the very end of the mass loading test. About 11 h after the start of the hydrated lime injection, the outlet opacity increased to the relatively high level. The outlet opacity remained at this level while the other mass loading test was conducted (the 0.068 lb/MBtu test). These data indicate that the performance of the ESP degrades with time when hydrated lime is injected at high injection rates.

Table 3. Summary Data from ESP Outlet Mass Load Tests

Sorbent type	Sorbent flow (lb/h/1000 acfm)	ESP outlet loading (lb/MBtu)
Fly ash only	—	0.027
Fly ash only	—	0.025
Fly ash only	—	0.048[a]
$Ca(OH)_2$ (1 h after sorbent flow initiated)	3.7	0.019
$Ca(OH)_2$ (48 h after sorbent flow initiated)	3.7	0.068
$Ca(OH)_2$	2.3	0.084[a]
$Ca(OH)_2$	1.0	0.065[a]
$NaHCO_3$	3.7	0.009

[a] Performed with only two fields energized.

The ESP outlet opacity returned to the baseline level soon after the sorbent injection was turned off. When the sorbent injection was restarted at a lower rate (1.0 lb/h/1000 acfm), the opacity did not increase. However, as shown in Table 3, the mass loading at the ESP outlet appeared to increase even at this low sorbent injection rate.

The sparking problems and the drastic altering of the voltage-current relationships were not apparent when either activated alumina or $NaHCO_3$ were injected into the flue gas stream. The outlet opacity also remained fairly constant while these sorbents were injected into the flue gas. In fact, the ESP outlet mass loading test which was performed while $NaHCO_3$ was injected showed that the efficiency of the ESP improved compared to that for fly-ash-only conditions. This result was somewhat expected because sodium compounds are known conditioning agents for ESPs due to their relatively low resistivity.

VII. FULL-SCALE IMPLEMENTATION

The data from this study suggest that it is possible to reduce SO_3 levels and plume opacity by injecting either hydrated lime or sodium bicarbonate into the flue gas. These sorbents were equivalent for SO_3 removal when compared on a mass basis. Costs for injecting these sorbents for SO_3 removal and plume opacity reduction were estimated for a 300-MW, base-loaded power plant. To achieve an 80% reduction in flue gas SO_3 levels from a moderate initial level (e.g., 20 ppm), about 7500 tons/year of either sorbent would have to be injected. For hydrated lime reagent at $65/ton, this would result in an annual sorbent cost of about $500,000, which is equivalent to about 0.2 to 0.3 mil/kWh. For sodium bicarbonate reagent at $200/ton, the annual sorbent cost would rise to about $1.5 million, which is equivalent to 0.7 to 0.8 mil/kWh. For either sorbent, a permanent sorbent storage and injection system would be estimated to cost between $500,000 and $750,000.

The injection of these sorbents would slightly increase the volume of solid waste produced by the plant. For a case with a 2.8% sulfur content and 8% ash content in the coal, the sorbent injected would represent about 4 to 5% of the dry weight of the

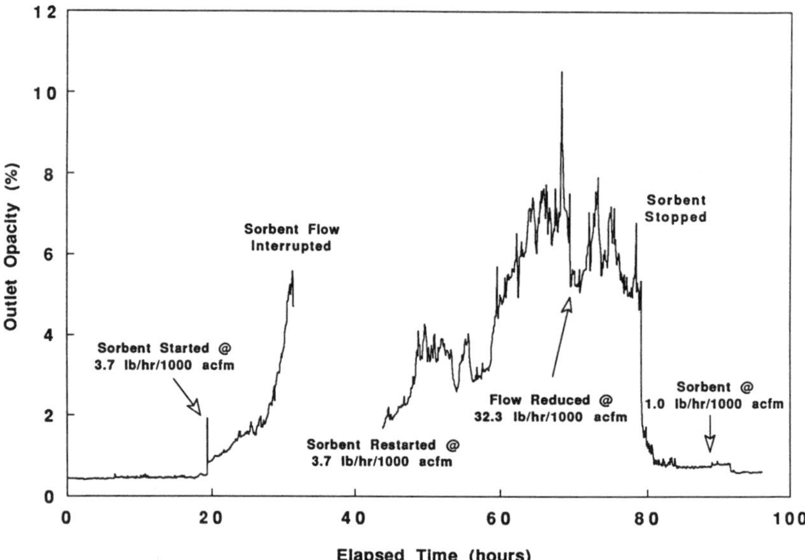

Figure 7. Effect of hydrated lime injection on flue gas opacity as measured at the ESP outlet.

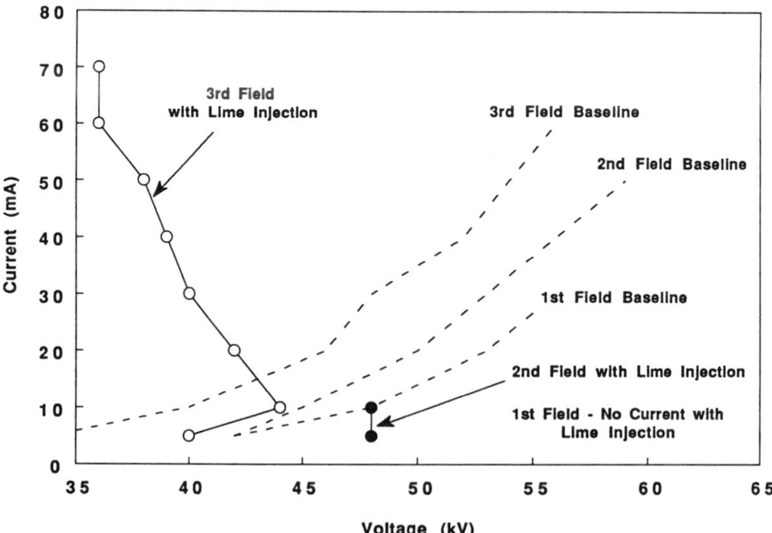

Figure 8. ESP current-voltage relationships for baseline and hydrated lime injection conditions. Data for hydrated lime conditions were taken after hydrated lime injection had been in progress for 48 h at 3.7 lb/h/1000 acfm.

combined ash and FGD sludge stream produced. Note that this represents about twice the amount of hydrated lime generally used in cases where it is added to the combined ash/FGD sludge stream for stabilization. In such cases, it may be possible to eliminate lime addition to the sludge if hydrated lime is used for SO_3 control. Thus, for this circumstance, the net reagent cost for SO_3 control by hydrated lime injection would be about half that of the estimate above, or only about \$250,000 per year for the example case.

The only drawback to using hydrated lime may be the potential adverse effects on ESP performance. The magnitude of these effects will likely be site specific and will depend greatly on the hydrated lime injection rate required. While sodium bicarbonate could be used instead to avoid any potential adverse effects on the ESP, for the example case described above, the sodium bicarbonate reagent would be at least three times more expensive than hydrated lime reagent. Also, the addition of highly water-soluble sodium salts to the solid waste stream from the plant may be undesirable.

VIII. CONCLUSIONS

Based on the results from the current test program, the following conclusions can be drawn:

- The injection of alkaline sorbents will remove SO_3 and, to a lesser extent, HCl from power plant flue gas streams. However, care must be taken to avoid ESP outlet particulate emission problems caused by certain sorbents.
- For all of the sorbents tested, the removal of HCl was much less than the removal of SO_3.
- For SO_3 removal, the commercial-grade $NaHCO_3$, the commercial-grade hydrated lime, and the high-surface-area hydrated lime all performed about the same when compared on a weight basis. The activated alumina was less reactive than these sorbents, even though it had a much higher specific surface area.
- The SO_3 removal results from the current study agree fairly well with the predictions of a simple gas-phase diffusion mathematical model. This model predicted that the most important parameters for SO_3 removal were the particle size of the sorbent, the sorbent injection rate, and the sorbent residence time in the flue gas.
- The injection of large amounts of hydrated lime caused the ESP outlet opacity and mass loading to increase. The voltage-current relationships for the ESP were also significantly altered.
- A permanent sorbent storage and injection system would cost between \$500,000 and \$750,000 for a 300-MW, base-loaded power plant. The annual sorbent costs for obtaining 80% removal of a 20-ppm SO_3 concentration in this plant would be about \$500,000 (0.2 to 0.3 mil/kWh) and \$1.5 million (0.7 to 0.8 mil/kWh) for hydrated lime and sodium bicarbonate, respectively.

ACKNOWLEDGMENTS

The work reported in this paper is the result of research carried out in part at EPRI's High Sulfur Test Center (HSTC) located near Barker, NY. We wish to acknowledge the support of the HSTC cosponsors: New York State Electric and Gas, Empire State Electric Energy Research Corporation, Electric Power Development Corporation, and the U.S. Department of Energy. In addition, partial funding of this project was provided by Kansas City Power and Light and by Louisville Gas and Electric.

We also wish to acknowledge the donations of the regular and high-surface-area hydrated lime reagents by the Chemical Lime Group of Fort Worth, TX.

REFERENCES

1. Fellows, K. T., and Pilat, M. J., *J. Air Waste Manage. Assoc.*, 40(6), 887, 1990.
2. Karlsson, H. T., Klingspor, J., and Bjerle, I., *APCA J.*, 31(11), 1177, 1981.
3. Uchida, S., Kageyama, S., Nogi, M., and Karakida, H., *J. Chin. Inst. Chem. Eng.*, 10, 45, 1979.
4. Reid, R. C., Prausnitz, J. M., and Poling, B. E., *The Properties of Gases and Liquids*, 4th ed., McGraw-Hill, New York, 1987.

Multipathway Risk Assessment at a Maryland Utility

Kenneth L. Zankel
Roger P. Brower
Jeroen Gerritsen
Sally A. Campbell

I. INTRODUCTION

A comprehensive, multipathway risk assessment for emissions from the routine operation of a complex of large combustion sources in a rural setting in west-central Maryland has been completed. As part of this project, the public health risk associated with ingestion of pollutants emitted from the sources, and subsequently incorporated into the human food chain, had to be estimated. In the course of this work, estimation of the ingestion risk was found to be more important and more difficult than anticipated. The results of this study and associated uncertainty analyses have elucidated several important considerations for performing multi-media food exposure assessments of airborne emissions from combustion sources. In this paper, lessons learned from this comprehensive exposure assessment are presented. Background analyses, data, and detailed discussions have been presented elsewhere.[1-4] In this paper, the overall design of the risk assessment and relative importance of the ingestion exposure pathway are discussed, followed by a discussion of major uncertainties and, hence, information needs in the exposure assessment process.

II. OVERALL DESIGN OF THE RISK ASSESSMENT

The sources addressed in the project included (1) three existing conventional pulverized coal boilers, totaling 558 MW at full load and requiring approximately 4500 ton/day of coal; (2) a proposed 768-MW combined-cycle gas turbine facility consisting of four gas/oil-fired combustion turbines, requiring approximately 680 gal/min of No. 2 fuel oil, in combination with two heat recovery steam generators; and (3) a proposed water-wall, mass-burn incinerator burning a maximum of 2250 ton/day of municipal refuse. This risk assessment was conducted for the Power Plant and Environmental Review Division of the Maryland Department of Natural Resources. It focused on the effects of routine airborne emissions of both toxic trace metals (such as As, Be, Cd, Cr, Cu, Pb, Hg, Ni) and organic compounds (such as formaldehyde, dioxins, PCBs, PAHs). Releases to the environment through wastewater or solid waste streams were not considered. The study examined both the inhalation and ingestion pathways to exposure. An integrated modeling system was employed to comprehensively address the transport and fate of the airborne emissions. For example, a state-of-the-art procedure was used to estimate the wet and dry deposition of both gases and particles to various surfaces; an improved food chain model[5] was used to project pollutant-specific ingestion exposures from the deposition estimates.

III. SIGNIFICANCE OF INGESTION EXPOSURE

Ingestion exposure is often neglected in risk assessment studies. However, because of the rural setting of this particular study, the potentially exposed population included local farm families who eat primarily products of their own farms. As a result, this study found that the indirect ingestion pathway was the dominant pathway of exposure of the most exposed individual (MEI) to the toxic emissions. The MEI was situated at the location of maximum joint (due to all sources) exposure to each pollutant. The risks associated with the combustion source complex were found to be dominated by carcinogenic emissions. Therefore, long-term exposure was the prime concern. The most significant emissions were arsenic, contributed mostly by the coal boilers, and dioxin and PCBs, contributed mostly by the incinerator. Within the food chain, the most significant route for arsenic exposure was ingestion of contaminated leafy vegetables, fruits, and fresh legumes, and incidental ingestion of soil. For the organic compounds, dioxin, and PCB, the most significant exposure route was ingestion of contaminated beef and dairy products.

This study examined both ingestion and inhalation exposures. For the defined MEI, projected ingestion exposure overwhelmed projected inhalation exposure. Table 1 presents the ratio of the maximum ingestion exposure to the maximum inhalation exposure of the MEI to the three critical pollutants. Ingestion exposure is more than two orders of magnitude higher than inhalation exposure. This is a significant result even acknowledging the large uncertainties associated with the ingestion exposure assessment (including uncertainties in deposition and in the food chain).

This study has shown that all pathways should be considered when developing a specific exposure assessment approach. In some circumstances the indirect ingestion exposure pathway can be an important and controlling pathway, and examining only the direct inhalation exposure pathway may lead to substantial underestimation of risk. Lifestyle and eating habits can have an enormous impact on potential ingestion exposure; thus, if ingestion exposure may be important, these habits should be carefully examined and specified. In particular, ingestion exposure cannot be neglected if the target population derives the bulk of its diet from contaminated homegrown or local products. In this study, the ingestion exposure to the general population would be far less than that to the most exposed individual whose diet consists of significant amounts of locally contaminated food. Also, if the emissions

Table 1. Comparison of Ingestion and inhalation Exposures

Pollutant	Maximum ingestion exposure / Maximum inhalation exposure
As	212
Dioxin	282
PCB	227

sources were located in more suburban or urban settings where diets largely consist of imported, nonlocal items, inhalation would most likely be the critical pathway of exposure of local populations to the toxic airborne emissions.

If ingestion exposure may be a critical pathway, the spatial distribution of the exposed population as well as the local climatology should be examined carefully. The specific location of home gardens or farms can be very important because wet deposition of particulate contaminants was found to be a key transport mechanism leading to ingestion exposure. Because of the unique pattern to wet deposition, farms close to the source generally receive the greatest deposition. Shifting the location by relatively short distances can result in large changes in exposure because of the large spatial gradients inherent in wet deposition.

Wet deposition is often thought to be slight in desert areas. However, food chain risk may not be directly related to annual precipitation. This study and other risk assessments have indicated that annual MEI risk may be dominated by very few rainfall events. In most areas, rainfall is distributed over a range of wind directions. In desert areas, however, although fewer events may occur, they may all affect the same downwind area.

IV. DIFFICULTIES IN INGESTION EXPOSURE ASSESSMENTS

A. Introduction

The methods for calculating inhalation exposure are much better understood than those associated with ingestion. Because exposure is proportional to emissions for both pathways, emissions estimates have the same effect on both. Methods of calculating transport through the atmosphere are common to both types of exposure. Ingestion exposure has the additional uncertainties and complexities associated with estimating deposition and transport from the ground to the food and from the food to human consumption. In the following, difficulties associated with ingestion exposure assessments and some of the ramifications to overall exposure estimates are discussed.

B. Emissions

As expected, the results of the study demonstrated the importance of obtaining good estimates of emissions. Uncertainties in the risk assessment were reduced considerably when direct measurements of emissions were possible. EPA has sponsored studies to accumulate existing data and EPRI is sponsoring studies to help better determine emission factors for power plant sources.

C. Atmospheric Transport

Deposition is affected by the transport and dispersion of the plume. Uncertainties in plume transport affect both wet and dry deposition. Plume dispersion or plume spread in the horizontal and vertical directions affects dry deposition (and also inhalation exposure); only spread in the horizontal direction affects wet deposition. More complicated dispersion scenarios such as plume transport within terrain, at a land/sea interface, or among structures and buildings can lead to much greater uncertainties. The impact of these features on wet deposition estimates may be far less than on dry deposition.

D. Deposition

In this study, wet deposition was key to ingestion exposure to arsenic and dioxins, which were treated as particles, while dry deposition was critical to the exposure to PCB. The study assumed washout of the pollutants below the clouds. The locations of maximum ingestion exposure for arsenic and dioxin differed greatly from the corresponding locations of maximum inhalation exposure because wet deposition dominated the ingestion exposure. Wet deposition is highest near the source where the elevated plume concentrations are the highest, and drops off rapidly in the crosswind and downwind directions, leading to large spatial gradients of deposition. This wet deposition pattern is consistent with measured patterns of rainwater concentrations beneath elevated plumes of toxic particulate metals.[6] Our analyses revealed that wet deposition of the gaseous contaminants of interest, such as PCB, was lower than dry gas deposition by about two orders of magnitude. The calculated maximum inhalation and ingestion exposure locations for gases such as PCB are coincident at the location of maximum ground-level concentration. Because it was most significant, attention was focused on difficulties in estimating wet deposition.

Currently, the general approach is to treat all wet deposition as washout. Wet deposition is commonly estimated by methods similar to those proposed by Bowman et al.[7] which incorporates precipitation scavenging of a below-cloud plume in the Industrial Source Complex (ISC) air quality model. The rate of deposition is controlled by specified scavenging coefficients which vary according to rainfall rate and the size of the scavenged particle. The most widely used coefficients are those computed by Bowman et al.[7] from data gathered by Radke et al.[8] These coefficients are only presented for three particle size categories and three general rainfall rate categories. Because of the relatively wide range of these few discrete particle size categories, deposition estimates based on the Bowman/Radke coefficients are fairly insensitive to the specifics of the particle size distribution of emissions from controlled combustion sources which consist mainly of fine particles. On the other hand, the specifics of the particle size distribution become more important for uncontrolled sources because of the wider spectrum of particle sizes.

Although the methods used assume below-cloud scavenging, the Bowman/Radke coefficients are based to a large extent on within-cloud plume scavenging data. Other scavenging coefficients were derived using limited data collected by Schumann et al.[9] for below-cloud scavenging (see Table 2). More field measurement programs, such as those performed by Schumann et al.,[9] are needed to improve methods to treat washout (i.e., below-cloud scavenging). Better data should help resolve the relationship of the scavenging coefficient with particle size and rainfall rate. As a result of the development of more appropriate coefficients that account for rapid changes in scavenging efficiencies in the 1-μm particle size range (the "Greenfield gap"), better specification of the size distribution of small particles may become more critical.

Ideally, actual particle sizes within the plume should be used to determine wet deposition; however, such measurements are seldom available. In developing estimates of particle size distributions it is important to assess such items as (1) whether the contaminants are present throughout the bulk of the particle or only on the surface; (2) whether the contaminants are present in hygroscopic particles that are likely to grow at high humidities; (3) whether the particles will undergo aggregation; and (4) whether contaminants are sufficiently volatile to redistribute to ambient particles (with a different size distribution) as the plume mixes with ambient air. The latter two considerations are more likely to be important farther downwind from the source.

The above considerations were based on the current practice of considering only washout. The results obtained using this assumption may be totally inappropriate for situations where the plume is entrained in the clouds. In this situation, plume material is incorporated into droplets as they form and may be transported tens of kilometers before being deposited. One could also envision situations, such as a convective system, for which plume material is entrained and deposited quite close to the source. Methods have not been developed for treating rainout (within-cloud scavenging) in deposition calculations.

E. Precipitation

The study investigated the impacts of using only 1 year of precipitation data instead of an average over a lifetime. Two types of errors could occur by the use of a single year of data in calculating the location and magnitude of maximum exposure over a lifetime: (1) the magnitude of maximum exposure in the year selected could be unusual and (2) because the location at which this maximum exposure occurs can change from year to year, the maximum exposure for a typical year will generally be greater than the maximum exposure over a lifetime; 7 years of meteorological data were used to determine year-to-year variability. There was about a twofold difference between the highest and lowest annual maximum deposition rates. The 7-year average maximum deposition was about one half the highest annual maximum deposition.

Table 2. Currently Available Scavenging Coefficients

Precipitation intensity	< 2 μm	Particle Size 2–10 μm	10–20 μm
		Coefficients Computed by Bowman[7] from Radke[8] Data (s^{-1})	
Light	2.20×10^{-4}	1.80×10^{-4}	9.69×10^{-3}
Moderate	5.60×10^{-4}	8.93×10^{-4}	9.69×10^{-3}
Heavy	1.46×10^{-3}	4.64×10^{-3}	9.69×10^{-3}
		Coefficients Computed by Zankel[1] from Schumann[9] Data (s^{-1})	
Light	4.0×10^{-5}	6.0×10^{-5}	6.0×10^{-4}
Moderate	1.6×10^{-4}	2.4×10^{-4}	2.4×10^{-3}
Heavy	4.0×10^{-4}	6.0×10^{-4}	6.0×10^{-3}

The impacts of year-to-year shifts in location of maximum wet deposition may be much more significant than that indicated above. For example, shifts in location of the maximum deposition could move the impacts to or from a target population such as a subsistence farmer.

F. Food Chain

Whereas exposure is directly proportional to emission rates and deposition, this is not the case for many of the input parameters used in the food chain analysis. In the food chain models, there is a large number of input parameters and assumptions, many of which are not well understood. Careful evaluation of parameters and assumptions are needed to reduce uncertainties in the estimates. To estimate the variabilities due to parameter uncertainties, exposure analyses were performed varying each input parameter singly between its upper and lower limits. The maximum variability in overall exposure was calculated to be less than about a factor of five for each of the single input parameters. The parameters that led to the highest variabilities to overall exposure were plant bioconcentration factors, soil loss rates, and the percent of food ingested that is home grown (assuming that a person who ate a large portion of home-grown food lived at the location of maximum deposition).

G. Overall Uncertainty

Uncertainty of the overall exposure is not simply the product of individual parameter uncertainties. Monte Carlo simulations using Latin Hypercube sampling[10] were used to furnish insight into how the individual uncertainties combine. Three sets of simulations, furnishing frequency distributions of overall exposure, were performed for arsenic and dioxin exposure: (1) variable emissions, and all other parameters used in the exposure assessments varied within their range; (2) constant (unit) emissions, and dispersion, deposition, and food chain parameters varied; and (3) constant (unit) deposition, and only the food chain parameters varied.

The simulations (Figures 1 and 2) showed that uncertainties in estimates of food chain parameters resulted in about a threefold variability about the geometric mean. Dispersion and scavenging uncertainties increased the total variability only slightly. Because there was a large uncertainty in dioxin emissions, inclusion of these emissions substantially increased the overall variability to about sevenfold from the geometric mean. Inclusion of uncertainties in arsenic emissions did not substantially change the overall variability.

These results indicate that, when emissions are fairly well characterized, the algorithms used to determine exposure can yield results with a threefold variability due to uncertainties in the input parameters. However, the simulations do not test the validity of the algorithms that were used. Questions concerning definition of the MEI and the validity of the "below-cloud" concept of wet deposition, as examples, could lead to orders of magnitude uncertainties in overall results.

V. SUMMARY AND CONCLUSIONS

The study indicated that, in this case, calculated risk due to ingestion via wet deposition was far greater than that due to inhalation, even acknowledging the large uncertainties in ingestion exposure. Although there are large uncertainties, it is possible to calculate upper-bound estimates of ingestion exposure. However, these estimates may differ greatly from actual exposure. It is difficult to analyze the uncertainties because the ramifications of some of the assumptions, such as the use of below-cloud scavenging, are not well understood. Further efforts are needed to improve, in descending order of importance, estimates of emissions, deposition, and transport through the food chain. Also, care must be taken in defining the target population.

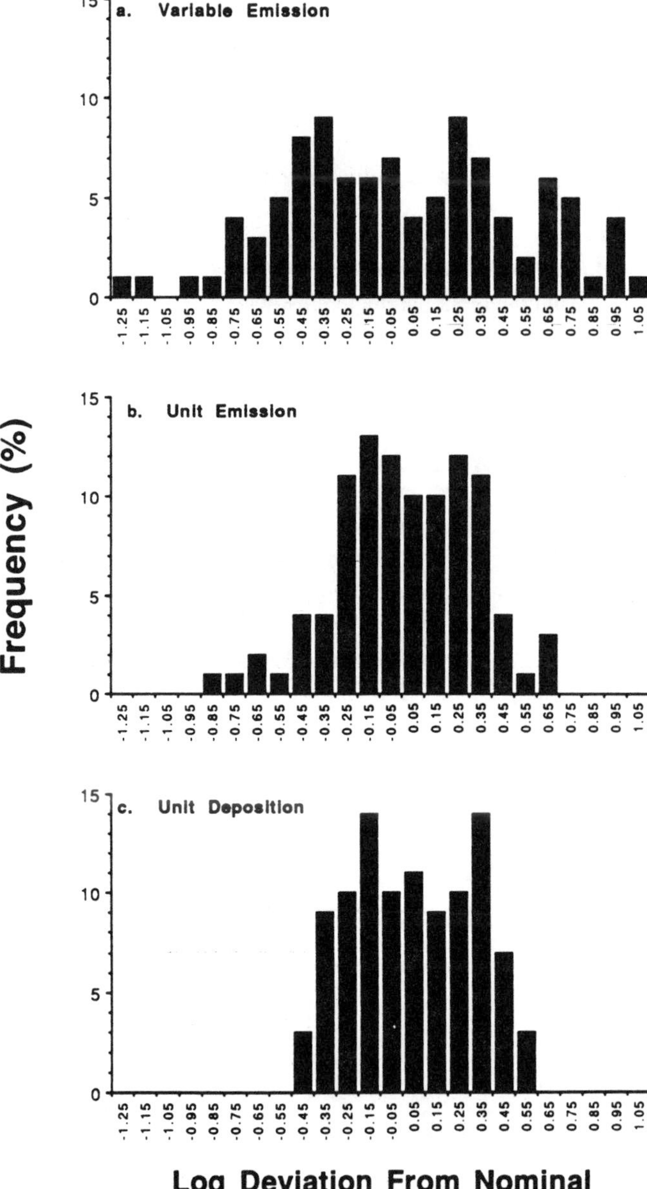

Figure 1. Frequency distributions of Monte Carlo-LHS results for dioxin exposure. Log nominal ratio is the log of the ratio of the Monte Carlo exposure to the nominal exposure [= log (Monte Carlo exp./Nominal exp.)]. (a) Results for variable emission case, with dioxin emission varying 25-fold. (b) Results for unit emission, where emission was fixed to unity but dispersion and scavenging were allowed to vary. (c) Variability due to the food chain module alone, where deposition was fixed to unity.

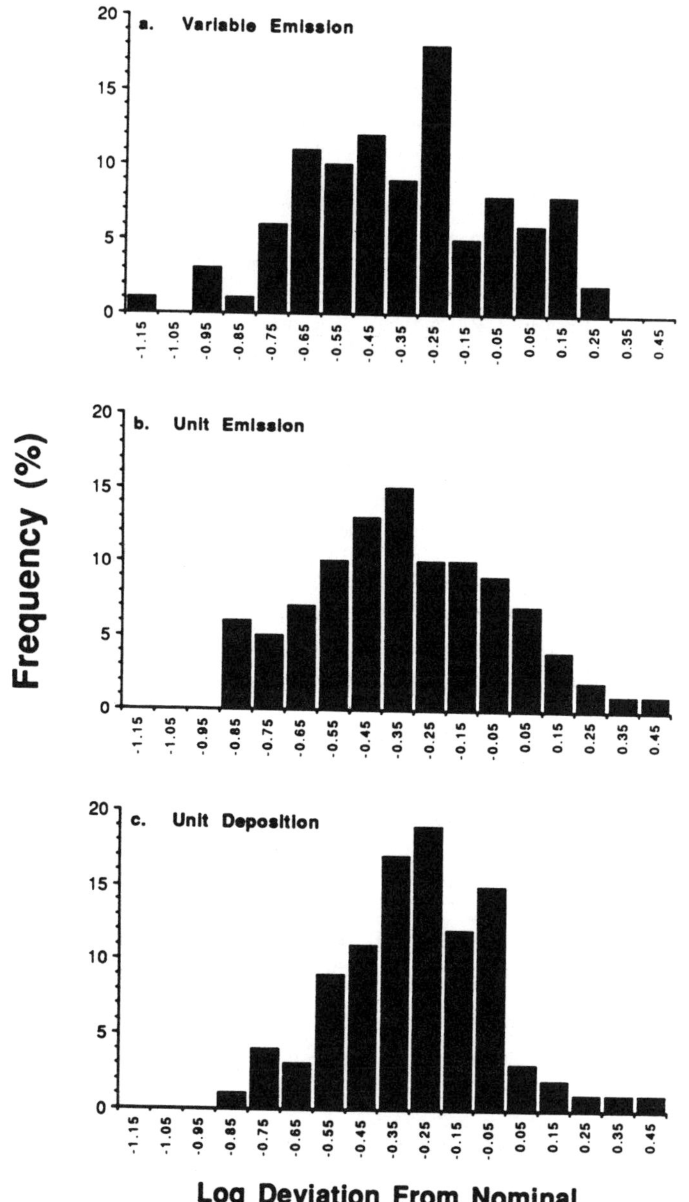

Figure 2. Frequency distributions of Monte Carlo-LHS results for arsenic exposure. See Figure 1 for explanation.

REFERENCES

1. Brower, R.P., Gerritsen, J., Zankel, K. L., Huggins, A., Peters, N., Campbell, S. A., and Nilsson, R., Risk Assessment Study of the Dickerson Site, Maryland Department of Natural Resources Power Plant and Environmental Review Division, Annapolis, MD, 1990.
2. Brower, R. P., Zankel, K. L., Gerritsen, J., Campbell, S. A., and Teitt, J. M., Importance of Indirect Exposure When Assessing Risk to Toxic Airborne Emissions, presented at Air and Waste Management Association 83rd Annual Meeting and Exhibition, Pittsburgh, PA, June 24–29, 1990.
3. Zankel, K. L., Brower, R. P., Campbell, S. A., and Teitt, J. M., Difficulties in Estimating Wet Deposition of Atmospheric Particulate Material, presented at Air and Waste Management Association 83rd Annual Meeting and Exhibition, Pittsburgh, PA, June 24–29, 1990.
4. Zankel, K. L., Gerritsen, J., Kou, J., Brower, R. P., and Teitt, J. M., Uncertainties in Multipathway Risk Assessment, presented at Air and Waste Management Association 83rd Annual Meeting and Exhibition, Pittsburgh, PA, June 24–29,1990.
5. Travis, C. C., Yambert, M. W., and Arms, A. D., *Food Chain Exposure from Municipal Waste Incineration*, Oak Ridge National Laboratory, Oak Ridge, TN, 1988.
6. Luecken, D. J., Laulainen, N. S., Payton, D. L., and Larson, T. V., Pollutant scavenging from plumes: a modeling case study from the ASARCO smelter, *Atmos. Environ.*, 23, 1063, 1989.
7. Bowman, C. R., Jr., Geary, H. V., Jr., and Schewe, G., Incorporation of Wet Deposition in the Industrial Source Complex Model, presented at Air Pollution Control Association 80th Annual Meeting, New York, 1987.
8. Radke, L. F., Hobbs, P. V., and Eltgroth, M. W., Scavenging of aerosol particles by precipitation, *J. Applied Meteor.*, 19, 715, 1980.
9. Schumann, T., Zinder, B., and Waldvogel, A., Aerosol and hydrometer concentrations and their chemical composition during winter precipitation along a mountain slope. I. Temporal evolution of the aerosol, microphysical and meteorological conditions, *Atmos. Environ.*, 22, 1443, 1988.
10. McKay, M. D., Conover, W. J., and Beckman, R. J., A comparison of three methods for selecting values of input variables in the analysis of output form a computer code, *Technometrics*, 21, 239, 1979.

Chapter 8
Conference Synthesis

Summary of the Conference on Managing Hazardous Air Pollutants Washington, D.C. November 4–6, 1991

Michael Miller

[Editor's note: Mike Miller heads the Waste & Water Management Program in EPRI's Environment Division. This program is responsible for the Institute's research activities in the areas of coal ash and scrubber by-product utilization and disposal, pollution prevention, aqueous discharge monitoring and treatment, analytical methods assessment, chemical and toxics waste management, and air toxics characterization. Miller's remarks are divided into three parts: general observations, a review of key information presented in each session, and conclusions.]

GENERAL OBSERVATIONS

About 225 people attended the 3-day conference, representing the following:

- 25 U.S. utilities
- 30 vendors or suppliers of technology and services
- 15 universities and nongovernmental research institutions
- 10 government agencies
- 22 international organizations (including overseas utilities)
- 10 other key industries (e.g., aluminum)

The moderate size of this conference indicates the importance yet relative newness of air toxics as an environmental issue. By contrast, approximately 800 people, chiefly from utilities and vendors, attended the December 1991 SO_2 Control Symposium, also held in Washington, D.C. As an environmental issue, acid rain is much further developed, in both scientific understanding and regulatory status, than air toxics. At this point, more is unknown than known about air toxics, but this should be considered a challenge to action rather than an excuse for inaction.

The diversity of attendees and the cooperative spirit of the conference, with EPA, DOE, EPRI, and the international community working together, is impressive. The international perspective of the conference is especially welcome.

PRESENTATION HIGHLIGHTS

U.S. and International Regulatory Overview and Implications

The 1990 Clean Air Act Amendments (CAAA) provide a very short timetable for studying air toxics emissions associated with the electric utility industry. The Administrator of the EPA is required to issue a report to Congress by November 1993 *[Editor's note: The deadline for this report has been postponed until at least September 1994]*. Meeting this deadline will require that all concerned parties work together to provide the input for this interim report.

In the CAAA, Congress asks EPA to regulate air toxics differently — on a source-by-source rather than a chemical-by-chemical basis. The Act's provisions call for a technology-based, rather than a risk-based, approach to air toxics control. Conversely, the Act calls for a study of the *risk* posed by electric utility emissions. EPA must evaluate the risks to public health and the environment from utility air toxics emissions, and then determine whether control technologies should be applied and, if so, what types of technologies should be used. The study of the utility industry is unique in this respect; for other industries, EPA will first assess available control technologies, and then conduct a residual risk analysis.

Several states are considering implementing their own air toxics studies/regulations. This possibility raises the question of how state initiatives will interface with EPA's efforts.

One question raised at this conference was, "How will compliance decisions made for the CAAA Acid Rain or Ozone Nonattainment provisions affect air toxics?" Concern has been expressed that some compliance strategies may increase trace element emissions. In conjunction with member utilities, EPRI recommends that air toxics considerations be factored into fuel/technology selections for SO_2 and NO_X compliance. When weighing the performance of two SO_2 and NO_X control technologies, for example, consider which technology may more effectively control air toxics.

The CAAA Title III (Air Toxics) provisions contain several unanswered questions, many of which Lee Zeugin of Hunton & Williams raised in his presentation. In order to determine whether and how to regulate the utility industry, EPA may choose to segregate power plants according to boiler/fuel/control technology subcategories.

Another regulatory issue raised at the conference was whether the 10- and 25-ton/year "major source" thresholds apply to the electric utility industry. EPA's Warren Peters asserted that, because the CAAA calls for a risk-based assessment of utility emissions, separate thresholds for the industry may be warranted on the basis of study results. However, if emissions risks exceed levels EPA deems acceptable, EPA may be compelled to apply some form of emission controls to electric utilities.

Concluding the discussion of U.S. regulatory issues was the question, "What is the definition of source?" Does the term refer to just the stack, or does it include the cooling tower, ash ponds, and landfill areas? Such semantics may determine whether power plants exceed emissions thresholds (if applicable) and may govern the range of control technologies to be considered.

What is the international perspective on controlling air toxics? Although the issue is being addressed in virtually all OECD countries, it appears that there is no uniform approach. For this reason, exchange of strategies and experiences at conferences like this one is especially important. The U.K., for example, is looking at the conversion of fossil steam plants from coal to natural gas — its per capita reserve base being greater than that of the U.S. — as the best solution for air toxics, global warming, and other emissions concerns. Incidentally, the award for "best acronym at the conference" goes to Hugh Evans of PowerGen and to the government of the U.K. for BATNEEC, meaning best available technology not entailing excessive costs.

Other countries, such as The Netherlands, have established extensive lists of environmentally based standards for hazardous air pollutants. In general, OECD member countries have been divided in their approach to setting standards — some are using health-based criteria whereas others are using ecological considerations, nor has there been a consensus on whether regulatory instruments should be source oriented or effect based. Concern has focused, however, on regional and global implications. International conventions have produced meaningful results, such as the North Sea Directive.

Emissions Sources

The session on emissions sources highlighted the extensive variability in data on trace element concentrations in coal. Often, the concentrations of these elements varied by a factor of three more than sulfur or ash concentrations. Coal cleaning, however, can play a significant role in removing trace elements, according to Ed Obermiller of Consolidation Coal R&D Center. Obermiller stated that coal cleaning is an important option for trace element emissions reduction.

Most available trace substance emissions data have been collected downstream of ESP. Relatively little information is available for fabric filter outlets, and even less for FGD system outlets. However, studies on FGD removal of trace elements indicate that SO_2 scrubbers may be partially effective in removing volatiles such as mercury. EPRI is having difficulty accurately determining mercury concentrations because the element is difficult to measure and is present in very low concentrations. Nevertheless, better data on FGD system removal of all trace elements need to be gathered.

Some conference participants expressed concern that fuel switching for CAAA SO_2 compliance could lead to increased trace element emissions. Data compiled to date are ambiguous on this point. For example, mercury is present in lower concentration in Powder River Basin and other western coals and in central and southern Appalachian low-sulfur coals than it is in eastern bituminous coals. Thus, mercury emissions may decrease, rather than increase, as SO_2 compliance options are implemented. Some low-sulfur coals, however, appear to have higher concentrations of other trace metals.

Power plant emissions of organics on the CAAA list have not been studied as extensively as those of trace metals. Data to date indicate that organic concentrations in flue gas are very low, except possibly during startup.

Advanced generating technologies, such as gasification-combined-cycle (GCC), have the potential for very low trace element emissions, according to Lee Clarke of IEA Coal Research and Dan Baker from Shell. Baker urged utilities to factor air toxics considerations into control technology decisions; although GCC may initially be more costly, low air toxics emissions may make it an overall better option.

Atmospheric Chemistry, Measurements, and Models

In his presentation on atmospheric chemistry, EPRI's Peter Mueller acknowledged that there are many uncertainties in the closure of mass balances for trace elements, and that measurement capabilities are still relatively primitive, especially for HCl and inorganics such as arsenic and mercury. Mueller suggested coordination between the sampling and analytical community and the risk assessment community to determine acceptable levels of uncertainty. Focusing efforts to improve measurement accuracy first on the trace elements that pose the greatest risk will allow the electric utility industry to maximize the effectiveness of its limited resources. For example, HCl does not seem to pose a significant risk, even though, of the chemicals on the list of 189 air toxics, it is probably emitted in the greatest quantity. Thus, measurement efforts should focus on other substances.

Possible methods for improved mercury sampling were discussed by Nicolas Bloom of Brooks Rand. Generally, these approaches rely on iodated activated charcoal or carbon.

Other speakers noted that plume sampling must also be improved. Representative samples are essential for health assessment purposes. Obtaining such samples has always been a challenge. The plume simulation dilution sampling (PSDS) system, introduced by Karl Simmons of Keystone/NEA, may provide representative samples at a much more reasonable cost than that of samples taken from an airplane while flying around a stack.

Iodated activated charcoal or carbon may also be the most effective sampling methods for organics in the atmosphere and can be measured to very low part-per-billion levels. However, as Bob Mann of Radian reiterated, effectively measuring organics in flue gas is still difficult.

Health and Environmental Information and Models; Models for Risk Assessment, Management, and Design

The session on environmental and health effects highlighted interesting models to help predict plant uptake of air toxics. One study, conducted by EPRI's Don Porcella and Doug Knauer of the Wisconsin Department of Natural Resources, modeled mercury balances in lakes. On the basis of their results, mercury cycling in the Great Lakes states is now fairly well understood. As expected, methyl mercury was found to be the species of greatest concern.

Bernard Weiss of the University of Rochester School of Medicine explained that, at relatively low levels, mercury is a toxin, not a carcinogen, although that distinction makes it no less of a health concern. According to Weiss, methyl mercury is a "superb human pesticide". It kills central nervous system cells and is especially damaging to fetal brain development.

Arsenic speciation and toxicity were discussed by EPRI's Ron Wyzga and Jeff Hicks of Radian. Trivalent arsenic (As^{+3}), the form likely to be of greatest concern for acute toxicity, is seldom found in significant quantities in coal ash. Utilities should not assume, however, that they need not consider arsenic a potential health risk. Although the predominant As species found in coal ash — pentavalent arsenic (As^{+5}) — is not as toxic as As^{+3}, long-term exposure to As^{+5} may be a potential health concern.

Interestingly, As^{+5} may be a carcinogen through ingestion, rather than inhalation, the exposure method studied previously. This possibility underscores the need to study different exposure routes, synergisms, bioavailability, and the mechanisms of action. All of these factors are essential to health risk assessment, and must be considered when studying the more-critical trace elements. On the basis of information from the risk assessment sessions, efforts in health risk assessment should focus on mercury, nickel, arsenic, and chromium. In terms of worker exposure, arsenic and silica are probably the most critical.

Ed Rubin of Carnegie-Mellon University and Warren Peters from the U.S. EPA Office of Air Quality Planning Studies presented an appraisal of advances in risk modeling. Risk models are changing from deterministic, single-point models to more stochastic and probabilistic models. With these newer models, it is possible to determine the probability of a certain event occurring, rather than simply using the worst-case scenario.

To be accurate, the risk assessment for some trace substances must consider the multi-media exposure pathways, even though this consideration makes the assessments much more complicated. Chromium exposure, for example, may be of more significant concern through the routes of ingestion and skin contact rather than through inhalation, according to Christian Seigneur of ENSR Consulting.

One issue from the risk assessment discussion that could be considered somewhat counterintuitive is the potential risk from oil-fired power plants. As Katy Connor of Decision Focus mentioned, oil-fired plants, because they typically are in urban areas and have shorter stacks, could pose a greater individual and population risk than coal-fired plants. Consequently, EPRI needs to gather better data on emissions from oil-fired power plants and conduct further risk assessments.

Control Strategies and Applicable Technologies

The session on control strategies re-emphasized the need for R&D in fine particle removal and vaporous emissions control. It covered mechanisms for the enrichment of trace elements on fine particles, and described technologies to control fine particles effectively, such as flue gas conditioning and improved fabric filtration. New methods for controlling mercury in flue gas appear to be promising.

One observation of positive synergism was the combined use of flue gas conditioning and fabric filtration. Using these technologies together can result in a decrease of fine particle emissions by three or four orders of magnitude *and* a reduction in pressure drop. This is an unusual win-win situation, because these benefits are seldom achievable in tandem.

CONCLUSIONS

Reviewing the presentations at the conference, the trace species of greatest concern seem to be arsenic, nickel, mercury, and chromium. At this time, HCl and organics do not seem to pose much risk. EPA's study is on a short timeframe — 3 years — and EPRI will do its best to provide the data necessary for the Agency to report to Congress.

Some analysts have said that the utility industry represents a large percentage of industrial trace element emissions. This statement is misleading in several ways:

- Some of the emissions factors used to determine utility discharges are out of date. EPRI is currently obtaining better emission factors and dispersion information.
- Estimates obtained using the toxics release inventory that industries report to EPA are inexact. Representatives from the chemical industry admit they are using estimates and they believe the data they are reporting are low.
- In reporting total releases, risk is not taken into consideration. Although industries can compare tonnages, risk is the relevant measure for health effects studies.

In other words, the electric utility industry might emit a large percentage of some trace elements; however, these elements and emissions patterns are likely not to be those that pose significant health risks.

In conclusion, cooperation is very important in learning more about air toxics. As a result of the positive feedback on this conference, EPRI will hold another in July 1993. Thank you for attending.

Index

WIDENER UNIVERSITY
WOLFGRAM
LIBRARY
CHESTER, PA.